MATRIX Book Series **8**

MATRIX is Australia's international and residential mathematical research institute. It facilitates new collaborations and mathematical advances through intensive residential research programs, each lasting 1–4 weeks.

David R. Wood (Editor-in-Chief) ·
Alison M. Etheridge · Jan de Gier · Nalini Joshi
Editors

2024 MATRIX Annals, Part II

Springer

Editors
David R. Wood *(Editor-in-Chief)*
School of Mathematics
Monash University
Melbourne, VIC, Australia

Alison M. Etheridge
Department of Statistics
University of Oxford
Oxford, UK

Jan de Gier
School of Mathematics and Statistics
University of Melbourne
Melbourne, VIC, Australia

Nalini Joshi
School of Mathematics and Statistics
The University of Sydney
Sydney, NSW, Australia

ISSN 2523-3041 ISSN 2523-305X (electronic)
MATRIX Book Series
ISBN 978-3-032-16205-2 ISBN 978-3-032-16206-9 (eBook)
https://doi.org/10.1007/978-3-032-16206-9

This Springer imprint is published by the registered company Springer Nature Switzerland AG
The registered company address is: Gewerbestrasse 11, 6330 Cham, Switzerland

Preface

MATRIX is Australia's residential mathematical research institute. It facilitates new collaborations and mathematical advances through intensive residential research programs, which are held in Creswick, a small town nestled in the beautiful forests of the Macedon Ranges. MATRIX was established in 2015, and is a joint partnership between Monash University, The University of Melbourne, and The University of Queensland, with The Australian National University and The University of New South Wales as associate members. The ARC Centre of Excellence for Mathematical and Statistical Frontiers (ACEMS) provided seed funding for MATRIX and was an associate member until the end of 2021. MATRIX is supported by the Simons Foundation via its *Targeted Grants to Institutes* scheme.

The following research programs were held at MATRIX in the second half of 2024. Each program lasted 1–4 weeks, and included ample unstructured time to encourage collaborative research. Some of the longer programs had an embedded lecture series.

- *Mathematics and Physics of Integrability*
  Guest editors: Patrick Dorey and Paul Pearce
- *Multivariate Dependence Modelling: Theory and Applications*
  Guest editor: Laleh Tafakori
- *Mathematical Models for Lipids and Cells in Atherosclerotic Plaques*
  Guest editor: Mary Myerscough
- *The Geometry of Moduli Spaces in String Theory*
  Guest editors: Sebastien Picard and Jock McOrist
- *Parameter Identifiability in Mathematical Biology*
- *Applications of Topological Data Analysis to Mathematical Biology*
- *Tensor Categories, Quantum Symmetries, and Mathematical Physics*
  Guest editor: James Tener
- *Algebraic Geometry*
- *Harmonic and Microlocal Analysis in Partial Differential Equations*

MATRIX also ran the following tandem workshop in partnership with the Mathematisches Forschungsinstitut Oberwolfach (MFO):

- *Invariants in Low-Dimensional Topology: Combinatorics, Geometry, and*

Computation

In this workshop, researchers meet simultaneously in Australia and in Germany to work on the joint topic.

The MATRIX Scientific Committee selected these programs based on scientific excellence and the participation rate of high-profile international participants. The following people served on the MATRIX Scientific Committee during 2024: Ben Andrews (ANU), Santiago Badia (Monash), Zdravko Botev (UNSW Sydney), Peter Bouwknegt (ANU), Ricardo Ruiz Baier (Monash), Peter Bühlmann (ETH Zurich), Alison Etheridge (Oxford), Yanan Fan (UNSW Sydney), Jan de Gier (Melbourne), Cecilia González Tokman (Queensland), Catherine Greenhill (UNSW Sydney), James McCaw (Melbourne), Marcy Robertson (Melbourne), Ngamta Thamwattana (Newcastle), Geordie Williamson (Sydney), and David Wood (Monash, Chair). The MFO and RIMS events were selected by sub-committees of the MATRIX, MFO and RIMS scientific committees.

MATRIX supports the entire Australian mathematical sciences research community. Indeed, the 2024 MATRIX programs included organisers from Australian National University, La Trobe University, Macquarie University, Monash University, Queensland University of Technology, University of Melbourne, University of New England, University of New South Wales, University of Queensland, University of Sydney, University of Western Australia, and University of Wollongong, along with numerous international organisers.

This volume is a scientific record of the above events. All participants were encouraged to submit articles to the MATRIX Annals. The articles are grouped into refereed contributions and other contributions. Refereed articles contain original results or surveys on a topic related to the MATRIX program. The other contributions are typically lecture notes or short articles based on talks or activities at MATRIX. A guest editor organised appropriate refereeing and ensured the scientific quality of submitted articles arising from each program. The Editors (Alison Etheridge, Jan de Gier, Nalini Joshi, and myself) finally evaluated and approved the papers.

Many thanks to the authors and to the guest editors for their wonderful work.

David Wood
MATRIX Annals Editor-in-Chief

MATRIX Program: Mathematics and Physics of Integrability

1 – 19 July 2024

Organisers

Changrim Ahn
Ewha Womans University

Murray Batchelor
Australian National University

Anastasia Doikou
Heriot-Watt University

Patrick Dorey
Durham University

Paul Pearce
Melbourne University
University of Queensland

Véronique Terras
Paris-Saclay University

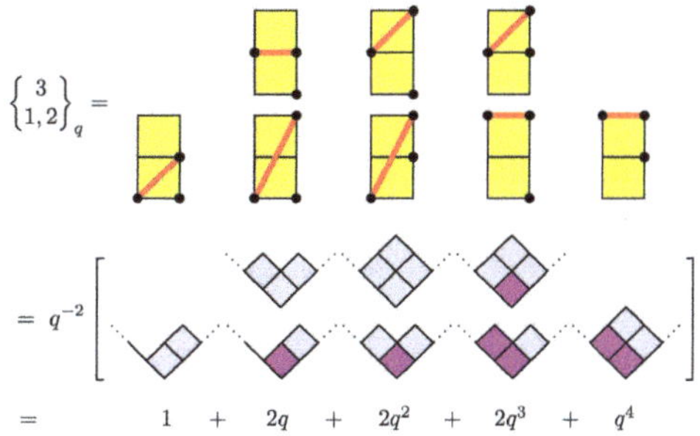

Participants

Murray Batchelor (ANU), Paul Pearce (Melbourne & Queensland), Changrim Ahn (Ewha, Seoul), Anastasia Doikou (Heriot-Watt), Patrick Dorey (Durham), Véronique Terras (Paris-Saclay), David Ridout (Melbourne), Andreas Klümper (Wuppertal), Rafael Nepomechie (Miami), Didina Serban (CEA Saclay), Roberto Tateo (Torino), Francisco Alcaraz (Sao Carlos), Zoltan Bajnok (Wigner Research Center), Junpeng Cao (IOP, CAS), Clare Dunning (Kent), Angela Foerster (Porto Alegre), Xiwen Guan (Wuhan), Marton Kormos (Budapest), Charlotte Kristjansen (Niels Bohr Inst), Atsuo Kuniba (Tokyo), Chihiro Matsui (Tokyo), Francesco Ravanini (Bologna), Ana Retore (Durham), Junji Suzuki (Shizuoka), Gabor Takacs (Budapest), Wen-Li Yang (Xian), Robert Weston (Heriot-Watt), Roger Behrend (Cardiff), Jean-Emile Bourgine (Melbourne), Peter Bouwknegt (ANU), Jan de Gier (Melbourne), Jon Links (Queensland), Thomas Quella (Melbourne), William Mead (Melbourne), Alexandr Garbali (Melbourne), Remy Adderton (ANU)

This research program focused on the mathematics and physics of integrability in classical and quantum physical systems in 1 and 2 dimensions. Following on from our 2017 MATRIX program, this program was organised around currently active hot topics and open problems with 50% of the program reserved for group collaboration.

The emphasis was on focused research and interaction in small groups to achieve real collaboration. The research topics included:

- integrability, symmetries and quantum algebras,
- exactly solvable lattice models,
- Bethe ansatz and quantum spin chains,
- diagram algebras, geometric phase transitions and logarithmic CFT,
- bulk and boundary CFT, AdS/CFT, QFT and their perturbations,
- quantum entanglement, quantum quenches and generalised hydrodynamics,
- cold atoms and strongly correlated systems.

The overarching motivation of this program was to muster all the disparate tools pure mathematics has to offer in disciplines such as affine Lie algebras, vertex algebras, planar algebras, quantum algebras and representation theory and apply them to understand, in a deep and precise manner, the behaviour of a diverse range of physical systems arising in disciplines such as classical and quantum statistical mechanics, condensed matter, quantum entanglement, quantum quenches and particle physics. The physical systems of interest include gases, fluids, ferromagnets, ferroelectrics, cold atoms, correlated electron systems, dimers, polymers and percolation among others.

The workshop brought together leading world experts to advance the frontier of knowledge on these specialized topics. Indeed, a significant proportion of the active world-wide community working on classical and quantum integrability participated in this program. There were 15 countries represented including Australia, Brazil, China, Denmark, England, France, Germany, Greece, Hungary, Italy, Japan, Romania, Scotland, South Korea and the USA. The program consisted of 40 one-hour talks with afternoons free for discussion and collaborative research.

The impact of this workshop will be of long-lasting benefit to the community in Mathematical and Theoretical Physics particularly in Australia and the region. With a series of lectures on Hot Topics and advanced seminars, it offered training to a number of PhD students and Early Career Researchers working in integrability.

Some highlights, among many of the meeting, include (i) a series of 4 lectures on the latest developments in the new area of integrable $T\overline{T}$ deformations and the Hagedorn transition, (ii) a beautiful lecture on the significant recent advances on solving tetrahedron and 3D reflections algebras by quantum cluster algebras, (iii) 3 lectures related to the hot topic of quantum computing, (iv) a striking overview of progress on the dynamics of confining spin chains connecting topics in condensed matter physics and high energy physics and (iv) a tour de force exposition on alternating sign matrices and integrability.

The work done at the workshop is well represented in the following articles published in this volume:

- Adel Ben Moussa, Jules Lamers and Didina Serban “The q-deformed Haldane–Shastry chain at $q = i$ with even length”
- Nicolò Brizio, Tommaso Morone and Roberto Tateo “A note on $T\bar{T}$ deforma-

tions and boundaries"
- Natalie Chung and Rafael I. Nepomechie "Quantum counting, and a relevant sign"
- Anastasia Doikou "Self-distributivity, braces, and the Yang-Baxter equation""
- Patrick Dorey and Brandon Morrison, "Analytic continuation in the sinh-Gordon model"
- Angela Foerster and Jon Links, "Superintegrable bosonic star networks"
- Andreas Klümper and Xin Zhang" "On the XXX spin-1/2 quantum chain with non-diagonal boundary fields"
- Paul A. Pearce and Jørgen Rasmussen "Ocneanu algebra of seams: Critical unitary E_6 RSOS lattice model'

We gratefully acknowledge the generous support of our sponsors — MATRIX, the Simons Foundation, the Asia Pacific Center for Theoretical Physics (APCTP), and the Australian Mathematical Sciences Institute (AMSI). We particularly thank Jan de Gier for his encouragement in bringing this program together. We also thank the very helpful MATRIX staff at Melbourne and Creswick campuses, as well as the outstanding chef Adam, for his contributions to the success of this meeting. Lastly, we thank the authors who kindly took the time and made the effort to contribute to this volume.

Patrick Dorey and Paul Pearce
Guest editors

MATRIX
MATHEMATICS AND
PHYSICS OF
INTEGRABILITY
(MPI2024)
2 - 19 JULY 2024
CRESWICK AUSTRALIA
MATRIX

MATRIX
MATRIX

MATRIX Program: Multivariate Dependence Modelling: Theory and Applications

22 July – 2 August 2024

Organisers

Pavel Krupskiy
University of Melbourne

Tingjin Chu
University of Melbourne

Laleh Tafakori
RMIT University

Bouchra Nasri
University of Montreal

MULTIVARIATE
DEPENDENCE
MODELLING
THEORY AND APPLICATIONS

Participants

Gery Geenens (UNSW), Clara Grazian (Sydney), Mohamed Chaouch (Qatar), Hans Manner (Graz), Harry Joe (British Columbia), Ying Chen (National University of Singapore), Arnab Hazra (Indian Institute of Technology Kanpur), Guilherme Ludwig (Campinas, Brazil), Rachael Quill (State Electricity Commission Victoria, Weatherzone), Yan Wang (RMIT), Jaehong Jeong (Hanyang), Jenny Wadsworth (Lancaster), Guoqi Qian (Melbourne), Huiyan Sang (Texas A&M), Radu Craiu (Toronto), Luciana Dalla Valle (Exeter), Kate Saunders (Monash), Elif Acar (Manitoba), Bouchra Nasri (Montreal), Bruno Remillard (HEC Montreal), Dietrich von Rosen (Linköping), Grace Y. Yi (Western Ontario), Wenqing He (Western Ontario), Sévérien Nkurunziza (Windsor), Archer Yang (McGill), Karim Oualkacha (Université du Québec à Montréal), Bahareh Ghanbari (RMIT), Pavel Krupskiy (Melbourne), Laleh Tafakori (RMIT), Jarryd Chapman (Monash), Tingjin Chu (Melbourne), Xueming Li (Melbourne), Haibo Li (Melbourne), Jialing Han (York), Xiaoting Li (British Columbia), Thong Nguyen-Huy (Southern Queensland), LuYi Shen (Melbourne), Reza Nosratpour (RMIT), Michael Smith (Melbourne Business School), Lin Deng (Melbourne)

The program brought together a diverse group of internationally recognized experts and early-career researchers working across various areas of multivariate dependence modelling. It served as a platform for exchanging ideas and strengthening interdisciplinary collaborations.

The first week focused on advances in copula modelling and inference, multivariate extremes, and their applications in climate science, environmental studies, and finance. Sessions also covered inference tools for spatio-temporal data, with applications in geology and environmental monitoring. The second week explored copula-based models for multivariate time series, Bayesian approaches for complex dependence structures, nonparametric methods, and variable selection techniques for high-dimensional data, with applications in biology, signal processing, and climate modelling.

Throughout the program, collaborative groups engaged in active, productive discussions. The workshop created valuable opportunities for theoretical and applied researchers to forge new interdisciplinary partnerships bridging methodology and real-world applications.

The program featured the following talks:

- Bruno Rémillard "Copula models for time series"
- Dietrich von Rosen "Safety belt regression, a mixture of classical regression and ridge regression"
- Huiyan Sang "Graph-Split-Based Decision Trees and Their Ensembles for Spatial and Network Clustering, and Classification"
- Grace Y. Yi "Estimation and variable selection under the function-on-scalar linear model with covariate measurement error"

This volume includes the following three papers arising from the program. They were the direct result of vibrant collaborative discussions, joint problem-solving sessions, and the dynamic exchange of ideas throughout the workshop.

- Arghadeep Basu and Arnab Hazra "A High-dimensional Bayesian Skew-Gaussian Spatial Model for Analysing Rainfall Trends"
- Xinyao Fan and Harry Joe "Estimation of Latent Variables in High-dimensional Factor Models with Weak Residual Dependence"
- Katarzyna Filipiak, Dietrich von Rosen, Martin Singull "Maximum likelihood estimation under inequality constraints in univariate linear models"

The program successfully achieved its goals through sustained engagement and interdisciplinary collaboration. By bringing together global experts, it sparked new research initiatives across statistics, climate science, and health studies. MATRIX provided an ideal setting for these interactions—a supportive, intellectually rich environment that fostered meaningful dialogue and laid the groundwork for future collaborative projects.

Laleh Tafakori
Guest editor

MATRIX
MULTIVARIATE
DEPENDENCE
MODELLING:
MATRIX

MATRIX
MULTIVARIATE
DEPENDENCE
MODELLING:
MATRIX

MATRIX Program: Mathematical models for lipids and cells in atherosclerotic plaques

19 – 30 August 2024

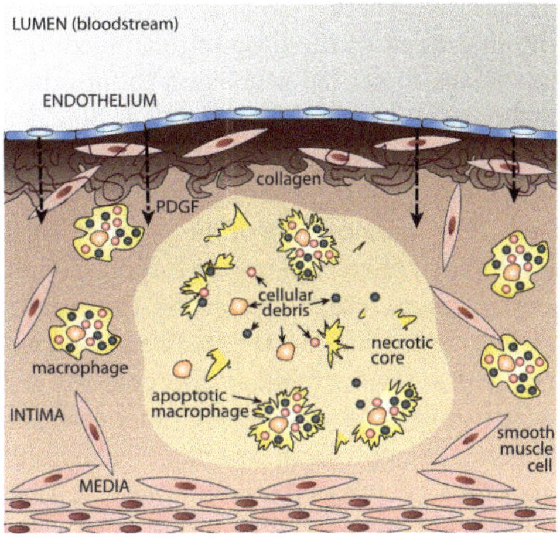

Organisers

Mary Myerscough
University of Sydney

Michael Watson
University of New South Wales

Helen Byrne
University of Oxford

Participants

Mary Myerscough (Sydney), Helen Byrne (Oxford), Michael Watson (UNSW), Christina Bursill (SAHMRI and Adelaide), Andrew Murphy (Baker Institute), Jennifer Flegg (Melbourne), Angela Reynolds (Virginia), Zhiyong Li (QUT), Pak-Wing Fok (Delaware), Vijay Rajagopal (Melbourne), Adrianne Jenner (QUT), Ishraq Ahmed (Sydney), Joseph Ndenda (Sydney), Keith Chambers (Oxford), Ryan Murphy (Melbourne), Jichao Pan (Southeast), Alex Tam (South Australia), Edward Green (Adelaide), Adelle Coster (UNSW)

Atherosclerosis is a chronic disease characterised by the growth of fatty plaques in the major arteries. It is a major cause of death in the developed world, and yet mathematical modelling of atherosclerosis is still in its infancy. By contrast, mathematical modelling of cancer, another major cause of death, is a mature field which contributes to the design of therapy and the delivery of personalised medicine. This program brought together researchers from the fields of atherosclerosis and cancer, in order to identify common modelling approaches and challenges, and catalyse new multidisciplinary collaborations in the growing field of mathematical modelling of atherosclerosis. The program focused on:

(i) Modelling approaches—how should we account for spatial and phenotypic heterogeneity of immune cells? How should mechanical effects be incorporated? How should the multiple timescales over which atherosclerosis develops be treated?

(i) Incorporating data—can we recover model parameters given perfect noise-free data (structural identifiability) and given limited, noisy data (practical identifiability)? How can we use synthetic data to help relate models to experiments?
(i) Discrete-to-continuum analysis—what are the relationships between models developed using different approaches? How do models at different spatial scales inform one another?

The first week of the program included a series of talks from scientists and mathematicians to set the scene and to introduce theory from complementary areas of mathematical modelling. There was time available for detailed discussions between experimental scientists, and mathematical biologists and engineers. We particularly valued the input of experimentalists Chris Bursill and Andrew Murphy during this week. The second week of the program focussed on continuing collaborative work, developing concrete plans for future research, and generating preliminary results to support future research projects. During the program, several collaborative groups formed, with some overlap between groups. Two groups worked on fitting experimental data on macrophage efferocytosis to a mathematical model, one using model identifiability analysis and the other using model fitting to obtain information about dynamic changes in macrophage behaviour that were evident in the dataset. Another group developed a preliminary, cell-based model using the PhysiCell package written by Paul Macklin's group in the US. Each of these initiatives has been included in this volume. A fourth group worked on an ODE model that describes how macrophage phenotype changes in developing atherosclerotic plaques. This has not yet been formally written up. On the last day of the workshop, all participants reflected on what we had individually and collectively learnt. A summary of these reflections forms a fourth article in this volume. Additionally, during the second week, a working group focused on learning the theory of morphoelasticity which is relevant to atherosclerosis.

The program also featured the following talks:

- Christina Bursill "Introduction to atherosclerosis"
- Mary Myerscough "A brief review of models for plaque development"
- Andrew Murphy "Topics in the science of plaques"
- Michael Watson "Models for macrophages and lipids in plaques"
- Angela Reynolds "What can we learn from macrophage dynamics elsewhere in the body?"
- Keith Chambers, Ishraq Ahmed, Joseph Ndenda "ECR talks on macrophage dynamics"
- Pak-Wing Fok "Incorporating mechanics into immunological models"
- Zhiyong Li and Jichao Pan "Agent-based models and models for the late-stage plaque"
- Vijay Rajagopal "Bioengineering perspective on plaques and cell mechanics"
- Ryan Murphy "Model identifiability and parameter estimation"
- Adelle Coster "Using models for synthetic data"
- Helen Byrne "Models for cancer. What can we learn?"

- Alex Tam and Ed Green “Models for growing yeast. What can we learn?”
- Jen Flegg “Models for wound healing. What can we learn?”

The aim of the program, to bring together researchers of different backgrounds to share expertise while working on new models for the development of atherosclerotic plaque, was met. The workshop provided a significant impulse to stimulate atherosclerosis modelling and increase the momentum, diversity and productivity of research into models for the immunology of atherosclerosis through collaborations in Australia and internationally. The aim to write a comprehensive review of the field has not been met. It became apparent that this is a more substantial undertaking than we had thought. Several attendees at the workshop are still discussing this and are likely to start work on it soon.

The workshop initiated several ongoing collaborations. Ed Green (Adelaide) has established a collaboration with Chris Bursill (SAHMRI) to model plaque calcification and has attracted a student to work on this. Angela Reynolds and Helen Byrne plan to continue to work with Adrianne Jenner’s draft PhysiCell model for plaques with the aim of producing synthetic data to use to fit differential equation models. A zoom-based workshop on morphoelasticity, led by Pak-Wing Fok was established immediately after the workshop to continue exploration of the possibilities of tissue mechanic models. Andrew Murphy invited Mary Myerscough to give an informal talk at the Baker Institute in September, 2024. Mary Myerscough and Helen Byrne continue to work on the ODE macrophage phenotype in November 2024. There is still more work to be done there.

Mary Myerscough
Guest editor

MATRIX Program:
The Geometry of Moduli Spaces in String Theory

2 – 13 September 2024

Organisers

Philip Candelas
University of Oxford

Xenia de la Ossa
University of Oxford

Johanna Knapp
University of Melbourne

Jock McOrist
University of New England

Sebastien Picard
University of British Columbia

Gabriele Tartaglino-Mazzucchelli
University of Queensland

Participants

Johanna Knapp (Melbourne), Jock McOrist (New England), Gabriele Tartaglino-Mazzucchelli (Queensland), Sebastien Picard (British Columbia), Philip Candelas (Oxford), Xenia dela Ossa (Oxford), Mario Garcia-Fernendez (Instituto de Ciencias Matematicas (ICMAT)), Kentaro Hori (Kavli IPMU, University of Tokyo), Eric Sharpe (Virginia Tech), Luca Cassia (Melbourne), Peter Bouwknegt (ANU), Joseph McGovern (Melbourne), Robert Pryor (Melbourne), Paul Norbury (Melbourne), Bryan Wang (ANU), Mateo Galdeano (Hamburg), Shinobu Hosono (Gakushuin), Martin Sticka (New England), Mohammad Elmi (Rutgers), Pyry Kuusela (Johannes Guttenberg), Emanuel Scheidegger (Beijing), Ruxandra Moraru (Waterloo), Thorsten Hertl (Freiburg), Thalia Greinke (ANU)

The program focused on moduli spaces in string theory. String theory unifies all known fundamental forces at the quantum level. For consistency, it requires a higher-dimensional spacetime. String compactification is the idea that there are four large non-compact dimensions that describe our world, plus extra compact ones. The mathematical structure of the extra dimensions determines the physics in four dimensions. The extra dimensions come with free parameters whose numbers are determined by the topology of the internal space. The configuration spaces of these parameters determine the moduli spaces that were the program's main theme. Quantum corrections in string compactifications depend on the moduli which makes their study essential for string theory. Another key concept of string theory is the idea of a duality. The statement is that different incarnations of string theory, compactified on different internal spaces can lead to the same physics in four dimensions. A classic example that shaped string theory and the associated mathematics is mirror symmetry for Calabi-Yau manifold. One definition of a Calabi-Yau is a Ricci-flat Kahler manifold. The moduli space is a product of the Kahler and complex structure moduli space whose dimensions are counted by the Hodge numbers of the Calabi-Yau. Mirror symmetry states that string theory on mirror pairs (X,Y), characterised by the fact that the Kahler moduli space of X is isomorphic to the complex structure moduli space of Y and vice versa, leads to the same physics in four dimensions. This has far-reaching consequences for mathematics and physics. In physics language, classical physics gets mapped to quantum physics in an a priori non-obvious way. Mathematically, it establishes highly non-trivial connections between areas of algebraic, differential and enumerative geometry. If one reduces the amount of supersymmetry of a string compactification, additional structures emerge that also influence the moduli spaces. The related technical terms are heterotic string theory and type II string theory with D-branes. The structure of the associated moduli spaces and the related string dualities are far less understood and were the focus of the workshop. Concretely, the following topics were covered all of which are active research areas in physics and mathematics:

- Moduli of non-Kahler 3-folds, G_2-manifolds and D-branes
- Generalised geometry and localisation of sigma models
- Topological invariants in string theory

The program was attended by expert mathematicians and physicists who have made ground-breaking contributions to these topics. The goal was to build bridges between mathematics and physics as well as different sides of dualities aiming to make progress on open research questions, to find new dualities and to forge new collaborations. There were 2-3 one-hour talks per day, leaving ample time for discussions, collaborations, and exchange of ideas. The program featured the following talks:

- Peter Bouwknegt "Introduction to Differential Cohomology"
- Luca Cassia "A-branes and symplectic cuts of quantum volumes"
- Mohamed Elmi "Hadamard Products and BPS Networks"
- Mateo Galdeano "The heterotic G_2 system on 3-(α,δ)-Sasaki manifolds"
- Mario Garcia-Fernandez "On the special geometry of the Hull-Strominger system"
- Mariana Graña "Moduli stabilisation and tadpole conjecture in non-geometric backgrounds"
- Thorsten Hertl "On the global topology of G_2 moduli spaces"
- Kentaro Hori "D-brane transport and monodromy in strong/weak CY/CY correspondence"
- Shinobu Hosono "Mirror symmetry of abelian fibered Calabi-Yau manifolds with $\rho = 2$"
- Pyry Kuusela "Calabi-Yau Compactifications and Modularity"
- Magdalena Larfors "Machine Learning Models for Calabi-Yau Geometry"

- Joseph McGovern "Misadventures in Moduli Spaces of Quotient Calabi-Yau Threefolds"
- Ruxandra Moraru "New examples of compact holomorphic symplectic manifolds"
- Paul Norbury "Volumes of moduli spaces of super Riemann surfaces"
- Sara Pasquetti "The mirror dualization algorithm at work"
- Sebastien Picard "Non-Kahler Degenerations of Calabi-Yau Threefolds"
- Robert Pryor "B-branes in $N = (2,2)$ Hybrid Models"
- Emanuel Scheidegger "Hodge theory and modular forms (and Calabi-Yau)"
- Eric Sharpe "The Bagger-Witten line bundle on moduli spaces of Calabi Yaus"
- Martin Sticka "Three Descriptions of the Hull-Strominger System"
- Bryan Wang "Branes, K-homology and Elliptic homology"

This volume includes the following refereed contributions arising from the program:

- A. de Arriba de la Hera, M. Galdeano and M. Garcia-Fernandez "G_2-structures with torsion and the deformed Shatashvili–Vafa vertex algebra"
- X. de la Ossa, M. Larfors, M. Magill and E. Svanes "Quantum aspects of heterotic G_2 systems"
- E. Sharpe "An overview of Bagger-Witten line bundles"
- S. Picard "Calabi-Yau Threefolds Across Quadratic Singularities"

and the following non-refereed contributions:

- J. Knapp, J.. McGovern "A single point as a Calabi-Yau zerofold"
- S. Hosono, H. Takagi "Mirror symmetry from families of Calabi-Yau manifolds"
- X. de la Ossa, M. Elmi "Some Identities For Periods of Hulek-Verrill Threefolds"
- M. Elmi, J. McGovern, E. Scheidegger "Classically modular points in moduli spaces of Calabi-Yau threefolds"

Furthermore, we know of the following publications and preprints to which the program made significant contributions:

- J. Knapp, J. McGovern "Noncommutative resolutions and CICY quotients from a non-abelian GLSM" arXiv 2504.06147
- J McOrist, M Sticka, E Svanes "The heterotic G_2 moduli space metric" arXiv: 2502.16093
- J. McOrist, M Sticka, E Svanes "The moduli of the universal geometry of heterotic moduli"
- J McOrist, M Sticka, E Svanes "The physical moduli of heterotic G_2 string compactifications"
- B Chisamanga, J McOrist, S Picard, E Svanes, "The decoupling of moduli about the standard embedding"
- P Candelas, X de la Ossa, J McGovern "Classical Weight-Four L-value Ratios as Sums of Calabi–Yau Invariants"

- X de la Ossa, M Galdeano, E Marchetto "SW-algebras and strings with torsion"
- L Cassia, P Longhi, M Zabzine "Symplectic cuts and open/closed strings II" arXiv:2410.10960
- L Cassia, K Hristov "Constant maps in equivariant topological strings and geometric modeling of fluxes"

The aims of the program were met in various ways. There was significant progress on existing research projects that have since led to publications. New collaborations between participants have been formed. The program has led to research visits and conference invitations of Australian participants to overseas institutions.

Sebastien Picard and Jock McOrist
Guest editors

MATRIX Program:
Parameter Identifiability in Mathematical Biology

16 – 20 September 2024

Organisers

Matthew Simpson
Queensland University of Technology

Ruth Baker
University of Oxford

Oliver Maclaren
University of Auckland

Jennifer Flegg
University of Melbourne

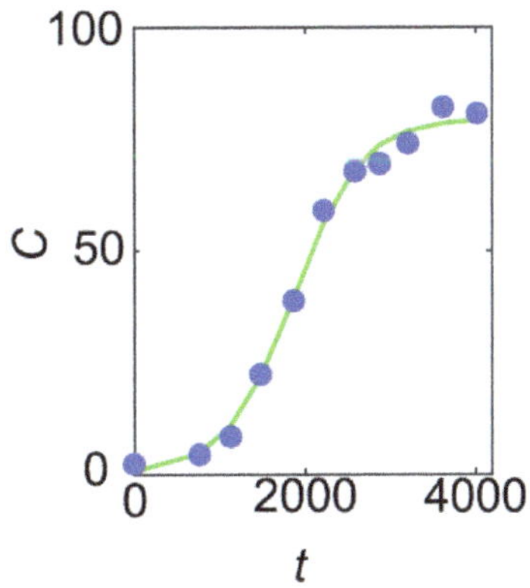

Participants

Ruth Baker (Oxford), Oliver Maclaren (Auckland), Matthew Simpson (QUT), Alejandro Villaverde (Vigo), Tyler Cassidy (Leeds), Alexander Browning (Oxford), Michael Plank (Canterbury), Stuart Johnston (Melbourne), Jennifer Flegg (Melbourne), Ryan Murphy (Melbourne), Sara Hamis (Uppsala), Carles Falcó (Oxford), Veronica Ciocanel (Duke), Jaekyoung Kim (KAIST), Isabel Cowlishaw (Auckland), Imke Botha (Melbourne), Adriana Zanca (Melbourne), Ivo Siekmann (Liverpool), Yihan Liu (QUT), Yong See Foo (Melbourne), Alexander Johnston (QUT), Ziwen Zhong (Monash), Nicholas Lam (Canterbury)

The program focused on methodologies for parameter identifiability in the field of mathematical biology. Some of the work presented at the workshop focused on developing new methodologies, whereas other work considered novel implementations of existing methodologies on new biological problems. The workshop included scientific presentations in two different formats. Firstly, the workshop included a series of broader presentations by Michael Plank who spoke about disease transmission modelling; Veronica Ciocanel who spoke about partial differential equation models of developmental biology with a focus on fluorescence recovery after photobleaching experiments; Ivo Siekmann spoke about identifiability using Markov Chain Monte Carlo approaches; Alex Villaverde spoke about novel methods for structural identifiability analysis; and Jaekyoung Kim spoke about parameter estimation problems in cell biology and biological signalling. These longer talks were presented alongside shorter focused presentations given by all remaining participants other than the organising committee, where participants had an opportunity to give a brief overview of a project they are interested in, or a recent piece of work that they have been working on.

The main activity of the workshop was a series of group work projects. Ahead of the meeting, several participants were given an opportunity to propose new research projects that participants could explore during the afternoons (and evenings) at the workshop. These research projects included projects that relate to identification of chemical reaction networks using sparse, noisy data; understanding the relationship between tools of structural identifiability and practical identifiability; and the role of identifiability in mixed effects modelling. These research projects continue to be explored by the participants, and we expect additional scientific results to appear in the literature as a result.

The aims of the program were met through initiating new collaborative research relationships that would never have been realised without the workshop. New scientific results have appeared in the literature that are related to the workshop with more to come. At least two collections of articles are being published, in the *Bulletin of Mathematical Biology*, and *Current Opinion in Systems Biology*, where workshop participants played a role as either author or editor.

The organisers

MATRIX-MFO Tandem Workshop: Invariants in Low-Dimensional Topology: Combinatorics, Geometry, and Computation

23 – 28 September 2024

Organisers

Joan Licata
Australian National University

Stephan Tillmann
University of Sydney

Stefan Friedl
University of Regensburg

Paula Truöl
ETH Zürich

Participants (MATRIX)

Stephan Tillmann (Sydney), Joan Licata (ANU), Agnese Barbensi (Queensland), Ben Burton (Queensland), Daniele Celoria (Queensland), Diarmuid Crowley (Melbourne), Zsuzsanna Dancso (Sydney), Adele Jackson (Sydney), James Morgan (Sydney), Jonathan Spreer (Sydney), Lucy Tobin (Sydney), Justin Lanier (Sydney), Dionne Ibarra (Monash), Jessica Purcell (Monash), Josh Howie (Monash), Bea Bleile (New England), Jayden Hammets (Melbourne), Adam Wood (Melbourne), Em Thompson (Monash)

Participants (MFO)

Valentina Bais (SISSA), Fathi Ben Aribi (Sorbonne), Vitalijs Brejevs (Vienna), Alessio Di Prisa (Pisa), Livio Ferretti (Geneve), Stefan Friedl (Regensburg), Daniel Galvin (Max-Planck), Tobias Hirsch (Regensburg), Chun-Sheng Hsueh (Humboldt), Christopher Johnson (Durham), Marc Kegel (Humboldt), Takahiro Kitayama (Tokyo), Naageswaran Manikandan (Humboldt), Alice Merz (Lille), Filip Misev (Regensburg), Mark Pencovitch (Glasgow), José Pedro Quintanilha (Heidelberg), Arunima Ray (Max-Planck), Diego Santoro (Wien), Lisa Schambeck (Regensburg), Saul Schleimer (Warwick), Annika Thiele (Humboldt), Paula Truöl (Max-Planck), Laura Wakelin (Max-Planck)

This workshop brought together researchers from different subfields of topology with a high potential for productive exchange. Specialists in knot theory, 3-manifolds, and 4-manifolds gathered at two locations, Creswick and Oberwolfach, for a program mixing in-person and distance collaboration on open problems of common interest. This is the second MATRIX-MFO tandem workshop in low-dimensional topology. The first, held in 2021, was particularly hybrid due to nation-wide lockdowns in Australia. This time, both institutes were able to host vibrant in-person sessions, and participants enthusiastically embraced the opportunities for mathematical and social interactions. (There may even have been karaoke.)

The workshop was organised jointly by Australian and German researchers: Stefan Friedl (Regensburg), Joan Licata, (ANU), Stephan Tillmann (Sydney), and Paula Truöl (Bonn).

The format of the MATRIX workshop was designed to support both high-value local discussion and international collaboration. In advance of the workshop, participants on both continents were invited to propose topics and vote on the suggestions. At the beginning of the week, we selected the following topics with a critical mass of interest in both locations:

1. Computable invariants of colored spatial graphs
2. Random knots
3. Small clasp number knots
4. Poincaré Duality Complexes in Dimension 4
5. Topological volume of the three-torus

Each topic had two discussion leaders, one at MFO and at MATRIX.

In the first tandem session, every participant gave a 1-minute, 1-slide introduction before initial working groups were established. The MATRIX-based researchers worked in groups during the day, with breaks for long walks and larger talks. The tandem portion of the schedule started at 17:00 after breakfast at MFO. The session began by gathering collectively for a check-in with summaries, and then teams joined separate Zoom meetings to work with their counterparts at MFO. We often reconvened after dinner at MATRIX for mini-talks from junior researchers or intensive discussions in working groups. The MATRIX participants then slept while work continued overseas. The MFO teams recorded Zoom summaries on their progress or new insights for us to watch when we started again the next day. Each working group also had a shared Google folder for notes, ideas, papers and pictures. Participants were encouraged to switch between groups during the week as they wished; most remained with their original choices, but there was some movement as the problems under discussion evolved, and in fact, distinct working groups converged on common topics.

In summary we feel that the collaboration across time zones and continents was a great success. The technical equipment at MATRIX and MFO worked extremely well and the big screens and the excellent cameras available in multiple rooms made it easy to share ideas between groups. The eight hour time difference gave each location the opportunitye to work on the problems separately and in smaller groups.

Workshop participants were enthusiastic about the format and have expressed a strong desire to hold a similar even in the near future. The organisers also wish to thank the staff at MATRIX and MFO for making this meeting possible and for providing us with all the required technical support.

We now summarise the work of each of the research groups.

Invariants of spatial graphs: Let G be an abstract graph, possibly with decorations. Given an embedding $\phi : G \to S^3$, the image Γ is called a spatial graph. A basic problem is to classify spatial graphs for a fixed G up to ambient isotopy. This should be viewed as a generalization of the classification of knots and links, and indeed the robust toolkit for analyzing knots and links can be leveraged effectively in studying more general spatial graphs. Our collaboration sought new invariants of spatial graphs. We considered Blanchfield pairings on Alexander modules and sets of representations of $\pi_1(S^3 \setminus \Gamma)$ into finite groups. We also worked on extending the notion of constituent knots and links using an "unzipping" construction.

Models for random knots: Several fundamental conjectures in low-dimensional topology are verified in many cases but have resisted rigorous proofs. Such problems admit a probabilistic point of view: given a property P of knots, what is the probability that a random knot K satisfies the property P? The answer will in general depend very heavily on the chosen model and, in particular, it will provide information on the fixed probability measure rather than on the set of knots itself. We examined grid diagrams and 1-vertex triangulations as models of knots suited to this approach. We performed explicit computer experiments to create random knots for both models, studying the prevelance of unknots, hyperbolicity, and invariants associated to sliceness.

Poincaré duality 4-complexes: Problems concerning 4-dimensional Poincaré duality complexes (PD_4-complexes) were considered. These are 4-dimensional CW-complexes equipped with a top-dimensional homology class which induces Poincaré duality isomorphisms via the cap product. We studied the equivariant intersection form $\lambda\colon H^2(X;\mathbb{Z}\pi) \times H^2(X;\mathbb{Z}\pi) \to \mathbb{Z}\pi$ which is defined using the cap product with the fundamental class $[X]$. Among other topics, we considered realisation problems and studied what data is required to recover the homotopy type of a complex.

Obstructing ribbon clasp number one: Clasp number is a natural measure of complexity for a knot $K \subset S^3$. We examined the relationship between different notions of clasp number for certain classes of knots. For example, a knot K with $g_4(K) = 0$ and $c_r^*(K) > 0$ would be a counterexample to the slice-ribbon conjecture; instead of looking for such an example, we considered the problem of finding an obstruction to a knot K from having $c_r^*(K) = 1$; we also discussed associated problems related to lattice embeddings and double branched covers.

Computing the topological volume of some three-manifolds: The *topological volume* of a compact orientable three-manifold with empty or toroidal boundary is

$$\mathrm{volt}(M) = \min\{\mathrm{vol}(M - L) \mid L \text{ is a link in } M \text{ such that } M - L \text{ is hyperbolic}\}.$$

Work of Kegel, Ray, Spreer, Thompson, and Tillmann that arose from the MATRIX-

MFO tandem workshop in 2021 gave general upper and lower bounds for the topological volume and determined the complements of volt-realising links for all but finitely many lens spaces. In this working group we tried to get a better understanding of the topological volume of some manifolds that were not already covered by previous work. In particular, we focused on product three-manifolds of the type $F \times \mathbb{S}^1$, where F is a compact orientable surface.

The organisers

MATRIX Program: Applications of Topological Data Analysis to Mathematical Biology

11 – 15 November 2024

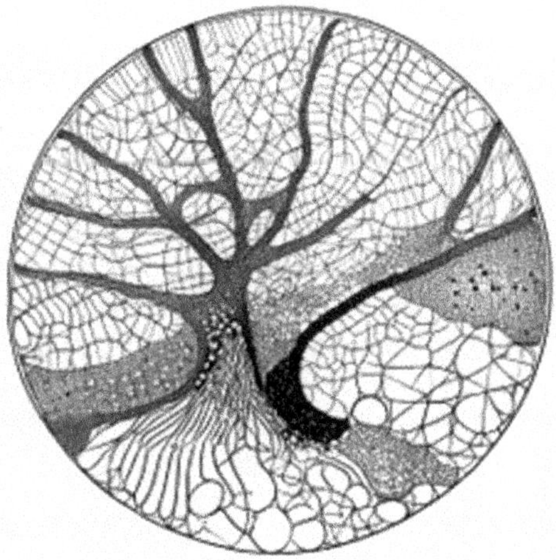

Organisers

Marcy Robertson
University of Melbourne

Katherine Turner
Australian National University

Yossi Bokor Bleile
Aalborg University, Denmark

Participants

Kate Turner (ANU), Marcy Robertson (Melbourne), Yossi Bokor Bleile (Aalborg), Heather Harrington (Max Planck & Oxford), Vanessa Robins (ANU), Facundo Memoli (Ohio State), Freya Jensen (Heidelberg), Musashi Koyama (ANU), Eve Cheng (ANU), James Nichols (ANU), Howard Bondell (Melbourne), Brittany Fasy (Montana State), Bea Bleile (New England), Agnese Barbensi (Queensland), Sophie Raynor (James Cook), Shmuel Weinberger (Chicago), Theresa Knudsen (Aalborg), Sayan Mukherjee (Max Planck & Leipzig), Thorsten Hertl (Melbourne), Thomas Burnett (ANU), Kelly Maggs (Max Planck), Renata Turkes (INRIA)

The program focused on uniting researchers in algebraic topology, topological data analysis (TDA), and mathematical biology. The primary goal was to foster collaboration among these diverse fields, bridging gaps between pure mathematics and applied biological research.

The first week of the program included lectures by Vanessa Robins, Heather Harrington, Facundo Memoli, and Agnese Barbensi.

During the program there were several groups of collaborators, which have continued discussions since, in particular using tools from TDA to study coral taxonomy. Coral taxonomy has traditionally been based on morphological features (at various scales) but, over the last decades, advances in genetic sequencing have thrown the existing taxonomy into disarray. This project considers whether applied topology, and TDA in particular, can help us understand the relationships between genetics, environment and morphology in hard corals.

The formation of research collaborations between attendees is the main marker of the success of the workshop.

The organisers

MATRIX Program: Tensor Categories, Quantum Symmetries, and Mathematical Physics

18 – 29 November 2024

Organisers

Kevin Coulembier
University of Sydney

Justine Fasquel
University of Melbourne

Zachary Fehily
University of Melbourne

Julia Plavnik
Indiana University Bloomington

David Ridout
University of Melbourne

James Tener
Australian National University

Participants

Iván Angiono (Córdoba), Peter Bouwknegt (ANU), Johannes Flake (Bonn), Terry Gannon (Alberta), Pinhas Grossman (UNSW), Mee Seong Im (Johns Hopkins), Shashank Kanade (Denver), Simon Lentner (University Hamburg), Peter McNamara (Melbourne), Robert McRae (Tsinghua), Cris Negron (Southern California), Yoshiko Ogata (Kyoto), Flor Orosz Hunziker (Colorado), Bregje Pauwels (Macquarie), Thomas Quella (Melbourne), Jorgen Rasmussen (Queensland), Eric Rowell (Texas A&M), Christoph Schweigert (Hamburg), Alex Sherman (Sydney), Kenichi Shimizu (Shibaura), Daniel Tubbenhauer (Sydney), Kevin Coulembier (Sydney), Justine Fasquel (Melbourne), Zachary Fehily (Melbourne), Julia Plavnik (Indiana), David Ridout (Melbourne), James Tener (ANU), Daniel Dunmore (UNSW), Thomas Dunmore (UNSW), Tyler Franke (Melbourne), Ethan Fursman (Melbourne), Jared Heymann (Melbourne), Joe Newton (Sydney), Tao Qin (Sydney), Jakob Sloan (Melbourne), Will Stewart (TU Munich), Brae Vaughan-Hankinson (Melbourne), Shigenori Nakatsuka (Alberta), Ana Kontrec (Kyoto), Monique Muller Lopes Rocha (Sao Joao del-Rei), Taiki Shibata (Okayama), Simon Wood (Cardiff), John Huerta (Lisbon & ANU)

This program focused on the theory of tensor categories and its recent applications to representation theory and mathematical physics. Participants learned of the latest progress on understanding symmetric and modular tensor categories, while enjoying an assortment of links to quantum groups, subfactors, topological field theories and vertex algebras. The workshop gathered leading experts on all these topics to facilitate collaborations between different groups working on similar themes.

The first week of the program featured five introductory 90-minute lectures, five 60-minute research talks and four 30-minute contributed talks. There were an additional eight 60-minute research talks, and six 30-minute contributed talks in the second week. The schedule was deliberately kept light to leave plenty of time for clarifying discussions and collaborative interaction. This was particularly appreciated by participants.

One paper arising from the program is included in this volume. Authored by Eric Rowell (Texas A&M) and Yuze Ruan (Tsinghua), it concerns the classification of indecomposable yet reducible representations of the braid group B_3.

James Tener
Guest editor

MATRIX Program: Algebraic Geometry

9 – 13 December 2024

Organisers

Daniel Chan
University of New South Wales

Ivan Cheltsov
University of Edinburgh

Behrouz Taji
University of New South Wales

Claire Voisin
Centre National de la Recherche Scientifique, IMJ Paris

Geordie Williamson
University of Sydney

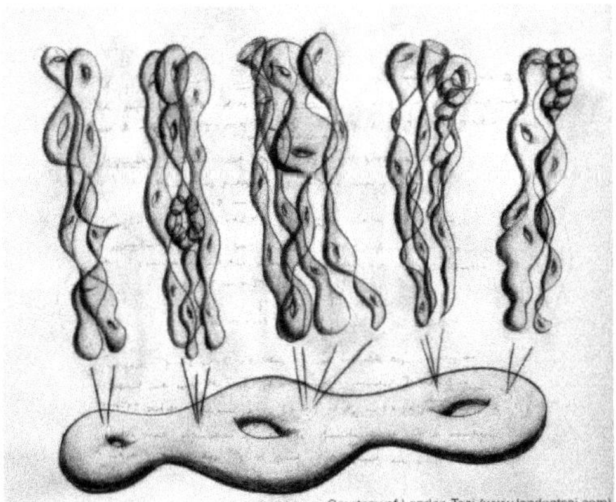

Courtesy of London Tsai (www.londontsai.com)

Participants

Paolo Cascini (Imperial), Mark De Cataldo (Stony Brook), Jack Hall (Melbourne), Elham Izadi (UC San Diego), James McKernan (UC San Diego), Yujiro Kawamata (Tokyo), Sándor Kovács (Washington), Amnon Neeman (Milano), Jihun Park (IBS / POSTECH), Frédéric Mangolte (Aix-Marseille), Kento Fujita (Osaka), Ziquan Zhuang (Johns Hopkins), Kenneth Ascher (UC Irvine), Kristin DeVleming (Washington), De-Qi Zhang (Singapore), Gavril Farkas (Humboldt), Philip Engel (Illinois), Susanna Zimmermann (Paris-Saclay), Evgeny Shinder (Sheffield), Hamid Abban (Nottingham), Erwan Rousseau (Brest), Eduard Looijenga (Chicago), Benjamin Bakker (UIC), Tiago Duarte Guerreiro (Paris-Saclay), Dougal Davis (Melbourne), Daniel Chan (UNSW), Ivan Cheltsov (Edinburgh), Behrouz Taji (UNSW), Ekaterina Amerik (HSE Moscow), Tyson Klingner (Washington), Fei Peng (Melbourne), Davood Nejaty (Melbourne), Erroxe Alberdi (Warwick), Joseph Malbon (Edinburgh), Svetlana Makarova (ANU), Antoine Pinardin (Edinburgh), Julia Schneider (Sheffield), Justin Sawon (North Carolina), Oliver Li (Melbourne), Dominic Matan (UNSW)

The week-long program focused on the following three key active areas of research in algebraic geometry:

- moduli, stability and K-stability,
- Hodge theory, and

- classification theory in higher dimensions (including MMP for Kähler geometry and foliations).

With research talks only being scheduled in mornings, the program devoted all afternoons to open discussion and problem-solving sessions around these topics. The program featured the following talks:

- James McKernan "Kähler versus projective"
- Jihun Park "Simply connected positive Sasakian 5-manifolds and log del Pezzo surfaces"
- Kenneth Ascher "Boundedness results for families of varieties"
- Ziquan Zhuang "Boundedness of singularities and discreteness of local volumes"
- Evgeny Shinder "Twists of intermediate Jacobian fibrations"
- Sándor Kovács "KSB stability is automatic in codimension 3"
- Mark De-Cataldo "The decomposition theorem for the logarithmic Hitchin fibration"
- Paolo Cascini "On the minimal model program for foliated varietie"
- Phillip Engel "The non-abelian Hodge locus"
- Eduardo Looijenga "A ball quotient of dimension 13, cubic threefold moduli and the Monster group"
- Gavril Farkas "Koszul modules and resonance in algebraic geometry"
- Jack Hall "Finiteness of integral transforms"
- Benjamin Bakker "The linear Shafarevich conjecture for quasiprojective varieties"
- Kristin DeVleming "The Hassett-Keel program in genus 4"
- Yujiro Kawamata "Non-commutative twisted deformations of varieties and the derived McKay correspondence"

The organisers

MATRIX Program: Harmonic and Microlocal Analysis in Partial Differential Equations

16 – 20 December 2024

Organisers

Zihua Guo
Monash University

Andrew Hassell
Australian National University

Qiuye Jia
Australian National University

Monica Visan
UCLA

Participants

Rowan Killip (UCLA), Kenji Nakanishi (Kyoto), Gong Chen (Georgia Tech), XueYing Yu (Oregon State), Xiaoxu Wu (Toronto), Jesse Gell-Redman (Melbourne), Moritz Doll (Melbourne), Pierre Portal (ANU), Todd Oliynyk (Monash), Adam Sikora (Macquarie), Xuan Thinh Duong (Macquarie), Melissa Tacy (Auckland), Tim Candy (Otago), Zihua Guo (Monash), Andrew Hassell (ANU), Qiuye Jia (ANU), Monica Visan (UCLA), Junyong Zhang (Beijing Tech), Jason Murphy (Oregon), Andreia Chapouto (CNRS Versailles), Yuzhou Zou (Northwestern), Yilin Ma (The Australian National), Alex Rose (ANU), Madelyne Brown (Auckland), Jared Wunsch (Northwestern), Satoshi Masaki (Hokkaido), Rana Badreddine (UCLA), Justin Forlano (Edinburgh), Alice Savvinos (Melbourne),

The mathematical theory of partial differential equations (PDE) draws on many subfields of analysis. Two of the most significant branches of analysis that have contributed to PDE theory are harmonic analysis and microlocal analysis. Harmonic analysis, in the tradition of Calderón and Stein, has been an indispensable tool for studying nonlinear dispersive equations such as nonlinear Schrödinger (NLS) and nonlinear wave equations, KdV and related equations, general relativity, fluid flow and many other types of PDE. Remarkable results on the threshold between scattering vs soliton behaviour in focusing NLS has been obtained over the past 20 years.

More recently, microlocal analysis has been fashioned into a powerful tool to study both linear and nonlinear PDE, based on the "Fredholm method". Successes include precise results in scattering theory (linear and nonlinear), including for Einstein's equations, nonlinear wave equations and NLS. Hintz and Vasy's celebrated work on the nonlinear stability of the Kerr-de Sitter family of black hole spacetimes provided great momentum for further development of this approach.

At the same time, a fusion between harmonic and microlocal analysis is being developed by several Australian researchers, by developing the theory of Hardy spaces associated to Fourier Integral Operators, function spaces on which Fourier Integral Operators (FIOs) act boundedly. The goal of the workshop was to foster and further develop this synergy between harmonic and microlocal analysis.

The program was held over one week, directly after the AustMS 2025 annual conference (which was held jointly with the US and NZ mathematical societies). There were 10 talks during the workshop, with ample time (more than 16 hours) reserved for collaboration in small groups.

A major topic of the workshop was recent developments in analysis of NLS and nonlinear Klein-Gordon equations. These recent developments include new methods for analyzing completely integrable equations (Badreddine, Chapouto and Killip), statistical or almost-sure properties of solutions (Forlano and Chapouto), new results for perturbed soliton and multi-soliton solutions (Nakanishi and Chen), and a novel microlocal approach to such equations (Gell-Redman and Zou). We also had lectures on unique continuation for Schrödinger equations (Yu) and local energy decay for the Schrödinger equation (Wu). These lectures stimulated lively conversations at MATRIX House and during meal times.

The organisers were thrilled with the quality of the talks and the conversations that they engendered. They look forward to new research arising from the discussions during the collaborative part of the workshop.

The organisers

Contents

Part I

Refereed Articles

Chapter 1

MATRIX Program: Mathematics and Physics of Integrability

The q-deformed Haldane–Shastry chain at $q = i$ with even length

Adel Ben Moussa, Jules Lamers and Didina Serban

Abstract In this note we announce some results extending our recent work with A. Toufik on the free-fermion point $q = i$ of the Haldane–Shastry chain to the case with an even number N of sites. The resulting long-range version of the Heisenberg XX chain may be viewed as a model of fermions with extended $\mathfrak{gl}(1|1)$ symmetry. Unlike for odd N, the conserved charges are nilpotent and exhibit Jordan blocks.

1 Introduction

The advent of logarithmic conformal field theories at the end of the last century, with applications in condensed matter theory and high-energy physics, has led to a growing interest in models whose algebraic structures are not completely reducible with respect to their algebras of symmetries [4, 10, 11]. More concretely, this property is related to conserved charges that are not diagonalisable, exhibiting 'non-trivial Jordan blocks'. An example of such a structure arises in the quantum-group invariant open XX chain with an even number of sites. Its symmetries make it a non-semisimple representation for a specialisation of $U_q(\mathfrak{sl}_2)$ at $q = i$ and, in the continuum limit, yields a well-known logarithmic CFT called the free symplectic fermions theory [10].

In this note, we announce new results for the free-fermion point of the Haldane–Shastry chain, extending the recent work [1] to the case with an even number of sites. The Haldane–Shastry model [5, 6] is an exactly solved spin chain with long-range

Adel Ben Moussa, Didina Serban
Université Paris–Saclay, CNRS, CEA, Institut de Physique Théorique, 91191 Gif-sur-Yvette, France
e-mail: {adel.benmoussa,didina.serban}@ipht.fr

Jules Lamers
Deutsches Elektronen-Synchrotron DESY, Notkestraße 85, 22607 Hamburg, Germany
e-mail: jules.lamers@desy.de

D. R. Wood et al. (eds.), *2024 MATRIX Annals, Part II*, MATRIX Book Series 8,
https://doi.org/10.1007/978-3-032-16206-9_1

interactions and remarkable properties including explicit eigenvectors and extended (Yangian) spin symmetry [2, 7]. It admits an integrable q-deformation, similar to how the Heisenberg XXZ chain generalises the isotropic Heisenberg XXX chain, called the q-deformed Haldane–Shastry (qHS) chain [8, 9, 12]. Amongst others it has a hierarchy of commuting charges, and q-deformed (quantum-affine) extended spin symmetry; see the Supplemental Material of [1] for a summary. While the properties of the qHS chain mimic that of HS at generic values of q, new phenomena emerge when q is a root of unity. We consider the case of $q = i$. Just like the Heisenberg XXZ chain reduces to the nearest-neighbour XX model at this point, the $q = i$ HS (iHS) chain is a model of free fermions, now with long-range hopping. Its properties depend critically on the parity of the number N of sites.

In [1], we studied the iHS chain with odd N. Following [3], the spin-chain representation of the Temperley–Lieb algebra can at $q = i$ be related to quadratic combinations of non-unitary versions of Jordan–Wigner fermions. We exhibited the first few commuting charges inherited from qHS explicitly in the fermionic language. This includes a chiral (parity-odd) hamiltonian that is quadratic in fermions, and a parity-even hamiltonian with quartic interactions implementing a statistical repulsion on the lattice. Both hamiltonians have a linear dispersion relation. The model is not translationally invariant in the traditional sense, yet both hamiltonians commute with a suitably modified 'quasi-translation' operator playing a pivotal role in [1]. Although the model is non-unitary, by combined parity and time-reversal symmetry, it has a real spectrum.

In this note we focus on an even number N of sites. The main difference with odd N is that the commuting charges of the qHS chain now exhibit singularities at $q = i$ that require regularisation. The renormalised conserved charges all become nilpotent operators. In Section 2 we give the explicit form of the first few of them in fermionic language and discuss some symmetries. In Section 3 we discuss examples for low N, illustrating that these operators have a non-trivial Jordan-block structure reminiscent of the $U_q(\mathfrak{sl}_2)$-invariant XX chain at $q = i$. In the present case, the Jordan blocks are larger, and we attribute this phenomenon to the extended $U_q(\widehat{\mathfrak{sl}}_2)$ spin symmetry of the qHS chain. We will elaborate on the extended symmetry its representation in a forthcoming work.

2 The conserved quantities

The conserved quantities of the parent model are rational in q. In the case of even length $N = 2L$, their expressions are singular at $q \to i$, but well-defined conserved charges can be extracted by carefully renormalising the original quantities. In this section, we define these quantities as elements of the Temperley–Lieb algebra, which can be represented as fermionic operators on the Fock space with N fermions, and consider their algebraic properties.

Algebraic setting We define the *free-fermion Temperley–Lieb algebra* by generators $e_1, \dots e_{n-1}$ and relations

$$\begin{aligned} e_i^2 &= 0\,, \\ e_i e_j e_i &= e_i\,, && \text{if } |i-j| = 1\,, \\ \left[e_i, e_j\right] &= 0\,, && \text{if } |i-j| > 1\,. \end{aligned} \tag{1}$$

This is the usual Temperley–Lieb algebra with 'loop fugacity' $\beta = 0$. Following [3], we realise this abstract algebra on a fermionic Fock space using *non-unitary* fermionic operators with the following anticommutation relations

$$\{f_i, f_j^+\} = (-1)^i \delta_{ij}\,, \quad \{f_i, f_j\} = \{f_i^+, f_j^+\} = 0\,. \tag{2}$$

They provide a basis of Fock vectors $f_{i_1}^+ \cdots f_{i_M}^+ |\varnothing\rangle$ for the Hilbert space of states. The free-fermion Temperley–Lieb algebra (1) can then be represented as quadratic expressions in two-site fermionic operators

$$e_k \equiv g_k^+ g_k\,, \qquad g_k \equiv f_k + f_{k+1}\,, \quad g_k^+ = f_k^+ + f_{k+1}^+\,, \qquad 1 \leqslant k < N\,. \tag{3}$$

As in [1], the building blocks of our conserved charges are nested commutators of *adjacent* Temperley–Lieb generators and their (disjoint) anticommutators

$$\begin{aligned} e_{[k,l]} &\equiv \left[e_k, \left[e_{k+1}, \dots \left[e_{l-1}, e_l\right] \dots\right]\right] = \left[\left[\dots\left[e_k, e_{k+1}\right], \dots e_{l-1}\right], e_l\right], \quad k < l\,, \\ \left\{e_{[i,j]}, e_{[k,l]}\right\} &= e_{[i,j]}\, e_{[k,l]} + e_{[k,l]}\, e_{[i,j]}\,, \qquad i \leqslant j < k \leqslant l\,. \end{aligned} \tag{4}$$

under the convention $e_{[k,k]} \equiv e_k$. It can be shown [1] that the nested commutators (4) can be expressed as

$$e_{[k,l]} = s_{kl} \left(g_l^+ g_k + (-1)^{l-k} g_k^+ g_l\right), \qquad k < l\,, \tag{5}$$

where $s_{kl} \equiv (-1)^{(k-l)(k+l-1)/2}$. In particular, all these quantities are quadratic in fermions f_k. It follows that the anticommutators of nested commutators $\left\{e_{[k,l]}, e_{[m,n]}\right\}$ that we will consider in what follows are at most quartic in fermions.

Definition of the hamiltonians The conserved charges at $q = i$ can be obtained, similarly to the odd case [1], in terms of the numbers $t_k \equiv \tan \frac{\pi k}{N}$. However, in this case the interaction potentials contain singularities (poles) from the antipodal points with $k = i - j = N/2 \equiv L$. To regularise these poles we introduce a parameter α and let

$$t_k \equiv \tan \frac{\pi k}{N}\,, \quad k \neq \frac{N}{2}\,, \qquad t_{N/2} = t_{N/2}(\alpha) \equiv \tan \frac{\pi\,(N/2 + \alpha)}{N} = \cot \frac{\pi\,\alpha}{N}\,. \tag{6}$$

The shift by α regularises the simple pole of $t_{N/2}$ at $\alpha = 0$. From the products

$$t_{k,l} \equiv \prod_{i=k}^{l-1} t_i\,, \quad k < l\,, \qquad t_{k,k} \equiv 1\,, \tag{7}$$

define

$$h^{\mathrm{L}}_{ij}(\alpha) \equiv \sum_{n=j+1}^{N} t_{n-j,n-i}\big(1-(-1)^i t^2_{n-i,n}\big)\,, \qquad h^{\mathrm{R}}_{ij}(\alpha) \equiv (-1)^{i-j}\, h^{\mathrm{L}}_{N-j,N-i}(\alpha)\,, \tag{8}$$

as well as

$$\begin{aligned} h^{\mathrm{L}}_{ij;kl}(\alpha) &= (-1)^{k-j} \sum_{n=l+1}^{N} t_{n-l,n-j}\, t_{n-k,n-i}\big(1-(-1)^i t^2_{n-i,n}\big)\,, \\ h^{\mathrm{R}}_{ij;kl}(\alpha) &= (-1)^{l-j+k-i}\, h^{\mathrm{L}}_{N-l,N-k;N-j,N-i}(\alpha)\,. \end{aligned} \tag{9}$$

These coefficients have at most double poles in α that we remove by taking their residues,

$$\begin{aligned} h^{+}_{ij} &\equiv \frac{\pi^2}{2N^2} \lim_{\alpha\to 0} \alpha^2 \Big(h^{\mathrm{L}}_{ij}(\alpha) + h^{\mathrm{R}}_{ij}(\alpha)\Big)\,, \\ h^{-}_{ij;kl} &\equiv \frac{\pi^2}{2N^2} \lim_{\alpha\to 0} \alpha^2 \Big(h^{\mathrm{L}}_{ij;kl}(\alpha) - h^{\mathrm{R}}_{ij;kl}(\alpha)\Big)\,. \end{aligned} \tag{10}$$

With these notations, two of the conserved Hamiltonians are given by

$$\begin{aligned} \mathrm{H}^{+} &= \sum_{1\leqslant i\leqslant j<N} h^{+}_{ij}\, e_{[i,j]}\,, \\ \mathrm{H}^{-} &= \sum_{1\leqslant i\leqslant j<k\leqslant l<N} h^{-}_{ij;kl}\, \{e_{[i,j]}, e_{[k,l]}\}\,. \end{aligned} \tag{11}$$

We can prove that H^{+} and H^{-} are non-vanishing and satisfy

$$\big[\mathrm{H}^{+},\mathrm{H}^{-}\big] = 0\,. \tag{12}$$

This follows from the commutativity of the 'chiral' hamiltonians $H^{\mathrm{L,R}}$ of the parent model [9] by arguments that are similar to [1]: in short, up to overall factors, H^{-} is the limit of $H^{\mathrm{L}} - H^{\mathrm{R}}$ as $q \to i$, while H^{+} is the residue of $H^{\mathrm{L}} + H^{\mathrm{R}}$ at $q = i$.

Like in [1], we have a quadratic and a quartic hamiltonian, both expressed via nested commutators (4) with coefficients that are products of tangents. Unlike for odd N, however, the hamiltonians (11) are not diagonalisable, and they are nilpotent: all their eigenvalues vanish. This feature, which is not obvious from the above expressions, will be made more explicit in the examples below.

Other conserved charges In addition to the two Hamiltonians, we get two further commuting quantities by taking the residue of the quasi-translation operator $\mathrm{G}(q)$ and its inverse [9] as $\mathrm{G}^{\pm} \equiv \lim_{q\to i}(q-i)\,\mathrm{G}(q)^{\pm 1}$. Explicitly, we have

$$\begin{aligned} \mathrm{G}^{+} &= (1+t_{N-1}\,e_{N-1})\cdots(1+t_{L+1}\,e_{L+1})\,e_L\,(1+t_{L-1}\,e_{L-1})\cdots(1+t_1\,e_1)\,, \\ \mathrm{G}^{-} &= \qquad (1-t_1\,e_1)\cdots(1-t_{L-1}\,e_{L-1})\,e_L\,(1-t_{L+1}\,e_{L+1})\cdots(1-t_{N-1}\,e_{N-1})\,, \end{aligned} \tag{13}$$

where we abbreviate $L \equiv N/2$. It follows from the definition of $\mathrm{G}^{\pm}$ as a limit and the properties of $\mathrm{G}(q)$ that

$$G^+ G^- = G^- G^+ = 0. \tag{14}$$

We can show that these charges are conserved

$$\left[G^\pm, H^\pm\right] = \left[G^\pm, H^\mp\right] = 0. \tag{15}$$

Spectrum Note that G^+ and G^- are here unsuited for the role of the 'quasi-translation' that $\lim_{q\to i} G(q)$ assumes for odd N [1], since neither of them is invertible. Accordingly, the basis of Fourier transformed fermions $\tilde{\Psi}_i$ in [1] cannot be defined in the same way here. All the conserved charges that we have defined here are nilpotent operators, and their spectrum will be given in terms of Jordan blocks. Numerical evidence indicated that at each length there is a single Jordan block of size $L+1 \equiv N/2+1$, and collections of blocks of smaller size.

Symmetries As in the case of the odd-length chain [1], we can define a parity transformation by $P(e_i) = e_{N-i}$. Under this transformation we have

$$P(H^\pm) = \pm H^\pm, \quad P(G^\pm) = G^\mp. \tag{16}$$

Note that as opposed to the odd-length case, the quadratic hamiltonian is here the one that is parity invariant while the quartic hamiltonian is anti-invariant.

It follows from the presentation of the model in terms of the Temperley–Lieb algebra that the model possesses a global $\mathfrak{gl}(1|1)$-symmetry given by

$$F_1 = \sum_{i=1}^{N} f_i, \quad F_1^+ = \sum_{i=1}^{N} f_i^+, \quad N = \sum_{i=1}^{N} (-1)^i f_i^+ f_i, \quad E = \sum_{i=1}^{N} (-1)^i = 0. \tag{17}$$

The vanishing of the central charge E is specific to even N and signals the occurrence of indecomposable features in the representations, hence the Jordan blocks [3]. The $\mathfrak{gl}(1|1)$ symmetry can only account for Jordan blocks of size two, and is only part of the symmetry, as can be concluded from the $U_q(\widehat{\mathfrak{sl}}_2)$ action for the parent model with generic q. The limit to $q = i$ of this algebraic structure will be considered in more detail in a future paper.

In the next section we will give examples of the action of the conserved charges and the symmetry algebra for systems of small sizes.

3 Small-size examples

In this section we illustrate the limiting procedure $q \to i$ on the smallest even size $N = 2$ and we give the conserved quantities and the structure of the Hilbert space for the smallest few sizes.

Origin from qHS for $N = 2$ It is instructive to consider $N = 2$ in some detail. Here, the hamiltonians $H^L(q)$ and $H^R(q)$ defined in [9] for general q are equal, up to normalisation, to the Temperley–Lieb generator

$$e(q) = \begin{pmatrix} 0 & 0 & 0 & 0 \\ 0 & q^{-1} & -1 & 0 \\ 0 & -1 & q & 0 \\ 0 & 0 & 0 & 0 \end{pmatrix}, \tag{18}$$

here expressed in the computational basis $|\uparrow\uparrow\rangle, |\uparrow\downarrow\rangle, |\downarrow\uparrow\rangle, |\downarrow\downarrow\rangle$ of $\mathbb{C}^2 \otimes \mathbb{C}^2$. More precisely

$$\mathrm{H}^+(q) = \mathrm{H}^{\mathrm{L}}(q) + \mathrm{H}^{\mathrm{R}}(q) = \frac{e(q)}{q+q^{-1}}, \qquad \mathrm{H}^-(q) = \mathrm{H}^{\mathrm{L}}(q) - \mathrm{H}^{\mathrm{R}}(q) = 0. \tag{19}$$

$\mathrm{H}^+(q)$ has eigenvalues $\{0^3, 1\}$ with eigenbasis basis formed by the q-symmetric states $|\uparrow\uparrow\rangle$, $|\uparrow\downarrow\rangle + q^{-1}|\downarrow\uparrow\rangle$, $|\downarrow\downarrow\rangle$ and the q-antisymmetric state $|\uparrow\downarrow\rangle - q\,|\downarrow\uparrow\rangle$. When $q \to i$ the matrix elements of of $H^+(q)$ diverge while the energies remain finite, and the q-symmetric and q-antisymmetric one-magnon states become collinear. In order to remove this pathology one can rescale the hamiltonian as

$$\mathrm{H}^+ = \lim_{q\to i} (q+q^{-1})\,\mathrm{H}^+(q) = \lim_{q\to i} e(q), \tag{20}$$

which sends all the eigenvalues to 0.

Results for low *N* After rescaling and sending q to i we find it more convenient to use the Fock basis associated to the fermions defined in (2). Write M for the number of fermions.

***N* = 2.** In the fermionic description the hamiltonian (20) for the smallest length is

$$\mathrm{H}^+ = \mathrm{G}^\pm = e_1 = (f_1^+ + f_2^+)(f_1 + f_2), \qquad \mathrm{H}^- = 0. \tag{21}$$

The vectors in the representation are given by

$$|\varnothing\rangle, \quad (f_1^+ + f_2^+)\,|\varnothing\rangle, \quad (f_1^+ - f_2^+)\,|\varnothing\rangle, \quad f_1^+ f_2^+\,|\varnothing\rangle. \tag{22}$$

which form a four-dimensional indecomposable representation of $\mathfrak{gl}(1|1)$. H^+ has a 2×2 Jordan block in the $M=1$ sector.

***N* = 4.** The conserved quantities are given by

$$\begin{aligned} \mathrm{G}^\pm &= (1 - t_1\, e_{2\pm 1})\, e_2\, (1 + t_1\, e_{2\mp 1}), \\ \mathrm{H}^+ &= [e_1 + e_3, e_2] + e_1 + e_3 - 2e_2, \\ \mathrm{H}^- &= -2\{e_1 - e_3, e_2\} - \{e_1, [e_2, e_3]\} - \{[e_1, e_2], e_3\} \\ &= -2\big(\{e_1 - e_3, e_2\} + e_1\, e_2\, e_3 - e_3\, e_2\, e_1\big). \end{aligned} \tag{23}$$

Both H^+ and H^- have the following Jordan-block structure: a 1×1 block at $M=0$ (and $M=4$), two 2×2 blocks at $M=1$ (and at $M=3$), and a 3×3 block plus three 1×1 blocks at the equator $M=2$.

In general, $\mathrm{G}^{\pm}$ have Jordan blocks up to size $M+1$ at each $M \leqslant N/2$. For $N \geqslant 6$ the Jordan-block structures of $\mathrm{H}^{\pm}$ do not necessarily coincide.

Acknowledgements We thank A. Toufik for collaboration at an early stage of this work. JL was funded by ERC-2021-CoG – BrokenSymmetries 101044226. DS thanks the organisers of the program *Mathematics and Physics of Integrability* (MPI2024) held in July 2024 at Matrix in Creswick, Victoria, Australia, for hospitality, and SMRI for financial support.

References

1. Ben Moussa, A., Lamers, J., Serban, D., Toufik, A.: A solvable non-unitary fermionic long-range model with extended symmetry. arXiv preprint arXiv:2404.10164 (2024)
2. Bernard, D., Gaudin, M., Haldane, F.D.M., Pasquier, V.: Yang–Baxter equation in long range interacting systems. J. Phys. A **26**, 5219–5236 (1993). DOI 10.1088/0305-4470/26/20/010
3. Gainutdinov, A.M., Read, N., Saleur, H.: Continuum limit and symmetries of the periodic $gl(1|1)$ spin chain. Nucl. Phys. B **871**, 245–288 (2013). DOI 10.1016/j.nuclphysb.2013.01.018
4. Gurarie, V.: Logarithmic operators and logarithmic conformal field theories. Journal of Physics A: Mathematical and Theoretical **46**(49), 494,003 (2013)
5. Haldane, F.D.M.: Exact Jastrow–Gutzwiller resonating-valence-bond groundstate of the spin-1/2 antiferromagnetic Heisenberg chain with $1/r^2$ exchange. Phys. Rev. Lett. **60**, 635 (1988). DOI 10.1103/PhysRevLett.60.635
6. Haldane, F.D.M.: Physics of the ideal semion gas: spinons and quantum symmetries of the integrable Haldane–Shastry spin chain. In: A. Okiji, N. Kawakami (eds.) Correlation effects in low-dimensional electron systems, vol. 118. Springer (1994). DOI 10.1007/978-3-642-85129-2_1
7. Haldane, F.D.M., Ha, Z.N.C., Talstra, J.C., Bernard, D., Pasquier, V.: Yangian symmetry of integrable quantum chains with long-range interactions and a new description of states in conformal field theory. Phys. Rev. Lett. **69**, 2021 (1992). DOI 10.1103/PhysRevLett.69.2021
8. Lamers, J.: Resurrecting the partially isotropic Haldane–Shastry model. Phys. Rev. B **97**, 214,416 (2018). DOI 10.1103/PhysRevB.97.214416
9. Lamers, J., Pasquier, V., Serban, D.: Spin-Ruijsenaars, q-deformed Haldane–Shastry and Macdonald polynomials. Commun. Math. Phys. **393**, 61–150 (2022). DOI 10.1007/s00220-022-04318-9
10. Read, N., Saleur, H.: Enlarged symmetry algebras of spin chains, loop models, and S-matrices. Nucl. Phys. B **777**, 263 (2007). DOI 10.1016/j.nuclphysb.2007.03.007
11. Rozansky, L., Saleur, H.: Quantum field theory for the multi-variable alexander-conway polynomial. Nuclear Physics B **376**(3), 461–509 (1992)
12. Uglov, D.: The trigonometric counterpart of the Haldane–Shastry model. arXiv preprint arXiv:hep-th/9508145 (1995)

A note on $T\overline{T}$ deformations and boundaries

Nicolò Brizio, Tommaso Morone and Roberto Tateo

Abstract The irrelevant composite operator $T\overline{T}$, constructed from components of the stress-energy tensor, exhibits unique properties in two-dimensional quantum field theories and represents a distinctive form of integrable deformation. Significant progress has been made in understanding the bulk aspects of the theory, including its interpretation in terms of coordinate transformations and its connection to topological gravity models. However, the behavior of $T\overline{T}$-deformed theories in the presence of boundaries and defects remains largely unexplored. In this note, we review analytical results obtained through various techniques. Specifically, we study the $T\overline{T}$-deformed exact g-function within the framework of the Thermodynamic Bethe Ansatz and show that the results coincide with those obtained by solving the corresponding Burgers-type flow equation. Finally, we highlight some potentially significant open problems.

Nicolò Brizio
Dipartimento di Fisica, Università di Torino and
INFN, Sezione di Torino, Via P. Giuria 1, 10125, Torino, Italy
e-mail: nicolo.brizio@unito.it

Tommaso Morone
Dipartimento di Fisica, Università di Torino and
INFN, Sezione di Torino, Via P. Giuria 1, 10125, Torino, Italy
e-mail: tommaso.morone@unito.it

Roberto Tateo
Dipartimento di Fisica, Università di Torino and
INFN, Sezione di Torino, Via P. Giuria 1, 10125, Torino, Italy
e-mail: roberto.tateo@unito.it

D. R. Wood et al. (eds.), *2024 MATRIX Annals, Part II*, MATRIX Book Series 8,
https://doi.org/10.1007/978-3-032-16206-9_2

1 Introduction and Overview

The investigation of thermodynamics in one-dimensional integrable models with factorized scattering originates from the pioneering work of Yang and Yang [1, 2]. Their formulation, now referred to as the Thermodynamic Bethe Ansatz (TBA), has proven to be remarkably flexible, finding application across a wide range of models [3–9], and leading to exact equations for the free energy.

In realistic physical systems, impurities, domain walls, and defects that break translational and rotational symmetries into subgroups are ubiquitous [10]. Methods from integrability provide robust tools for addressing such complexities. A particularly relevant notion is the g-function, which extends the notion of non-integer ground state degeneracy linked to critical boundary conditions, initially introduced by Affleck and Ludwig for the Kondo model [11, 12]. The g-function quantifies the contribution of a single boundary to the free energy, and represents a pivotal tool in analyzing renormalization group-type flows in boundary field theories.

Two distinct Hamiltonian representations exist for the partition function on a finite cylinder. In the first formulation, the Hamiltonian governs the time evolution of states within a Hilbert space defined on an interval of length L, with boundary conditions w_1 and w_2 imposed at the endpoints. Alternatively, the time evolution can be understood as corresponding to the height L of the cylinder, involving boundary states $|w_1\rangle$ and $|w_2\rangle$, along with the eigenstates $\{|\psi_n\rangle\}$ of the Hamiltonian on a circle of radius R. The equivalence of these two formulations yields the following fundamental identity:

$$\sum_{n\geq 0} e^{-RE_n^{\text{strip}}(L)} = \sum_{n\geq 0} \frac{\langle w_1|\psi_n\rangle\langle\psi_n|w_2\rangle}{\langle\psi_n|\psi_n\rangle} e^{-LE_n^{\text{circ}}(R)}, \tag{1}$$

and

$$\ln g_{n,|w_1\rangle} = \ln \frac{\langle w_1|\psi_n\rangle}{\langle\psi_n|\psi_n\rangle^{1/2}}. \tag{2}$$

In particular $\ln g_{|w_1\rangle} := \ln g_{0,|w_1\rangle}$ is the so-called ground-state g-function. For integrable models, in the Thermodynamic Bethe Ansatz framework, an exact expression for the g-function can be obtained for large families of quantum field theories [13–18].

Besides boundary conditions, studying renormalization group flows from non-trivial fixed points remains a primary area of interest.

Deformations of quantum field theories (QFTs) are classified based on the nature of the perturbing operators: relevant, marginal, or irrelevant [19]. While relevant and marginal deformations have been extensively studied, irrelevant deformations are less explored due to their corresponding challenges. Renormalization theory shows that irrelevant deformations require an infinite number of counterterms for the computation of physical quantities, resulting in Lagrangian densities, including an infinite series of terms, with significant ambiguities arising in the process. Nevertheless, recent advances show that these issues can be overcome for certain special classes of irrelevant deformations, making the problem more manageable. This class of deformations is known as $\mathrm{T\overline{T}}$ deformations [20, 21], which can be defined in terms of the Lagrangian flow equation

$$\frac{\partial \mathscr{L}^{(\tau)}}{\partial \tau} = -\frac{1}{\pi^2}\mathrm{T\overline{T}}, \tag{3}$$

where $\mathscr{L}^{(\tau)}$ denotes the Lagrangian density of the theory, τ is a flow parameter, and the $\mathrm{T\overline{T}}$ operator is defined as

$$\mathrm{T\overline{T}} = -\pi^2 \det T_{\mu\nu} = -\pi^2\left(T_{xx}T_{yy} - T_{xy}^2\right) = T\overline{T} - \Theta^2, \tag{4}$$

where $T_{\mu\nu}$ is the energy-momentum tensor. At the quantum level, denoting by $|n\rangle$ an eigenstate of both energy and momentum, such that

$$\mathscr{H}|n\rangle = E_n|n\rangle, \quad \mathscr{P}|n\rangle = P_n|n\rangle, \tag{5}$$

the following factorization property holds [22]:

$$\langle n|\mathrm{T\overline{T}}|n\rangle = \langle n|T|n\rangle\langle n|\overline{T}|n\rangle - \langle n|\Theta|n\rangle^2. \tag{6}$$

Previous research [23–26] has emphasized the special role of the $\mathrm{T\overline{T}}$ operator in integrable perturbations of conformal field theories (CFTs), suggesting connections with the exact S-matrix theory and the TBA. Within the context of 2D integrable models, $\mathrm{T\overline{T}}$-deformed field theories retain an infinite number of conserved charges, allowing for the exact computation of scattering amplitudes and other observables. In particular, the scattering matrix $\mathbf{S}(\theta)$ is modified via the introduction of a CDD (Castillejo-Dalitz-Dyson) factor, such that

$$\mathbf{S}^{(\tau)}(\theta) = \mathbf{S}^{(0)}(\theta) \times e^{\imath\tau m^2 \sinh\theta}, \tag{7}$$

where m is a mass scale, and θ is the relative rapidity. From the TBA equations, it is possible to show that the $\mathrm{T\overline{T}}$-deformed energy spectrum on the cylinder follows the inviscid Burgers equation of hydrodynamics [20, 21],

$$\frac{\partial E_n^{(\tau)}(R)}{\partial \tau} = E_n^{(\tau)}(R)\frac{\partial E_n^{(\tau)}(R)}{\partial R} + \frac{P_n^2(R)}{R}, \tag{8}$$

where $P_n = 2\pi n/R$ is the (quantized) total momentum. Notice, however, that in a field theory confined to a finite cylinder geometry with Neumann boundary conditions, only the states with $P_n(R) = 0$ contribute to the partition function. In the zero-momentum sector, equation (8) can be implicitly solved by the following equation:

$$E_n^{(\tau)}(R) = E_n^{(0)}(\tilde{R}), \quad \text{with} \quad \tilde{R} = R + \tau E_n^{(\tau)}. \tag{9}$$

Equation (9) establishes an explicit connection between the deformation parameter τ and an auxiliary deformed geometry through the dynamical map $R \mapsto \tilde{R}$.

When the un-deformed theory is a conformal field theory (CFT), a general solution to the Burgers equation (8) for zero-momentum states is provided by

$$E_n^{(\tau)}(R) = -\frac{R}{2\tau} \pm \sqrt{\left(\frac{R}{2\tau}\right)^2 - \frac{2\pi}{\tau}\left(\frac{c_{\text{eff}}}{12} - \Delta\right)}, \tag{10}$$

where c_{eff} denotes the effective central charge, and Δ is the scaling dimension of the corresponding CFT operator on the plane. The physical states of the deformed CFT correspond to those with the positive sign in (10). The non-physical branch (negative sign) merges with the physical one at a branch point, which, for the ground state, corresponds to the well-known Hagedorn-type singularity appearing in string theory. For $c_{\text{eff}} = D - 2$, with $D = 26$, the physical spectrum coincides with the spectrum of the Nambu-Goto string in a 26-dimensional target space obtained through light-cone quantization.[1] A key subtlety arises concerning the sign of the deformation parameter τ: different choices lead to significantly different physical behaviors.

- For negative τ, an analysis of the density of states in the deep UV regime reveals behavior characteristic of a class of non-local theories known as little string theory, which is dual to gravitational theories on a linear dilaton background [31]. The deformation effectively acts as a UV cutoff, softening high-energy divergences. However, this can be problematic from the perspective of causality.
- For positive τ, the deformed energy levels can become complex at high energies, signaling a breakdown of unitarity or the presence of a Hagedorn-type transition (an upper limit to the temperature of a system due to an exponential growth in the number of states), arising from a pair of square-root branch points in the complex R-plane (see Figure 1). This phenomenon is anticipated in theories of two-dimensional quantum gravity, where a minimum length scale exists [32,33].

By exploiting modular invariance, the flow equations for the $T\overline{T}$-deformed g-function can be derived [34,35]:

$$\frac{\partial s_n^{(\tau)}(R)}{\partial \tau} = E_n^{(\tau)}(R)\frac{\partial s_n^{(\tau)}(R)}{\partial R} + R\frac{\partial}{\partial R}\frac{E_n^{(\tau)}(R)}{R}, \quad s_n^{(\tau)} = \ln\left(g_{n,|w_1\rangle}^{(\tau)} g_{n,|w_2\rangle}^{(\tau)}\right). \tag{11}$$

The interpretation of this flow in the classical limit and its potential applications to the Nambu-Goto spectrum remain important areas for further investigation. Starting from a conformal field theory, the exact form of the $T\overline{T}$-deformed boundary entropy $s_n^{(\tau)}$ is obtained by incorporating an extra factor which depends on the energy $E_n^{(\tau)}(R)$ [20]:

---

[1] It must be noted that the spectrum of the form $E_n^{(\tau)}(R) + R/2\tau$ is known to be incompatible with target space Poincaré symmetry, except for $D = 26$, where the Virasoro algebra is free of anomalies, and possibly $D = 3$ (see, for example, the related comments in [27], showing that deviations from the Nambu-Goto action can only occur at eight-derivative order — or higher — in three space-time dimensions, as well as [28,29]). However, $T\overline{T}$ deformations of generic CFTs can be interpreted as non-critical string models [30].

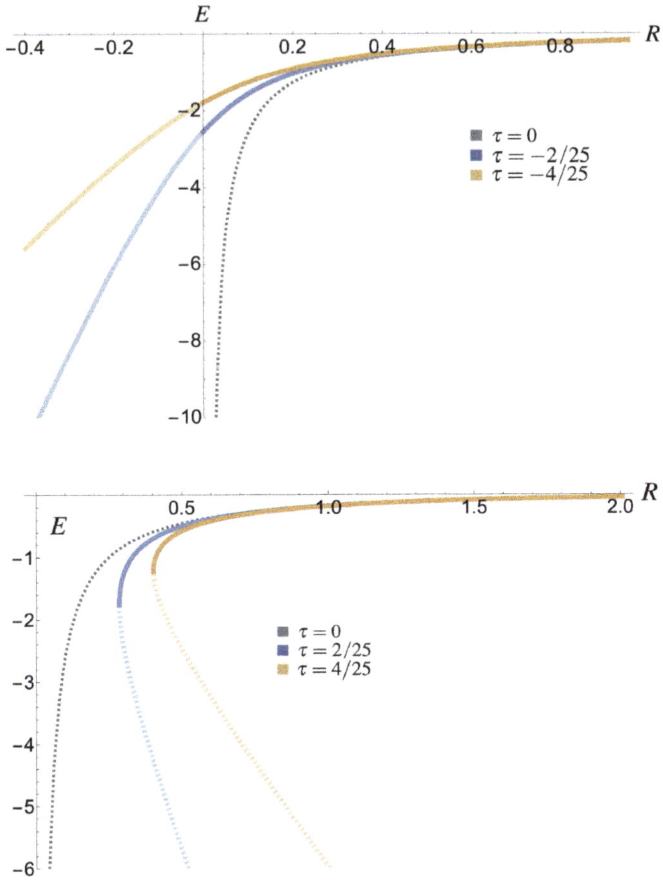

Fig. 1 The $T\overline{T}$-deformed ground-state spectrum for the free fermion case is presented. In the first panel, the lighter-colored lines correspond to the forbidden region where $R < 0$. In the second panel, the dashed-colored lines represent the analytic continuation around the solution branch. This non-physical solution depicts the behavior of the ground-state spectrum when extended to the additional branch in the solution space.

$$s^{(\tau)}_{n,\mathrm{CFT}}(R) = \ln\left(\langle w_1|\psi_n\rangle^{\mathrm{CFT}}\right) + \ln\left(\langle w_2|\psi_n\rangle^{\mathrm{CFT}}\right) + \ln\left(\frac{R}{R+2\tau E^{(\tau)}_n(R)}\right). \qquad (12)$$

The origin of the last term on the right-hand side of the equation in (12) can be traced back to the additional contribution, compared to the initial proposal of [13], identified in [14] (see also [36, 37]), which is required for the equality between the closed and open channel representations of the cylinder partition function. More details on the TBA-related approach to the g-function will be given in section 2 below.

On the other hand, one can employ the *method of characteristics* to derive the *general solution* to the flow equation (11) [35]

$$s_n^{(\tau)}(R) = s_n^{(0)}\left(R+\tau E_n^{(\tau)}(R)\right) + \ln\left(\frac{R+\tau R\partial_R E_n^{(\tau)}(R)}{R+\tau E_n^{(\tau)}(R)}\right), \tag{13}$$

where the first term on the right-hand side represents the un-deformed entropy on the deformed geometry $\tilde{R} = R+\tau E_n^{(\tau)}(R)$. The expressions (13) and (12) coincide when the Nambu-Goto type spectrum (10) is involved.

Beginning from the observation that, within the TBA framework, the deformation is equivalent to the modification of the S-matrix through a CDD factor (7), the objective of this note is to derive the expression (13) exclusively by employing the TBA definition of the exact g-function for a generic massive theory (see Section 2). While the calculations can be performed analytically in general, in this brief review, we focus on the specific case of the Ising model (massive Majorana fermion).

2 The Exact g-function

In this section, to simplify the discussion and avoid unnecessary complications, we will restrict our analysis to a purely elastic scattering theory involving a single particle species of mass m. The discussion can be easily extended to more general scattering theories and configurations [16,38–44].

It was first proposed by [13], that the boundary entropy should be given by an expression of the form:

$$\int_{\mathbb{R}} \frac{d\theta}{4\pi}\Theta(\theta)\ln\left(1+e^{-\varepsilon(\theta|R)}\right), \tag{14}$$

where $\varepsilon(\theta|R)$, satisfying a non-linear integral equation from the TBA, is given by

$$\varepsilon(\theta|R) = mR\cosh\theta - \frac{1}{2\pi}\phi * \ln\left(1+\mathrm{e}^{-\varepsilon}\right)(\theta), \tag{15}$$

where we introduced the compact notation

$$f*g(\theta) = \int_{\mathbb{R}} d\theta' f(\theta-\theta')g(\theta'). \tag{16}$$

The function $\Theta(\theta)$ is defined as

$$\Theta(\theta) = -\imath\frac{d}{d\theta}\ln(\mathbf{R}_{|w_1\rangle}(\theta)\mathbf{R}_{|w_2\rangle}(\theta)) + \imath\frac{d}{d\theta}\ln\mathbf{S}(2\theta) - 2\pi\delta(\theta), \tag{17}$$

where $\mathbf{S}(\theta)$ and $\mathbf{R}_{|w\rangle}(\theta)$ denote the scattering and reflection matrices, respectively. The kernel $\phi(\theta)$ is directly related to the scattering matrix through

$$\phi(\theta) = -\imath \frac{\mathrm{d}}{\mathrm{d}\theta} \ln \mathbf{S}(\theta). \tag{18}$$

However, it was later observed with a careful numerical analysis in [39] that the expression (14) is incomplete due to the exclusion of a boundary-independent term. By employing an n-particle cluster expansion, reminiscent of form-factor techniques for correlation functions, the missing term was finally recovered in [14]. The complete expression for the boundary entropy can be written in the form

$$s^{(0)}(R) = \int_{\mathbb{R}} \frac{\mathrm{d}\theta}{4\pi} \Theta(\theta) \ln\left(1 + e^{-\varepsilon(\theta|R)}\right) + \sum_{k>0} \mathbf{I}_k(R), \tag{19}$$

where

$$\mathbf{I}_1(R) = \int_{\mathbb{R}} \frac{\mathrm{d}\theta}{2\pi} \frac{\phi(2\theta)}{1 + e^{\varepsilon(\theta|R)}}, \tag{20}$$

$$\mathbf{I}_{k>1}(R) = \frac{1}{k} \left(\prod_{p=1}^{k} \int_{\mathbb{R}} \frac{\mathrm{d}\theta_p}{2\pi} \right) \frac{\phi(\theta_1 + \theta_2)}{1 + e^{\varepsilon(\theta_1|R)}} \left(\prod_{q=2}^{k} \frac{\phi(\theta_q - \theta_{q+1})}{1 + e^{\varepsilon(\theta_q|R)}} \right) \delta_{\theta_{k+1},\theta_1}. \tag{21}$$

We also recall that the ground-state energy of the theory on a circle of circumference R is given in terms of the pseudo-energy by [3]

$$E_0^{(0)}(R) = -m \int_{\mathbb{R}} \frac{\mathrm{d}\theta}{2\pi} \cosh\theta \ln\left(1 + e^{-\varepsilon(\theta|R)}\right). \tag{22}$$

3 Deforming the Entropy

We can implement the deformation by modifying the scattering properties of the theory [20,21,26,45],

$$\begin{aligned} \mathbf{S} &\longrightarrow \mathbf{S}^{(\tau)}, \\ \mathbf{R} &\longrightarrow \mathbf{R}^{(\tau)}. \end{aligned} \tag{23}$$

In this framework, the infinite sum of bulk contributions is always deformed, even in cases such as the free fermion, where $\mathbf{S}^{(0)} = -1$, and the contribution was originally absent.

As discussed in the introduction, the $T\overline{T}$-deformed scattering matrix is purely a phase, with no poles or zeros,

$$\mathbf{S}^{(\tau)}(\theta) = \mathbf{S}^{(0)}(\theta) \times e^{\imath\tau m^2 \sinh\theta}, \tag{24}$$

so that

$$\phi^{(\tau)}(\theta) = \phi^{(0)}(\theta) + \tau m^2 \cosh\theta. \tag{25}$$

By using this expression for the kernel in the TBA equation (15), and after straightforward manipulations, we obtain [20]:

$$\varepsilon^{(\tau)}(\theta|R) = \varepsilon^{(0)}(\theta|\tilde{R}), \tag{26}$$

where $\tilde{R} = R + \tau E_0^{(\tau)}$. This general property is central to the remarkable result (9), allowing us to derive the deformation by explicitly modifying the scale R in accordance with (26).

The bulk corrections are modified in a non-trivial manner, as they depend on the pseudo-energy and the kernel itself. To clarify this concept, let us consider the first contribution (20) for simplicity. Explicitly, one has

$$\mathbf{I}_1^{(0)}(R) = \int_{\mathbb{R}} \frac{d\theta}{2\pi} \frac{\phi^{(0)}(2\theta)}{1+e^{\varepsilon(\theta|R)}} \longrightarrow \mathbf{I}_1^{(\tau)}(R) = \int_{\mathbb{R}} \frac{d\theta}{2\pi} \frac{\phi^{(\tau)}(2\theta)}{1+e^{\varepsilon^{(\tau)}(\theta|R)}}, \tag{27}$$

where, using definition (25) and property (26),

$$\mathbf{I}_1^{(\tau)}(R) = \int_{\mathbb{R}} \frac{d\theta}{2\pi} \frac{\phi^{(0)}(2\theta)}{1+e^{\varepsilon^{(0)}(\theta|\tilde{R})}} + \tau m^2 \int_{\mathbb{R}} \frac{d\theta}{2\pi} \frac{\cosh 2\theta}{1+e^{\varepsilon^{(\tau)}(\theta|R)}} \tag{28}$$

$$= \mathbf{I}_1^{(0)}(\tilde{R}) + \mathbf{F}_1^{(\tau)}(R). \tag{29}$$

In particular,

$$\mathbf{F}_1^{(\tau)}(R) \equiv \tau m^2 \int_{\mathbb{R}} \frac{d\theta}{2\pi} \frac{\cosh 2\theta}{1+e^{\varepsilon^{(\tau)}(\theta|R)}} = \tau m^2 \int_{\mathbb{R}} \frac{d\theta}{2\pi} \frac{\cosh 2\theta}{1+e^{\varepsilon^{(0)}(\theta|\tilde{R})}}, \tag{30}$$

where we have used once again the property (26) in the last identification. Similarly, we can define $\forall k > 0$, functions $\mathbf{F}_k^{(\tau)}$ encoding the whole explicit τ-dependence. Compactly, one can write

$$\mathbf{I}_k^{(0)}(R) \longrightarrow \mathbf{I}_k^{(\tau)}(R) = \mathbf{I}_k^{(0)}(\tilde{R}) + \mathbf{F}_k^{(\tau)}(R). \tag{31}$$

Meanwhile, the general form of the boundary-perturbing reflection factor is given by [20, 26]:

$$\mathbf{R}^{(\tau)}(\theta) = \mathbf{R}^{(0)}(\theta) \times e^{\imath\tau m^2 \frac{\sinh(2\theta)}{2}} \prod_{a\geq 1} e^{\delta_a (\imath m \sinh\theta)^{2a-1}}. \tag{32}$$

Each factor in the product on the right-hand side of (32) corresponds to a specific boundary perturbation with coupling δ_a [20]. Combining these expressions, we obtain:

$$\Theta(\theta) = \Theta^{(0)}(\theta) + 2\sum_{a\geq 1}(2a-1)(-1)^{a+1}\delta_a m^{2a-1}\cosh\theta(\sinh\theta)^{2a-2}. \tag{33}$$

Interestingly, the function $\Theta(\theta)$ is explicitly independent of τ, due to a cancellation mechanism between the scattering factor and the fundamental boundary reflection factor. Using this property, we can express the deformed entropy as

$$
\int_{\mathbb{R}} \frac{d\theta}{4\pi} \Theta(\theta) \ln\left(1+e^{-\varepsilon^{(\tau)}(\theta|R)}\right) + \sum_{k>0} \mathbf{I}_k^{(\tau)}(R) + \int_{\mathbb{R}} \frac{d\theta}{2\pi} \sum_{a\geq 1} (2a-1)(-1)^{a+1} \delta_a m^{2a-1} \cosh\theta (\sinh\theta)^{2a-2} \ln\left(1+e^{-\varepsilon^{(\tau)}(\theta|R)}\right). \tag{34}
$$

When setting $\delta_a = 0$,[2] we can apply the property (26) to express

$$
s^{(\tau)}(R) = s^{(0)}(\tilde{R}) + \sum_{k>0} \mathbf{F}_k^{(\tau)}(R). \tag{35}
$$

Even in this relatively simple setup, where only a single particle species is involved, it is far from obvious how the infinite corrections in τ can be reduced to the expected result (13), i.e.

$$
\sum_{k>0} \mathbf{F}_k^{(\tau)}(R) = \ln\left(\frac{R+\tau R\partial_R E_0^{(\tau)}(R)}{R+\tau E_0^{(\tau)}(R)}\right). \tag{36}
$$

This holds in the massless limit, as demonstrated in [20], but becomes highly non-trivial in the general case.

4 Deforming the Free Fermion

The aim of this section is to present a concrete example of how the exact expression (13) can be systematically reconstructed within the context of the free fermion case.

The challenge arises from the infinite sum of bulk corrections that contribute to the exact entropy as a result of the deformation. Using the kernel (25), where

$$
\phi^{(0)}_{\text{Fermion}} \equiv 0, \tag{37}
$$

we can express these corrections ($\mathbf{I}_k^{(0)} \equiv 0$) as:

$$
\mathbf{F}_1^{(\tau)} = \tau m^2 \int_{\mathbb{R}} \frac{d\theta}{2\pi} \frac{1}{1+e^{\tilde{r}\cosh\theta}} \cosh(2\theta), \tag{38}
$$

$$
\mathbf{F}_{k>1}^{(\tau)} = \frac{(\tau m^2)^k}{k} \left(\prod_{p=1}^{k} \int_{\mathbb{R}} \frac{d\theta_p}{2\pi}\right) \frac{\cosh(\theta_1+\theta_2)}{1+e^{\tilde{r}\cosh\theta_1}} \left(\prod_{q=2}^{k} \frac{\cosh(\theta_q-\theta_{q+1})}{1+e^{\tilde{r}\cosh\theta_q}}\right) \delta_{\theta_{k+1},\theta_1}, \tag{39}
$$

where we have used the property (26) to represent the deformed pseudo-energy as $\varepsilon^{(\tau)}(\theta)|_{\text{Fermion}} = \tilde{r}(\tau)\cosh\theta$, with $\tilde{r}/m \equiv \tilde{R}$.

[2] We focus on cases where no additional irrelevant fields act on the boundary. Although nonzero values of some δ_a with $a \geq 1$ may correspond to more general and physically interesting scenarios [26], we concentrate here on the contributions (20)–(21), which are independent of these parameters, and, as already mentioned, on what appears to us as a rather surprising match with the second term in equation (13).

Remarkably, much like in the case of CFTs [20], we are able to *resum* these contributions. Through explicit calculations involving trigonometric functions, we have derived a fundamental identity:

$$\sum_{k>0} \mathbf{F}_k^{(\tau)} = \sum_{n\geq 0} \mathbf{M}_n^{(\tau)}, \tag{40}$$

where

$$\begin{aligned}\mathbf{M}_n^{(\tau)} =& \frac{2^{-2n-3}\tau^{2n+1}}{(n+1)(2n+1)} \Big\{ (V+U)^{2n+1} \left[(2n+1)\tau(V+U)+4(n+1)\right] \\ & - (V-U)^{2n+1} \left[(2n+1)\tau(V-U)+4(n+1)\right] \Big\},\end{aligned} \tag{41}$$

and

$$U = m^2 \int_{\mathbb{R}} \frac{\mathrm{d}\theta}{2\pi} \frac{\cosh 2\theta}{e^{\tilde{r}\cosh\theta}+1}, \quad V = m^2 \int_{\mathbb{R}} \frac{\mathrm{d}\theta}{2\pi} \frac{1}{e^{\tilde{r}\cosh\theta}+1}. \tag{42}$$

This identity can be straightforwardly verified by applying Werner's second formula,

$$\cosh(x+y)\cosh(x-y) = \frac{1}{2}(\cosh 2x + \cosh 2y), \tag{43}$$

and using the following relation:

$$\left(\prod_i \int_{-c}^{c} dx_i\right) \cosh \sum_i x_i = \left(\prod_i \int_{-c}^{c} dx_i\right) \prod_i \cosh x_i. \tag{44}$$

The sum in (40) is well-defined, and we get

$$\sum_{k>0} \mathbf{F}_k^{(\tau)}(R) = \frac{1}{2}\ln\left[\left(\frac{(U-V)\tau+2}{(U+V)\tau-2}\right)^2\right] = \frac{1}{2}\ln\left[\left(\frac{R+\tau R\partial_R E_0^{(\tau)}(R)}{R+\tau E_0^{(\tau)}(R)}\right)^2\right]. \tag{45}$$

The final identification in (45) follows from straightforward manipulations of the total energy for the deformed fermion, i.e.,

$$\begin{aligned}E_0^{(\tau)} &= -m \int_{\mathbb{R}} \frac{\mathrm{d}\theta}{2\pi} \cosh\theta \ln\left(1+e^{-\tilde{r}\cosh(\theta)}\right) \\ &= -m^2 R \int_{\mathbb{R}} \frac{\mathrm{d}\theta}{2\pi} \sinh^2\theta \frac{1}{e^{\tilde{r}\cosh\theta}+1} - m^2\tau E_0^{(\tau)} \int_{\mathbb{R}} \frac{d\theta}{2\pi} \sinh^2\theta \frac{1}{e^{\tilde{r}\cosh\theta}+1}\end{aligned} \tag{46}$$

and

$$\begin{aligned}\frac{\partial}{\partial R} E_0^{(\tau)} =& m^2 \int_{\mathbb{R}} \frac{\mathrm{d}\theta}{2\pi} \cosh^2\theta \frac{1}{e^{\tilde{r}\cosh\theta}+1} \\ & + m^2 \tau \frac{\partial}{\partial R} E_0^{(\tau)} \int_{\mathbb{R}} \frac{\mathrm{d}\theta}{2\pi} \cosh^2\theta \frac{1}{e^{\tilde{r}\cosh\theta}+1}.\end{aligned} \tag{47}$$

In particular, we can express

$$-\frac{E_0^{(\tau)}}{R} = \frac{u}{1+\tau u}, \quad \frac{\partial}{\partial R} E_0^{(\tau)} = \frac{v}{1-\tau v}, \tag{48}$$

with

$$U = u + v, \quad V = u - v. \tag{49}$$

By assembling all the contributions, one precisely recovers the result (13) for the ground state. The explicit calculation provides substantial control over each contribution. The generalization to excited states is obtained by replacing the integration contour $\mathbb{R}$ with a contour encircling zeros of $1 + e^{-\varepsilon(\theta)}$ [46], this means $E_0^{(\tau)} \to E_n^{(\tau)}$. It has thus been shown that the solution obtained via the method of characteristics is entirely consistent with the one derived using the more traditional TBA-related approach.

5 Discussion & Numerics

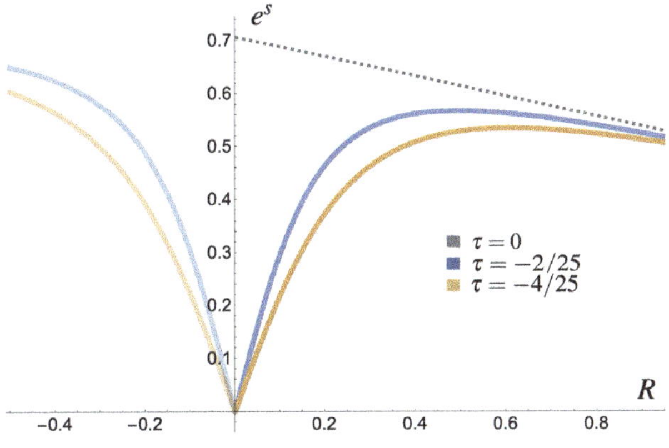

Fig. 2 The $T\overline{T}$-deformed ground state entropy in the Majorana free fermion model, under "fixed-free" boundary conditions [47], is examined here with $\delta_a = 0$ for simplicity. In the panel, the lighter-colored lines indicate the forbidden region corresponding to $R < 0$.

In this note, we investigate the effects of $T\overline{T}$ deformation on a simple model with boundaries. Our methodology can be easily generalized. The *resummation procedure* employed here may also be extended to incorporate more general scattering matrices than that specified in (24). Let us notice that the flow equation (11) can also be proven true in the case $\delta_a \neq 0$.

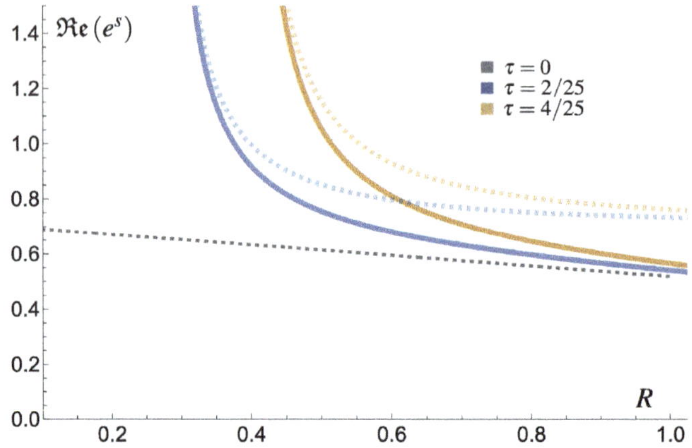

Fig. 3 In the panel, the deformed ground state entropy in the free fermion model, under "fixed-free" boundary conditions, is examined ($\delta_a = 0$). The analytic continuation around the solution branch is highlighted by dashed-colored lines.

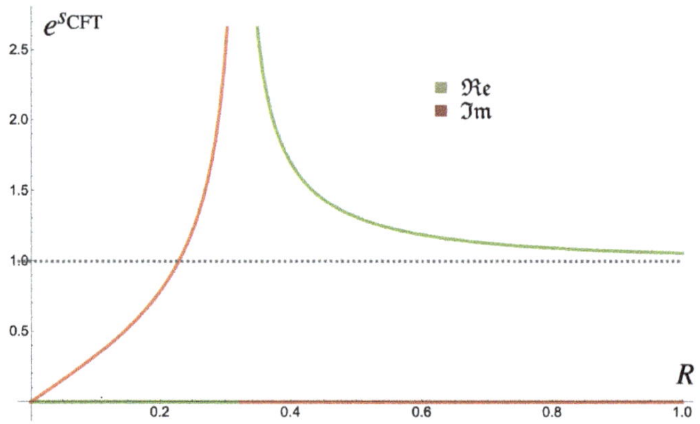

Fig. 4 The $T\overline{T}$-deformed ground state entropy in the CFT model, under "free" boundary conditions [15], is examined here at $\tau = 1/10$. The un-deformed value is reported in dashed-gray

A numerical evaluation of the flow equations reveals striking similarities between the deformed spectrum and the entropy. For the sake of simplicity, we consider "fixed-free" boundary conditions in the un-deformed theory [47],

$$\mathbf{R}_{|w_1\rangle}(\theta) = \mathbf{R}_{|\text{fixed}\rangle}(\theta) = i\tan(i\pi/4 - \theta/2)\,, \tag{50}$$

$$\mathbf{R}_{|w_2\rangle}(\theta) = \mathbf{R}_{|\text{free}\rangle}(\theta) = -i\,\text{cotan}(i\pi/4 - \theta/2) \tag{51}$$

which simplifies to

$$\Theta(\theta) = -2\pi\delta(\theta)\,, \tag{52}$$

assuming then $\delta_a = 0$ in (33).[3]

For $\tau < 0$, the ground state function $\exp s_0$ exhibits no singularities, but deviates from its behavior at $\tau = 0$, vanishing linearly in the ultraviolet (UV) regime as $R \to 0$ (see Figure 2). This behavior, typically manifesting at macroscopic scales, can be expressed as:

$$\lim_{R\to 0} \frac{\langle w_1 | \psi_0 \rangle^{(\tau<0)}}{\sqrt{\langle \psi_0 | \psi_0 \rangle^{(\tau<0)}}} = 0 . \tag{53}$$

This result is straightforwardly recovered in the massless limit, where the relevant expressions are explicit.

For $\tau > 0$, $\exp s_0^{(\tau)}$ reveals the existence of a minimum length scale R_c, below which such quantity becomes complex. In Figures 3 and 4, this corresponds to a singularity of the form $(R - R_c)^{-1/2}$. To probe the behavior for $R < R_c$, we can consider the CFT case, where $\exp s_{0,\mathrm{CFT}}^{(\tau)}$ becomes purely imaginary, and again tends to zero linearly as $R \to 0$. For $R > R_c$, in addition to the physical solution, $\exp s_0^{(\tau)}$ has a second real branch solution (depicted by the dashed-colored lines in Figure 3). This non-physical branch arises from the analytic continuation of the spectrum around the branch point (see Figure 1). In the massless limit, where $R_c = \sqrt{2\pi\tau c_{\mathrm{eff}}/3}$, the two branches (for $R > R_c$) collapse into a single one.

In general, unlike the $\tau = 0$ cases, the $T\overline{T}$-deformed quantities $\exp s_{n,\mathrm{CFT}}^{(\tau)}$ vanish linearly as $R \to 0$, suggesting the need for an appropriate renormalization of the partition function. Moreover, when transitioning from the ground state to the excited states, the qualitative behavior for $\tau < 0$ and $\tau > 0$ is exchanged.

A comprehensive numerical investigation could provide further insights into these aspects (see, for example, [48]).

6 Open Problems and Future Directions

The effect of the $T\overline{T}$ deformation on boundary conditions, particularly on the boundary term of the classical action, as well as the nature and properties of the associated deforming operators, remains a critical open question in this research field (see, e.g., [35, 49]). A thorough investigation of these aspects is a crucial first step toward understanding the implications of the broader class of $T\overline{T}$-like deformations [21, 50–53], including the more recently introduced classically marginal root-$T\overline{T}$ deformation [54–60], in systems with boundaries. The geometric interpretation of these more general deformations is particularly intriguing, as it may uncover further connections to topological gravity and random geometry [34, 61, 62].

An important open question worth highlighting concerns the so-called ODE/IM correspondence (see [63–67]). This framework establishes a precise dictionary between classical and quantum quantities in 2D integrable field theories. As shown

[3] We work in natural units where $m = 1$.

in [68], this quantum/classical correspondence is preserved under $T\overline{T}$ deformation. However, the extension of the ODE/IM correspondence to the case of boundaries and crosscaps [18, 69] in the presence of $T\overline{T}$-like deformations has not yet been explored and remains an intriguing open problem.

Notably, boundary problems—together with the simple Ising model example discussed in this note—naturally link with recent advances in the study of non-invertible symmetries, a topic that has garnered considerable interest in the scientific community (see [70] for a comprehensive review and [71] for a recent perspective on the topic). See also [72] for a discussion related to non-invertible symmetries emerging in deformed Ising models on the lattice, which share remarkable analogies with $T\overline{T}$ flows in the continuous limit. In particular, the analytical tools offered by the $T\overline{T}$ formalism may provide valuable insights into the space of topological defects, which can be understood as the insertion of boundary conditions.

Finally, extending the current framework to field theories in higher spacetime dimensions represents a promising direction for future research (see, for instance, [54, 73–75]).

Acknowledgements We thank Changrim Ahn, Zoltan Bajnok, Patrick Dorey, Clare Dunning, Paul Pearce, Francesco Ravanini, Marton Kormos, and Gabor Takacs for insightful discussions. This work received partial support from the INFN project SFT and the PRIN Project No. 2022ABPBEY, "Understanding quantum field theory through its deformations".

Roberto Tateo thanks the organizers of the MATRIX Research Program "Mathematics and Physics of Integrability (MPI2024)" for the invitation and its participants for the stimulating atmosphere and discussions. He also expresses his gratitude to the Mathematical Research Institute MATRIX and the Australian Mathematical Sciences Institute for their financial support during his visit to Australia.

References

1. C.-N. Yang and C. P. Yang, "One-dimensional chain of anisotropic spin spin interactions. 1. Proof of Bethe's hypothesis for ground state in a finite system," *Phys. Rev.* **150** (1966) 321–327.
2. C.-N. Yang and C. P. Yang, "Thermodynamics of one-dimensional system of bosons with repulsive delta function interaction," *J. Math. Phys.* **10** (1969) 1115–1122.
3. A. B. Zamolodchikov, "Thermodynamic Bethe Ansatz in Relativistic Models. Scaling Three State Potts and Lee-yang Models," *Nucl. Phys. B* **342** (1990) 695–720.
4. T. R. Klassen and E. Melzer, "Purely Elastic Scattering Theories and their Ultraviolet Limits," *Nucl. Phys. B* **338** (1990) 485–528.
5. T. R. Klassen and E. Melzer, "The Thermodynamics of purely elastic scattering theories and conformal perturbation theory," *Nucl. Phys. B* **350** (1991) 635–689.
6. A. Klumper, M. T. Batchelor, and P. A. Pearce, "Central charges of the 6- and 19- vertex models with twisted boundary conditions," *J. Phys. A* **24** (1991) 3111.
7. C. Destri and H. J. de Vega, "New thermodynamic Bethe ansatz equations without strings," *Phys. Rev. Lett.* **69** (1992) 2313–2317.
8. C. Destri and H. J. De Vega, "Unified approach to thermodynamic Bethe Ansatz and finite size corrections for lattice models and field theories," *Nucl. Phys. B* **438** (1995) 413–454, hep-th/9407117.

9. P. Dorey and R. Tateo, "Excited states by analytic continuation of TBA equations," *Nucl. Phys. B* **482** (1996) 639–659, hep-th/9607167.
10. N. Andrei *et al.*, "Boundary and Defect CFT: Open Problems and Applications," *J. Phys. A* **53** (2020), no. 45, 453002, 1810.05697.
11. I. Affleck and A. W. W. Ludwig, "Universal noninteger 'ground state degeneracy' in critical quantum systems," *Phys. Rev. Lett.* **67** (1991) 161–164.
12. I. Affleck and A. W. W. Ludwig, "Critical theory of overscreened Kondo fixed points," *Nucl. Phys. B* **360** (1991) 641–696.
13. A. LeClair, G. Mussardo, H. Saleur, and S. Skorik, "Boundary energy and boundary states in integrable quantum field theories," *Nucl. Phys. B* **453** (1995) 581–618, hep-th/9503227.
14. P. Dorey, D. Fioravanti, C. Rim, and R. Tateo, "Integrable quantum field theory with boundaries: The Exact g function," *Nucl. Phys. B* **696** (2004) 445–467, hep-th/0404014.
15. P. Dorey, C. Rim, and R. Tateo, "Exact g-function flow between conformal field theories," *Nucl. Phys. B* **834** (2010) 485–501, 0911.4969.
16. P. Dorey, A. Lishman, C. Rim, and R. Tateo, "Reflection factors and exact g-functions for purely elastic scattering theories," *Nucl. Phys. B* **744** (2006) 239–276, hep-th/0512337.
17. P. Dorey, R. Tateo, and R. Wilbourne, "Exact g-function flows from the staircase model," *Nucl. Phys. B* **843** (2011) 724–752, 1008.1190.
18. J. Caetano and S. Komatsu, "Functional equations and separation of variables for exact g-function," *JHEP* **09** (2020) 180, 2004.05071.
19. K. G. Wilson, "The renormalization group: Critical phenomena and the Kondo problem," *Rev. Mod. Phys.* **47** (Oct, 1975) 773–840.
20. A. Cavaglià, S. Negro, I. M. Szécsényi, and R. Tateo, "$T\bar{T}$-deformed 2D Quantum Field Theories," *JHEP* **10** (2016) 112, 1608.05534.
21. F. A. Smirnov and A. B. Zamolodchikov, "On space of integrable quantum field theories," *Nucl. Phys. B* **915** (2017) 363–383, 1608.05499.
22. A. B. Zamolodchikov, "Expectation value of composite field T anti-T in two-dimensional quantum field theory," hep-th/0401146.
23. A. B. Zamolodchikov, "From tricritical Ising to critical Ising by thermodynamic Bethe ansatz," *Nucl. Phys. B* **358** (1991) 524–546.
24. G. Mussardo and P. Simon, "Bosonic type S matrix, vacuum instability and CDD ambiguities," *Nucl. Phys. B* **578** (2000) 527–551, hep-th/9903072.
25. S. Dubovsky, R. Flauger, and V. Gorbenko, "Solving the Simplest Theory of Quantum Gravity," *JHEP* **09** (2012) 133, 1205.6805.
26. M. Caselle, D. Fioravanti, F. Gliozzi, and R. Tateo, "Quantisation of the effective string with TBA," *JHEP* **07** (2013) 071, 1305.1278.
27. O. Aharony and N. Klinghoffer, "Corrections to Nambu-Goto energy levels from the effective string action," *JHEP* **12** (2010) 058, 1008.2648.
28. S. Dubovsky and V. Gorbenko, "Towards a Theory of the QCD String," *JHEP* **02** (2016) 022, 1511.01908.
29. C. Chen, P. Conkey, S. Dubovsky, and G. Hernández-Chifflet, "Undressing Confining Flux Tubes with $T\bar{T}$," *Phys. Rev. D* **98** (2018), no. 11, 114024, 1808.01339.
30. N. Callebaut, J. Kruthoff, and H. Verlinde, "$T\overline{T}$ deformed CFT as a non-critical string," *JHEP* **04** (2020) 084, 1910.13578.
31. A. Giveon, N. Itzhaki, and D. Kutasov, "$T\overline{T}$ and LST," *JHEP* **07** (2017) 122, 1701.05576.
32. L. McGough, M. Mezei, and H. Verlinde, "Moving the CFT into the bulk with $T\overline{T}$," *JHEP* **04** (2018) 010, 1611.03470.
33. M. Guica and R. Monten, "$T\bar{T}$ and the mirage of a bulk cutoff," *SciPost Phys.* **10** (2021), no. 2, 024, 1906.11251.
34. J. Cardy, "The $T\overline{T}$ deformation of quantum field theory as random geometry," *JHEP* **10** (2018) 186, 1801.06895.
35. Y. Jiang, F. Loebbert, and D.-l. Zhong, "Irrelevant deformations with boundaries and defects," *J. Stat. Mech.* **2204** (2022), no. 4, 043102, 2109.13180.
36. B. Pozsgay, "On O(1) contributions to the free energy in Bethe Ansatz systems: The Exact g-function," *JHEP* **08** (2010) 090, 1003.5542.

37. I. Kostov, D. Serban, and D.-L. Vu, "Boundary TBA, trees and loops," *Nucl. Phys. B* **949** (2019) 114817, 1809.05705.
38. P. Dorey, A. Pocklington, R. Tateo, and G. Watts, "TBA and TCSA with boundaries and excited states," *Nucl. Phys. B* **525** (1998) 641–663, hep-th/9712197.
39. P. Dorey, I. Runkel, R. Tateo, and G. Watts, "g function flow in perturbed boundary conformal field theories," *Nucl. Phys. B* **578** (2000) 85–122, hep-th/9909216.
40. Z. Bajnok, L. Palla, and G. Takács, "Finite size effects in boundary sine-Gordon theory," *Nucl. Phys. B* **622** (2002) 565–592, hep-th/0108157.
41. E. Corrigan and A. Taormina, "Reflection factors and a two parameter family of boundary bound states in the Sinh-Gordon model," *J. Phys. A* **33** (2000) 8739–8754, hep-th/0008237.
42. C. Ahn, M. Bellacosa, and F. Ravanini, "Excited states NLIE for sine-Gordon model in a strip with Dirichlet boundary conditions," *Phys. Lett. B* **595** (2004) 537–546, hep-th/0312176.
43. C. Ahn, Z. Bajnok, R. I. Nepomechie, L. Palla, and G. Takacs, "NLIE for hole excited states in the Sine-Gordon model with two boundaries," *Nucl. Phys. B* **714** (2005) 307–335, hep-th/0501047.
44. M. Kormos, "Boundary renormalisation group flows of the supersymmetric Lee-Yang model and its extensions," *Nucl. Phys. B* **772** (2007) 227–248, hep-th/0701061.
45. S. Dubovsky, V. Gorbenko, and M. Mirbabayi, "Natural Tuning: Towards A Proof of Concept," *JHEP* **09** (2013) 045, 1305.6939.
46. P. Dorey and R. Tateo, "Excited states in some simple perturbed conformal field theories," *Nucl. Phys. B* **515** (1998) 575–623, hep-th/9706140.
47. S. Ghoshal and A. B. Zamolodchikov, "Boundary S matrix and boundary state in two-dimensional integrable quantum field theory," *Int. J. Mod. Phys. A* **9** (1994) 3841–3886, hep-th/9306002. [Erratum: Int.J.Mod.Phys.A 9, 4353 (1994)].
48. G. Camilo, T. Fleury, M. Lencsés, S. Negro, and A. Zamolodchikov, "On factorizable S-matrices, generalized TTbar, and the Hagedorn transition," *JHEP* **10** (2021) 062, 2106.11999.
49. J. P. Babaro, V. F. Foit, G. Giribet, and M. Leoni, "$T\overline{T}$ type deformation in the presence of a boundary," *JHEP* **08** (2018) 096, 1806.10713.
50. A. Bzowski and M. Guica, "The holographic interpretation of $J\bar{T}$-deformed CFTs," *JHEP* **01** (2019) 198, 1803.09753.
51. M. Guica, "On correlation functions in $J\bar{T}$-deformed CFTs," *J. Phys. A* **52** (2019), no. 18, 184003, 1902.01434.
52. R. Conti, S. Negro, and R. Tateo, "Conserved currents and $\mathrm{T\bar{T}}_s$ irrelevant deformations of 2D integrable field theories," *JHEP* **11** (2019) 120, 1904.09141.
53. B. Le Floch and M. Mezei, "KdV charges in $T\bar{T}$ theories and new models with super-Hagedorn behavior," *SciPost Phys.* **7** (2019), no. 4, 043, 1907.02516.
54. R. Conti, J. Romano, and R. Tateo, "Metric approach to a $\mathrm{T\overline{T}}$-like deformation in arbitrary dimensions," *JHEP* **09** (2022) 085, 2206.03415.
55. C. Ferko, A. Sfondrini, L. Smith, and G. Tartaglino-Mazzucchelli, "Root-$T\bar{T}$ Deformations in Two-Dimensional Quantum Field Theories," *Phys. Rev. Lett.* **129** (2022), no. 20, 201604, 2206.10515.
56. H. Babaei-Aghbolagh, K. Babaei Velni, D. Mahdavian Yekta, and H. Mohammadzadeh, "Marginal $T\bar{T}$-like deformation and modified Maxwell theories in two dimensions," *Phys. Rev. D* **106** (2022), no. 8, 086022, 2206.12677.
57. R. Borsato, C. Ferko, and A. Sfondrini, "Classical integrability of root-$T\bar{T}$ flows," *Phys. Rev. D* **107** (2023), no. 8, 086011, 2209.14274.
58. S. Ebert, C. Ferko, and Z. Sun, "Root-$T\bar{T}$ deformed boundary conditions in holography," *Phys. Rev. D* **107** (2023), no. 12, 126022, 2304.08723.
59. H. Babaei-Aghbolagh, S. He, T. Morone, H. Ouyang, and R. Tateo, "Geometric Formulation of Generalized Root-$T\bar{T}$ Deformations," *Phys. Rev. Lett.* **133** (2024), no. 11, 111602, 2405.03465.

60. E. Tsolakidis, "Massive gravity generalization of $T\overline{T}$ deformations," *JHEP* **09** (2024) 167, 2405.07967.
61. S. Dubovsky, V. Gorbenko, and G. Hernández-Chifflet, "$T\overline{T}$ partition function from topological gravity," *JHEP* **09** (2018) 158, 1805.07386.
62. S. Hirano and M. Shigemori, "Random boundary geometry and gravity dual of $T\overline{T}$ deformation," *JHEP* **11** (2020) 108, 2003.06300.
63. P. Dorey and R. Tateo, "Anharmonic oscillators, the thermodynamic Bethe ansatz, and nonlinear integral equations," *J. Phys. A* **32** (1999) L419–L425, hep-th/9812211.
64. V. V. Bazhanov, S. L. Lukyanov, and A. B. Zamolodchikov, "Spectral determinants for Schrodinger equation and Q operators of conformal field theory," *J. Statist. Phys.* **102** (2001) 567–576, hep-th/9812247.
65. P. Dorey, C. Dunning, and R. Tateo, "The ODE/IM Correspondence," *J. Phys. A* **40** (2007) R205, hep-th/0703066.
66. S. L. Lukyanov and A. B. Zamolodchikov, "Quantum Sine(h)-Gordon Model and Classical Integrable Equations," *JHEP* **07** (2010) 008, 1003.5333.
67. P. Dorey, C. Dunning, S. Negro, and R. Tateo, "Geometric aspects of the ODE/IM correspondence," *J. Phys. A* **53** (2020), no. 22, 223001, 1911.13290.
68. F. Aramini, N. Brizio, S. Negro, and R. Tateo, "Deforming the ODE/IM correspondence with $\mathrm{T}\overline{\mathrm{T}}$," *JHEP* **03** (2023) 084, 2212.13957.
69. J. Caetano and S. Komatsu, "Crosscap States in Integrable Field Theories and Spin Chains," *J. Statist. Phys.* **187** (2022), no. 3, 30, 2111.09901.
70. S.-H. Shao, "What's Done Cannot Be Undone: TASI Lectures on Non-Invertible Symmetries," 2308.00747.
71. Y. Choi, B. C. Rayhaun, and Y. Zheng, "Noninvertible Symmetry-Resolved Affleck-Ludwig-Cardy Formula and Entanglement Entropy from the Boundary Tube Algebra," *Phys. Rev. Lett.* **133** (2024), no. 25, 251602, 2409.02806.
72. F. Yan, R. Konik, and A. Mitra, "Duality defect in a deformed transverse-field Ising model," 2410.17317.
73. M. Taylor, "$T\bar{T}$ deformations in general dimensions," *Adv. Theor. Math. Phys.* **27** (2023), no. 1, 37–63, 1805.10287.
74. T. Morone, S. Negro, and R. Tateo, "Gravity and $T\overline{T}$ flows in higher dimensions," *Nucl. Phys. B* **1005** (2024) 116605, 2401.16400.
75. C. D. A. Blair, J. Lahnsteiner, N. A. Obers, and Z. Yan, "Matrix Theory Reloaded: A BPS Road to Holography," 2410.03591.

Quantum counting, and a relevant sign

Natalie Chung and Rafael I. Nepomechie

Abstract Two indispensable algorithms in an introductory course on Quantum Computing are Grover's search algorithm and quantum phase estimation. Quantum counting is a simple yet beautiful blend of these two algorithms, and it is therefore an attractive topic for a student project in such a course. However, a sign that is irrelevant when implementing Grover's algorithm becomes relevant. We briefly review these algorithms, highlighting the aforementioned sign.

1 Introduction

Two indispensable algorithms in an introductory undergraduate course on Quantum Computing are Grover's search algorithm [4, 5] and quantum phase estimation (QPE) [6]. Quantum counting [2, 3] is a simple yet beautiful blend of these two algorithms, and it is therefore an attractive topic for a student project in such a course. However, a sign that is irrelevant when implementing Grover's algorithm becomes relevant.

We start by briefly reviewing in Sec. 2 Grover's algorithm for a single marked element, and its extension [2] to the case of multiple marked elements. QPE is briefly reviewed in Sec. 3. The key part of this paper is Sec. 4, where we review quantum counting and highlight the aforementioned sign. We finish in Sec. 5 with a brief conclusion.

Natalie Chung
Doral Academy Preparatory High School, 11100 NW 27th St, Doral, FL 33172 USA
e-mail: nataliechung05@gmail.com

Rafael I. Nepomechie
Department of Physics, P.O. Box 248046, University of Miami, Coral Gables, FL 33124 USA
e-mail: nepomechie@miami.edu

D. R. Wood et al. (eds.), *2024 MATRIX Annals, Part II*, MATRIX Book Series 8,
https://doi.org/10.1007/978-3-032-16206-9_3

2 Grover's algorithm

Grover's search algorithm, one of the "crown jewels" of Quantum Computing, has been widely described in detail, see e.g. [7]. We content ourselves here with reminding the reader of the main steps, which also serves to set out our notations and conventions. We first treat in Sec. 2.1 the familiar case of a single marked element, and then consider in Sec. 2.2 the perhaps less-familiar case of multiple marked elements.

2.1 One marked element

Let x and a be n-bit integers, and $f(x)$ the search function

$$f(x) = \begin{cases} 0 & \text{if} \quad x \neq a \\ 1 & \text{if} \quad x = a \end{cases} \tag{1}$$

The "marked element" a is the object of our search. We define the corresponding unitary operator U_f, the so-called standard protocol acting on the n-qubit "input register" $|x\rangle$ and the 1-qubit "output register" $|y\rangle$, by

$$U_f\left(|y\rangle|x\rangle\right) = |y \oplus f(x)\rangle|x\rangle\,. \tag{2}$$

The initial state of the output register is taken to be

$$|y\rangle = H|1\rangle = \frac{1}{\sqrt{2}}\left(|0\rangle - |1\rangle\right), \tag{3}$$

and therefore

$$U_f\left(H|1\rangle|x\rangle\right) = (-1)^{f(x)} H|1\rangle|x\rangle\,. \tag{4}$$

We define the operator V on the input register by

$$V|x\rangle = (-1)^{f(x)}|x\rangle = \begin{cases} |x\rangle & \text{if} \quad x \neq a \\ -|x\rangle & \text{if} \quad x = a \end{cases}, \tag{5}$$

which can be realized as

$$V = 1 - 2|a\rangle\langle a|\,, \tag{6}$$

so that (4) can re-expressed as

$$U_f\left(H|1\rangle|x\rangle\right) = H|1\rangle V|x\rangle\,. \tag{7}$$

The output register remains in the state $H|1\rangle$ throughout the algorithm, and so we henceforth focus only on the input register.

The initial state of the input register is taken to be

$$|\phi\rangle \equiv H^{\otimes n}|0\rangle^{\otimes n} = \frac{1}{2^{n/2}} \sum_{x=0}^{2^n-1} |x\rangle \,. \tag{8}$$

The kets $|a\rangle$ and $|\phi\rangle$ determine a (real) plane; we define a normalized ket $|a_\perp\rangle$ in this plane that is perpendicular to $|a\rangle$, i.e. $\langle a|a_\perp\rangle = 0$. We define θ as the angle between $|a_\perp\rangle$ and $|\phi\rangle$, as in Fig. 1. We therefore have

$$|\phi\rangle = \cos\theta\, |a_\perp\rangle + \sin\theta\, |a\rangle \,. \tag{9}$$

We observe from (8) that $\langle a|\phi\rangle = \frac{1}{2^{n/2}}$; and from (9) that $\langle a|\phi\rangle = \sin\theta$. Hence,

$$\sin\theta = \frac{1}{2^{n/2}} \,. \tag{10}$$

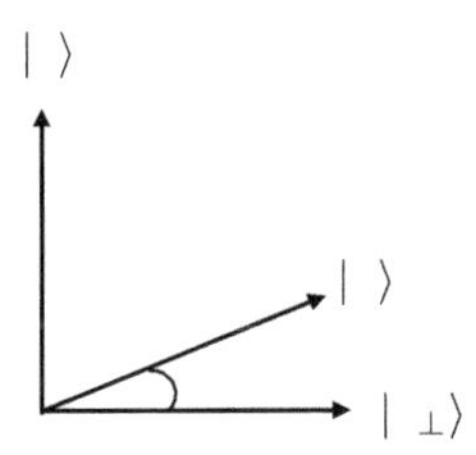

Fig. 1: The plane spanned by $|a\rangle$ and $|\phi\rangle$

We define the unitary operator W, the so-called diffuser, by

$$W = 2|\phi\rangle\langle\phi| - 1 \,. \tag{11}$$

By explicit computation using (6), (9) and (11), we obtain

$$\begin{aligned} (WV)\, |a_\perp\rangle &= \cos(2\theta)\, |a_\perp\rangle + \sin(2\theta)\, |a\rangle \,, \\ (WV)\, |a\rangle &= -\sin(2\theta)\, |a_\perp\rangle + \cos(2\theta)\, |a\rangle \,. \end{aligned} \tag{12}$$

We see that WV can be represented in the basis $|a_\perp\rangle\,, |a\rangle$ by the matrix

$$WV = \begin{pmatrix} \cos(2\theta) & -\sin(2\theta) \\ \sin(2\theta) & \cos(2\theta) \end{pmatrix} = e^{-i2\theta Y} \,, \tag{13}$$

which rotates any ket in the plane by the angle 2θ counterclockwise (i.e, from $|a_\perp\rangle$ to $|a\rangle$). After k such Grover rotations, the input register is in the state

$$(WV)^k\, |\phi\rangle = \cos\left((2k+1)\theta\right)\, |a_\perp\rangle + \sin\left((2k+1)\theta\right)\, |a\rangle \,. \tag{14}$$

The number k is chosen such that the result (14) is the sought-after ket $|a\rangle$; measuring the input register will then give a with probability 1. To this end, we set $(2k+1)\theta = \pi/2$, and thus $k = \frac{\pi}{4\theta} - \frac{1}{2}$. We see from (10) that $\theta \approx \sin\theta = \frac{1}{2^{n/2}}$ for large n. Hence, the required number of Grover rotations is

$$k \approx \frac{\pi}{4}\sqrt{2^n}\,. \tag{15}$$

It is not difficult to show that the W operator (11) can be re-expressed as

$$W = -H^{\otimes n} X^{\otimes n} \left(c^{n-1} Z\right) X^{\otimes n} H^{\otimes n}\,, \tag{16}$$

where $c^{n-1}Z$ denotes the $(n-1)$-fold controlled Z gate. This form for W is useful for implementing Grover's algorithm on, say, a quantum simulator. For the purpose of determining a, the minus sign in (16) is irrelevant, so it is typically dropped. (Note that multiplication by -1 is not a standard gate. Nevertheless, this sign could be implemented using a product of standard 1-qubit gates, for example $ZXZX = -\mathbb{I}$.) That is, when simulating Grover's algorithm, typically one uses

$$\tilde{W} \equiv -W = H^{\otimes n} X^{\otimes n} \left(c^{n-1} Z\right) X^{\otimes n} H^{\otimes n} \tag{17}$$

instead of (16). We will revisit this sign in Sec. 4.

2.2 *Multiple marked elements*

Suppose we want to search now for a set $S = \{a_1, a_2, \ldots, a_m\}$ of m marked elements. The standard protocol U_f is still given by (2), except that the search function is now given by

$$f(x) = \begin{cases} 0 & \text{if} \quad x \notin S \\ 1 & \text{if} \quad x \in S \end{cases}. \tag{18}$$

The operator V is therefore now defined by

$$V|x\rangle = (-1)^{f(x)}|x\rangle = \begin{cases} |x\rangle & \text{if} \quad x \notin S \\ -|x\rangle & \text{if} \quad x \in S \end{cases}, \tag{19}$$

and is given by

$$V = 1 - 2\sum_{j=1}^{m} |a_j\rangle\langle a_j|\,. \tag{20}$$

The key insight is to consider the plane spanned by $|s\rangle$ and $|\phi\rangle$, where $|s\rangle$ is an equal-weight superposition of all the $|a_j\rangle$'s

$$|s\rangle = \frac{1}{\sqrt{m}}\sum_{j=1}^{m} |a_j\rangle\,. \tag{21}$$

We define a vector $|s_\perp\rangle$ in this plane that is perpendicular to $|s\rangle$, i.e. $\langle s|s_\perp\rangle = 0$; moreover, we define θ as the angle between $|s_\perp\rangle$ and $|\phi\rangle$, see Fig. 2. We therefore have

$$|\phi\rangle = \cos\theta\,|s_\perp\rangle + \sin\theta\,|s\rangle\,. \tag{22}$$

We observe from (21) and (8) that $\langle s|\phi\rangle = \sqrt{\frac{m}{2^n}}$; and from (22) that $\langle s|\phi\rangle = \sin\theta$. Hence,

$$\sin\theta = \sqrt{\frac{m}{2^n}}\,. \tag{23}$$

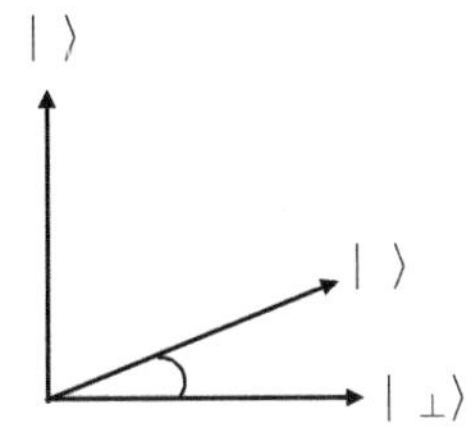

Fig. 2: The plane spanned by $|s\rangle$ and $|\phi\rangle$

It is easy to check that $V|s\rangle = -|s\rangle$; furthermore, $\langle a_j|s_\perp\rangle = 0$, and hence $V|s_\perp\rangle = |s_\perp\rangle$. We see that, in the plane of $|s_\perp\rangle$ and $|s\rangle$, V acts as

$$V = 1 - 2|s\rangle\langle s|\,, \tag{24}$$

cf. (6). The computation of WV on $|s_\perp\rangle$ and $|s\rangle$ is therefore the same as the previous computation of WV on $|a_\perp\rangle$ and $|a\rangle$, respectively (12); and in the basis $|s_\perp\rangle\,, |s\rangle$, WV is given by the same matrix (13). After k Grover rotations, the input register is in the state

$$(WV)^k\,|\phi\rangle = \cos\left((2k+1)\theta\right)\,|s_\perp\rangle + \sin\left((2k+1)\theta\right)\,|s\rangle\,. \tag{25}$$

Choosing again $(2k+1)\theta = \pi/2$, so that

$$k \approx \frac{\pi}{4}\sqrt{\frac{2^n}{m}}\,, \tag{26}$$

the result (25) is the ket $|s\rangle$ (21); measuring the input register will then give any one of the $a_j \in S$ with equal probability. Of course, (26) reduces to (15) for $m = 1$.

In order to use this algorithm, we must know the number of marked elements (m), which enters into the formula (26) for the number of Grover rotations. What do we do if we do *not* know m ahead of time? One way to proceed, the so-called quantum counting considered in Sec. 4, involves using QPE.

3 Quantum phase estimation

QPE is another important quantum algorithm with many uses, including an elegant formulation of Shor's celebrated period-finding algorithm [11]. Given a unitary operator $\mathscr{U}$ and an eigenvector $|\psi\rangle$, whose corresponding eigenvalue necessarily has the form $e^{i\alpha}$ with α real,

$$\mathscr{U}\,|\psi\rangle = e^{i\alpha}|\psi\rangle\,, \tag{27}$$

QPE provides an estimate for α. A key ingredient of QPE is the quantum Fourier transform, which we recall (see e.g. [7]) is defined by

$$U_{FT}\,|x\rangle = \frac{1}{2^{\frac{n}{2}}}\sum_{y=0}^{2^n-1} e^{\frac{2\pi ixy}{2^n}}\,|y\rangle\,, \qquad 0 \le x < 2^n\,, \tag{28}$$

and which can be shown to be written more explicitly as

$$U_{FT}\,|x\rangle = \frac{1}{2^{\frac{n}{2}}}\left(|0\rangle + e^{\frac{2\pi ix}{2}}\,|1\rangle\right)\otimes\ldots\otimes\left(|0\rangle + e^{\frac{2\pi ix}{2^{n-1}}}\,|1\rangle\right)\otimes\left(|0\rangle + e^{\frac{2\pi ix}{2^n}}\,|1\rangle\right). \tag{29}$$

The QPE circuit diagram, with t auxiliary qubits, is shown in Fig. 3.

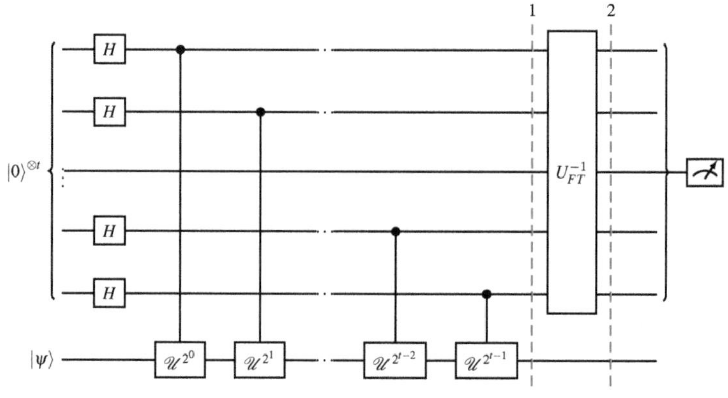

Fig. 3: Circuit diagram for QPE

At slice 1, one can check with the help of (29) that the circuit is in the state

$$|\psi\rangle\,U_{FT}\,|\frac{2^t\alpha}{2\pi}\rangle\,. \tag{30}$$

At slice 2, after applying the inverse quantum Fourier transform, the state becomes simply

$$|\psi\rangle\,|\frac{2^t\alpha}{2\pi}\rangle\,. \tag{31}$$

Thus, the measurement of the ancillary qubits gives an integer j that is within 1/2 of $\frac{2^t\alpha}{2\pi}$. This estimate can be improved by increasing the value of t.

4 Quantum counting

We are now in position to address the question raised at the end of Sec. 2.2: what can we do if we are searching for multiple marked elements, but we do not know their number m ?

An answer, known as quantum counting, is to apply QPE to the Grover rotation operator; that is, recalling (13), set

$$\mathscr{U} = WV = e^{-i2\theta Y}, \tag{32}$$

and implement the QPE circuit in Fig. 3 using $|\psi\rangle = |\phi\rangle$ (8). Evidently, $\mathscr{U}$ has eigenvalues $e^{\pm 2i\theta}$. Although $|\phi\rangle$ is not an eigenket of $\mathscr{U}$, $|\phi\rangle$ lies in the subspace spanned by $|s_\perp\rangle$ and $|s\rangle$ (22); hence, QPE provides estimates for both $\pm 2\theta$. More precisely, QPE gives integers j_+ and j_- that are closest to $\frac{2^t(2\theta)}{2\pi}$ and $\frac{2^t(2\pi-2\theta)}{2\pi}$, respectively, see (31). Solving for θ, we see that

$$\theta \approx \frac{\pi j_+}{2^t}, \quad \pi - \frac{\pi j_-}{2^t}. \tag{33}$$

Using (23), we conclude that the number of marked elements is given by

$$m = 2^n \sin^2\theta \approx 2^n \sin^2\left(\frac{\pi j_\pm}{2^t}\right) \qquad (\text{ implementing } W). \tag{34}$$

Using $\tilde{W}$ (17) instead of W (that is, dropping the sign in (16)) leads to a result different from (34). Indeed, a simple way to account for the different sign is to observe from (32) that changing $W \mapsto -W = \tilde{W}$ implies $\mathscr{U} \mapsto -\mathscr{U}$; and the corresponding eigenvalues transform accordingly $e^{\pm 2i\theta} \mapsto -e^{\pm 2i\theta} = e^{\pm 2i(\theta \pm \frac{\pi}{2})}$. Performing the shift $\theta \mapsto \theta - \frac{\pi}{2}$ in (34), we obtain

$$m = 2^n \sin^2\left(\theta - \frac{\pi}{2}\right) \approx 2^n \sin^2\left[\pi\left(\frac{j_\pm}{2^t} - \frac{1}{2}\right)\right] \qquad (\text{ implementing } \tilde{W} = -W). \tag{35}$$

In our simulations, we implemented $\tilde{W}$, and the number of marked elements indeed matched with (35). For example, we searched for the set of **three** 3-bit integers $S = \{2,4,6\}$ using 5 ancillas, so that $n = m = 3$ and $t = 5$. In 1024 shots, the two most frequent integers obtained from measuring the ancillas were $j = 9$ and $j = 23$. Eq. (35) then correctly gave $m \approx 3.22 \approx 3$, while (34) would imply the incorrect result $m \approx 4.78 \approx 5$.

5 Conclusion

Although quantum counting is not considered part of the standard undergraduate introductory Quantum Computing curriculum (see e.g. [1, 8]), it entails only modest – yet very interesting – extensions of two topics that *are* part of the standard repertoire, namely Grover's algorithm and QPE. Hence, coding and simulating quantum counting, using e.g. qiskit [9] or cirq [10], is an attractive topic for a student project in such a course, especially for students who have already coded and simulated Grover and QPE.[1] However, the sign of the diffuser W (16) should not be neglected.

Acknowledgements RN was supported in part by the National Science Foundation under Grant No. PHY 2310594 and by a Cooper fellowship.

References

1. Asfaw, A., et al.: Building a Quantum Engineering Undergraduate Program. IEEE Trans. Educ. **65**, 220 (2022). DOI 10.1109/TE.2022.3144943
2. Boyer, M., Brassard, G., Hoeyer, P., Tapp, A.: Tight bounds on quantum searching. Fortsch. Phys. **46**, 493–506 (1998). DOI 10.1002/(SICI)1521-3978(199806)46:4/5⟨493::AID-PROP493⟩3.0.CO;2-P
3. Brassard, G., Hoyer, P., Tapp, A.: Quantum counting. In: 25th Intl. Colloquium on Automata, Languages, and Programming (ICALP), LNCS 1443, pp. 820–831 (1998). DOI 10.48550/arXiv.quant-ph/9805082
4. Grover, L.K.: A Fast quantum mechanical algorithm for database search. In: Proceedings, 28th Annual ACM Symposium on the Theory of Computing (STOC), pp. 212–219 (1996). DOI 10.48550/arXiv.quant-ph/9605043
5. Grover, L.K.: Quantum mechanics helps in searching for a needle in a haystack. Phys. Rev. Lett. **79**, 325–328 (1997). DOI 10.1103/PhysRevLett.79.325
6. Kitaev, A.Y.: Quantum measurements and the Abelian stabilizer problem (1995). DOI 10.48550/arXiv.quant-ph/9511026
7. Mermin, N.D.: Quantum computer science, an introduction. Cambridge University Press (2007)
8. Meyer, J.C., Passante, G., Pollock, S.J., Wilcox, B.R.: Introductory quantum information science coursework at US institutions: Content coverage (2023). DOI 10.48550/arXiv.2308.12929
9. Qiskit contributors: Qiskit: An open-source framework for quantum computing (2023). DOI 10.5281/zenodo.2573505
10. Quantum AI team and collaborators: qsim (2020). DOI 10.5281/zenodo.4023103. URL `https://doi.org/10.5281/zenodo.4023103`
11. Shor, P.W.: Polynomial time algorithms for prime factorization and discrete logarithms on a quantum computer. SIAM J. Sci. Statist. Comput. **26**, 1484 (1997). DOI 10.1137/S0097539795293172

[1] Sample qiskit code is included as Supplementary Material; it can be found at `https://arxiv.org/abs/2310.07428`.

Self-distributivity, braces, and the Yang-Baxter equation

Anastasia Doikou

Abstract The theory of the set-theoretic Yang-Baxter equation is reviewed from a purely algebraic point of view. We recall certain algebraic structures called shelves, racks and quandles. These objects satisfy a *self-distributivity* condition and lead to solutions of the Yang-Baxter equation. The quantum algebra as well as the integrability associated to Baxterized involutive set-theoretic solutions is briefly discussed. We then present the theory of the universal algebras associated to rack and general set-theoretic solutions. We show that these are quasi-triangular Hopf algebras and we derive the universal set-theoretic Drinfel'd twist. It is shown that this is an admissible twist allowing the derivation of the universal set-theoretic $\mathscr{R}$-matrix.

1 Introduction

The Yang-Baxter equation (YBE) was first introduced in a purely physical context in [64] as the main mathematical tool for the investigation of quantum systems with many particle interactions, and in [5] for the study of statistical model known as the anisotropic Heisenberg magnet. The idea of set-theoretic solutions to the Yang-Baxter equation was suggested in early 90's by Drinfel'd [27] and since then, set-theoretic solutions have been extensively investigated primarily by means of representations of the braid group, but almost exclusively for the parameter free Yang-Baxter equation (see for instance [30, 39, 55, 56]). The investigation of set-theoretic solutions of the Yang-Baxter equation and the associated algebraic structures is a highly active research field that has been particularly prolific, given that a significant number of related studies has been produced over the past several years (see for instance [6]–[10] [13]–[25],[35]–[38], [41, 42, 49, 50, 51, 59]). The

A. Doikou
Department of Mathematics, Heriot-Watt University, Edinburgh EH14 4AS, UK; and
Maxwell Institute for Mathematical Sciences, Edinburgh EH8 9BT, UK
e-mail: a.doikou@hw.ac.uk

D. R. Wood et al. (eds.), *2024 MATRIX Annals, Part II*, MATRIX Book Series 8,
https://doi.org/10.1007/978-3-032-16206-9_4

study of the set-theoretic Yang-Baxter equation has produced numerous significant connections to distinct mathematical areas, such as group theory, algebraic number theory, Hopf-Galois extensions, non-commutative rings, knot theory, Hopf algebras and quantum groups, universal algebras, groupoids, heaps and trusses, pointed Hopf algebras, Yetter-Drinfel'd modules and Nichols algebras (see for instance among [3, 4, 6, 8, 9, 10, 30], [41]-[43],[46]-[50], [56, 59]). Moreover, interesting links between the set-theoretic Yang-Baxter equation and geometric crystals [7, 31], or soliton cellular automatons [40, 61] have been shown. Concrete connections with quantum spin-chain like systems were also made in [18, 19].

We note that set-theoretic solution for the parametric Yang-Baxter equation (Yang-Baxter maps) have been primarily studied up to date only in the context of classical discrete and continuous integrable systems connected also to the notion of Darboux-Bäcklund transformation or the discrete zero curvature condition in the Lax pair formulation and the refactorization problem and soliton interactions (see for instance [2, 53, 54, 63], [1, 12, 62]). The refactorization is also synonymous to the so called Bianchi's permutability, which describes the exchange of two consecutive Bäcklund transformations. Specifically, Figure 1 describes graphically the construction of say a two-soliton solution for some integrable non-linear ODE or PDE via two consecutive Bäcklund transformations and the use of Bianchi's permutability.

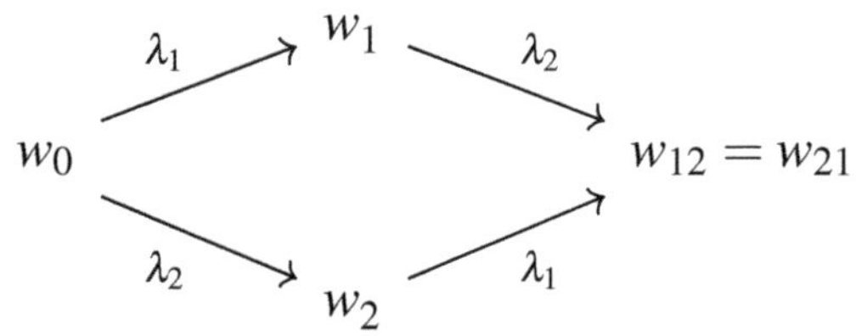

1. Bianchi Permutability

The set-theoretic Yang-Baxter equation may be also seen as a cube or 3D consistency condition in classical integrable discrete systems (discrete time evolution), see Figure 2.

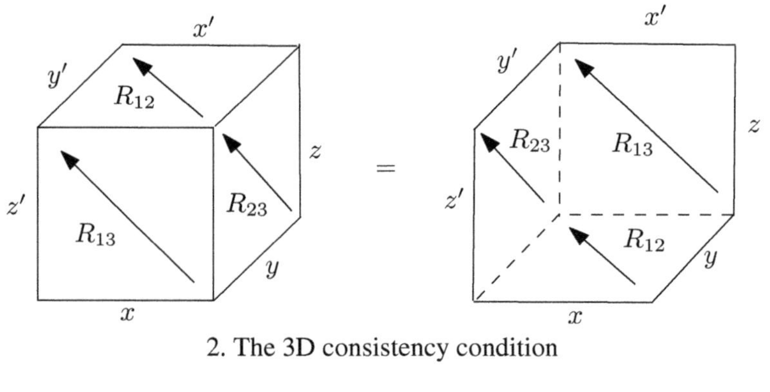

2. The 3D consistency condition

In [25] an entirely algebraic analysis for the parametric set-theoretic Yang-Baxter equation was undertaken and purely algebraic solutions were produced. Earlier works on the algebraic structures as well as the associated admissible Drinfel'd twists of the non-parametric, set-theoretic Yang-Baxter equation provided a basic algebraic blueprint [20, 21, 24].

We introduce now the *parameter free* set-theoretic braid equation. Following [27], given a non-empty set X, a map $\check{r}: X \times X \to X \times X$ is said to be a *set-theoretic solution of the braid equation*, if $\check{r}$ satisfies the *braid identity*

$$(\check{r} \times \mathrm{id}_X)(\mathrm{id}_X \times \check{r})(\check{r} \times \mathrm{id}_X) = (\mathrm{id}_X \times \check{r})(\check{r} \times \mathrm{id}_X)(\mathrm{id}_X \times \check{r}). \tag{1}$$

We call such a map $\check{r}$ simply a *solution* and write $(X, \check{r})$ to denote a solution $\check{r}$ on a set X. Besides, if we write $\check{r}(a,b) = (\sigma_a(b), \tau_b(a))$, with σ_a, τ_a maps from X into itself, then $\check{r}$ is said to be *left non-degenerate* if σ_a is bijective for every $a \in X$, *right non-degenerate* if τ_a is bijective for every $a \in X$, and *non-degenerate* if $\check{r}$ is both left and right non-degenerate. Furthermore, if $\check{r}$ is a solution such that $\check{r}^2 = \mathrm{id}_{X \times X}$, then $\check{r}$ is said to be *involutive*.
It is also worth recalling the connection between the set-theoretic braid equation (1) and the set-theoretic Yang-Baxter equation (see also e.g. [44, 45]). We introduce the map $r: X \times X \to X \times X$, such that $r = \check{r}\pi$, where $\pi: X \times X \to X \times X$ is the flip map: $\pi(x,y) = (y,x)$. Hence, $r(y,x) = (\sigma_x(y), \tau_y(x))$, and it satisfies the set-theoretic Yang-Baxter equation:

$$r_{12}\, r_{13}\, r_{23} = r_{23}\, r_{13}\, r_{12}, \tag{2}$$

where we denote $r_{12}(y,x,w) = (\sigma_x(y), \tau_y(x), w)$, $r_{23}(w,y,x) = (w, \sigma_x(y), \tau_y(x))$ and $r_{13}(y,w,x) = (\sigma_x(y), w, \tau_y(x))$. If $\check{r}$ is involutive then r satisfies $r_{12}r_{21} = \mathrm{id}_{X \times X}$ and is called *reversible*.

After having introduced the set-theoretic braid or Yang-Baxter equation, which is the main mathematical object in the present work, we may state the main aim of this paper, which is the review of basic algebraic structures associated to solutions of the set-theoretic Yang-Baxter equation. Specifically, the key objectives of each of the subsequent sections of this article are presented below.

- In Section 2 we present basic definitions and examples of fundamental algebraic structures associated to solutions of the set-theoretic Yang-Baxter equation. These structures are self distributive and are known as shelves, racks and quandles [46, 52]. They also satisfy axioms analogous to the Reidemeister moves used to manipulate knot diagrams and are associated to link invariants. Other fundamental structures introduced more recently are the so-called (skew) braces invented precisely for the study of generic solutions of the set-theoretic Yang-Baxter equation [55, 56, 57, 39].
- In Section 3 we focus on Baxterized involutive set-theoretic solutions of the braid equation coming from involutive set-theoretic solutions. We present some basic properties of these solutions and we then construct the associated quantum algebra via the FRT (Faddeev-Reshetikhin-Takhtajan) construction [32]. The Yangian is a special case within this class of quantum algebras. We also present

information on the construction of a new class of integrable quantum spin chain-like systems associated to set-theoretic solutions (see also [18, 19] for a more detailed exposition).

- In Section 4 we present the new findings on Drinfel'd twists for involutive, non-degenerate, set-theoretic solutions. We focus on a simple, but characteristic example of set-theoretic solution of the YBE known as Lyubashenko's solution. We show that Lyubashenko's solutions can be obtained from the permutation operator via a simple Drinfel'd twist. We derive the simple twist as well as explicit expressions for the n-fold twist. We also present the action of the twist on the Yangian coproducts and the derivation of the twisted co-products associated to the Baxterized Lyubashenko solution.
- In Section 5, which is divided in three subsections, we focus on the Hopf algebras associated to set-theoretic solutions [20, 21, 22, 24]. Specifically, in the first subsection, we introduce the Yang-Baxter algebras of rack/quandle type as quasi-triangular Hopf algebras (see also related results in [30, 3, 49, 4]). We first introduce and study the rack and quandle algebras $\mathscr{A}$, we then show that these are Hopf algebras and we systematically construct the associated universal invertible (or non-degenerate) $\mathscr{R}$-matrix (i.e. $\mathscr{R}^{-1}$ exists) [24]. In the second subsection, we suitably extend the quandle algebra and present the set-theoretic Yang-Baxter algebra, which is also a Hopf algebra. A suitable universal Drinfel'd twist is introduced in the third subsection and it is shown to be admissible. Then the universal set-theoretic $\mathscr{R}$-matrices are derived as twists, and we conclude that invertible, universal set-theoretic $\mathscr{R}$-matrices are coming from universal rack $\mathscr{R}$-matrices via the admissible Drinfel'd twist. The fundamental representation of the universal $\mathscr{R}$-matrices are the linearized versions of rack and general set-theoretic solutions, and it is consequently shown that all involutive set-theoretic solutions of the braid equation are coming from the permutation operator via the set-theoretic Drinfel'd twist. A detailed analysis on these results is presented in [20, 24].
- In Section 6 using the notion of the admissible Drinfel'd twist as well as a certain algebraic structure associated to any generic set-theoretic solution called the *structure group* we are able to extract explicit invertible set-theoretic solutions from quandles. We provide several explicit examples of solutions emerging from distinct quandles.

2 Preliminaries

Before we present the fundamental algebraic structures associated to set-theoretic solutions we first introduce some simple examples of finite set-theoretic solutions and the notion of linearization.

Example 2.1

1. **Flip map.** *Let* $X = \{1,2,\dots,n\}$, *the map* $\check{r} : X \times X \to X \times X$, *such as* $\check{r}(a,b) = (b,a)$ *is set-theoretic solution of the braid equation. This is the simplest set-theoretic solution.*
2. **Lyubashenko's solution.** *Let* $X = \{1,2,\dots,n\}$, *and the map* $\check{r} : X \times X \to X \times X$, *such as* $\check{r}(a,b) = (b+c, a-c)$, *where the addition is defined mod n and* $c \in \{1,2,\dots,n-1\}$. *Then* $\check{r}$ *is a solution of the set-theoretic braid equation.*

Notice that both solutions above are involutive, i.e. $\check{r}^2 = id$.

Linearization. *Via the linearization process we will be able to express the maps* $\check{r} : X \times X \to X \times X$ *as* $n^2 \times n^2$ *matrices* $\check{r} \in End(\mathbb{C}X^{\otimes 2})$ *(we slightly abuse the notation and use the same symbol* $\check{r}$ *for the matrices). Specifically, consider a free vector space* $V = \mathbb{C}X$ *of dimension equal to the cardinality of* X. *Let* $\mathbb{B} = \{\hat{e}_a\}_{a \in X}$ *be the basis of* V *and* $\mathbb{B}^* = \{\hat{e}^*_a\}_{a \in X}$ *be the dual basis:* $\hat{e}^*_a \hat{e}_b = \delta_{a,b}$, *also* $e_{a,b} := \hat{e}_a \hat{e}^*_b$. *Then any set-theoretic solution of the braid equation is expressed as an* $n^2 \times n^2$ *matrix:*

$$\check{r} = \sum_{a,b \in X} e_{a,\sigma_a(b)} \otimes e_{b,\tau_b(a)} \tag{1}$$

Specifically, for a set $X = \{x_1, x_2, \dots, x_n\}$, *the canonical basis of the n-dimensional vector space reads as expected:* $\hat{e}_{x_1} = \begin{pmatrix} 1 \\ 0 \\ 0 \\ \vdots \\ 0 \end{pmatrix}, \hat{e}_{x_2} = \begin{pmatrix} 0 \\ 1 \\ 0 \\ \vdots \\ 0 \end{pmatrix}, \dots, \hat{e}_{x_n} = \begin{pmatrix} 0 \\ 0 \\ \vdots \\ 0 \\ 1 \end{pmatrix}$.

Remark 2.2 *The action of the* $\check{r}$*-matrix on the tensor product of any two elements of the basis* $\hat{e}_x, \hat{e}_y \in V$, *for all* $x, y \in X$ *reads as:*

$$\check{r}\ \hat{e}_{\sigma_x(y)} \otimes \hat{e}_{\tau_y(x)} = \hat{e}_x \otimes \hat{e}_y \quad \& \quad \check{r}^T\ \hat{e}_x \otimes \hat{e}_y = \hat{e}_{\sigma_x(y)} \otimes \hat{e}_{\tau_y(x)}, \tag{2}$$

where ^{T} *denotes total transposition. Or equivalently for any test function* $f(x,y)$, $x, y \in X$: $(\check{r}f)(x,y) = f(\sigma_x(y), \tau_y(x))$. *Similarly, the action of* $r = \mathscr{P}\check{r}$, *where* $\mathscr{P}$ *is the permutation operator* $\mathscr{P} = \sum_{x,y \in X} e_{x,y} \otimes e_{y,x}$, *on* $\hat{e}_x \otimes \hat{e}_y \in V \otimes V$ *is*

$$r\ \hat{e}_{\sigma_x(y)} \otimes \hat{e}_{\tau_y(x)} = \hat{e}_y \otimes \hat{e}_x \quad \& \quad r^T\ \hat{e}_y \otimes \hat{e}_x = \hat{e}_{\sigma_x(y)} \otimes \hat{e}_{\tau_y(x)}. \tag{3}$$

We consider for the purposes of this article only finite sets. The linerization process can be formally generalized in the case of infinite countable sets as above. In the case of compact sets the use of functional analysis and the study of kernels of integral operators that represent the solution $\check{r}$ is required.

We consider the simple, but non-trivial example of the Lyubashenko solution [27].

Example 2.3 *The Lyubashenko solution may be expressed as a* $n^2 \times n^2$ *matrix:*

$$\check{r}^{(c)} = \sum_{a,b=1}^{n} e_{a,b+c} \otimes e_{b,a-c}. \tag{4}$$

The addition is defined mod n, and for all $x,y \in \{1,2,\dots,n\}$ *and for a fixed* $c \in \{1,2,\dots,n-1\}$,

$$\check{r}^{(c)}\ \hat{e}_{y+c} \otimes \hat{e}_{x-c} = \hat{e}_x \otimes \hat{e}_y, \quad r^{(c)}\ \hat{e}_{y+c} \otimes \hat{e}_{x-c} = \hat{e}_y \otimes \hat{e}_x. \tag{5}$$

We focus here on the case $n = 3$ *and present the two distinct solutions below as* 9×9 *matrices:*

1. $c = 1$, $\check{r}^{(1)} = \sum_{a,b} e_{a,b+1} \otimes e_{b,a-1}$. *We write all the nine non-zero terms:*

$$\begin{aligned}\check{r}^{(1)} &= e_{1,2} \otimes e_{1,3} + e_{2,3} \otimes e_{2,1} + e_{3,1} \otimes e_{3,2} + e_{1,3} \otimes e_{2,3} + e_{1,1} \otimes e_{3,3} \\ &+ e_{2,1} \otimes e_{3,1} + e_{2,2} \otimes e_{1,1} + e_{3,2} \otimes e_{1,2} + e_{3,3} \otimes e_{2,2}.\end{aligned}$$

2. $c = 2$, $\check{r}^{(2)} = \sum_{a,b} e_{a,b+2} \otimes e_{b,a-2}$, *and explicitly,*

$$\begin{aligned}\check{r}^{(2)} &= e_{1,3} \otimes e_{1,2} + e_{2,1} \otimes e_{2,3} + e_{3,2} \otimes e_{3,1} + e_{1,1} \otimes e_{2,2} \\ &+ e_{1,2} \otimes e_{3,1} + e_{2,2} \otimes e_{3,3} + e_{2,3} \otimes e_{1,3} + e_{3,3} \otimes e_{1,1} + e_{3,1} \otimes e_{2,1}.\end{aligned}$$

The explicit 9×9 *matrices are,*

$$\check{r}^{(1)} = \begin{pmatrix} 0&0&0&0&0&1&0&0&0 \\ 0&0&0&0&0&0&0&0&1 \\ 0&0&1&0&0&0&0&0&0 \\ 0&0&0&1&0&0&0&0&0 \\ 0&0&0&0&0&0&1&0&0 \\ 1&0&0&0&0&0&0&0&0 \\ 0&0&0&0&1&0&0&0&0 \\ 0&0&0&0&0&0&0&1&0 \\ 0&1&0&0&0&0&0&0&0 \end{pmatrix}, \quad \check{r}^{(2)} = \begin{pmatrix} 0&0&0&0&0&0&0&1&0 \\ 0&1&0&0&0&0&0&0&0 \\ 0&0&0&0&1&0&0&0&0 \\ 0&0&0&0&0&0&0&0&1 \\ 0&0&1&0&0&0&0&0&0 \\ 0&0&0&0&0&1&0&0&0 \\ 0&0&0&0&0&0&1&0&0 \\ 1&0&0&0&0&0&0&0&0 \\ 0&0&0&1&0&0&0&0&0 \end{pmatrix}.$$

Notice that $\check{r}^{(2)} = \mathscr{P}\check{r}^{(1)}\mathscr{P}$, *where* $\mathscr{P}$ *is the permutation operator. In general, for any* $n \in \{1,2,\dots\}$ *we observe that* $\check{r}^{(n-k)} = \mathscr{P}\check{r}^{(k)}\mathscr{P}$, $k \in \{1,2,\dots,[\frac{n}{2}]\}$.

2.1 Self distributivity & special non-involutive solutions

To describe non-involutive solutions of the braid equation we introduce certain algebraic structures that satisfy a self distributivity condition. Self-distributive structures, such as shelves, racks & quandles [11, 46, 52] satisfy axioms analogous to the Reidemeister moves used to manipulate knot diagrams and are associated to link invariants (see also biracks, biquandles), and the coloring of links, i.e. a knot is tri-colored or not. According to Alexander's theorem all links are closed braids, hence these self distributive structures lead naturally to special non-involutive, set-theoretic solutions of the braid equation. Other algebraic structures naturally connected to bijective, non-degenerate solutions are the *bi-racks*. These are algebraic structures that appear in low-dimensional topology that are associated to link diagrams, and are invariant under the generalized Reidemeister moves for virtual knots and links, see [33]. We provide in what follows some preliminaries on left shelves, racks and quandles. For the first systematic study of shelves, we refer the interested reader to [11]. For recent reviews on self-distributive structures the interested reader is referred to [48, 58, 29] and [34] for potential physical applications.

Definition 2.4 *Let X be a non-empty set and $\triangleright$ a binary operation on X. Then, the pair $(X,\triangleright)$ is said to be a* left shelf *if $\triangleright$ is left self-distributive, namely, the identity*

$$a\triangleright(b\triangleright c)=(a\triangleright b)\triangleright(a\triangleright c) \tag{6}$$

is satisfied, for all $a,b,c\in X$. Moreover, a left shelf $(X,\triangleright)$ is called

1. *a* left spindle *if $a\triangleright a=a$, for all $a\in X$;*
2. *a* left rack *if $(X,\triangleright)$ is a* left quasigroup, *i.e., the maps $L_a: X\to X$ defined by $L_a(b):=a\triangleright b$, for all $b\in X$, are bijective, for every $a\in X$.*
3. *a* quandle *if $(X,\triangleright)$ is both a left spindle and a left rack.*

We are mostly interested in racks and quandles here, given that we always require invertible solutions of the Yang-Baxter equation. We provide below some fundamental known cases of quandles and racks (see also [48, 58, 29]):

(a) **Conjugate quandle.** Let $(X,\cdot)$ be a group and define $\triangleright: X\times X\to X$, such that $a\triangleright b=a^{-1}\cdot b\cdot a$. Then $(X,\triangleright)$ is a quandle.
(b) **Core quandle.** Let $(X,\cdot)$ be a group and $\triangleright: X\times X\to X$, such that $a\triangleright b=a\cdot b^{-1}\cdot a$. Then $(X,\triangleright)$ is a quandle.
(c) **Alexander (affine) quandle.** Let Q be a $\mathbb{Z}[t,t^{-1}]$ ring module and $\triangleright: Q\times Q\to Q$, $a\triangleright b=(1-t)a+bt$, then $(Q,\triangleright)$ is a quandle.
Or, let X be a non empty set equipped with two group operations, $+$ and $\circ$, such that $a\circ(b+c)=a\circ b-a+a\circ c$. This is a so-called left skew brace. The precise definition is given later in the text. Then define $\triangleright: X\times X\to X$, such that for a fixed $z\in X$ and for all $a,b\in X$, $a\triangleright b=z-a\circ z+b\circ z-z+a$ and $(a+b)\circ z=a\circ z-z+a\circ b$.
(d) **Rack, but not quandle.** Let $(G,\cdot)$ be a group and define $\triangleright: G\times G\to G$, such that $a\triangleright b=b\cdot a^{-1}\cdot x\cdot a$, where $x\in G$ is fixed. Then $(G,\triangleright)$ is a rack, but not a quandle.

We also present below some concrete examples of finite quandles:

Example 2.5

1. **The dihedral quandle.** *Let $i,j\in X=\mathbb{Z}_n$ and define $\triangleright: X\times X\to X$, such that $i\triangleright j=2i-j \bmod n: (X,\triangleright)$ is a quandle. This is a core quandle with an abelian group. An explicit table of the action $\triangleright$ is presented below for $n=3$ and $X=\{x_1=0,x_2=1,x_3=2\}:$*

$\triangleright$	x_1	x_2	x_3
x_1	x_1	x_3	x_2
x_2	x_3	x_2	x_1
x_3	x_2	x_1	x_3

Table 1

Although the table above comes from the dihedral quandle, another simple representation exists, $\rho : X \to End(\mathbb{C}X)$, such that

$$x_1 \mapsto \begin{pmatrix} 1\,0\,0 \\ 0\,0\,1 \\ 0\,1\,0 \end{pmatrix}, \quad x_2 \mapsto \begin{pmatrix} 0\,0\,1 \\ 0\,1\,0 \\ 1\,0\,0 \end{pmatrix}, \quad x_3 \mapsto \begin{pmatrix} 0\,1\,0 \\ 1\,0\,0 \\ 0\,0\,1 \end{pmatrix}.$$

$\rho(x \triangleright y) = \rho(x)^{-1} \bullet \rho(y) \bullet \rho(x)$ *for all* $x, y \in X$ *and* $\bullet$ *is the usual matrix multiplication.*

2. **The tetrahedron quandle.** *Let* $X = \{1,2,3,4\}$ *and define* $\triangleright : X \times X \to X$, *such that* $1\triangleright = (234)$, $2\triangleright = (143)$, $3\triangleright = (124)$ *and* $4\triangleright = (132)$. *This is also a cyclic quandle. We construct below the explicit table of the action* $\triangleright$:

$\triangleright$	x_1	x_2	x_3	x_4
x_1	x_1	x_3	x_4	x_2
x_2	x_4	x_2	x_1	x_3
x_3	x_2	x_4	x_3	x_1
x_4	x_3	x_1	x_2	x_4

Table 2

3. **Example of affine quandle.** *Let* $X = U_m$ *denote the set of odd integers mod* 2^m, $m \in \mathbb{N}$, *and the two operation be* $+_1$, *such that* $a +_1 b = a - 1 + b$ *and* $\circ$, *where* $+$ *and* $\circ$ *are the usual addition and multiplication. Then a quandle is obtained as in case (c) (Alexander's quandle), i.e. for a fixed* $z \in U_m$ *define,* $a \triangleright b = z -_1 a \circ z +_1 b \circ z -_1 z +_1 a$. *Specifically, the sets are: 1. for* $m = 1$, $U_1 = \{1\}$, *2. for* $m = 2$, $U_2 = \{1,\ 3\}$, *3. for* $m = 3$, $U_3 = \{1,\ 3,\ 5,\ 7\}$, *etc.*
Recalling that $a +_1 b = a - 1 + b$ *and that* $(X, +)$ *in an abelian group we conclude that* $a \triangleright b = -a \circ z + b \circ z + a$. *For instance for* $m = 3$ *(*$X = \{x_1 = 1,\ x_2 = 3,\ x_3 = 5,\ x_4 = 7\}$*) and by choosing for example* $z = 3$ *we obtain the following table*

$\triangleright$	x_1	x_2	x_3	x_4
x_1	x_1	x_4	x_3	x_2
x_2	x_3	x_2	x_1	x_4
x_3	x_1	x_4	x_3	x_2
x_4	x_3	x_2	x_1	x_4

Table 3

Notice here the distributivity rule between $+_1$ and $\circ$ in case (3) of Example 2.5 of an affine quandle. Indeed, we observe that for all $a,b,c \in X$ $a\circ(b+_1 c) = a\circ b -_1 a +_1 a\circ c$ (see also later in the text the definition of skew braces, Definition 2.9).

We recall now a fundamental statement regarding shelves and solutions of the set-theoretic Yang-Baxter equation.

Proposition 2.6 *We define the binary operation* $\triangleright : X\times X \to X$, $(a,b)\mapsto a\triangleright b$. *Then* $\check{r}: X\times X \to X\times X$, *such that for all* $a,b\in X$, $\check{r}(a,b) = (b, b\triangleright a)$ *is a solution of the set-theoretic braid equation if and only if* $(X,\triangleright)$ *is a shelf.*

Proof. The proof is straightforward by direct substitution in the Yang-Baxter equation and comparison between LHS and RHS (a graphical depiction of the proof is given below in Figure 4).

Remark 2.7 *If* $\check{r}: X\times X \to X\times X$, *such that for all* $a,b\in X$, $\check{r}(a,b) = (b,b\triangleright a)$ *is an invertible braid solution then* $(X,\triangleright)$ *is a rack (or a quandle).*

The graphical representation of the shelve solution $\check{r}(a,b) = (b,b\triangleright a)$:

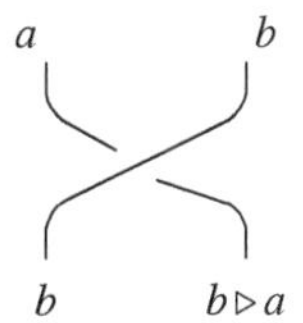

We are interested here in invertible solutions of the braid equation, so we are focusing on rack solutions. We note that the inverse of $\check{r}$ above is $\check{r}^{-1}: X\times X \to X\times X$, $\check{r}^{-1}(a,b) = (a\triangleright^{-1} b, a)$, such that $a\triangleright(a\triangleright^{-1} b) = a\triangleright^{-1}(a\triangleright b) = b$ for all $a,b\in X$. Notice also that a different map denoted as $\check{r}': X\times X\to X\times X$, such that $\check{r}'(a,b) = (a\triangleright b, a)$ is also a solution of the braid equation. $\check{r}'$ is graphically depicted below in Figure 3 together with the braid relation in Figure 4.
The expressions in blue in Figure 4 indicate the left and right hand side of the self-distributivity condition.

Example 2.8 *We express the solution of the braid equation associated to the dihedral and tetrahedron quandle of Example 2.5 as* 9×9 *and* 16×16 *matrices respectively. Recall from linearization we obtain,* $\check{r}$ *as a matrix,* $\check{r} = \sum_{x,y\in X} e_{x,y}\otimes e_{y,y\triangleright x}$, *where* $e_{x,y}$ *are the elementary* $n\times n$ *matrices* $e_{x,y} = e_x e_y^T$ *(*^{T} *denotes transposition). In our examples* $n = 3,4$.

1. *(A non-involutive solution from the dihedral quandle). More specifically, from Table 1*

$$\check{r} = \sum_{j=1}^{3} e_{x_j,x_j}\otimes e_{x_j,x_j} + e_{x_1,x_2}\otimes e_{x_2,x_3} + e_{x_2,x_1}\otimes e_{x_1,x_3} + e_{x_2,x_3}\otimes e_{x_3,x_1}$$
$$+ e_{x_3,x_2}\otimes e_{x_2,x_1} + e_{x_1,x_3}\otimes e_{x_3,x_2} + e_{x_3,x_1}\otimes e_{x_1,x_2}.$$

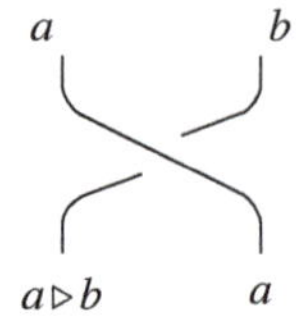

3. The shelve braiding

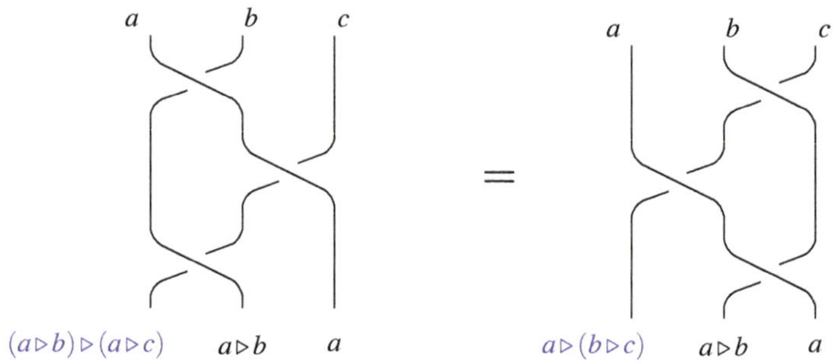

4. Shelve solution of the braid equation

Then $\check{r}$ is explicitly expressed as a 9×9 matrix,

$$\check{r} = \begin{pmatrix} 1&0&0&0&0&0&0&0&0\\ 0&0&0&0&0&1&0&0&0\\ 0&0&0&0&0&0&0&1&0\\ 0&0&1&0&0&0&0&0&0\\ 0&0&0&0&1&0&0&0&0\\ 0&0&0&0&0&0&1&0&0\\ 0&1&0&0&0&0&0&0&0\\ 0&0&0&1&0&0&0&0&0\\ 0&0&0&0&0&0&0&0&1 \end{pmatrix}.$$

2. *(A non-involutive solution from the tetrahedron quandle). More specifically, from Table 2*

$$\begin{aligned}\check{r} = &\sum_{j=1}^{4} e_{x_j,x_j}\otimes e_{x_j,x_j} + e_{x_1,x_2}\otimes e_{x_2,x_4} + e_{x_2,x_1}\otimes e_{x_1,x_3} + e_{x_2,x_3}\otimes e_{x_3,x_4}\\ &+ e_{x_3,x_2}\otimes e_{x_2,x_1} + e_{x_1,x_3}\otimes e_{x_3,x_2} + e_{x_3,x_1}\otimes e_{x_1,x_4} + e_{x_2,x_4}\otimes e_{x_4,x_1}\\ &+ e_{x_4,x_2}\otimes e_{x_2,x_3} + e_{x_3,x_4}\otimes e_{x_4,x_2} + e_{x_4,x_3}\otimes e_{x_3,x_1} + e_{x_1,x_4}\otimes e_{x_4,x_3} + e_{x_4,x_1}\otimes e_{x_1,x_2}.\end{aligned}$$

2.2 Skew braces & generic set-theoretic solutions

It is useful to recall at this point the definition of (skew) braces [55]-[57], [14, 39] as this will allow us to derive generic solutions of the set-theoretic Yang-Baxter equation of the type $r : X \times X \to X \times X$, $r(b,a) = (\sigma_a(b), \tau_b(a))$.

Definition 2.9 *A left skew brace is a set B together with two group operations* $+, \circ :$ $B \times B \to B$*, the first is called addition and the second is called multiplication, such that for all* $a, b, c \in B$*,*

$$a \circ (b + c) = a \circ b - a + a \circ c. \tag{7}$$

If $+$ *is an abelian group operation, then B is called a left brace. Moreover, if B is a left skew brace and for all* $a, b, c \in B$ $(b + c) \circ a = b \circ a - a + c \circ a$*, then B is called a two sided skew brace. Analogously if* $+$ *is abelian and B is a skew brace, then B is called a two sided brace.*

The additive identity of a skew brace B will be denoted by 0 and the multiplicative identity by 1. In every skew brace $0 = 1$.

From now on when we say skew brace we mean left skew brace. Some useful examples of braces are presented below:

Example 2.10 (See [9] Corollary 3.14) *Let* $\mathrm{U}(\mathbb{Z}/2^m\mathbb{Z})$ *denote a set of invertible integers modulo* 2^m*, for some* $m \in \mathbb{N}$*. Then a triple* $(\mathrm{U}(\mathbb{Z}/2^m\mathbb{Z}), +_1, \circ)$ *is a brace, where* $a +_1 b = a - 1 + b$*, for all* $a, b \in \mathrm{U}(\mathbb{Z}/2^m\mathbb{Z})$*,* $+$ *and* $\circ$ *are addition and multiplication of integer numbers modulo* 2^m*, respectively.*

Example 2.11 (See [10] Example 5.7) *Consider a ring* $\mathbb{Z}/8\mathbb{Z}$*. A triple*

$$\left(\mathrm{OM} := \left\{ \begin{pmatrix} a & b \\ c & d \end{pmatrix} \mid a, d \in \{1,3,5,7\},\ b, c \in \{0,2,4,6\} \right\}, +_{\mathbb{I}}, \circ \right)$$

is a brace, where $(A, B) \stackrel{+_{\mathbb{I}}}{\longmapsto} A - \mathbb{I} + B$*,* $(A, B) \stackrel{\circ}{\longmapsto} A \cdot B$*, and* $+, \cdot$ *are addition and multiplication of two by two matrices over* $\mathbb{Z}/8\mathbb{Z}$*, respectively.*

Example 2.12 (See [10] Example 5.6 or [9] Example 3.15) *Consider a set* $\mathrm{Odd} := \left\{ \frac{2n+1}{2k+1} \mid n, k \in \mathbb{Z} \right\}$ *together with two binary operations* $(a, b) \stackrel{+_1}{\longmapsto} a - 1 + b$ *and* $(a, b) \stackrel{\circ}{\longmapsto} a \cdot b$*, where* $+, \cdot$ *are addition and multiplication of rational numbers, respectively. The triple* $(\mathrm{Odd}, +_1, \circ)$ *is a brace.*

We recall now the basic conditions associated to any generic solution of the set-theoretic Yang-Baxter equation, as they will be used in our analysis here.

Proposition 2.13 *Let X be a nonempty set and define for all* $a, b \in X$*, the maps* $\sigma_a, \tau_b : X \to X$*,* $b \mapsto \sigma_a(b)$ *and* $a \mapsto \tau_b(a)$*. Then* $r : X \times X \to X \times X$*, such that for all* $a, b \in X$*,* $r(b, a) = (\sigma_a(b), \tau_b(a))$ *is a solution of the set-theoretic Yang-Baxter equation if and only if for all* $a, b, c \in X$*,*

$$\sigma_a(\sigma_b(c)) = \sigma_{\sigma_a(b)}(\sigma_{\tau_b(a)}(c)) \tag{8}$$
$$\tau_c(\tau_b(a)) = \tau_{\tau_c(b)}(\tau_{\sigma_b(c)}(a)) \tag{9}$$
$$\sigma_{\tau_{\sigma_b(c)}(a)}(\tau_c(b)) = \tau_{\sigma_{\tau_b(a)}(c)}(\sigma_a(b)). \tag{10}$$

Proof. Let r be a solution of the Yang-Baxter equation. We explicitly compute the LHS and RHS of the parametric Yang-Baxter equation. The LHS of the Yang-Baxter equation gives, $a,b,c \in X$,

$$r_{12}\, r_{13}\, r_{23}(c,b,a) = \big(\sigma_{\sigma_a(b)}(\sigma_{\tau_b(a)}(c)),\ \tau_{\sigma_{\tau_b(a)}(c)}(\sigma_a(b)),\ \tau_c(\tau_b(a))\big), \tag{11}$$

whereas the RHS gives

$$r_{23}\, r_{13}\, r_{12}(c,b,a) = \big(\sigma_a(\sigma_b(c)),\ \sigma_{\tau_{\sigma_b(c)}(a)}(\tau_c(b)),\ \tau_{\tau_c(b)}(\tau_{\sigma_b(c)}(a))\big). \tag{12}$$

By equating (11) and (12) we arrive at (8)-(10). And conversely, if conditions (8)-(10) are satisfied then r automatically satisfies the Yang-Baxter equation.

We may now prove a key proposition on generic set-theoretic solutions coming from skew braces [55, 56, 14, 39].

Proposition 2.14 *Let $(X,+,\circ)$ be a skew brace and let $\sigma_a(b) := -a + a\circ b$ and $a\circ b = \sigma_a(b)\circ\tau_b(a)$ for all $a,b \in X$. Then the map $r : X\times X \to X\times X$, $r(b,a) = (\sigma_a(b), \tau_b(a))$ is a solution of the set theoretic Yang-Baxter equation.*

Proof. In order to prove that r is a solution of the Yang-Baxter equation, it suffices to show that the three conditions (8)-(10) hold.

Before we proceed with the proof, we observe that the distributivity condition in skew braces for all $a,b,c \in X$, $a\circ(b+c) = a\circ b - a + a\circ c$ is equivalent to $a\circ(b-c+d) = a\circ b - a\circ c + a\circ d$ (see also [22]) for a detailed proof).

We first show condition (8), indeed for $a,b,c \in X$,

$$\begin{aligned}\sigma_a(\sigma_b(c)) &= -a + a\circ\sigma_b(c) = -a + a\circ(-b + b\circ c) = -a\circ b + a\circ b\circ c = \sigma_{a\circ b}(c)\\ &= \sigma_{\sigma_a(b)\circ\tau_b(a)}(c) = \sigma_{\sigma_a(b)}(\sigma_{\tau_b(a)}(c)).\end{aligned}$$

We show now condition (9),

$$\tau_c(\tau_b(a)) = \sigma_{\tau_b(a)}(c)^{-1}\circ\tau_b(a)\circ c = \sigma_{\tau_b(a)}(c)^{-1}\circ\sigma_a(b)^{-1}\circ a\circ b\circ c.$$

But,

$$\begin{aligned}\sigma_a(b)\circ\sigma_{\tau_b(a)}(c) &= \sigma_a(b)\circ(-\tau_b(a) + \tau_b(a)\circ c) = \sigma_a(b) - a\circ b + a\circ b\circ c\\ &= -a + a\circ b\circ c = \sigma_a(b\circ c)\end{aligned} \tag{13}$$

hence,

$$\tau_c(\tau_b(a)) = \tau_{b\circ c}(a) = \tau_{\sigma_b(c)\circ\tau_c(b)} = \tau_{\tau_c(b)}(\tau_{\sigma_b(c)}(a)). \tag{14}$$

Condition (10) follows from (8), (9) using $a\circ b = \sigma_a(b)\circ\tau_b(a)$, for all $a,b \in X$. And, this concludes our proof.

The group $(X,\circ)$, such that for all $a,b \in X$, $a \circ b = \sigma_a(b) \circ \tau_b(a)$ is called the *structure group* of a set-theoretic solution. In the case that $(X,+,\circ)$ is a brace we obtain involutive solutions [55, 56, 57]. According to Rump all involutive solutions are obtained from braces, therefore in this article we often call the involutive set-theoretic solutions *brace solutions*.

The important fact that will be discussed in Section 5, is that all involutive solutions are obtained from the permutation operator by a suitable Drinfel'd twist (see also [20, 60, 50]), whereas all the non-involutive but invertible solutions are obtained from rack solutions via a Drinfel'd twist (degenerate solutions are coming from shelves) [22, 24].

3 Involutive solutions & Baxterization

We focus in this section on involutive set-theoretic solutions and derive Baxterized solutions of the braid and Yang-Baxter equations. We then identify the quantum algebras associated to these solutions (see also [18, 19]).

Recall the Yang-Baxter equation in the braid form in the presence of spectral parameters λ_1, λ_2 ($\delta = \lambda_1 - \lambda_2$):

$$\check{R}_{12}(\delta)\, \check{R}_{23}(\lambda_1)\, \check{R}_{12}(\lambda_2) = \check{R}_{23}(\lambda_2)\, \check{R}_{12}(\lambda_1)\, \check{R}_{23}(\delta), \tag{1}$$

where $\check{R} : V \otimes V \to V \otimes V$, ($V$ is an n dimensional space) and let in general $\check{R} = \sum_j a_j \otimes b_j$, then in the index notation $\check{R} = \sum_j a_j \otimes b_j \otimes 1_V$, $\check{R}_{23} = \sum_j 1_V \otimes a_j \otimes b_j$ and $\check{R}_{13} = \sum_j a_j \otimes 1_V \otimes b_j$.

We focus here on Baxterized solutions of the form

$$\check{R}(\lambda) = \lambda \check{r} + 1_V^{\otimes 2}, \tag{2}$$

where $\check{r}$ is an involutive solution of the braid equation. Let also, $R = \mathscr{P}\check{R}$, then

$$R(\lambda) = \lambda r + \mathscr{P}, \tag{3}$$

and R is a solution of the Yang-Baxter equation,

$$R_{12}(\delta)\, R_{13}(\lambda_1)\, R_{23}(\lambda_2) = R_{23}(\lambda_2)\, R_{13}(\lambda_1)\, R_{12}(\delta). \tag{4}$$

Remark 3.1 *It would be useful for the following Proposition to define the partial transposition. Let $A \in End\big(\mathbb{C}^n \otimes \mathbb{C}^n\big)$ expressed as: $A = \sum_{i,j,k,l=1}^n A_{ij,kl}\, e_{i,j} \otimes e_{k,l}$. We define the partial transposition as follows (in the index notation):*

$$A_{12}^{t_1} = \sum_{i,j,k,l=1}^{n} A_{ij,kl}\, e_{i,j}^t \otimes e_{k,l}, \quad A_{12}^{t_2} = \sum_{i,j,k,l=1}^{n} A_{ij,kl}\, e_{i,j} \otimes e_{k,l}^t \tag{5}$$

where $e_{i,j}^t = e_{j,i}$.

Proposition 3.2 *The brace R-matrix satisfies the following fundamental properties:*

$$R_{12}(\lambda)\, R_{21}(-\lambda) = (-\lambda^2+1)1_V^{\otimes 2}, \qquad \textit{Unitarity} \tag{6}$$
$$R_{12}^{t_1}(\lambda)\, R_{12}^{t_2}(-\lambda-n) = \lambda(-\lambda-n)1_V^{\otimes 2}, \qquad \textit{Crossing-unitarity} \tag{7}$$
$$R_{12}^{t_1 t_2}(\lambda) = R_{21}(\lambda), \tag{8}$$

where $^{t_{1,2}}$ *denotes transposition on the first, second space respectively.*

Proof. The proof follows immediately after a straightforward computation [19].

3.1 The quantum algebra associated to braces

We recall the definitions of two quadratic algebras $\mathscr{A}$ and $\mathscr{Q}$ associated to solutions of the Yang-Baxter equation, which arise from the FRT (Faddeev, Reshetikhin and Takhtajan) construction [32]. Indeed, from the FRT construction we recall that given a solution of the braid equation $\check{r}: V\times V \to V\times V$ (henceforth we consider $V=\mathbb{C}^n$) the associated quantum algebra $\mathscr{A}$ is a quotient of a free associative $\mathbb{C}$-algebra, generated by $\{L_{z,w}|\ x,w\in X\}$, and relations

$$\check{r}_{12}\, L_1\, L_2 = L_1\, L_2\, \check{r}_{12}, \tag{9}$$

where $L=\sum_{x,y\in X} e_{x,y}\otimes L_{x,y}\in \mathrm{End}(V)\otimes\mathscr{A}$. Recall the *index notation*: $\check{r}_{12}=\check{r}\otimes 1_{\mathscr{A}}$ and $L_1=\sum_{z,w\in X} e_{z,w}\otimes 1_V\otimes L_{z,w},\ L_2=\sum_{z,w\in X} 1_V\otimes e_{z,w}\otimes L_{z,w}$.

From the fundamental relation (9) and by considering the set-theoretic solution of the braid equation

$$\check{r} = \sum_{x,y\in X} e_{x,\sigma_x(y)}\otimes e_{y,\tau_y(x)}, \tag{10}$$

we obtain [30]:

$$L_{x,\hat{x}}L_{y,\hat{y}} = L_{\sigma_x(y),\sigma_{\hat{x}}(\hat{y})}L_{\tau_y(x),\tau_{\hat{y}}(\hat{x})}. \tag{11}$$

Recall that sometimes we call the involutive, set-theoretic solutions, brace solutions, because they are all obtained from braces [55, 56]. Given a solution of the braid equation $\check{r}: V\otimes V\to V\otimes V$, the quadratic algebra $\mathscr{Q}$ is generated by $\{q_x|\ x\in X\}$ and relations

$$\check{r}_{12}\, q_1\, q_2 = q_1\, q_2, \tag{12}$$

where $q=e_x\otimes q_x\in V\otimes\mathscr{Q}$. Also, $\check{r}_{12}=\check{r}\otimes 1_{\mathscr{A}},\ q_1=\sum_{z,w\in X} e_x\otimes 1_V\otimes q_x,\ q_2=\sum_{x\in X} 1_V\otimes e_x\otimes q_x$. The quadratic relation (12) for the set-theoretic solution implies

$$q_x q_y = q_{\sigma_x(y)}q_{\tau_y(x)}, \tag{13}$$

also obtained in [30].

We now consider the Baxterized solution $\check{R}(\lambda)=\lambda\check{r}+1_{V\otimes V}$, where $\check{r}$ in our analysis here is the set-theoretic solution of the braid equation (10). Given a parametric solution of the Yang-Baxter equation, the quantum algebra is defined via the fundamental relation [32]:

$$\check{R}_{12}(\lambda_1-\lambda_2)\, L_1(\lambda_1)\, L_2(\lambda_2) = L_1(\lambda_2)\, L_2(\lambda_1)\, \check{R}_{12}(\lambda_1-\lambda_2). \tag{14}$$

$\check{R}(\lambda) \in \mathrm{End}(\mathbb{C}^n \otimes \mathbb{C}^n)$, $L(\lambda) \in \mathrm{End}(\mathbb{C}^n) \otimes \mathfrak{A}$, where $\mathfrak{A}$[1] is the quantum algebra defined by (14). We shall focus henceforth on Baxterized solutions coming from involutive, set-theoretic solutions. The defining relations of the corresponding quantum algebra were derived in [18, 19].

The quantum algebra associated to Baxterized solutions coming from braces is defined by generators $L^{(m)}_{z,w}$, $z, w \in X$, and defining relations (see also [19])

$$\begin{aligned} L^{(n)}_{z,w} L^{(m)}_{\hat{z},\hat{w}} - L^{(m)}_{z,w} L^{(n)}_{\hat{z},\hat{w}} &= L^{(m)}_{z,\sigma_w(\hat{w})} L^{(n+1)}_{\hat{z},\tau_{\hat{w}}(w)} - L^{(m+1)}_{z,\sigma_w(\hat{w})} L^{(n)}_{\hat{z},\tau_{\hat{w}}(w)} \\ &\quad - L^{(n+1)}_{\sigma_z(\hat{z}),w} L^{(m)}_{\tau_{\hat{z}}(z),\hat{w}} + L^{(n)}_{\sigma_z(\hat{z}),w} L^{(m+1)}_{\tau_{\hat{z}}(z),\hat{w}}. \end{aligned} \tag{15}$$

The proof is based on the fundamental relation (14) and the form of the Baxterized brace R-matrix (for the detailed proof see [18, 19]). Recall also that in the index notation we define $\check{R}_{12} = \check{R} \otimes \mathrm{id}_{\mathfrak{A}}$:

$$L_1(\lambda) = \sum_{z,w\in X} e_{z,w} \otimes 1_V \otimes L_{z,w}(\lambda), \qquad L_2(\lambda) = \sum_{z,w\in X} 1_V \otimes e_{z,w} \otimes L_{z,w}(\lambda). \tag{16}$$

The exchange relations among the various generators of the affine algebra are derived below via (14). Let us express L as a formal power series expansion $L(\lambda) = \sum_{n=0}^{\infty} \frac{L^{(n)}}{\lambda^n}$. Substituting expressions (2), and the λ^{-1} expansion in (14) we obtain the defining relations of the quantum algebra associated to a brace R-matrix (we focus on terms $\lambda_1^{-n}\lambda_2^{-m}$):

$$\begin{aligned} &\check{r}_{12} L_1^{(n+1)} L_2^{(m)} - \check{r}_{12} L_1^{(n)} L_2^{(m+1)} + L_1^{(n)} L_2^{(m)} \\ &= L_1^{(m)} L_2^{(n+1)} \check{r}_{12} - L_1^{(m+1)} L_2^{(n)} \check{r}_{12} + L_1^{(m)} L_2^{(n)}. \end{aligned} \tag{17}$$

The latter relations immediately lead to the quantum algebra relations (15), after recalling: $L_1^{(k)} = \sum_{x,y\in X} e_{x,y} \otimes 1_V \otimes L^{(k)}_{x,y}$, $L_2^{(k)} = \sum_{x,y\in X} 1_V \otimes e_{x,y} \otimes L^{(k)}_{x,y}$, and $\check{r}_{12} = \check{r} \otimes \mathrm{id}_{\mathfrak{A}}$, $L^{(k)}_{x,y}$ are the generators of the associated quantum algebra. The quantum algebra is also equipped with a co-product $\Delta : \mathfrak{A} \to \mathfrak{A} \otimes \mathfrak{A}$ [32, 28]. Indeed, we define

$$(\mathrm{id} \otimes \Delta) L(\lambda) := L_{13}(\lambda) L_{12}(\lambda), \tag{18}$$

which satisfies (14) and is expressed as $(\mathrm{id} \otimes \Delta) L(\lambda) = \sum_{x,y\in X} e_{x,y} \otimes \Delta(L_{x,y}(\lambda))$.

Remark 3.3 *In the special case $\check{r} = \mathscr{P}$ the $\mathscr{Y}(\mathfrak{gl}_n)$ algebra is recovered:*

---

[1] Notice that in L in addition to the indices 1 and 2 in (14) there is also an implicit "quantum index" n associated to $\mathfrak{A}$, which for now is omitted, i.e. one writes L_{1n}, L_{2n}.

$$\left[L^{(n+1)}_{i,j},\, L^{(m)}_{k,l}\right] - \left[L^{(n)}_{i,j},\, L^{(m+1)}_{k,l}\right] = L^{(m)}_{k,j} L^{(n)}_{i,l} - L^{(n)}_{k,j} L^{(m)}_{i,l}. \tag{19}$$

The next natural step is the classification of solutions of the fundamental relation (14), for the brace quantum algebra. A first step towards this goal will be to examine the fundamental object $L(\lambda) = L_0 + \frac{1}{\lambda} L_1$, and search for finite and infinite representations of the respective elements. The classification of L-operators will allow the identification of new classes of quantum integrable systems, such as the analogues of Toda chains or deformed boson models. A first obvious example to consider is associated to Lyubashenko's solution, which is further discussed later in the manuscript.

3.2 Integrability: local Hamiltonians

Given any solution of the Yang-Baxter equation we define the so-called monodromy matrix $T_{0,12\dots N}(\lambda) \in \mathrm{End}\big(\mathbb{C}^n \otimes (\mathbb{C}^n)^{\otimes N}\big)$, which is a tensor representation of the quantum group (14), [32]

$$T_{0,12\dots N}(\lambda) := R_{0N}(\lambda)\dots R_{02}(\lambda)\, R_{01}(\lambda), \tag{20}$$

recall $R = \mathscr{P}\check{R}$. We define also the transfer matrix $\mathrm{t}_{12\dots N}(\lambda) = tr_0\big(T_{0,12\dots N}(\lambda)\big) \in \mathrm{End}\big((\mathbb{C}^n)^{\otimes N}\big)$. The monodromy matrix T satisfies (14), and hence one can show that the transfer matrix provides mutually commuting quantities [32]: ($\mathrm{t}(\lambda) = \lambda^N \sum_k \frac{\mathrm{t}^{(k)}}{\lambda^k}$)

$$\left[\mathrm{t}(\lambda),\, \mathrm{t}(\mu)\right] = 0 \Rightarrow \left[\mathrm{t}^{(k)},\, \mathrm{t}^{(l)}\right] = 0. \tag{21}$$

Note that historically the index 0 is called "auxiliary", whereas the indices $1, 2, \dots, N$ are called "quantum", and they are usually suppressed for simplicity, i.e. we simply write $T_0(\lambda)$ and $\mathrm{t}(\lambda)$.

The following Proposition 3.4 is quite general and holds for any $R(\lambda) = \lambda \mathscr{P}\check{r} + \mathscr{P}$, where $\check{r}$ is an involutive solution of the braid equation and $\mathscr{P}$ is the permutation operator. The key property that allows the derivation of local Hamiltonians is $R(\lambda = 0) = \mathscr{P}$.

Proposition 3.4 *Consider the λ-series expansion of the monodromy matrix: $T(\lambda) = \lambda^N \sum_{k=0}^N \frac{T^{(k)}}{\lambda^k}$ for any $R(\lambda) = \lambda \mathscr{P}\check{r} + \mathscr{P}$, where $\check{r}$ is an involutive solution of the braid equation. Let also $H^{(k)} = \mathrm{t}^{(k)}(\mathrm{t}^{(N)})^{-1}$, $k = 0, \dots, N-1$ and $H^{(N)} = \mathrm{t}^{(N)}$, where $\mathrm{t}^{(k)} = tr_0(T_0^{(k)})$. Then the commuting quantities, $H^{(k)}$ for $k = 1, \dots, N-1$, are expressed exclusively in terms of the elements $\check{r}_{n n+1}$, $n = 1, \dots, N-1$, and $\check{r}_{N1}$.*

Proof. We refer the interested reader to [19] for the detailed proof.

The generic first neighbor Hamiltonian is given as $\mathscr{H} = \sum_{j=1}^N \check{r}_{jj+1}$. Higher commuting quantities can be found in [19] for periodic spin chains and in [18] for open spin chains. In the special case of set-theoretic solutions (10) the local Hamiltonian becomes

$$\mathscr{H} = \sum_{j=1}^{N} \sum_{a,b\in X} e^{(j)}_{a,\sigma_a(b)} e^{(j+1)}_{b,\tau_b(a)}. \tag{22}$$

A simple example within this class is the Lyubashenko solution (see Example 2.1), then the Hamiltonian (22) takes the simple form,

$$\mathscr{H}_c = \sum_{j=1}^{N} \sum_{a,b=1}^{n} e^{(j)}_{a,b+c} e^{(j+1)}_{b,a-c},$$

where recall $c \in \{1,2,\ldots,n-1\}$ is fixed.

The ultimate goal in the context of quantum integrable systems, or any quantum system for that matter, is the identification of the eigenvalues and eigenvectors of the corresponding Hamiltonian. In the frame of quantum integrable systems there exists a set of mutually commuting "Hamiltonians", guaranteed by the existence of a quantum R-matrix that satisfies the Yang-Baxter equation. An exhaustive analysis of the symmetries of periodic and open quantum spin chains constructed from Baxterized involutive set-theoretic solutions is presented in [18, 19]. The hierarchy of periodic and open mutually commuting Hamiltonians is also explicitly derived in [18, 19] exclusively in terms of the elements of the symmetric group.

4 Set-theoretic solutions as Drinfel'd twists: an example

We recall in this section the Drinfel'd twist for set-theoretic solutions. It was shown in [20] that all involutive, set-theoretic solutions can be obtained from the permutation operator via suitable twists (see also [60] for an analogous non-local map), whereas non-involutive, invertible solutions are obtained via the same twist from rack/quandle solutions [22, 24]. We consider here a simple example of set-theoretic solution of the braid equation obtained as a simple twist of the permutation to provide a key motivation for the general results presented in the subsequent section.

Simple non-trivial case: Lyubashenko's solution

We recall the Lyubashenko solution and show that is immediately obtained from the permutation operator as a simple twist. Although the construction is simple it has significant implications on the associated symmetries of the braid solutions. Inspired by the isotropic case a similar construction for the q-deformed analogue of Lyubashenko's solution is provided in [19, 18].

Before we derive the Lyubashenko solution as a suitable twist we first introduce a useful Lemma.

Lemma 1. *Let $\check{r}' : V \otimes V \to V \otimes V$ (V is an n dimensional vector space) satisfy the braid relation and $(\check{r}')^2 = 1_V^{\otimes 2}$. Let also $u : V \to V$ be an invertible map, such that $(u \otimes u)\check{r}' = \check{r}'(u \otimes u)$. We define $\check{r} = (u \otimes 1_V)\check{r}'(u^{-1} \otimes 1_V) = (1_V \otimes u^{-1})\check{r}'(1_V \otimes u)$, then:*

1. $\check{r}^2 = 1_V^{\otimes 2}$
2. *$\check{r}$ satisfies the braid relation.*

Proof. The proof is straightforward [18].

Proposition 4.1 *Let $\tau,\ \sigma : X \to X$, $X = \{1,\dots,n\}$ be isomorphisms, such that $\sigma(\tau(x)) = \tau(\sigma(x)) = x$ and let $u = \sum_{x\in X} e_{x,\tau(x)}$ and $u^{-1} = \sum_{x\in X} e_{\tau(x),x}$. Then any solution of the type (Lyubashenko's solution)*

$$\check{r} = \sum_{x,y\in X} e_{x,\sigma(y)} \otimes e_{y,\tau(x)}, \tag{1}$$

is obtained from the permutation operator $\mathscr{P} = \sum_{x,y\in X} e_{x,y} \otimes e_{y,x}$ as

$$\check{r} = (u \otimes 1_V)\mathscr{P}(u^{-1} \otimes 1_V) = (1_V \otimes u^{-1})\mathscr{P}(1_V \otimes u). \tag{2}$$

Proof. The proof relies on the definitions of $\mathscr{P}$, u, u^{-1}, the fundamental property $e_{x,y}e_{z,w} = \delta_{y,z}e_{x,w}$ and by straightforward computation.

Note that $r = \mathscr{P}\check{r}$, and consequently $R = \mathscr{P}\check{R}$ take a simple form for this class of solutions:

$$r = u^{-1} \otimes u \;\Rightarrow\; R(\lambda) = \lambda u^{-1} \otimes u + \mathscr{P}. \tag{3}$$

Before we present our findings on the symmetry of Lyubashenko's $\check{r}$-matrix we first introduce a useful Lemma [19, 18].

Lemma 2. *Let $\mathfrak{l}_{x,y}$ be the generators of the $\mathfrak{gl}_n$ algebra:*

$$\left[\mathfrak{l}_{x,y}, \mathfrak{l}_{z,w}\right] = \delta_{y,z}\mathfrak{l}_{x,w} - \delta_{x,w}\mathfrak{l}_{z,y}. \tag{4}$$

The $\mathfrak{gl}_n$ algebra is equipped with a coproduct $\Delta : \mathfrak{gl}_n \to \mathfrak{gl}_n \otimes \mathfrak{gl}_n$, such that

$$\Delta(\mathfrak{l}_{x,y}) = \mathfrak{l}_{x,y} \otimes id + id \otimes \mathfrak{l}_{x,y}. \tag{5}$$

The N-coproduct is obtained by iteration $\Delta^{(N)} = (\Delta^{(N-1)} \otimes id)\Delta = (id \otimes \Delta^{(N-1)})\Delta$ and is given as $\Delta^{(N)}(\mathfrak{l}_{x,y}) = \sum_{n=1}^{N} id \otimes \dots \otimes \underbrace{\mathfrak{l}_{x,y}}_{n^{th}\ position} \otimes \dots \otimes id$.

Let also $\mathscr{F}^{(N)} \in \mathfrak{gl}_n^{\otimes N}$ be an invertible element ($\mathscr{F}^{(2)} =: \mathscr{F}$), and define $\Delta_T^{(N)}(\mathfrak{l}_{x,y}) := \mathscr{F}^{(N)}\Delta^{(N)}(\mathfrak{l}_{x,y})(\mathscr{F}^{(N)})^{-1}$, then $\Delta_T^{(N)}(\mathfrak{l}_{x,y})$ also satisfy the $\mathfrak{gl}_n$ algebraic relations.

Proof. The N-coproducts satisfy the $\mathfrak{gl}_n$ relations (4), i.e. $\left[\Delta^{(N)}(\mathfrak{l}_{x,y}), \Delta^{(N)}(\mathfrak{l}_{z,w})\right] = \delta_{y,z}\Delta^{(N)}(\mathfrak{l}_{x,w}) - \delta_{x,w}\Delta^{(N)}(\mathfrak{l}_{z,y})$. By acting from the left with $\mathscr{F}^{(N)}$ and with $(\mathscr{F}^{N})^{-1}$

from the right in the latter commutator we immediately obtain $\left[\Delta_T^{(N)}(\mathfrak{l}_{x,y}), \Delta_T^{(N)}(\mathfrak{l}_{z,w})\right] = \delta_{y,z}\Delta_T^{(N)}(\mathfrak{l}_{x,w}) - \delta_{x,w}\Delta_T^{(N)}(\mathfrak{l}_{z,y})$.

Corollary 4.2 *Let $\rho : \mathfrak{gl}_n \to End(\mathbb{C}^n)$ be the fundamental representation of $\mathfrak{gl}_n$, such that $\mathfrak{l}_{x,y} \mapsto e_{x,y}$, where recall $e_{x,y}$ are $n \times n$ matrices with elements $(e_{x,y})_{z,w} = \delta_{x,z}\delta_{y,w}$. The special solution $\check{r}$ (1) is $\mathfrak{gl}_n$ symmetric, i.e.*

$$\left[\check{r},\ \Delta_i(e_{x,y})\right] = 0, \quad x,\ y \in X, \tag{6}$$

where we define the "twisted" co-products ($i = 1,2$):

$$\begin{aligned} \Delta_1(e_{x,y}) &= e_{\sigma(x),\sigma(y)} \otimes 1_V + 1_V \otimes e_{x,y}, \\ \Delta_2(e_{x,y}) &= e_{x,y} \otimes 1_V + 1_V \otimes e_{\tau(x),\tau(y)}, \end{aligned} \tag{7}$$

$(\Delta_1(e_{\tau(x),\tau(y)}) = \Delta_2(e_{x,y}))$.

Proof. This can be shown using the form of the special class of solutions (1). The permutation operator is $\mathfrak{gl}_n$ symmetric, i.e.

$$\left[\mathscr{P},\ \Delta(e_{x,y})\right] = 0, \tag{8}$$

where the co-products $\Delta(e_{x,y})$ are defined in Lemma (2) ($\mathfrak{l}_{x,y} \mapsto e_{x,y}$).

Let $u = \sum_{x \in X} e_{x,\tau(x)}$, then (6) immediately follows from (8) and (2) after acting (8) from the left and right with $u \otimes 1_V$, $u^{-1} \otimes 1_V$ or $1_V \otimes u^{-1}$, $1_V \otimes u$ respectively. $\Delta_i(e_{x,y})$ are then defined as

$$\begin{aligned} \Delta_1(e_{x,y}) &= u e_{x,y} u^{-1} \otimes 1_V + 1_V \otimes e_{x,y}, \\ \Delta_2(e_{x,y}) &= e_{x,y} \otimes 1_V + 1_V \otimes u^{-1} e_{x,y} u \end{aligned} \tag{9}$$

and explicitly given by (7). Indeed, $u e_{x,y} u^{-1} = e_{\sigma(x),\sigma(y)}$ and $u^{-1} e_{x,y} u = e_{\tau(x),\tau(y)}$.

According to Lemma 2 $\Delta_i(e_{x,y})$ also satisfy the $\mathfrak{gl}_n$ algebra relations, thus $\check{r}$ (1) is $\mathfrak{gl}_n$ symmetric. In this particular case, as is clear from the computation above, two invertible linear maps are involved, $F_i \in \mathrm{End}(\mathbb{C}^n \otimes \mathbb{C}^n), i \in \{1,2\}$, such that $F_1 := u \otimes 1_V$ and $F_2 := 1_V \otimes u^{-1}$ and $F_i \Delta(e_{x,y}) F_i^{-1} = \Delta_i(e_{x,y})$.

By iteration one derives the N co-products: $\Delta_1^{(N)} = (\Delta_1^{(N-1)} \otimes \mathrm{id})\Delta_1$ and $\Delta_2^{(N)} = (\mathrm{id} \otimes \Delta_2^{(N-1)})\Delta_2$, which explicitly read as

$$\Delta_1^{(N)}(e_{x,y}) = \sum_{k=1}^{N} 1_V \otimes \ldots \otimes e_{\sigma^{N-k}(x),\sigma^{N-k}(y)} \otimes \ldots \otimes 1_V, \tag{10}$$

$$\Delta_2^{(N)}(e_{x,y}) = \sum_{k=1}^{N} 1_V \otimes \ldots \otimes e_{\tau^{k-1}(x),\tau^{k-1}(y)} \otimes \ldots \otimes 1_V. \tag{11}$$

The above expressions can be written in a compact form as:
$\Delta_i^{(N)}(e_{x,y}) = F_i^{(N)} \Delta^{(N)}(e_{x,y}) (F_i^{(N)})^{-1}$, where

$\Delta^{(N)}(e_{x,y}) = \sum_{k=1}^{N} \mathrm{id} \otimes \ldots \otimes \underbrace{e_{x,y}}_{k^{th}\ \text{position}} \otimes \ldots \otimes \mathrm{id}$, and we define $F_1^{(N)} := u^{N-1} \otimes u^{N-2} \otimes \ldots \otimes u \otimes 1_V$ and $F_2^{(N)} := 1_V \otimes u^{-1} \otimes u^{-2} \otimes \ldots \otimes u^{-(N-1)}$ (see also relevant findings in [20, 21]), where $F_i^{(2)} =: F_i$, $i \in \{1,2\}$. Notice that co-associativity does not hold in this case. In fact, it turns out that the quantum algebra associated to generic set-theoretic solutions is a quasi-bialgebra (see e.g. [20, 21]).

We note that local open Hamiltonians for generic Baxterized solutions coming from involutive solutions were systematically derived in [19]. Specifically, the open Hamiltonian associated to Lyubashenko's solution with special boundary conditions is givens as

$$\mathscr{H}_c = \sum_{j=1}^{N-1} \sum_{a=1}^{n} e_{a,b+c}^{(j)} e_{b,a-c}^{(j+1)}, \tag{12}$$

$c \in \{1,2,\ldots,n-1\}$. This is a $\mathfrak{gl}_n$ symmetric Hamiltonian, i.e. $\left[\mathscr{H}_c,\ \Delta_i^{(N)}(y)\right] = 0$, $y \in \mathfrak{gl}_n$, $i \in \{1,2\}$, as opposed to the periodic one, which is not $\mathfrak{gl}_n$ symmetric. Recall $\Delta_i^{(N)}$ are defined in (10),(11).

Motivated by the case of Lyubashenko's solution and the suitable twist of $\mathfrak{gl}_n$ (see also [21]) we present in the subsequent section the full theory of quantum algebras associated to set-theoretic solutions. These are new types of quasi-triangular Hopf algebras introduced in [20, 22, 24]. The quantum algebras associated to the set-theoretic solutions of the parametric Yang-Baxter equation were also introduced in [25].

5 Set-theoretic quasi-triangular Hopf algebras & Drinfel'd twists

In this section we discuss the Yang-Baxter algebras associated to rack type and set-theoretic solutions as quasi-triangular Hopf algebras [20, 24]. The Hopf algebra theory associated to set-theoretic solutions has been developed in [20, 22, 24] (see also [30] and [3] in connection with pointed Hopf algebras and [49, 4] on Hopf algebras in connection to braces [55]–[57], [39]). The associated universal $\mathscr{R}$-matrices are also derived in this frame. A full analysis of set-theoretic twists in the Yangian $\mathfrak{gl}_n$ is presented in [20, 21, 26].

We start our analysis with the rack and quandle algebras and the construction of the associated universal $\mathscr{R}$-matrix. We then extend the algebra to the *decorated rack algebra* and via a suitable admissible Drinfel'd twist we construct the corresponding universal $\mathscr{R}-$matrix. For a more detailed exposition on the proofs presented in this section the interested reader is referred to [20, 24].

5.1 Rack & quandle algebras

We first define the *rack* and *quandle* algebras [24].

Definition 5.1 *(Rack algebra) Let $(X,\triangleright)$ be a finite magma, or such that $a\triangleright$ is surjective, for every $a \in X$. We say that the unital, associative algebra $\mathscr{A}$, over a field k generated by indeterminates $1_{\mathscr{A}}$ (the unit element), q_a, q_a^{-1}, $h_a \in \mathscr{A}$ $(h_a = h_b \Leftrightarrow a = b)$ and relations, $a,b \in X$:*

$$q_a^{-1}q_a = q_a q_a^{-1} = 1_{\mathscr{A}}, \quad q_a q_b = q_b q_{b\triangleright a}, \quad h_a h_b = \delta_{a,b} h_a, \quad q_b h_{b\triangleright a} = h_a q_b, \quad (1)$$

is a rack algebra.

Definition 5.2 *(Quandle algebra) A rack algebra is called a quandle algebra if there is a magma $(X,\bullet)$, such that for all $a,b \in X$, $a\bullet : X \to X$ is a bijection and $a \bullet b = b \bullet (b \triangleright a)$.*

The following proposition fully justifies the names rack and quandle algebras.

Proposition 5.3 *Let $\mathscr{A}$ be a rack algebra, then for all $a,b,c \in X$, $c \triangleright (b \triangleright a) = (c \triangleright b) \triangleright (c \triangleright a)$, and $a\triangleright$ is bijective, i.e. $(X,\triangleright)$ is a rack. If $\mathscr{A}$ is a quandle algebra, then in addition, for all $a \in X$, $a \triangleright a = a$, i.e. $(X,\triangleright)$ is a quandle.*

Proof. We compute $h_a q_b q_c$ using the associativity of the rack algebra and due to invertibility of q_a for all $a \in X$ we conclude for all $a,b,c \in X$,

$$h_{c\triangleright(b\triangleright a)} = h_{(c\triangleright b)\triangleright(c\triangleright a)} \Rightarrow c \triangleright (b \triangleright a) = (c \triangleright b) \triangleright (c \triangleright a).$$

We assume $c \triangleright a = c \triangleright b$, then $q_c h_{c\triangleright a} = q_c h_{c\triangleright b}$, by the fourth relation in 1, we get $h_a q_c = h_b q_c$ and by the invertibility of q_c, $h_a = h_b$, hence $a = b$, i.e. $a\triangleright$ is bijective for all $a \in X$ and thus $(X,\triangleright)$ is a rack.

Moreover, we recall for the quandle algebra $a \bullet b = b \bullet (b \triangleright a)$, then for $a = b$ and by recalling bijectivity and hence left cancellation in $(X,\bullet)$ we conclude that $a \triangleright a = a$ for all $a \in X$, i.e. $(X,\triangleright)$ is a quandle.

Lemma 3. *Let $c = \sum_{a\in X} h_a$, then c is a central element of the rack algebra $\mathscr{A}$.*

Proof. The proof is direct by means of the definition of the algebra $\mathscr{A}$ and Proposition 5.3. Without loss of generality let $c = 1_{\mathscr{A}}$.

Having defined the rack algebra we are now in the position to identify the associated universal $\mathscr{R}$-matrix (solution of the Yang-Baxter equation).

Proposition 5.4 *Let $\mathscr{A}$ be a rack algebra and $\mathscr{R} \in \mathscr{A} \otimes \mathscr{A}$ be an invertible element, such that $\mathscr{R} = \sum_{a\in X} h_a \otimes q_a$. Then $\mathscr{R}$ satisfies the Yang-Baxter equation*

$$\mathscr{R}_{12}\mathscr{R}_{13}\mathscr{R}_{23} = \mathscr{R}_{23}\mathscr{R}_{13}\mathscr{R}_{12},$$

where $\mathscr{R}_{12} = \sum_{a\in X} h_a \otimes q_a \otimes 1_{\mathscr{A}}$, $\mathscr{R}_{13} = \sum_{a\in X} h_a \otimes 1_{\mathscr{A}} \otimes q_a$, and $\mathscr{R}_{23} = \sum_{a\in X} 1_{\mathscr{A}} \otimes h_a \otimes q_a$.

Proof. The proof is a direct computation of the two sides of the Yang-Baxter equation (and use of the fundamental relations (1)):

$$\begin{aligned}
\text{LHS}: \quad & \sum_{a,b,c\in X} h_a h_b \otimes q_a h_c \otimes q_b q_c = \sum_{a,b,c\in X} h_a \otimes q_a h_c \otimes q_a q_c \\
& = \sum_{a,b,c\in X} h_a \otimes q_a h_{a\triangleright c} \otimes q_a q_{a\triangleright c} \\
\text{RHS}: \quad & \sum_{a,b,c\in X} h_b h_a \otimes h_c q_a \otimes q_c q_b = \sum_{a,b,c\in X} h_a \otimes q_a h_{a\triangleright c} \otimes q_c q_a,
\end{aligned}$$

where we have used that $a\triangleright$ is bijective. Then due to the basic relation $q_a q_b = q_b q_{b\triangleright a}$, we show that LHS=RHS, and this concludes our proof.

Remark 5.5 *The universal $\mathscr{R}-$matrix is invertible, in fact, by Lemma 3, $\sum_{a\in X} h_a = 1_{\mathscr{A}}$, hence $\mathscr{R}^{-1} = \sum_{a\in X} h_a \otimes q_a^{-1}$.*

Remark 5.6 Fundamental representation: *Let $\mathscr{A}$ be a rack algebra and $\rho : \mathscr{A} \to End(V)$ be the map defined by*

$$q_a \mapsto \sum_{x\in X} e_{x,a\triangleright x}, \qquad h_a \mapsto e_{a,a}, \tag{2}$$

where $(X,\triangleright)$ is a rack. Then $\mathscr{R} \mapsto r := \sum_{a,b\in X} e_{b,b} \otimes e_{a,b\triangleright a}$, is the linearized version of a rack solution of the Yang-Baxter equation. We note that r is invertible, because $a\triangleright : X \to X$ is a bijection for all $a \in X$, then $r^{-1} = \sum_{a,b\in X} e_{b,b} \otimes e_{b\triangleright a,a}$.

Let $\mathscr{P} = \sum_{a,b\in X} e_{a,b} \otimes e_{b,a}$ be the permutation (flip) operator, then the solution of the braid equation is the linearized version of rack solution, $\check{r} = \mathscr{P} r = \sum_{a,b} e_{a,b} \otimes e_{b,b\triangleright a}$.

Note that in the special case where $\check{r}$ is involutive, i.e. $\check{r}^2 = id$, then $a\triangleright b = b$, for all $a,b \in X$, which means that the rack algebra (Definition 5.1) becomes a commutative algebra. In this case $\check{r} = \sum_{a,b\in X} e_{a,b} \otimes e_{b,a}$, i.e. $\check{r}$ reduces to the permutation operator, whereas r reduces to the identity

Theorem 5.7 *Let $\mathscr{A}$ be a quandle algebra (Definition 5.2). Let also $\mathscr{R} = \sum_{a\in X} h_a \otimes q_a$ and for all $a,b \in X$, $q_a q_b = q_{a\bullet b}$. If $(X,\bullet,e)$ is a group, then $(\mathscr{A},\Delta,\varepsilon,S,\mathscr{R})$ is a quasi-triangular Hopf algebra:*

- *Co-product.* $\Delta : \mathscr{A} \to \mathscr{A} \otimes \mathscr{A}$, $\Delta(q_a^{\pm 1}) = q_a^{\pm 1} \otimes q_a^{\pm 1}$ *and* $\Delta(h_a) = \sum_{b,c\in X} h_b \otimes h_c\big|_{b\bullet c = a}$.
- *Co-unit.* $\varepsilon : \mathscr{A} \to k$, $\varepsilon(q_a^{\pm 1}) = 1$, $\varepsilon(h_a) = \delta_{a,e}$.
- *Antipode.* $S : \mathscr{A} \to \mathscr{A}$, $S(q_a^{\pm 1}) = q_a^{\mp 1}$, $S(h_a) = h_{a^*}$, *where a^* is the inverse in $(X,\bullet)$ for all $a \in X$.*

Proof. It is straightforward to check that Δ is an algebra homomorphism. Indeed, this can be explicitly checked via the distributivity condition $a\triangleright(b\bullet c) = (a\triangleright b)\bullet(a\triangleright c)$, which readily follows from, $a\triangleright b = a^* \bullet b \bullet a$.

We are now going to show that all the axioms of a quasi-triangular Hopf algebra hold.

Given the co-products of the generators we have to check co-associativity and also uniquely derive the counit $\varepsilon : \mathscr{A} \to k$ (homomorphism) and antipode $S : \mathscr{A} \to \mathscr{A}$ (anti-homomorphism).

(i) Co-associativity.:

$$(\mathrm{id} \otimes \Delta)\Delta(q_a) = (\Delta \otimes \mathrm{id})\Delta(q_a) = q_a \otimes q_a \otimes q_a,$$
$$(\mathrm{id} \otimes \Delta)\Delta(h_a) = (\Delta \otimes \mathrm{id})\Delta(h_a) = \sum_{b,c,d \in X} h_b \otimes h_c \otimes h_d\big|_{b \bullet c \bullet d = a}.$$

(ii) Counit: $(\varepsilon \otimes \mathrm{id})\Delta(x) = (\mathrm{id} \otimes \varepsilon)\Delta(x) = x$, for all $x \in \{q_a,\ q_a^{-1}, h_a\}$. The generators q_a are group-like elements, so $\varepsilon(q_a) = 1$, and

$$\sum_{a,b \in X} \varepsilon(h_a) h_b = \sum_{a,b} h_a \varepsilon(h_b)\big|_{a \bullet b = c} = h_c \Rightarrow \varepsilon(h_a) = \delta_{a,e}. \tag{3}$$

(iii) Antipode: $m\big((S \otimes \mathrm{id})\Delta(x)\big) = m\big((\mathrm{id} \otimes S)\Delta(x)\big) = \varepsilon(x) 1_{\mathscr{A}}$ for all $x \in \{q_a,\ q_a^{-1}, h_a\}$. For q_a, we immediately have $S(q_a) = q_a^{-1}$ and (recall $h_a h_b = \delta_{a,b} h_a$ and $\sum_{a \in X} h_a = 1_{\mathscr{A}}$)

$$\sum_{a,b \in X} S(h_a) h_b\big|_{a \bullet b = c} = \sum_{a,b \in X} h_a S(h_b)\big|_{a \bullet b = c} = \delta_{c,e} 1_{\mathscr{A}} \ \Rightarrow\ S(h_a) = h_{a^*},$$

where a^* is the inverse in $(X, \bullet)$ for all $a \in X$. So $(\mathscr{A}, \Delta, \varepsilon, S)$ is a Hopf algebra.

(iv) Moreover,

$$\mathscr{R}_{13}\mathscr{R}_{12} = \sum_{a \in X} h_a \otimes q_a \otimes q_a = \sum_{a \in X} h_a \otimes \Delta(q_a) = (\mathrm{id} \otimes \Delta)\mathscr{R}, \tag{4}$$
$$\mathscr{R}_{13}\mathscr{R}_{23} = \sum_{a,b \in X} h_a \otimes h_b \otimes q_c\big|_{a \bullet b = c} = \sum_{c \in X} \Delta(h_c) \otimes q_c = (\Delta \otimes \mathrm{id})\mathscr{R}. \tag{5}$$

(v) It is also readily shown from the relations of the rack algebra that

$$\Delta^{(op)}(q_a)\mathscr{R} = \mathscr{R}\Delta(q_a) \quad \Delta^{(op)}(h_a)\mathscr{R} = \mathscr{R}\Delta(h_a), \tag{6}$$

where $\Delta^{(op)} = \pi \circ \Delta$, π is the flip map.
We conclude that $(\mathscr{A}, \Delta, \varepsilon, S, \mathscr{R})$ is a quasi-triangular Hopf algebra.

5.2 *Set-theoretic YB algebras*

In this subsection we suitably extend the rack and quandle algebras in order to construct the universal $\mathscr{R}-$matrix associated to general set-theoretic solutions of the Yang-Baxter equation.

We define the *decorated rack* and the set-theoretic Yang-Baxter algebras.

Definition 5.8 *(Decorated rack algebra.) Let $\mathscr{A}$ be a rack algebra (Definition 1). Let also $\sigma_a,\ \tau_b : X \to X$, and σ_a be bijective for all $a \in X$. We say that the unital, associative algebra $\hat{\mathscr{A}}$ over k, generated by indeterminates $1_{\hat{\mathscr{A}}}, q_a, q_a^{-1}, h_a \in \mathscr{A}$ ($h_a = h_b \Leftrightarrow a = b$) and $w_a, w_a^{-1} \in \hat{\mathscr{A}}$, $a \in X$, and relations, $a, b \in X$:*

$$q_a^{-1} q_a = q_a q_a^{-1} = 1_{\hat{\mathscr{A}}},\ \ q_a q_b = q_b q_{b\triangleright a},\ \ h_a h_b = \delta_{a,b} h_a,\ \ q_b h_{b\triangleright a} = h_a q_b,$$
$$w_a^{-1} w_a = w_a w_a^{-1} = 1_{\hat{\mathscr{A}}},\ \ w_a w_b = w_{\sigma_a(b)} w_{\tau_b(a)}\ \ w_a h_b = h_{\sigma_a(b)} w_a,\ \ w_a q_b = q_{\sigma_a(b)} w_a \tag{7}$$

is a decorated rack algebra.

Lemma 4. *Let $c = \sum_{a\in X} h_a$, then c is a central element of the decorated rack algebra $\hat{\mathscr{A}}$.*

Proof. The proof is straightforward by means of the definition of the algebra $\hat{\mathscr{A}}$. We consider henceforth, without loss of generality, $c = 1_{\hat{\mathscr{A}}}$ (see also Lemma 3).

Proposition 5.9 *Let $\hat{\mathscr{A}}$ be a decorated rack algebra, then for all $a, b, c \in X$,*

$$\sigma_a(\sigma_b(c)) = \sigma_{\sigma_a(b)}(\sigma_{\tau_b(a)}(c)) \quad \& \quad \sigma_c(b) \triangleright \sigma_c(a) = \sigma_c(b \triangleright a).$$

Proof. We compute $w_a w_b h_c$ using the associativity of the algebra and the invertibility of w_a for all $a \in X$ and we deduce for all $a, b, c \in X$,

$$h_{\sigma_{\sigma_a(b)}(\sigma_{\tau_b(a)}(c))} = h_{\sigma_a(\sigma_b(c))} \ \Rightarrow\ \sigma_{\sigma_a(b)}(\sigma_{\tau_b(a)}(c)) = \sigma_a(\sigma_b(c)).$$

We also compute $h_a q_b w_c$, via associativity and the invertibility of $q_a,\ w_a$ for all $a \in X$, we obtain for all $a, b, c \in X$,

$$h_{\sigma_c^{-1}(b) \triangleright \sigma_c^{-1}(a)} = h_{\sigma_c^{-1}(b\triangleright a)} \ \Rightarrow\ \sigma_c^{-1}(b) \triangleright \sigma_c^{-1}(a) = \sigma_c^{-1}(b \triangleright a),$$

from the latter it immediately follows, $\sigma_c(b) \triangleright \sigma_c(a) = \sigma_c(b \triangleright a)$.

Definition 5.10 *(Set-theoretic Yang-Baxter algebra.) Let $\mathscr{A}$ be a quandle algebra. Let also $\sigma_a,\ \tau_b : X \to X$, and σ_a be bijective for all $a \in X$. We say that the unital, associative algebra $\hat{\mathscr{A}}$ over k, generated by indeterminates $1_{\hat{\mathscr{A}}}, q_a, q_a^{-1}, h_a \in \mathscr{A}$ ($h_a = h_b \Leftrightarrow a = b$) and $w_a, w_a^{-1} \in \hat{\mathscr{A}}$, $a \in X$, and relations, (7) is a set-theoretic Yang-Baxter algebra.*

Proposition 5.11 *(Hopf algebra) Let $\hat{\mathscr{A}}$ be a set-theoretic Yang-Baxter algebra, $\mathscr{R} = \sum_{b\in X} h_b \otimes q_b$ and $(\mathscr{A}, \Delta, \varepsilon, S, \mathscr{R})$ be the quasi-triangular Hopf algebra of Theorem 5.7. If for all $a, b, x \in X$,*

$$\sigma_x(a) \bullet \sigma_x(b) = \sigma_x(a \bullet b), \tag{8}$$

then,

1. $(\hat{\mathscr{A}},\Delta,\varepsilon,S)$ *is a Hopf algebra with* $\Delta(w_a) = w_a \otimes w_a$, *for all* $a \in X$.
2. $\Delta(w_a)\mathscr{R} = \mathscr{R}\Delta(w_a)$, *for all* $a \in X$.

Proof. In our proof below we are using the Definition 5.10 and (8).

1. The coproduct Δ is an algebra homomorphism. It is sufficient to check below the consistency of all algebraic relations of Definition 5.10 for the corresponding coproducts and (8). Then we have for $Y_b \in \{h_b,\ q_b\}$ and for all $a,b \in X$,
$$\Delta(w_a)\Delta(w_b) = \Delta(w_{\sigma_a(b)})\Delta(w_{\tau_b(a)}), \quad \Delta(w_a)\Delta(Y_b) = \Delta(Y_{\sigma_a(b)})\Delta(w_a).$$
Also, w_a is a group-like element, thus the counit and antipode are given as: $\varepsilon(w_a) = 1$ and $S(w_a) = w_a^{-1}$. Recall that the coproducts, counits and antipodes of the generators $h_a,\ q_a$ are given in Theorem 5.7.
2. By a direct computation and using the algebraic relations of the Definition 5.10 we conclude for all $a \in X,\ \Delta(w_a)\mathscr{R} = \mathscr{R}\Delta(w_a)$.

5.3 Set-theoretic Drinfel'd twist

In this subsection we introduce the universal set-theoretic (or combinatorial) Drinfel'd twist ([20, 21, 24] (see also, relevant construction in [60, 50]). Using the twist, we will be able to obtain the universal $\mathscr{R}$-matrix associated with the set-theoretic Yang-Baxter (YB) algebra.

Before we introduce the set-theoretic twist, we recall a general statement [28].

Proposition 5.12 *(Admissible Drinfel'd twist [28]) Let* $\mathscr{A}$ *be a unital, associative algebra,* $\mathscr{F},\mathscr{R} \in \mathscr{A}\otimes\mathscr{A}$ *be invertible elements and* $\mathscr{R}$ *satisfies the Yang-Baxter equation. Let also* $\mathscr{F}_{1,23},\mathscr{F}_{12,3} \in \mathscr{A}^{\otimes 3}$, *such that*

1. $\mathscr{F}_{23}\mathscr{F}_{1,23} = \mathscr{F}_{12}\mathscr{F}_{12,3}$, *where recall* $\mathscr{F}_{12} = \mathscr{F}\otimes 1_{\mathscr{A}}$ *and* $\mathscr{F}_{23} = 1_{\mathscr{A}}\otimes\mathscr{F}$.
2. $\mathscr{F}_{1,32}\mathscr{R}_{23} = \mathscr{R}_{23}\mathscr{F}_{1,23}$ *and* $\mathscr{F}_{21,3}\mathscr{R}_{12} = \mathscr{R}_{12}\mathscr{F}_{12,3}$.

That is, $\mathscr{F}$ *is an admissible Drinfel'd twist. Define also* $\mathscr{R}^F := \mathscr{F}^{(op)}\mathscr{R}\mathscr{F}$, $\mathscr{F}^{(op)} = \pi(\mathscr{F})$ *where* $\pi : \mathscr{A}\otimes\mathscr{A} \to \mathscr{A}\otimes\mathscr{A}$ *is the flip map. Then* $\mathscr{R}^F$ *also satisfies the Yang-Baxter equation.*

Proof. It is convenient to introduce some handy notation that can be used in the following. First, let $\mathscr{F}_{123} := \mathscr{F}_{12}\mathscr{F}_{1,23} = \mathscr{F}_{23}\mathscr{F}_{1,23}$. Let also $i,j,k \in \{1,2,3\}$, then $\mathscr{F}_{jik} = \pi_{ij}(\mathscr{F}_{ijk})$ and $\mathscr{F}_{ikj} = \pi_{jk}(\mathscr{F}_{ijk})$, where π is the flip map. This notation describes all possible permutations of the indices 1, 2, 3.

The proof is quite straightforward, [28], we just give a brief outline here. We first prove that $\mathscr{F}_{jik}\mathscr{R}_{ij}\mathscr{F}_{ijk}^{-1} = \mathscr{R}_{ij}^F$, indeed via condition (2) of the proposition the definition of $\mathscr{R}^F$ and the notation introduced above we have

$$\mathscr{F}_{jik}\mathscr{R}_{ij}\mathscr{F}_{ijk}^{-1} = \mathscr{F}_{ji}\mathscr{F}_{ji,k}\mathscr{R}_{ij}\mathscr{F}_{ijk} = \mathscr{F}_{ji}\mathscr{R}_{ij}\mathscr{F}_{ij,k}\mathscr{F}_{ijk}^{-1} = \mathscr{R}_{ij}^F\mathscr{F}_{ij}\mathscr{F}_{ij,k}\mathscr{F}_{ijk}^{-1} = \mathscr{R}_{ij}^F. \tag{9}$$

Similarly, it is shown that $\mathscr{F}_{ikj}\mathscr{R}_{jk}\mathscr{F}_{ijk}^{-1} = \mathscr{R}_{jk}^F$.

Then from the YBE we have,

$$\mathscr{F}_{321}\mathscr{R}_{12}\mathscr{R}_{13}\mathscr{R}_{23} = \mathscr{F}_{321}\mathscr{R}_{23}\mathscr{R}_{13}\mathscr{R}_{12} \Rightarrow \mathscr{R}_{12}^F\mathscr{R}_{13}^F\mathscr{R}_{23}^F\mathscr{F}_{123} = \mathscr{R}_{23}^F\mathscr{R}_{13}^F\mathscr{R}_{12}^F\mathscr{F}_{123}.$$

But $\mathscr{F}_{123}$ is invertible, hence $\mathscr{R}^F$ indeed satisfies the Yang-Baxter equation.

Below we prove the main theorem on the set-theoretic Drinfel'd twist (see more details on the proof in [20, 24]).

Theorem 5.13 *(Set-theoretic Drinfel'd twist [20, 24]) Let* $\mathscr{R} = \sum_{a\in X} h_a \otimes q_a \in \mathscr{A} \otimes \mathscr{A}$ *be the universal rack* $\mathscr{R}$*−matrix. Let also* $\hat{\mathscr{A}}$ *be a decorated rack algebra,* $\mathscr{F} \in \hat{\mathscr{A}} \otimes \hat{\mathscr{A}}$*, such that* $\mathscr{F} = \sum_{b\in X} h_b \otimes w_b^{-1}$ *and* $\mathscr{R}_{ij}^F := \mathscr{F}_{ji}\mathscr{R}_{ij}\mathscr{F}_{ij}^{-1}$*,* $i,j \in \{1,2,3\}$*. We also define:*

$$\mathscr{F}_{1,23} := \sum_{a\in X} h_a \otimes w_a^{-1} \otimes w_a^{-1} =, \quad \mathscr{F}_{12,3}^* := \sum_{a,b\in X} h_a \otimes h_{\sigma_a(b)} \otimes w_b^{-1} w_a^{-1}. \quad (10)$$

Let also for every $a,b \in X$, $b \triangleright a = \sigma_b(\tau_{\sigma_a^{-1}(b)}(a))$*. Then, the following statements are true:*

1. $\mathscr{F}_{12}\mathscr{F}_{12,3}^* = \mathscr{F}_{23}\mathscr{F}_{1,23} =: \mathscr{F}_{123}$.
2. *For* $i,j,k \in \{1,2,3\}$*: (i)* $\mathscr{F}_{ikj}\mathscr{R}_{jk} = \mathscr{R}_{jk}^F\mathscr{F}_{ijk}$ *and (ii)* $\mathscr{F}_{jik}\mathscr{R}_{ij} = \mathscr{R}_{ij}^F\mathscr{F}_{ijk}$.

That is, $\mathscr{F}$ *is an admissible Drinfel'd twist.*

Proof. The proof is straightforward based on the underlying algebra $\hat{\mathscr{A}}$.

1. Indeed, this is proved by a direct computation and use of the decorated rack algebra. In fact, $\mathscr{F}_{123} = \sum_{a,b\in X} h_a \otimes h_b w_a^{-1} \otimes w_b^{-1} w_a^{-1}$.
2. Given the notation introduced in the proof of Proposition 5.12 it suffices to show that $\mathscr{F}_{132}\mathscr{R}_{23} = \mathscr{R}_{23}^F\mathscr{F}_{123}$ and $\mathscr{F}_{213}\mathscr{R}_{12} = \mathscr{R}_{12}^F\mathscr{F}_{123}$.
   (i) Due to the fact that for all $a \in X$, $\Delta(w_a)\mathscr{R} = \mathscr{R}\Delta(w_a)$ (see Proposition 5.11) we arrive at $\mathscr{F}_{1,32}\mathscr{R}_{23} = \mathscr{R}_{23}F_{1,23}$, then
   $$\mathscr{F}_{132}\mathscr{R}_{23} = \mathscr{F}_{32}\mathscr{F}_{1,32}\mathscr{R}_{23} = \mathscr{F}_{32}\mathscr{R}_{23}\mathscr{F}_{1,23} = \mathscr{R}_{23}^F\mathscr{F}_{123}.$$
   (ii) By means of the relations of the decorated rack algebra $\hat{\mathscr{A}}$ we compute:
   $$\mathscr{F}_{21,3}^*\mathscr{R}_{12} = \sum_{a,c\in X} h_a \otimes q_a h_{a\triangleright c} \otimes (w_c w_{\sigma_c^{-1}(a)})^{-1},$$
   $$\mathscr{R}_{12}\mathscr{F}_{12,3}^* = \sum_{a,b\in X} h_a \otimes q_a h_{\sigma_a(b)} \otimes (w_a w_b)^{-1}.$$
   Due to the fact that $b \triangleright a = \sigma_b(\tau_{\sigma_a^{-1}(b)}(a))$ and $w_a w_b = w_{\sigma_a(b)} w_{\tau_b(a)}$ we conclude that $\mathscr{F}_{21,3}^*\mathscr{R}_{12} = \mathscr{R}_{12}\mathscr{F}_{12,3}^*$ and consequently (recall $\mathscr{F}_{213} = \mathscr{F}_{21}\mathscr{F}_{21,3}^*$)
   $$\mathscr{F}_{213}\mathscr{R}_{12} = \mathscr{F}_{21}\mathscr{F}_{21,3}^*\mathscr{R}_{12} = \mathscr{F}_{21}\mathscr{R}_{12}\mathscr{F}_{12,3}^* = \mathscr{R}_{12}^F\mathscr{F}_{123}.$$

Corollary 5.14 *Let $\hat{\mathscr{A}}$ be a decorated rack algebra, $\mathscr{F} \in \hat{\mathscr{A}} \otimes \hat{\mathscr{A}}$, such that $\mathscr{F} = \sum_{b\in X} h_b \otimes w_b^{-1}$ and $\mathscr{R}^F_{ij} := \mathscr{F}_{ji}\mathscr{F}_{ij}^{-1}$, $i,j \in \{1,2,3\}$. We also define:*

$$\mathscr{F}_{1,23} := \sum_{a\in X} h_a \otimes w_a^{-1} \otimes w_a^{-1} =, \quad \mathscr{F}^*_{12,3} := \sum_{a,b\in X} h_a \otimes h_{\sigma_a(b)} \otimes w_b^{-1} w_a^{-1}. \tag{11}$$

Let also for every $a,b \in X$, $\sigma_{\sigma_a(b)}(\tau_b(a)) = a$. Then, the following statements are true:

1. $\mathscr{F}_{12}\mathscr{F}^*_{12,3} = \mathscr{F}_{23}\mathscr{F}_{1,23} =: \mathscr{F}_{123}$.
2. *For $i,j,k \in \{1,2,3\}$: (i) $\mathscr{F}_{ikj} = \mathscr{R}^F_{jk}\mathscr{F}_{ijk}$ and (ii) $\mathscr{F}_{jik} = \mathscr{R}^F_{ij}\mathscr{F}_{ijk}$.*
3. $\mathscr{R}^F_{12}\mathscr{R}^F_{21} = 1_{\hat{\mathscr{A}}^{\otimes 2}}$, *i.e. $\mathscr{R}^F$ is reversible.*

Proof. This is an immediate consequence of Theorem 5.13. $\mathscr{R}^F$ is reversible by construction.

We now examine the twisted $\mathscr{R}-$matrix as well as the twisted co-products of the set-theoretic Yabg-Baxter algebra (see also [20, 22]).

Remark 5.15 *(Twisted universal $\mathscr{R}$-matrix) We extract now the explicit expressions of the twisted universal $\mathscr{R}-$matrix and the twisted coproducts of the algebra. We recall the admissible twist $\mathscr{F} = \sum_{b\in X} h_b \otimes w_b^{-1}$.*

- *The twisted $\mathscr{R}-$matrix:*

$$\mathscr{R}^F = \mathscr{F}^{(op)}\mathscr{R}\mathscr{F}^{-1} = \sum_{a,b\in X} h_b w_a^{-1} \otimes h_a q_{\sigma_a(b)} w_{\sigma_a(b)}.$$

- *The twisted coproducts: $\Delta_F(y) = \mathscr{F}\Delta(y)\mathscr{F}^{-1}$, $y \in \hat{\mathscr{A}}$ (in this Remark $\hat{\mathscr{A}}$ denotes the set-theoretic Yang-Baxter algebra) and we recall, for $a \in X$,*

$$\Delta(w_a) = w_a \otimes w_a, \quad \Delta(h_a) = \sum_{b,c\in X} h_b \otimes h_c\big|_{b\bullet c=a}, \quad \Delta(q_a) = q_a \otimes q_a.$$

Then, the twisted coproducts read as: $\Delta_F(w_a) = \sum_{b\in X} h_{\sigma_a(b)} w_a \otimes w_{\tau_b(a)}$, $\Delta_F(h_a) = \sum_{b\in X} h_b \otimes w_b^{-1} h_c w_b\big|_{b\bullet c=a}$, $\Delta_F(q_a) = \sum_{b\in X} q_a h_{a\triangleright b} \otimes w_b^{-1} q_a w_{a\triangleright b}$,

and it immediately follows that $\mathscr{R}^F \Delta_F(Y) = \Delta_F^{(op)}(Y)\mathscr{R}^F$, $Y \in \hat{\mathscr{A}}$.

Remark 5.16 Fundamental representation & the set-theoretic solution:
Let $\rho : \hat{\mathscr{A}} \to End(V)$, such that

$$q_a \mapsto \sum_{x\in X} e_{x,a\triangleright x}, \quad h_a \mapsto e_{a,a}, \quad w_a \mapsto \sum_{b\in X} e_{\sigma_a(b),b}, \tag{12}$$

then $\mathscr{F} \mapsto F := \sum_{a,b\in X} e_{a,a} \otimes e_{\sigma_a(b),b}$ and $\mathscr{R}^F \mapsto r^F := \sum_{a,b\in X} e_{b,\sigma_a(b)} \otimes e_{a,\tau_b(a)}$, where we recall that for all $a,b,c \in X$, $\sigma_{\sigma_a(b)}(\sigma_{\tau_b(a)}(c)) = \sigma_a(\sigma_b(c)$ and $\tau_b(a) := \sigma^{-1}_{\sigma_a(b)}(\sigma_a(b) \triangleright a)$ and $(X,\triangleright)$ is a rack (see also [60, 50, 22]). This is the linearized version of the general set-theoretic solution of the Yang-Baxter equation.

Recall, $\mathscr{P} = \sum_{a,b\in X} e_{a,b} \otimes e_{b,a}$ *is the permutation (flip) operator, then the solution of the braid equation is the linearized set-theoretic solution,* $\check{r}^F = \mathscr{P} r^F = \sum_{a,b\in X} e_{a,\sigma_a(b)} \otimes e_{b,\tau_b(a)}$.

In the special case, where the set-theoretic solution of the braid equation is invertible, i.e. $(\check{r}^F)^2 = id$, *then* $\sigma_{\sigma_a(b)}(\tau_b(a)) = a$, *which leads to* $a \triangleright b = b$ *for all* $a,b \in X$ *(see also Remark 5.6). That is to say that all involutive set-theoretic* $\check{r}$*-matrices are coming from the permutation operator via the set-theoretic twist, i.e.* $\check{r}^F = F\mathscr{P}F^{-1}$ *(and* $r^F = F^{(op)}F^{-1}$*) where recall* $\mathscr{P}$ *is the permutation operator and* $F = \sum_{a,b\in X} e_{a,a} \otimes e_{\sigma_{a(b)},b}$, *such that for all* $a,b,c \in X$, $\sigma_a(\sigma_b(c)) = \sigma_{\sigma_a(b)}(\sigma_{\tau_b(a)}(c))$ *and* $\sigma_{\sigma_a(b)}(\tau_b(a)) = a$ *(see also Corollary 5.14 and [24, 26]).*

6 Solutions from Drinfel'd twists

The main aim of this section is the derivation of general invertible solutions of the set-theoretic Yang-Baxter equation via an admissible Drinfel'd twist.

From the analysis of Section 5 we conclude that any admissible set-theoretic twist satisfies two fundamental conditions, for all $a,b,c \in X$, $\sigma_a(\sigma_b(c)) = \sigma_{\sigma_a(b)}(\sigma_{\tau_b(a)}(c))$, $\sigma_{\sigma_a(b)}(\tau_b(a)) = \sigma_a(b) \triangleright a$ and $(X,\triangleright)$ is a rack (see also Theorem 5.13). It is thus convenient to introduce an alternative definition for the admissible set-theoretic twist at the fundamental representation (see also Remark 5.16 and a similar definition first introduced in [24]).

Definition 6.1 *Let* $(X,\triangleright)$ *be a rack and define* $F := \sum_{x,y\in X} e_{x,x} \otimes e_{\sigma_x(y),y}$, *such that* $\sigma_x : X \to X$ *is bijective. F is called an admissible set-theoretic Drinfel'd twist if for all* $a,b,c \in X$ *(see also Theorem 5.13)*

(a) $\sigma_a(\sigma_b(c)) = \sigma_{\sigma_a(b)}(\sigma_{\tau_b(a)}(c))$.
(b) $\sigma_{\sigma_a(b)}(\tau_b(a)) = \sigma_a(b) \triangleright a$.

Notice that in the involutive case the second condition above becomes $\sigma_{\sigma_a(b)}(\tau_b(a)) = a$ for all $a,b \in X$.

6.1 Involutive case

We first focus on the systematic derivation of involutive, set-theoretic solutions of the braid equation by exploiting the existence of an admissible Drinfel'd twist.

The following useful proposition can be now formulated (see also [56, 57, 23]).

Proposition 6.2 *Let* $(X,\circ,1)$ *be a group and let* $\sigma_a, \tau_b : X \to X$, *such that for all* $a,b \in X$, σ_a *is a bijection,* $a \circ b = \sigma_a(b) \circ \tau_b(a)$ *and* $\sigma_{\sigma_a(b)}(\tau_b(a)) = a$. *Moreover, define* $+ : X \times X \to X$, *such that* $a + b := a \circ \sigma_a^{-1}(b)$ *and assume that* $+$ *is associative and for all* $a,b,c \in X$, $a \circ (b+c) = a\circ b - a + a \circ c$. *Then for all* $a,b,c \in X$,

1. $(X,+,\circ)$ *is a brace and* $\sigma_a(b) = -a + a \circ b$.

2. $\sigma_a(\sigma_b(c)) = \sigma_{\sigma_a(b)}(\sigma_{\tau_b(a}(c))$.

Proof.

1. For the proof of $(X,+)$ being a group we refer the interested reader to [23] for a step by step constructive approach of the algebraic structure. Then due to distributivity condition $(X,+,\circ)$ is a skew brace (see the original works on (skew) braces [55, 56, 57, 39]). To prove that $(X,+,\circ)$ is a brace we need to show that $(X,+)$ is an abelian group.
Indeed, from condition $\sigma_{\sigma_a(b)}(\tau_b(a)) = a$ and the structure group condition $a \circ b = \sigma_a(b) \circ \tau_b(a)$ we obtain for all $a,b \in X$,
$$\sigma_a(b)^{-1} \circ a \circ b = \sigma^{-1}_{\sigma_a(b)}(a) \Rightarrow a \circ \sigma_a^{-1}(b) = b \circ \sigma_b^{-1}(a) \Rightarrow a+b = b+a,$$
i.e. $(X,+)$ is abelian. We used in the last part of the proof above the definition $a+b := a \circ \sigma_a^{-1}(b)$.
Moreover, for all $a,b \in X$
$$a \circ \sigma_a^{-1}(b) = a+b \Rightarrow \sigma_a^{-1}(b) = a^{-1} \circ (a+b)$$
and
$$\sigma_a^{-1}(\sigma_a(b)) = b \Rightarrow a^{-1} \circ (a+\sigma_a(b)) = b \Rightarrow \sigma_a(b) = -a + a \circ b.$$
2. Condition (2) is condition (8) and its proof is given in the proof of Proposition 2.14.

Notice that in Proposition 6.2 we assume a non-standard distributivity condition as the usual one does not apply (see for more details [23]). Indeed, let us assume that the usual distributivity condition holds, then
$$a \circ (1+0) = a \Rightarrow a \circ 1 + a \circ 0 = a \Rightarrow a \circ 0 = 0 \Rightarrow a = a \circ 0^{-1} = 1, \quad \forall a \in X.$$
The latter statement is false given that we consider sets with not just the unit element 1. In general, when a non-empty set X is equipped with two groups operation $\circ$ and $+$, then the non-standard distributivity condition of Proposition 6.2 (or slight variations of it [23]) applies.

We conclude from Proposition 6.2 that the maps $\sigma_a(b)$ and $\tau_b(a)$ provide an involutive solution $\check{r}(a,b) = (\sigma_a(b), \tau_b(a))$ of the set-theoretic braid equation (Rump's solution, Proposition 2.14).

6.2 *Non-involutive case*

We recall from Section 4 (see also [24]) that invertible, non-involutive, set-theoretic solutions, which are the main focus of this subsection, are constructed from the rack-quandle solutions via an admissible Drinfel'd twist.

The following proposition is useful in describing general set-theoretic solutions.

Proposition 6.3 *Let* $\sigma_a, \tau_b : X \to X$*, such that for all* $a,b \in X$*,* σ_a *is a bijection and* $\tau_b(a) = \sigma^{-1}_{\sigma_a(b)}(\sigma_a(b) \triangleright a)$*. Moreover, let* $(X,\circ,1)$ *be a group, such that for all* $a,b \in X$*,* $a \circ b = \sigma_a(b) \circ \tau_b(a)$ *and define* $\bullet : X \times X \to X$*, such that* $a \bullet b := a \circ \sigma_a^{-1}(b) \circ \xi$*,* $\xi \in X$ *is a fixed element. Then, for all* $a,b \in X$*,*

1. *(a)* $a \bullet b = b \bullet (b \triangleright a)$
   (b) $a \bullet \sigma_a(b) = a \circ b \circ \xi$*.*
2. *Assume that* $(X,\bullet)$ *is a group and set* $a \bullet b =: a + b$ *and* $\xi = 1$*.*
   a. *Then,* $b \triangleright a = -b + a + b$*, where* $-b$ *is the inverse of* b *in* $(X,+)$ *(conjugate quandle), and* $\sigma_a(b) = -a + a \circ b$*.*
   b. *If* $(X,+,\circ)$ *is a skew brace, then* $\sigma_a(\sigma_b(c)) = \sigma_{\sigma_a(b)}(\sigma_{\tau_b(a)}(c))$ *and* $\sigma_a(b \triangleright c) = \sigma_a(b) \triangleright \sigma_a(c)$*.*

Proof.

1. (a) Using the definitions $a \bullet b = a \circ \sigma_a^{-1}(b) \circ \xi$ and $b \triangleright a = \sigma_b(\tau_{\sigma_a^{-1}(b)}(a))$ for all $a,b \in X$, we compute

$$b \bullet (b \triangleright a) = b \bullet \sigma_b(\tau_{\sigma_a^{-1}(b)}(a)) = b \circ \sigma_b^{-1}(\sigma_b(\tau_{\sigma_a^{-1}(b)}(a))) \circ \xi = b \circ \tau_{\sigma_a^{-1}(b)}(a)) \circ \xi.$$

   But due to the condition $a \circ b = \sigma_a(b) \circ \tau_b(a)$ we conclude that $a \bullet b = b \bullet (b \triangleright a)$.
   (b) From the definition of $a \bullet b$, and the fact that σ_a is bijection:

$$a \bullet \sigma_a(b) = a \circ \sigma_a^{-1}(\sigma_a(b)) \circ \xi = a \circ b \circ \xi.$$

2. (a) The first part follows immediately from (1) (a). From (1) (b) we immediately conclude that $\sigma_a(b) = -a + a \circ b$.
   (b) This is condition (8) and it is shown in the proof of Proposal 2.14.

Notice that the binary operation $\bullet$ such that it satisfies condition (1) (a) in Proposition 6.3 is not uniquely defined (see [24] for a more detailed discussion). However, based on the definition of the operation $\bullet$ given in Proposition 6.3 we provide below a classification of non-involutive set-theoretic solutions given a specific rack/quandle.

In what follows we assume the existence of the map $\sigma_a : X \to X$ being a bijection and $(X,\circ)$ is a group.

1. **The conjugate quandle.** This case corresponds to Part (2) of Proposition 6.3. Recall that σ_a satisfies condition (a) of Definition 6.1 and provides a solution to the Yang-Baxter equation. We also confirm that condition (b) of Definition 6.1 is equivalent to $a \circ b = \sigma_a(b) \circ \tau_b(a)$. This corresponds to the Guarnieri-Vendramin solution [39].
2. **The affine quandle.** We generalize the definition of an affine quandle as follows. Let $f : X \to X$ be a a bijection and

$$f(-a+b+c) = -f(a) + f(b) + f(c) \tag{1}$$

for all $a,b,c \in X$. Define also $\triangleright : X \times X \to X$, such that $b \triangleright a = -f(b) + f(a) + b$ then $(X, \triangleright)$ is called an affine quandle. Note that, due to condition (1) it is shown that $(X, \triangleright)$ is a quandle. We define for all $a,b \in X$ $a \bullet b := f(a) + b$, and conclude that $f(a) + b = f(b) + (b \triangleright a)$. Then we obtain from the definition of σ_a^{-1} of Proposition 6.3, $\sigma_a^{-1}(b) = a^{-1} \circ (f(a) + b) \circ \xi^{-1}$ and consequently $\sigma_a(b) = -f(a) + a \circ b \circ \xi$. Note that σ_a satisfies condition (a) of Definition 6.1 if and only if

$$a \circ f(b) - a + f(a) \circ \xi^{-1} = \sigma_a(b) \circ f(\tau_b(a)) - \sigma_a(b) + f(\sigma_a(b)) \circ \xi^{-1}. \quad (2)$$

For instance, let $(X, +, \circ)$ be a skew brace and consider $f(a) := a \circ z - z$, $z \in X$ is a fixed element and $\xi = 1$, then conditions (1) and (2) are satisfied. We also confirm that condition (b) of Definition 6.1 is equivalent to $a \circ b = \sigma_a(b) \circ \tau_b(a)$, (see also [22, 24]).

3. **The core quandle.** Recall the core quandle. Recall that $(X, +)$ is a group and we define for all $a, b \in X$, $b \triangleright a = b - a + b$. We also define for all $a, b \in X$, $a \bullet b := -a + b$, then $-a + b = -b + (b \triangleright a)$. Then according to Proposition 6.3 for $\xi = 1$, $\sigma_a(b) = a + a \circ b$. We confirm that σ_a satisfies condition (a) and (b) of Definition 6.1. Notice, in particular that $\sigma_a(\sigma_b(c)) = \sigma_{a \circ b}(c)$, for all $a, b, c \in X$, if and only if $(X, +, \circ)$ is a brace, i.e. $(X, +)$ is abelian.

And with this we conclude our discussion on the derivation of generic solutions of the set-theoretic Yang-Baxter equation from admissible twists. A more exhaustive analysis of admissible twists and general set-theoretic solutions in accordance to Proposition 6.3 will be presented in a forthcoming publication.

Acknowledgements I am indebted to G. Papamikos for sharing Figure 2 on the 3D consistency condition. I am grateful for the support from the Matrix Institute, University of Melbourne, and in particular the program "Mathematics and Physics of Integrability", 1-19 July 2024 in Creswick, where part of this work was completed. Funding from the EPSRC research grant EP/V008129/1 is a also acknowledged.

References

1. M. J. Ablowitz, B. Prinari and A. D. Trubatch, *Soliton interactions in the vector NLS equation*, Inverse Problems, vol. 20, no. 4 (2004) 1217.
2. V.E. Adler, A.I. Bobenko and Yu.B. Suris, *Classification of integrable equations on quad-graphs. The consistency approach*, Comm. Math. Phys. 233 (2003) 513.
3. N. Andruskiewitsch and M. Graña, *From racks to pointed Hopf algebras*, Adv. Math., 178 (2003) 2, 177–243.
4. I. Angiono, C. Galindo and L. Vendramin, *Hopf braces and Yang-Baxter operators*, Proc. Amer. Math. Soc. 1 45 (2017) no. 5, 1981–1995.
5. R. Baxter, *Exactly solved models in statistical mechanics*, Academic Press (1982).
6. D. Bachiller, F. Cedó, E. Jespers and J. Okniński, *Iterated matched products of finite braces and simplicity; new solutions of the Yang-Baxter equation*, Trans. Amer. Math. Soc. 370 (2018) 4881–4907.

7. A. Berenstein and D. Kazhdan, Geometric and unipotent crystals, in Visions in 13 Mathematics: GAFA special volume (2000) 1p. 88.
8. T. Brzeziński, *Lie trusses and heaps of Lie affebras*, Proceedings of Corfu Summer Institute 2021 "School and Workshops on Elementary Particle Physics and Gravity" — PoS(CORFU2021), 406.
9. T. Brzeziński and B. Rybołowicz, *Congruence classes and extensions of rings with an application to braces*, Comm. Contemp. Math. **23** (2021) 2050010.
10. T. Brzeziński, S. Mereta and B. Rybołowicz, *From pre-trusses to skew bracess,* Publicacions Matemàtiques 66, Issue 2 (2022).
11. C. Burstinm and W. Mayer, *Distributive Gruppen von endlicher Ordnung*, J. Reine Angew. Math. **160** (1929) 111–130. (English translation arXiv:1403.6326)
12. V. Caudrelier and Q.C. Zhang, *Yang-Baxter and reflection maps from vector solitons with a boundary*, Nonlinearity 27 (2014) 1081.
13. M. Castelli, F. Catino and P. Stefanelli, *Left non-degenerate set-theoretic solutions of the Yang-Baxter equation and dynamical extensions of q-cycle sets*, J. Algebra Appl., 21 (2022) 8, Paper No. 2250154, 22.
14. F. Cedó, E. Jespers and J. Okniński, *Braces and the Yang-Baxter equation*, Comm. Math. Phys. 327 (2014) 1, 101–116.
15. F. Chouraqui, *On some Garsideness properties of structure groups of set-theoretic solutions of the Yang-Baxter equation*, Comm. in Alg., Volume 51 (2023) Issue 8.
16. I. Colazzo, E. Jespers, L. Kubat, A. Van Antwerpen and C. Verwimp, *Finite Idempotent Set-Theoretic Solutions of the Yang-Baxter Equation* , International Mathematics Research Notices (2023); rnad183.
17. I. Colazzo, E. Jespers, A. Van Antwerpen and C. Verwimp, *Left non-degenerate set-theoretic solutions of the Yang-Baxter equation and semitrusses*, J. Algebra, 610 (2022) 409–462.
18. A. Doikou and A. Smoktunowicz, *Set-theoretic Yang-Baxter & reflection equations and quantum group symmetries*, Lett. Math. Phys. 111 (2021) 105.
19. A. Doikou and A. Smoktunowicz, *From Braces to Hecke algebras & Quantum Groups*, J. of Algebra and its Applications (2022) 2350179.
20. A. Doikou, *Set-theoretic Yang-Baxter equation, braces and Drinfel'd twists*, J. Phys. A 54 (2021) 41.
21. A. Doikou, A. Ghionis and B. Vlaar, *Quasi-bialgebras from set-theoretic type solutions of the Yang-Baxter equation*, Lett. Math. Phys. 112 (2022) 4, Paper No. 78, 29.
22. A. Doikou and B. Rybołowicz, *Novel non-involutive solutions of the Yang-Baxter equation from (skew) braces*, Journal of the London Math. Soc. 110, 4 (2024) e12999.
23. A. Doikou and B. Rybołowicz, *Near braces and p-deformed braided groups*, Bull. London Math. Soc. 56 (2024) 124-139.
24. A. Doikou, B. Rybołowicz and P. Stefanelli, *Quandles as pre-Lie skew braces, set-theoretic Hopf algebras & universal R-matrices*, J. Phys. A: Math. Theor. 57 (2024) 405203.
25. A. Doikou, *Parametric set-theoretic Yang-Baxter equation: p-racks, solutions & quantum algebras*,
arXiv:2405.04088 [math-ph].
26. A. Doikou, *Combinatorial twists in* $\mathfrak{gl}_n$ *Yangian*, arXiv:2504.21690 [math-QA].
27. V.G. Drinfel'd, *On some unsolved problems in quantum group theory*, in: Quantum groups (Leningrad, 1990), vol. 1510 of Lecture Notes in Math., Springer, Berlin, (1992) pp. 1–8.
28. V.G. Drinfel'd, *Hopf algebras and the quantum Yang–Baxter equation*, Soviet. Math. Dokl. 32 (1985) 254.
29. M. Elhamdadi, *A Survey of Racks and Quandles: Some Recent Developments*, Algebra Colloquium, Vol. 27, No. 03, pp. 509-522 (2020).
30. P. Etingof, T. Schedler and A. Soloviev, *Set-theoretical solutions to the Quantum Yang-Baxter equation*, Duke Math. J. 100(2) (1999) 169–209.
31. P. Etingof, *Geometric crystals and set-theoretical solutions to the quantum Yang-Baxter equation*, Comm. Algebra 31 (2003) 1961.
32. L.D. Faddeev, N.Yu. Reshetikhin and L.A. Takhtajan, *Quantization of Lie groups and Lie algebras*, Leningrad Math. J. 1 (1990) 193.

33. R. Fenn, M. Jordan-Santana, and L. Kauffman, Biquandles and virtual links, Topology Appl., 145 (2004) pp. 157–175.
34. T. Fritz, *Self-distributive structures in physics*, arXiv:2403.14458 [quant-ph].
35. T. Gateva-Ivanova and S. Majid, *Matched pairs approach to set theoretic solutions of the Yang–Baxter equation*, J. Algebra 319 (2008) 1462–1529.
36. T. Gateva-Ivanova, *Set-theoretic solutions of the Yang–Baxter equation, braces and symmetric groups*, Adv. Math. 388(7):649–701 (2018).
37. T. Gateva-Ivanova, *Quadratic algebras, Yang-Baxter equation, and Artin- Schelter regularity*, Adv. in Math. 230 (2012) 2152–2175.
38. T. Gateva-Ivanova and S. Majid, *Quadratic algebras and idempotent braided sets*, arXiv:2409.02939.
39. L. Guarnieri, L. Vendramin, *Skew braces and the Yang-Baxter equation*, Math. Comput. 86(307) (2017) 2519–2534.
40. G. Hatayama, A. Kuniba and T. Takagi, *Soliton cellular automata associated with crystal bases*, Nucl. Phys. B577 (2000) 619.
41. P. Jedlicka, A. Pilitowska and A. Zamojska-Dzienio, *The retraction relation for biracks*, J. Pure Appl. Algebra 223 (2019) 3594–3610.
42. E. Jespers, E. Kubat and A. Van Antwerpen, *The structure monoid and algebra of a non-degenerate set-theoretic solution of the Yang-Baxter equation*, Transactions of the American Mathematical Society, 372, 10 (2019) 7191-7223.
43. E. Jespers, E. Kubat, A. Van Antwerpen and L. Vendramin, *Factorizations of skew braces*, Math. Ann. 375 (2019) no. 3-4, 1649–1663.
44. M. Jimbo, A q-difference analogue of U(g) and the Yang-Baxter equation, Lett. Math. Phys. 10 (1985) 63.
45. M. Jimbo, Quantum R-matrix for the generalized Toda system, Comm. Math. Phys. 102 (1986) 537–547
46. D. Joyce, *A classifying invariant of knots, the knot quandle*, J. Pure Appl. Algebra, 23 (1982) 1, 37–65.
47. L.H. Kauffman, Virtual knot theory, European Journal of Combinatorics 20 (1999) 663–691.
48. V. Lebed, *Applications of self-distributivity to Yang–Baxter operators and their cohomology*, Journal of Knot Theory and Its RamificationsVol. 27, No. 11, 1843012 (2018).
49. V. Lebed and L. Vendramin, *On structure groups of set-theoretical solutions to the Yang-Baxter equation*, Proc. Edinburgh Math. Soc., Volume 62, Issue 3 (2019) 683 -717.
50. V. Lebed and L. Vendramin, *Reflection equation as a tool for studying solutions to the Yang-Baxter equation*, J. Algebra, 607 (2022) 360–380.
51. J.-H. Lu, M. Yan, and Y.-C. Zhu, *On the set-theoretical Yang-Baxter equation*, Duke Math. J., 104 (2000) 1, 1–18.
52. S. V. Matveev, *Distributive groupoids in knot theory*, Mat. Sb. (N.S.) 119(161) (1982) no. 1, 78–88, 160 (Russian).
53. V.G. Papageorgiou, A.G. Tongas and A.P. Veselov, *Yang-Baxter maps and symmetries of integrable equations on quad-graphs*, J. Math. Phys. 47, 083502 (2006).
54. V.G. Papageorgiou, Yu.B. Suris, A.G. Tongas and A.P. Veselov, *On quadrirational Yang-Baxter Maps*, SIGMA 6 (2010) 033.
55. W. Rump, *A decomposition theorem for square-free unitary solutions of the quantum Yang–Baxter equation*, Adv. Math. 193 (2005) 40–55.
56. W. Rump, *Braces, radical rings, and the quantum Yang-Baxter equation*, J. Algebra 307 (1) (2007) 153–170.
57. W. Rump, *A covering theory for non-involutive set-theoretic solutions to the Yang–Baxter equation*, J. Algebra 520 (2019) 136–170.
58. J. Scott Carter, *A Survey of Quandle Ideas*, Introductory Lectures on Knot Theory, pp. 22-53 (2011).
59. A. Smoktunowicz, L. Vendramin, *On Skew Braces* (with an appendix by N. Byott and L. Vendramin), Journal of Combinatorial Algebra Volume 2, Issue 1, (2018) 47-86.
60. A. Soloviev, Non-unitary set-theoretical solutions to the quantum Yang-Baxter equation, Math. Res. Lett. 7(5-6) (2000) 577-596.

61. D. Takahashi and J. Satsuma, *A soliton cellular automaton*, J. Phys. Soc. Japan 59 (1990) 3514.
62. T. Tsuchida, *N-Soliton Collision in the Manakov Model*, Progress of Theoretical Physics, vol. 111 (2004) 151–182.
63. A.P. Veselov, *Yang-Baxter maps and integrable dynamics*, Phys. Lett. A314 (2003) 214.
64. C.N. Yang, *Some exact results for the many-body problem in one dimension with repulsive delta-function interaction*, Rev. Lett. 19 (1967) 1312.

Analytic continuation in the sinh-Gordon model

Patrick Dorey and Brandon Morrison

Abstract We study the analytic continuation of the finite-volume energy levels of the sinh-Gordon model. The structure is surprisingly rich, and in the small-coupling limit we find systematic patterns which suggest that a complete picture may be possible.

1 Introduction

The idea that different states in a quantum theory might be linked by a process of analytic continuation in a parameter is an old one, perhaps most spectacularly realised in Bender and Wu's analysis of the quantum anharmonic oscillator [2]. It is natural to ask whether a similar level of understanding could be attained for a quantum field theory, and in this short note we describe some steps in this direction.

Our main focus will be the quantum sinh-Gordon model, an integrable quantum field theory in 1+1 dimensions. But to illustrate the main idea we start with a much simpler example, discussed in [6]: the finite volume spectrum of the 1+1 dimensional Ising field theory. With the spatial dimension curled up to form a circle of circumference R, the ground state energy can be written as [9]

$$E_0(M,R) = E_{\text{bulk}}(M,R) - \frac{\pi}{6R} c_0^{(\text{Ising})}(MR) ,$$

where M the inverse of the bulk correlation length, $E_{\text{bulk}}(M,R)$ contains the bulk contribution, and $c_0^{(\text{Ising})}(MR)$, the so-called effective central charge for the Ising

Patrick Dorey
Durham University, UK
e-mail: p.e.dorey@durham.ac.uk

Brandon Morrison
Durham University, UK
e-mail: brandon.m.morrison@dunelm.org.uk

D. R. Wood et al. (eds.), *2024 MATRIX Annals, Part II*, MATRIX Book Series 8,
https://doi.org/10.1007/978-3-032-16206-9_5

field theory, is given in closed form by the following expression:

$$c_0^{(\text{Ising})}(r) = \frac{1}{2} - \frac{3r^2}{2\pi^2}\left(\log\frac{1}{r} + \frac{1}{2} + \ln\pi - \gamma_E\right) + \frac{6}{\pi}\sum_{k=1}^{\infty}\left(\sqrt{r^2+(2k-1)^2\pi^2} - (2k-1)\pi - \frac{r^2}{2(2k-1)\pi}\right)$$

with $r = MR$ and $\gamma_E = 0.57721566\ldots$. Apart from the logarithmic singularity at the origin, $c_0(r)$ has a sequence of square-root singularities in the complex r plane, at odd-integer multiples of $i\pi$. If R (or equivalently r) is continued along a closed path from the positive real axis, around the singularities with $k = k_1,\, k_2 \ldots k_n$ and back, then the corresponding square roots will flip signs from positive to negative, decreasing $c_0(r)$, and thereby increasing $E_0(R)$ to

$$E_{k_1,k_2\ldots k_n}(M,R) = E_0(M,R) + \frac{2}{R}\sum_{i=1}^{n}\sqrt{r^2+(2k-1)^2\pi^2}$$

which is the energy of an excited state of the model, containing $2n$ Bethe-quantised particles in the infrared.

In more general, interacting cases, an explicit representation of the finite-volume ground state energy is not available, and so the analysis above cannot be applied directly. However, when the quantum field theory in question is integrable with a known exact S-matrix, $E_0(R)$ can be expressed in terms of solutions the relevant thermodynamic Bethe Ansatz (TBA) equations [11]. For a theory with a single scalar particle of mass M and exact scattering matrix $S(\theta)$ (where θ is the rapidity), confined to live on a circle of circumference R, the recipe is as follows. First, find the so-called pseudoenergy $\varepsilon(\theta)$ by solving the following non-linear integral equation (the TBA equation)

$$\varepsilon(\theta) = r\cosh\theta - \phi * L(\theta)\,,$$

where $r = MR$ is the system size in units of the correlation length, and

$$L(\theta) = \log\left(1 + e^{-\varepsilon(\theta)}\right),$$
$$f * g(\theta) = \frac{1}{2\pi}\int_{\mathbb{R}} f(\theta-\theta')g(\theta')\,d\theta'\,,$$
$$\phi(\theta) = -i\frac{\partial}{\partial\theta}\log S(\theta)\,.$$

Next, evaluate the effective central charge, also sometimes called a scaling function:

$$c_0(r) = \frac{3}{\pi^2}\int_{\mathbb{R}} r\cosh\theta L(\theta)\,d\theta\,.$$

In terms of this scaling function we have

$$E_0(R) = E_{\text{bulk}}(M,R) - \frac{\pi}{6R} c_0(r)\,.$$

(Here $E_{\text{bulk}}(M,R)$ is a possible bulk term which will not play a role in the continuations we discuss below.) For the Ising case, $S(\theta) = -1$, so $\phi(\theta) = 0$ and the TBA equation is easily solved to yield the following alternative formula for $c_0^{(\text{Ising})}(r)$:

$$c_0^{(\text{Ising})}(r) = \frac{3}{\pi^2} \int_{\mathbb{R}} r\cosh\theta \log\left(1 + e^{-r\cosh\theta}\right) d\theta\,.$$

In the more interesting cases of interacting quantum field theories, $\phi(\theta)$ will be nontrivial and it won't be possible to solve for $\varepsilon(\theta)$ explicitly. Nevertheless, it is possible to track how the TBA equation itself changes under analytic continuation. This was studied in detail in [6], taking the Yang-Lee model (a perturbation of the $\mathcal{M}_{2,5}$ minimal model) as the initial example. The key is to keep track of the zeros of $1+e^{-\varepsilon(\theta)}$, the argument of the logarithm in the definition of $L(\theta)$, in the complex θ plane. (By a small abuse of language we will sometimes refer to such points as singularities of the pseudoenergies, even though they are really only singularities of the associated functions $L(\theta)$.) For general r these will all be clear of the real axis. But if a pair of these zeros, say at θ_0 and $-\theta_0$, approach the real axis as r varies along some path[1], the TBA convolution risks becoming ill-defined. The solution is to distort the contour away from the real axis and around the singularities to allow them to cross over, which yields the correct analytic continuation of the equation. Returning the contour to the real axis generates a pair of residue terms, one each for the singularities at θ_0 and $-\theta_0$, which can be found explicitly via an integration by parts, as explained in [6]. The net result is that the TBA equation becomes

$$\varepsilon(\theta) = r\cosh\theta + \log\frac{S(\theta-\theta_0)}{S(\theta+\theta_0)} - \phi * L(\theta)\,.$$

Likewise, the expression for $c_0(r)$ acquires an extra term, becoming

$$c(r) = \frac{12r}{\pi} i\sinh\theta_0 + \frac{3}{\pi^2}\int_{\mathbb{R}} r\cosh\theta L(\theta)\, d\theta\,.$$

This is no longer the effective central charge, but rather a scaling function associated with an excited state of the model. The new excited-state TBA equation might appear to have an extra unknown, θ_0, the location of the singularity which has crossed the integration contour. This is fixed by the self-consistency requirement that $1+e^{\varepsilon(\theta_0)} = 0$, or equivalently $\varepsilon(\theta_0) = (2N+1)\pi i$ with $N \in \mathbb{Z}$.

In principle, this process can now be repeated, varying r again in the complex plane and looking for occasions when θ-plane singularities for the excited-state TBA equation cross the real axis, which will generate more complicated TBA equations corresponding to further excited states. At each stage the TBA equations found are exact, and some of their asymptotics can be evaluated exactly in small and large

[1] Since the TBA equation is symmetric in θ, the zeros in question are always paired in this way.

r limits. However, the 'connectivity' of these equations depends on how, and in what order, singularities of the pseudoenergies cross the TBA integration contour. This relies on the behaviour of solutions of the TBA equations at finite (and indeed complex) values of r, something which needs to be found numerically. In [6] (for the Yang-Lee model) and then [7] (for a further series of perturbed minimal models), the first few sheets of the Riemann surface of energy levels and the structure of the branch points linking them were found. However, and in spite of some further work with Lorenz Hilfiker and István Szécsényi [5], a complete picture for these theories has proved elusive: the arrangements of the low-lying levels do not show any clear patterns, and gathering more extensive data is made difficult by the fact that, even with the addition of damping factors, the numerical iteration of the TBA equations tends not to converge near branch points.

Instead, in the following we will report on some work on a different theory with more structure in the form of a tunable coupling, which will allow us to see evidence of a more uniform behaviour in the limit where this coupling is small: the sinh-Gordon model.

2 The sinh-Gordon model and its TBA equation

The infrared description of the sinh-Gordon model as a scattering theory is particularly simple: it is a theory of a single massive scalar particle, with an exact scattering matrix $S(\theta)$ which depends on a real parameter p lying between 0 and 1:

$$S(\theta) = \frac{\sinh(\theta) - i\sin(\pi p)}{\sinh(\theta) + i\sin(\pi p)}$$

This has physical strip zeros at $i\pi p$ and $i\pi(1-p)$. In the conventions of [13], p is related to the coupling b in the Euclidean action

$$A_{\text{ShG}} = \int \left[\frac{1}{4\pi} (\partial_\mu \phi)^2 + 2\mu \cosh 2b\phi \right] d^2x$$

by

$$p = \frac{b^2}{1+b^2} .$$

The strong-weak coupling duality of the model under $b \to 1/b$ is reflected in the symmetry of $S(\theta)$ under $p \to 1-p$, allowing p to be restricted to $0 < p \leq 1/2$.

The TBA equation for this model assumes the standard form given in the last section:

$$\varepsilon(\theta) = r\cosh\theta - \phi * L(\theta) .$$

Since $S(\theta)$ has zeros at $\theta = i\pi p$ and $\theta = i\pi(1-p)$, and simple poles at $\theta = -i\pi p$ and $\theta = -i\pi(1-p)$, $\phi(\theta) = -i\frac{\partial}{\partial\theta}\log S(\theta)$ has simple poles at these same locations. Once the pseudoenergy has been found on the real axis, for example by numerical iteration of the TBA equation, allowing θ to become complex provides a representation for the same function in the complex plane. However, due to the poles in the kernel $\phi(\theta)$, this only holds on the strip $-\pi p < \mathrm{Im}\,\theta < \pi p$. To extend further, we can follow Aliosha Zamolodchikov [13] and define an associated pair of functions, $Y(\theta)$

$$Y(\theta) = e^{-\varepsilon(\theta)} = \exp\left(-r\cosh(\theta) + \frac{1}{2\pi}\int_{\mathbb{R}} \phi(\theta-\theta')L(\theta')\,d\theta'\right)$$

and $X(\theta)$

$$X(\theta) = \exp\left(-\frac{r}{2\sin(\pi p)}\cosh(\theta) + \frac{1}{2\pi}\int_{\mathbb{R}} \frac{1}{\cosh(\theta-\theta')}L(\theta')\,d\theta'\right).$$

The kernel in the convolution defining $X(\theta)$ is $1/\cosh\theta = -i\frac{\partial}{\partial\theta}\log S_X(\theta)$, where

$$S_X(\theta) = \frac{1}{i}\frac{1+ie^{\theta}}{1-ie^{\theta}}.$$

Note that $S_X(0) = 1$, $S_X(\theta)S_X(-\theta) = 1$, and its nearest poles and zeros to the real axis are at $\pm i\pi/2$. This means that while the initial definition of $Y(\theta)$ holds in the strip $-\pi p < \mathrm{Im}\,\theta < \pi p$, that of $X(\theta)$ holds in the wider strip $-\pi/2 < \mathrm{Im}\,\theta < \pi/2$.

In [13] Zamolodchikov further showed that the X and Y functions just defined satisfy a number of functional relations, most succinctly expressed using an alternative parameter $a = 1-2p$. These are an X-system:

$$X\left(\theta+\frac{i\pi}{2}\right)X\left(\theta-\frac{i\pi}{2}\right) = 1 + X\left(\theta+\frac{ia\pi}{2}\right)X\left(\theta-\frac{ia\pi}{2}\right);$$

a Y-system:

$$Y\left(\theta+\frac{i\pi}{2}\right)Y\left(\theta-\frac{i\pi}{2}\right) = \left(1+Y\left(\theta+\frac{ia\pi}{2}\right)\right)\left(1+Y\left(\theta-\frac{ia\pi}{2}\right)\right);$$

and a couple of 'X-Y-systems':

$$X\left(\theta+\frac{ia\pi}{2}\right)X\left(\theta-\frac{ia\pi}{2}\right) = Y(\theta);$$

$$X\left(\theta+\frac{i\pi}{2}\right)X\left(\theta-\frac{i\pi}{2}\right) = 1+Y(\theta).$$

Under duality, a maps to $-a$ and all of these relations map to themselves.

Unlike the more familiar Y-systems associated with relevant perturbations of many rational conformal field theories (see for example [12, 10]), these functional equations do not imply any straightforward periodicity properties for their solutions. As discussed in [13], this is associated with the subtleties of the UV behaviour of the sinh-Gordon model. Here we just add two further remarks. First, at the self-dual point $p=1/2$, the X-system reduces to $X\left(\theta+\frac{i\pi}{2}\right)X\left(\theta-\frac{i\pi}{2}\right)=1+X(\theta)^2$. After a trivial relabeling this is the recurrence relation associated with the simplest cluster algebra of infinite type, namely $A_1^{(1)}$, a (non-periodic) closed form solution to which was found in [14]. Second, in the small-coupling, $p\to 0$ limit we have $a=1-2p\to 1$ and the two X-Y systems might appear to come into conflict. However this is not the case: for $X(\theta)$, the factor of $\exp(-\frac{r}{2\sin(\pi p)}\cosh(\theta))$ in its initial definition comes to dominate in the $p\to 0$ limit, so that

$$X(\theta)\to\begin{cases}0 & |\mathrm{Im}\,\theta|<\pi/2\\ \infty & \pi/2<|\mathrm{Im}\,\theta|<3\pi/2\end{cases}$$

and so on. This explains the apparent disagreement between the two X-Y-systems at $p=0$: the left-hand sides of these equations cease to be well-defined in this limit. It is possible to get finer control of this limit, and get to an X-Y-system which continues to make sense when $p\to 0$, by defining, for $|\mathrm{Im}(\theta)|<\frac{\pi}{2}$,

$$\widetilde{X}(\theta)=\exp\left(\frac{1}{2\pi}\int_{\mathbb{R}}\frac{1}{\cosh(\theta-\theta')}L(\theta')\,d\theta'\right)$$

so that

$$X(\theta)=e^{-\frac{r}{2\sin(\pi p)}\cosh(\theta)}\widetilde{X}(\theta)\,.$$

The X and X-Y systems become the following $\widetilde{X}$ and $\widetilde{X}$-Y systems:

$$\widetilde{X}\left(\theta+\frac{i\pi}{2}\right)\widetilde{X}\left(\theta-\frac{i\pi}{2}\right)=1+e^{-r\cosh\theta}\widetilde{X}\left(\theta+\frac{ia\pi}{2}\right)\widetilde{X}\left(\theta-\frac{ia\pi}{2}\right)$$

$$\widetilde{X}\left(\theta+\frac{ia\pi}{2}\right)\widetilde{X}\left(\theta-\frac{ia\pi}{2}\right)=e^{r\cosh\theta}Y(\theta)$$

$$\widetilde{X}\left(\theta+\frac{i\pi}{2}\right)\widetilde{X}\left(\theta-\frac{i\pi}{2}\right)=1+Y(\theta)$$

For $p=0$, $a=1$ these equations become

$$\widetilde{X}\left(\theta+\frac{i\pi}{2}\right)\widetilde{X}\left(\theta-\frac{i\pi}{2}\right)=1+e^{-r\cosh\theta}\widetilde{X}\left(\theta+\frac{i\pi}{2}\right)\widetilde{X}\left(\theta-\frac{i\pi}{2}\right)$$

$$\widetilde{X}\left(\theta+\frac{i\pi}{2}\right)\widetilde{X}\left(\theta-\frac{i\pi}{2}\right)=e^{r\cosh\theta}Y(\theta)$$

$$\widetilde{X}\left(\theta+\frac{i\pi}{2}\right)\widetilde{X}\left(\theta-\frac{i\pi}{2}\right)=1+Y(\theta)$$

and are no longer inconsistent. Indeed, equating the right-hand sides of the second and third equations implies that, in this limit, $Y(\theta)=1/(e^{r\cosh\theta}-1)$, a result which will be found by other means in the next section. However we should add a caveat that this is only consistently the limiting form of $Y(\theta)$ in the region $|\mathrm{Im}\,\theta| < \pi/2$. For $\pi/2 < |\mathrm{Im}\,\theta| < \pi$, our numerical results show that $Y(\theta)$ develops singularities along a zig-zag pattern of lines as $p \to 0$, as illustrated in Figs. 3 and 4 below.

3 A subtlety in the small-p limit

As $p \to 0$ the kernel $\phi(\theta)$ becomes increasingly concentrated near $\theta = 0$. Taking the limit, $\phi(\theta-\theta')$ can be replaced by $2\pi\delta(\theta-\theta')$ in the convolution term in the TBA equation, which becomes

$$\varepsilon(\theta) = r\cosh\theta - \int_{\mathbb{R}} \delta(\theta-\theta')L(\theta')\,d\theta' = r\cosh\theta - L(\theta)\,.$$

Exponentiating and solving recovers the formula for $Y(\theta)$ given at the end of the previous section:

$$Y(\theta) = e^{-\varepsilon(\theta)} = 1/(e^{r\cosh\theta} - 1)\,.$$

Substituting into the formula for $c(r)$,

$$c(r) = -\frac{3}{\pi^2}\int_{\mathbb{R}} r\cosh\theta \log\left(1 - e^{-r\cosh\theta}\right) d\theta\,.$$

Apart from some signs, this is very similar to the integral formula for $c_0^{(\mathrm{Ising})}(r)$ reported in the introduction. In fact, and perhaps unsurprisingly, it precisely matches $c_0^{(\mathrm{boson})}(r)$, the effective central charge for a single free boson [9]. Just as for the Ising field theory, an alternative formula exists which makes its analytic structure clear:

$$c_0^{(\mathrm{boson})}(r) = 1 - \frac{3r}{\pi} + \frac{3r^2}{2\pi^2}\left(\log\frac{1}{r} + \frac{1}{2} + \ln 4\pi - \gamma_E\right) - \frac{6}{\pi}\sum_{k=1}^{\infty}\left(\sqrt{r^2 + (2k\pi)^2} - 2k\pi - \frac{r^2}{4k\pi}\right).$$

This time, the predicted branch points are at even integer multiples of $i\pi$, in contrast to the odd integer multiples seen for the Ising field theory. But more crucially, the *signs* of the square root terms are reversed. When continuing $c_0^{(\mathrm{Ising})}(r)$ around branch points from the ground state, the resulting flips in signs of square roots led to an *increase* in $E(m,R)$, and excited state energies which were higher than that of the ground state. But with the square roots in the formula for $c_0^{(\mathrm{boson})}(r)$ starting with the opposite signs, any flips will *decrease* the corresponding $E(m,R)$, leading to 'states' with lower energy than the ground state. Since this cannot be correct, the whole analytic continuation programme might be questioned. Fortunately, it can be argued that the case of the free boson is genuinely exceptional.

Consider first a (rather standard) toy example: the quantum mechanical simple harmonic oscillator, which can be thought of as a free boson in a universe with just one point. The eigenvalue problem is

$$\left(-\frac{d^2}{dx^2}+\nu^2x^2\right)\psi(x)\equiv\mathcal{H}_\nu\,\psi(x)=E\,\psi(x)$$

with ν a real parameter, and the energy levels, found by imposing $\psi\in L^2(\mathbb{R})$, are

$$E_n=(2n+1)\nu \quad n=0,1,2\ldots.$$

Now, in analogy with the continuations performed earlier, set $\nu=re^{i\phi}$, with r and ϕ real, and continue ϕ from 0 to π. Then ν continues to $-\nu$ and all the energies change sign (and thus decrease), even though under the continuation $\mathcal{H}_\nu\to\mathcal{H}_{-\nu}=\mathcal{H}_\nu$, so the eigenvalue problem might appear to be unchanged. The resolution of this apparent paradox comes from realising that eigenvalue problems associated with differential operators are specified not only by the differential operator, but also by the boundary conditions that are imposed. For the simple harmonic oscillator, the continuation just described rotates the associated Stokes sectors by ninety degrees, which in turn moves the contour along which the wavefunction must be square integrable from $\mathbb{R}$ to $i\mathbb{R}$ – and for this revised eigenvalue problem, the eigenvalues are indeed the negatives of their more usual values.

The same phenomenon occurs for two simple harmonic oscillators, coupled together, when the continuation is in their coupling [1]. This corresponds to a free boson in a universe with two points, with the 'size' of this universe relating to the strength of the coupling. Continuation on a closed path from the decoupling point and back can flip the boundary conditions and send energies to their negatives. However, this interference in the boundary conditions is only possible because the coupling between the oscillators is of the same quadratic order at infinity as the potential for each single oscillator. For more severely confining potentials such as the hyperbolic cosine found in the sinh-Gordon model, this is not the case and a continuation in system size would not be expected to flip boundary conditions from their starting configurations. In the absence of a detailed theory of the Stokes sectors for the relevant functions of many variables, the reasoning just given is far from being rigorous. However our numerical work has confirmed that for the sinh-Gordon model, continuation in r at fixed but nonzero p, no matter how small p is taken to be, always leads to states of higher, not lower, energies. This implies that the process of analytic continuation in r does not commute with the taking of the $p\to 0$ limit, and so the surprising (and unfortunate) continuation of the ground state energy when p is strictly zero should not deter us from its study for $p>0$.

As a last remark we note that the at first sight surprising continuation properties of the free boson ground state energy have an analogue in the behaviour of $T\bar{T}$-deformed conformal field theories with $c_{\text{eff}}>0$ [4]. In that context the branch point occurs for physical values of R and is interpreted as a Hagedorn singularity. It would be interesting to investigate whether the considerations just given might be relevant for understanding the otherwise mysterious level that is found on continuation around this point.

4 Numerical results

In this section, we outline some of the results we have obtained from numerical work on the sinh-Gordon TBA, focusing on small but non-zero values of p. The picture we find is not yet complete, and we hope to be able to give a more comprehensive description elsewhere.

The procedure we followed is as described in the introduction. First we solved the ground state TBA to find the effective central charge $c_0(r)$ in the complex r plane[2]. A typical plot, for $p = 0.04$, is shown in Fig. 1 (a). For sufficiently negative values of $\mathrm{Re}(r)$, iteration of the TBA equation failed to converge no matter how strongly it was damped, and these regions are left blank on the figure[3]. Along the edge of this region lies a sequence of branch points, labeled on the figure with the letters $A \dots D$ and $A' \dots D'$, all of which turn out to be of square root type. As $p \to 0$, the locations of these branch points tend to even multiples of $2\pi i$: those at A and A' to $2\pi i$, those at B and B' to $4\pi i$, and so on. As the TBA system is continued around these branch points, pairs of points in the complex θ plane at which $e^{-\varepsilon(\theta)} = -1$ cross the contour of integration, converting the equation into an excited-state equation in the manner reviewed in the introduction. Continuation back to the real axis results in excited states which in the infrared, that is for large real values of r, contain two particles with equal and opposite momenta, and Bethe quantum numbers ± 1 for A and A', ± 2 for B and B', and so on. It is worth noting that continuation about paired branch points such as A and A' leads to the *same* excited state; given that all branch points are square roots, this means that continuation along a loop encircling both but no others is monodromy free.

Once the relevant excited state TBA equations have been identified, plots can be made of the scaling functions implied by these equations. One such plot is shown in Fig. 1 (b), again for $p = 0.04$, and two more are in Figs. 2 (a) and (b). The branch points on these plots which connect back to the 'parent' ground state energy level are indicated with the corresponding letters; there are also further branch points, which connect not with the ground state, but rather with more highly-excited levels, with more than two particles in the infrared. Continuing around these points, we have been able to connect with states containing 4 and then 6 particles, and there seem to be no obstacles to going further. Given that the $\theta \to -\theta$ symmetry of the ground-state TBA is preserved under continuation, the states we find are always

[2] We should note that, as done in previous work [6, 7], even for the ground state it was necessary to deform the contour of integration in the convolution term to lie a little below the real axis at $+\infty$, and above it at $-\infty$, in order to extend to values of r with argument larger than $\pi/2$.

[3] In the $p \to 0$ limit this convergence failure can be understood analytically, as will be explained in the next section.

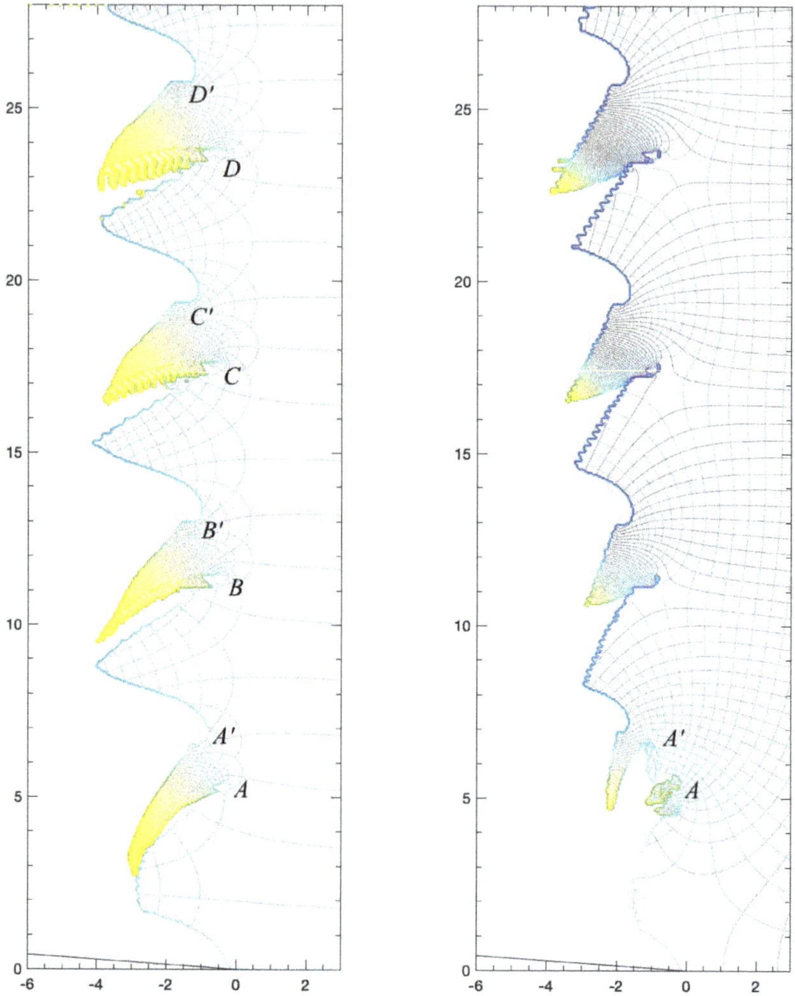

(a) (b)

Fig. 1 Contour plots of the real and imaginary parts of the $p = 0.04$ sinh-Gordon effective central charge $c_0(r)$ (a), and of the scaling function $c(r)$ for the lowest two-particle excited state (b). Horizontal and vertical axes correspond to the real and imaginary parts of r, while the letters $A \ldots D$ and $A' \ldots D'$ label selected square-root branch points, described further in the main text.

symmetrical with respect to a negation of all momenta, but we did not observe any other restrictions on the set of states that can be reached.

To perform these continuations, it is important to keep track of the places where singularities (in the sense given in the introduction) of the pseudoenegies approach the real axis. For this purpose we used the X-system to continue $X(\theta)$ beyond the strip $-\pi/2 < \operatorname{Im}\theta < \pi/2$ on which it is initially defined, and then the X-Y system

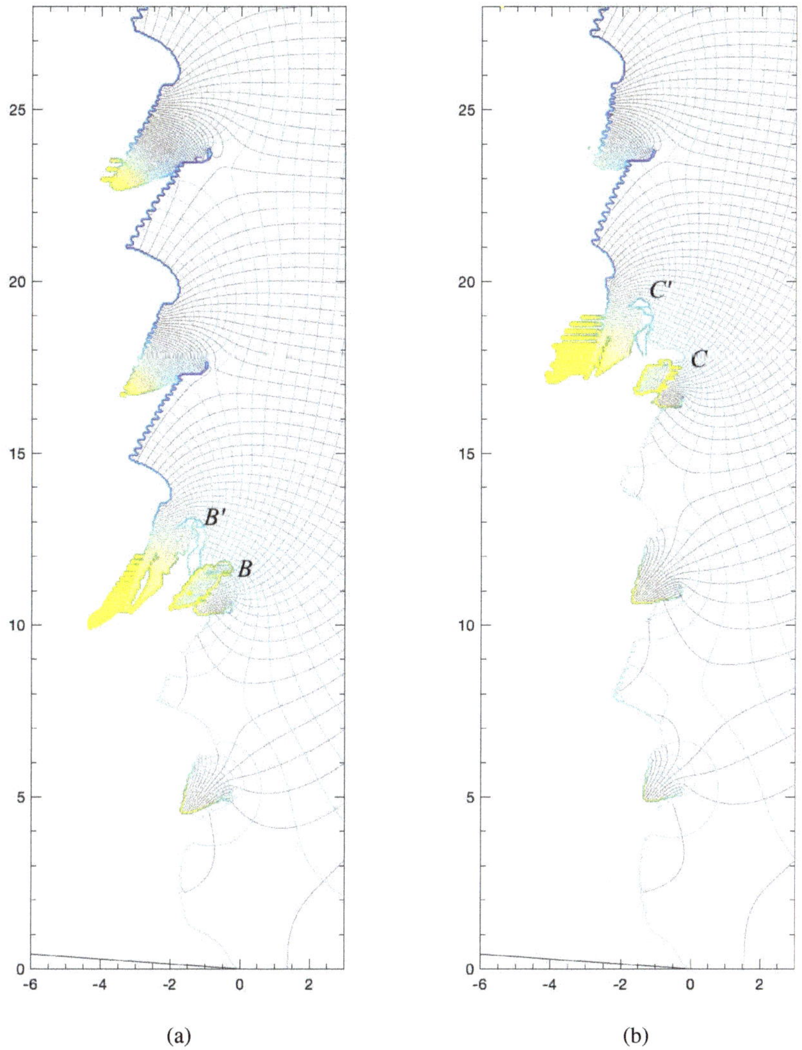

Fig. 2 Contour plots of the real and imaginary parts of the scaling function $c(r)$ for the second- and third-lowest symmetrical two-particle excited states of the $p = 0.04$ sinh-Gordon model, with Bethe quantum numbers $N = \pm 2$ (a) and $N = \pm 3$ (b).

to reconstruct $Y(\theta)$. For small values of p, high precision is needed in order to eliminate numerical artifacts; we used the Advanpix toolbox in Matlab [8].

The plots shown in Figs. 3 and 4, all computed for $r = 1$, illustrate the surprising behaviour of the ground state pseudoenergy as the small-p limit is taken at real values of r. They show contours of $Z(\theta)/(1+Z(\theta))$, where $Z(\theta) = |1+Y(\theta)|$. This function lies between zero and one and its zeros are the 'singularities' of the pseu-

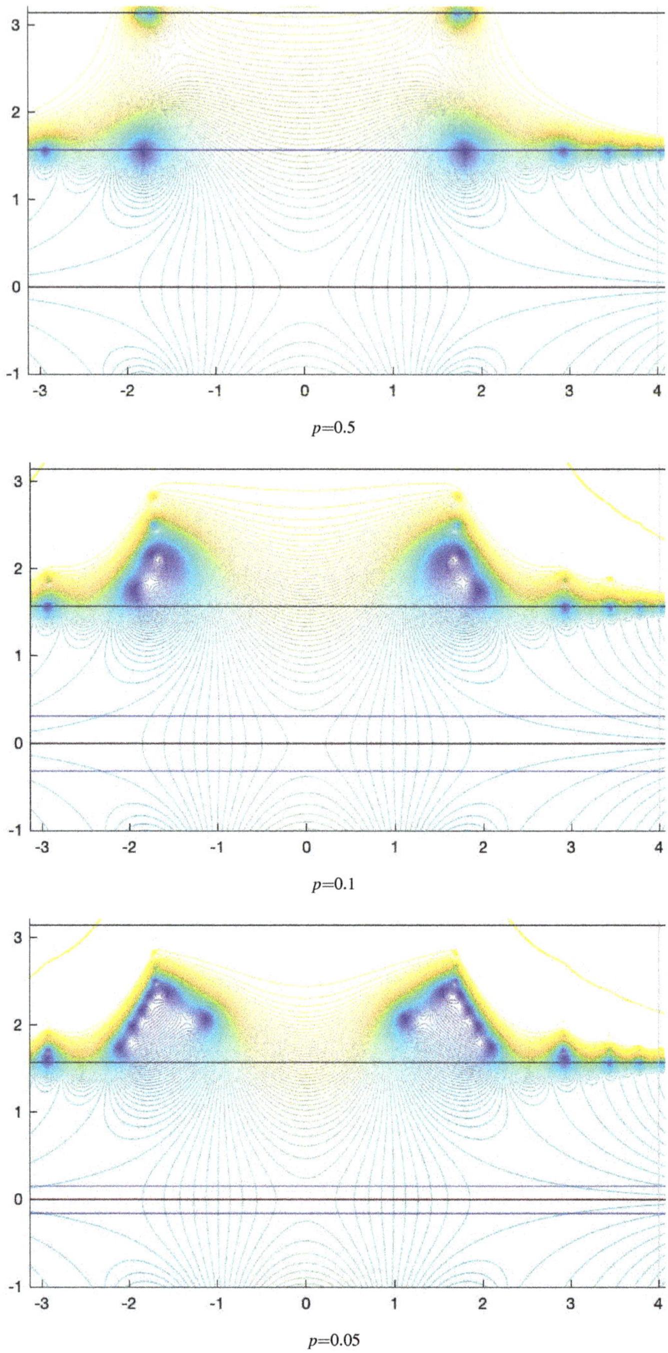

Fig. 3 Contour plots of $Z(\theta)/(1+Z(\theta))$ in the complex θ plane, where $Z(\theta) = |1+Y(\theta)|$, at various values of p, with $r = 1$. The horizontal lines lie at $\mathrm{Im}\,\theta = -p\pi$, 0, $p\pi$, $\pi/2$ and π .

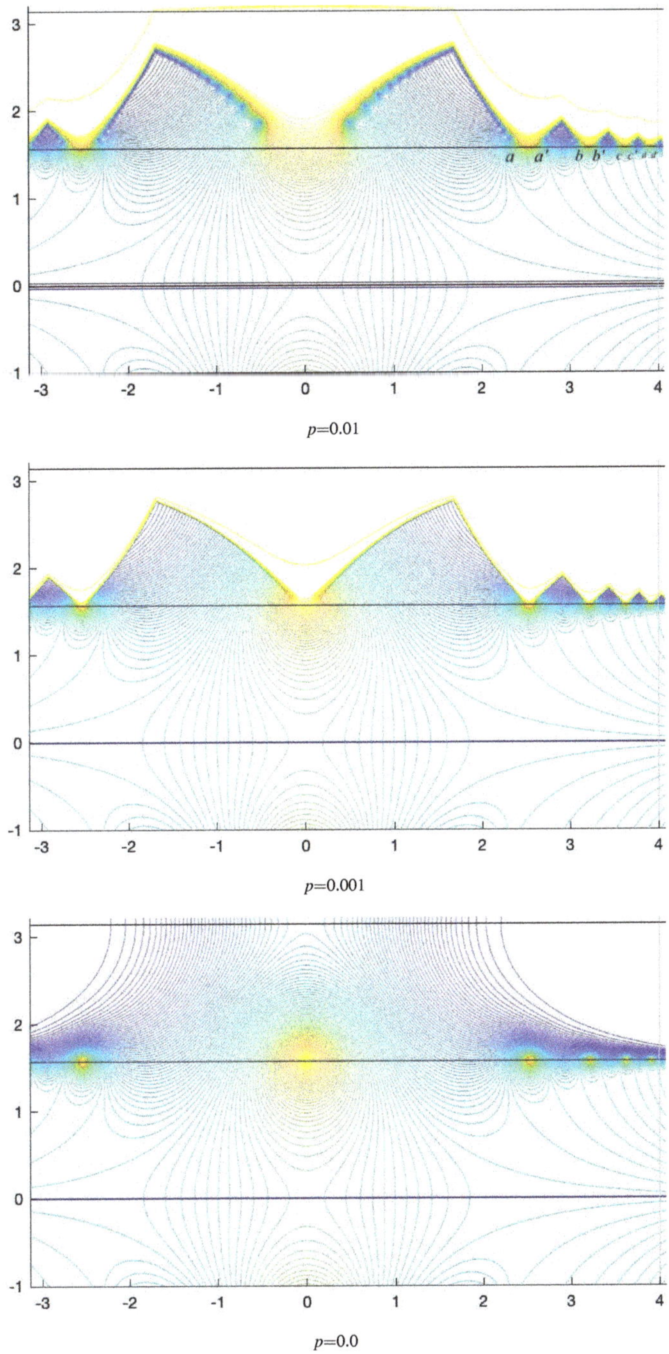

Fig. 4 Contour plots of $Z(\theta)/(1+Z(\theta))$ for further values of p, again with $r=1$. For the final plot, $Y(\theta)$ is the function $1/(e^{r\cosh\theta}-1)$ found in the formal $p\to 0$ limit, as discussed in the main text. The labels $a\ldots d$, $a'\ldots d'$ on the plot for $p=0.01$ indicate zeros of $Z(\theta)$ which cross the real axis under continuation around the branch points $A\ldots D$, $A'\ldots D'$, respectively.

doenergy, that is the points at which $Y(\theta) = -1$. In a central strip $-\pi/2 < \mathrm{Im}(\theta) < \pi/2$, we found that as $p \to 0$, the pseudoenergy tends smoothly to the $p = 0$ solution, shown in the final plot of Fig. 4, and $Z(\theta)$ is free of zeros. However, further from the real θ axis the situation becomes more complicated: zeros accumulate along a set of curves in the strip $\pi/2 < \mathrm{Im}(\theta) < \pi$, resulting in a sequence of inverted Y-like structures in the upper half plane which spread out from a central pair of 'cat ears', while their mirror images are formed in the lower half plane. Though not visible on the plots, this pattern repeats itself at larger values of $|\mathrm{Im}(\theta)|$.

During continuation around one of the ground state branch points shown in Fig. 1 (a), the pattern of zeros distorts, and a zero from the tip of one of the structures in the upper half plane crosses the integration contour into the lower half plane, while its negative makes the opposite journey. Together, they become the pair of 'activated' singularities at θ_0 and $-\theta_0$ in the excited-state TBA equation. The relevant singularities are indicated on the $p = 0.01$ plot of Fig. 4, and their pairing matches the pairing of ground state branch points seen in Fig. 1 (a): under continuation around branch point A, the singularity at a crosses to the lower half plane, while under continuation about A', it is its neighbour at a' which crosses, and likewise for b and b', c and c', and d and d'. A similar picture is found for continuation of the excited state TBA equations.

5 TBA iteration in the small-p limit

One feature of the scaling function plots shown in the last section was the presence of regions of the complex r plane where iteration of the TBA equation failed to converge. In the small-p limit their boundary acquires a characteristic, somewhat sinusoidal, shape, which can be understood analytically.

We begin by writing the TBA equation as $\varepsilon = F[\varepsilon]$ where

$$F[\varepsilon] = r\cosh\theta - \frac{1}{2\pi}\int_{\mathbb{R}} \phi(\theta-\theta')\log(1+e^{-\varepsilon(\theta')})\,d\theta' .$$

This can be solved numerically by iteration: $\varepsilon_{n+1} = F[\varepsilon_n]$. To help convergence it is useful to introduce a damping factor λ with $0 < \lambda \leq 1$ and instead iterate $\varepsilon_{n+1} = G_\lambda[\varepsilon_n]$ with $G_\lambda[\varepsilon] = (1-\lambda)\varepsilon + \lambda F[\varepsilon]$. However even this does not always converge. In the limit $p \to 0$, we have $F[\varepsilon] = r\cosh\theta - \log(1+e^{-\varepsilon(\theta)})$, the iterations at different values of θ decouple, and the lack of convergence can be understood as follows. Setting $r\cosh\theta = \hat{r}$, the fixed point ε_∞ for both the F and G iterations is the $p = 0$ TBA solution found earlier, satisfying $e^{\varepsilon_\infty} = e^{\hat{r}} - 1$. For this to be stable for $\varepsilon_{n+1} = G_\lambda[\varepsilon_n]$ we need $|G'_\lambda(\varepsilon_\infty)| < 1$, that is

$$|1-\lambda+\lambda e^{-\hat{r}}| < 1 ,$$

or, with $\hat{r} = x + iy$,

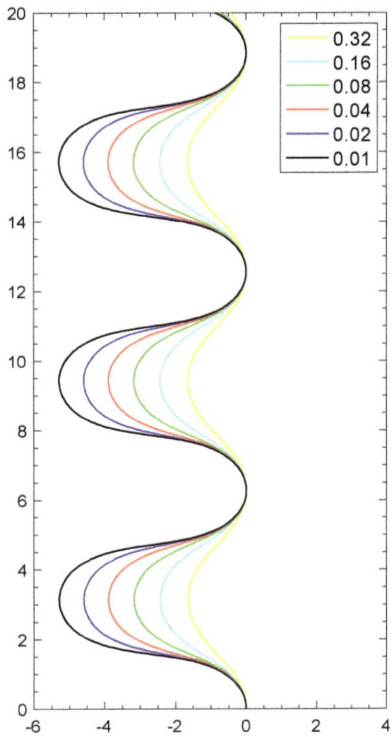

Fig. 5 Convergence regions for the $p = 0$ TBA equation in the complex $r = x + iy$ plane, for various values of the damping factor λ. For the leftmost curve, $\lambda = 0.01$; for the rightmost, $\lambda = 0.32$. In all cases the convergence region is to the right of the curve shown.

$$|1-\lambda+\lambda e^{-x-iy}|^2 = (1-\lambda)^2 + 2\lambda(1-\lambda)e^{-x}\cos(y) + \lambda^2 e^{-2x} < 1$$

which implies

$$\lambda e^{-2x} + 2(1-\lambda)e^{-x}\cos(y) - 2 + \lambda < 0 .$$

This is the region of the complex $x+iy$ plane to the right of the curve

$$x(y) = -\log\left(\frac{-(1-\lambda)\cos(y) + \sqrt{(1-\lambda)^2\cos^2(y) + (2-\lambda)\lambda}}{\lambda}\right).$$

The point $\theta = 0$, for which $\hat{r} = r$, always lies on the integration contour and so for convergence r must certainly lie in this region, which is shown for various values of λ in Fig. 5. The smaller the value of λ, the stronger the damping, and the larger the convergence region becomes. However, for $\cos(y) \neq 0$ and $\lambda \to 0$ we have

$$x(y) = -\log\left(\frac{(1-\lambda)|\cos(y)|}{\lambda}\left(\sqrt{1+\frac{(2-\lambda)\lambda}{(1-\lambda)^2\cos^2(y)}} - \sigma\right)\right)$$

$$\sim \begin{cases} \log(\cos(y)) & \sigma = +1 \\ \log(\lambda/2) - \log|\cos(y)| & \sigma = -1 \end{cases}$$

where $\sigma = \mathrm{sgn}(\cos(y))$. The fact that the limiting curve for $\sigma = +1$ is independent of λ means that in the $p \to 0$ limit there are 'forbidden zones' lying in the strips $|\,\mathrm{Im}(r) - 2n\pi| < \pi/2$, $n \in \mathbb{Z}$, where the iterative algorithm does not converge no matter how strong the damping. This picture is consistent with our numerical results at small but nonzero values of p.

6 Conclusions

In this article we have shown that the analytic continuation of energy levels in the finite volume sinh-Gordon model simplifies significantly in the $p \to 0$ limit. However, much work remains to be done before we can claim to have a complete picture. In particular, it turns out that the branch points labeled in Fig. 1 (a) are not the only ones which are accessible from the ground state: there are others, hidden in the region where the numerical methods we have presented so far do not converge. Some of these can be glimpsed by allowing the convergence factor λ to become complex[4], though we suspect that their detailed study will require a more sophisticated approach, possibly along the lines of that used in [3]. Our preliminary results indicate that these extra branch points lie on curves in the complex r plane terminating on the ground state branch points that were already discussed, condensing in the $p \to 0$ limit much as did the zeros of $Z(\theta)$ for the pseudoenergy. If so, this condensation should be relevant to a deeper understanding of the subtleties that were described in section 3, and shed light on the mechanism by which continuation to states of higher energy for nonzero values of p is flipped to a continuation to a shadow sector with lower energies when p is exactly equal to zero.

Acknowledgements We would like to thank David Meier for collaboration on early stages of this project, and Carl Bender, Lorenz Hilfiker, Didina Serban, István Szécsényi, Roberto Tateo and Pavel Tumarkin for helpful discussions and suggestions. PED thanks the MATRIX research institute, Australia, for hospitality, supported by a Simons Travel Grant. The work was also supported in part by a grant from the UK STFC, number ST/X000591.

References

1. Bender, C.M., Felski, A., Hassanpour, N., Klevansky, S.P., Beygi, A.: Analytic structure of eigenvalues of coupled quantum systems. Phys. Scripta **92**(1), 015,201 (2017). doi:10.1088/0031-8949/92/1/015201
2. Bender, C.M., Wu, T.T.: Anharmonic oscillator. Phys. Rev. **184**, 1231–1260 (1969). doi:10.1103/PhysRev.184.1231

[4] In the calculations of the last section, λ was assumed, as is standard, to be real. If λ becomes complex the curves plotted in Fig. 5 shift, and in some cases, even for nonzero values of p, the iterations then converge in previously inaccessible regions. It was this which allowed us to detect the additional branch points.

3. Camilo, G., Fleury, T., Lencsés, M., Negro, S., Zamolodchikov, A.B.: On factorizable S-matrices, generalized TTbar, and the Hagedorn transition. JHEP **10**, 062 (2021). doi: 10.1007/JHEP10(2021)062
4. Cavaglià, A., Negro, S., Szécsényi, I.M., Tateo, R.: $T\bar{T}$-deformed 2D Quantum Field Theories. JHEP **10**, 112 (2016). doi:10.1007/JHEP10(2016)112
5. Dorey, P., Hilfiker, L., Szécsényi, I.: Unpublished
6. Dorey, P., Tateo, R.: Excited states by analytic continuation of TBA equations. Nuclear Physics B **482**(3), 639–659 (1996)
7. Dorey, P., Tateo, R.: Excited states in some simple perturbed conformal field theories. Nuclear Physics B **515**(3), 575–623 (1998)
8. Holodoborodko, P.: Multiprecision Computing Toolbox for MATLAB 4.8.5.14569. Advanpix LLC., Yokohama, Japan (2021)
9. Klassen, T.R., Melzer, E.: The Thermodynamics of purely elastic scattering theories and conformal perturbation theory. Nucl. Phys. B **350**, 635–689 (1991). doi:10.1016/0550-3213(91) 90159-U
10. Ravanini, F., Tateo, R., Valleriani, A.: Dynkin TBAs. Int. J. Mod. Phys. A **8**, 1707–1728 (1993). doi:10.1142/S0217751X93000709
11. Zamolodchikov, Al. B.: Thermodynamic Bethe Ansatz in relativistic models: Scaling 3-state Potts and Lee-Yang models. Nuclear Physics B **342**(3), 695–720 (1990)
12. Zamolodchikov, Al. B.: On the thermodynamic Bethe ansatz equations for reflectionless ADE scattering theories. Phys. Lett. B **253**, 391–394 (1991). doi:10.1016/0370-2693(91)91737-G
13. Zamolodchikov, Al. B.: On the Thermodynamic Bethe Ansatz equation in sinh-Gordon model. Journal of Physics A: Mathematical and General **39**(41), 12,863 (2006)
14. Zelevinsky, A.: Semicanonical basis generators of the cluster algebra of type $a_1^{(1)}$. arXiv preprint math/0606775 (2006)

Superintegrable bosonic star networks

Angela Foerster and Jon Links

Abstract We introduce a class of bosonic star networks involving a central site interacting with the surrounding environment sites. These networks are shown to be *superintergable*. We present two forms of Bethe Ansatz solution providing expressions for the energy eigenvalues. A brief discussion is included on the potential applications.

1 Introduction

There has recently been significant interest in quantum models describing a central degree of freedom interacting with an environment modeled as an ensemble of surrounding degrees of freedom. This has been especially the case for spin degrees of freedom, giving rise to the class of *central spin models*, also known as *spin stars*. Many of these examples are integrable systems admitting Bethe Ansatz solutions, allowing for detailed analyses of their properties to be undertaken e.g. [4, 7, 9, 10].

In this note we provide an analogous class of models for bosonic systems, consisting of a single bosonic site coupled to environment sites through tunneling terms, as well as a global collective interaction. These models are defined in Sect. 2. Our interest in these is that they are superintegrable in the sense that there exists a larger number of conserved operators than degrees of freedom. We will explicitly construct the conserved operators in Sect. 3. Identifying these facilitates the calculation of a

Angela Foerster
Instituto de Física da UFRGS, Av. Bento Gonçalves 9500, Agronomia, Porto Alegre, RS 91501-970, Brazil
e-mail: `angela@if.ufrgs.br`

Jon Links
School of Mathematics and Physics, The University of Queensland, 4072, Australia
e-mail: `jrl@maths.uq.edu.au`

D. R. Wood et al. (eds.), *2024 MATRIX Annals, Part II*, MATRIX Book Series 8,
https://doi.org/10.1007/978-3-032-16206-9_6

Bethe Ansatz solution for determining the energy spectrum. Results on this aspect are collated in Sect 4. Concluding remarks are provided in Sect. 5.

2 The models

We introduce the star model Hamiltonian $H_{(p,1)}$ for $p+1$ sites in terms of a set of canonical boson operators a_i, $a_i^\dagger$, $N_i = a_i^\dagger a_i$, $i = 1, ..., p$, and another set b, $b^\dagger$, $N_b = b^\dagger b$. The Hamiltonian reads

$$H_{(n,1)} = U(N_a - N_b)^2 - J \sum_{j=1}^{p} (a_j b^\dagger + a_j^\dagger b), \tag{1}$$

where we have defined

$$N_a = \sum_{j=1}^{p} a_j^\dagger a_j.$$

The coupling parameter U provides the interaction strength between the central site and the surrounding environment sites, while the parameter J is the coupling strength for the hopping between sites. These star networks belong to a larger class of networks that are associated with complete bipartitite graphs [1]. A schematic representation of these systems is provided below in Fig. 1.

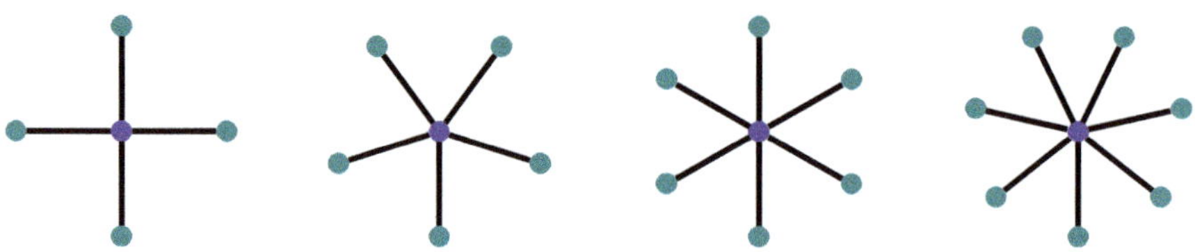

Fig. 1 Pictorial representation of the star network models $H_{(4,1)}$ on five sites (leftmost) through to $H_{(7,1)}$ on eight sites (rightmost).

The fundamental case of the $H_{(2,1)}$ Hamiltonian was first introduced in [12] as a model of cold dipolar bosonic atoms confined in a triple well potential. In that work, an analysis was undertaken to gain insights into the functionality of the system as an atomtronic switch. This same model has been subsequently studied in relation to generating and controlling entanglement [8, 14], as well as the transition to quantum chaos under the breaking of integrability through the inclusion of additional interactions terms [2, 3, 13].

3 Superintegrability

It is easily verified that the Hamiltonian (1) commutes with the total particle number operator given by $N = N_a + N_b$. In addition, the Hamiltonian commutes with the generators for a realisation of $gl(p-1)$. This result shows that for $p \geq 3$ the Hamiltonian is superintegrable - the number of degrees of freedom in the system, which is $p+1$, is less than the number of algebraically independent conserved operators, which is $2p-1$.

To identify superintegrability, a discrete Fourier transform is implemented via

$$c_k = \frac{1}{\sqrt{p}} \sum_{j=1}^{p} \exp\left(\frac{2\pi i j(k-p)}{p}\right) a_j, \qquad c_k^\dagger = \frac{1}{\sqrt{p}} \sum_{j=1}^{p} \exp\left(\frac{-2\pi i j(k-p)}{p}\right) a_j^\dagger,$$

such that the canonical bosonic commutation relations are preserved and

$$N_c = \sum_{j=1}^{p} c_j^\dagger c_j = N_a$$

The Hamiltonian (1) then assumes the form

$$H_{(n,1)} = U(N_b - N_c)^2 - J\sqrt{p}(b^\dagger c_p + c_p^\dagger b). \tag{2}$$

In this form it becomes apparent that the operators $\{E_k^j = c_j^\dagger c_k : k = 1, \ldots, p-1\}$ satisfy the $gl(p-1)$ commutation relations

$$[E_k^j, E_n^m] = \delta_k^m E_n^j - \delta_n^j E_k^m$$

and commute with the Hamiltonian (2), demonstrating superintegrability.

Moreover, the states

$$|r_1, \ldots, r_{p-1}\rangle = \prod_{k=1}^{p-1} (c_k^\dagger)^{r_k} |0\rangle, \tag{3}$$

where $|0\rangle$ is the Fock vacuum, are eigenstates of (2). The corresponding eigenvalues are given by Ur^2, where $r = \sum_{k=1}^{p-1} r_k$. The states (3) serve as pseudovacuum states in the derivation of the Bethe Ansatz results stated below.

4 Bethe Ansatz solutions

There are two known forms of Bethe Ansatz solution for the above star Hamiltonians. One is a specialisation of results derived in 2017 [15] through the standard formulation of the Algebraic Bethe Ansatz (ABA). Later, in 2024, an alternative form was obtained [1] via a different means. In the first instance the Bethe Ansatz Equations (BAE) have a multiplicative form, and are amenable for studies in the weak interaction regime following the approach of [16]. In the second case the BAE are additive, and are amenable for studies in the strong interaction regime as described in [1].

Set $\eta^2 = \frac{4U}{J\sqrt{p}}$. Using the ABA, the energies of the Hamiltonian (1) are provided via solutions of the Bethe Ansatz equations

$$\eta^2 v_i (v_i + \eta r) = \prod_{j\neq i}^{N-r} \frac{v_i - v_j - \eta}{v_i - v_j + \eta}, \quad r < N. \tag{4}$$

Given a solution to (4), the associated energy reads

$$E = -J\sqrt{p}\left(\lambda_r(u) - u^2 - \eta^{-2} - u\eta N - \frac{\eta^2 N^2}{4}\right), \tag{5}$$

where

$$\lambda_r(u) = u(u+\eta r)\prod_{j=1}^{N-r} \frac{u - v_j + \eta}{u - v_j} + \eta^{-2}\prod_{j=1}^{N-r} \frac{u - v_j - \eta}{u - v_j}.$$

Note that for $r = N$ the above is to be interpreted as $\lambda_N(u) = u(u+\eta r) + \eta^{-2}$ such that $E = Ur^2$.

The alternative exact-solution formulation developed in [1] yields the energy expression

$$E = U(N-2r)^2 - J\sqrt{p}\sum_{j=1}^{N-r} u_j, \tag{6}$$

where $\{u_n : n = 1, \ldots, N-r\}$ is the set of solutions to the BAE

$$\frac{J\sqrt{p}}{4U} u_k^{-2} + (1-N)u_k^{-1} - \frac{J\sqrt{p}}{4U} = \sum_{j\neq k}^{N-r} \frac{2}{u_j - u_k}. \tag{7}$$

We remark that, using (7), the formula (6) can also be expressed as

$$E = UN^2 - J\sqrt{p}\sum_{j=1}^{N-r} u_j^{-1}. \tag{8}$$

5 Discussion

We have introduced a class of bosonic superintegrable star networks and presented two distinct Bethe Ansatz solutions. In future work we will explore potential applications of these networks. Starting with the 3+1 model, we have found that this provides a framework for the design of a directional quantum switch and a diplexer-like atomic system.

Due to the dihedral symmetry of the star networks, a relevant question is: Can we enhance the design of quantum devices by rotating these systems? We have found that the superintegrability property plays a crucial role in this regard, which will be reported in upcoming work.

Acknowledgements This research was supported by the Australian Research Council through Discovery Project DP200101339. AF also acknowledges support from CNPq (Conselho Nacional de Desenvolvimento Científico e Tecnológico) - Edital Universal 406563/2021-7. We thank the mathematical research institute MATRIX in Australia where part of this research was performed. JL acknowledges the traditional owners of the land on which The University of Queensland (St. Lucia campus) operates, the Turrbal and Jagera people

References

1. Bennett, L., Foerster, A,. Isaac, P.S., Links, J.: Occupancy probabilities in superintegrable bosonic networks, Nucl. Phys. B **998**, 116406 (2024) doi:10.1016/j.nuclphysb.2023.116406
2. Castro, E.R., Chávez-Carlos, J., Roditi, I., Santos, L.F., Hirsch, J.G.: Quantum-classical correspondence of a system of interacting bosons in a triple-well potential, Quantum **5**, 563 (2021) doi:10.22331/q-2021-10-19-563
3. Castro, E.R,. Wittmann W., K., Chávez-Carlos, J., Roditi, I., Foerster, A., Hirsch J.G.: Quantum-classical correspondence in a triple-well bosonic model: From integrability to chaos, Phys. Rev. A **109**, 032225 (2024) doi:10.1103/PhysRevA.109.032225
4. Dimo, C., Faribault, A.: Strong-coupling emergence of dark states in *XX* central spin models, Phys. Rev. B **105**, L121404 (2022) doi:10.1103/PhysRevB.105.L121404
5. Faribault, A., Schuricht, D.: Integrability-based analysis of the hyperfine-interaction-induced decoherence in quantum dots, Phys. Rev. Lett. **110**, 040405 (2013) doi:/10.1103/PhysRevLett.110.040405
6. Jivulescu, M., Ferraro, E., Napoli, A., Messina, A.: Dynamical behaviour of an *XX* central spin model through Bethe Ansatz techniques, Rep. Math. Phys. **64**, 315 (2009) doi:10.1016/S0034-4877(09)90036-2
7. Nepomechie, R.I., Guan X.-W.: The spin-*s* homogeneous central spin model: exact spectrum and dynamics, J. Stat. Mech. Theory Exp. 103104 (2018) doi:10.1088/1742-5468/aae2d9
8. Tonel, A.P., Ymai, L.H., Wittmann W., K., Foerster, A., Links, J.: Entangled states of dipolar bosons generated in a triple-well potential, SciPost Phys. Core **2**, 003 (2020) doi:10.21468/SciPostPhysCore.2.1.003
9. van Tonder, J., Links, J.: Supersymmetry and integrability for a class of XY central spin models, J. Phys. A: Math. Theor. **56**, 47LT01 (2023) doi:10.1088/1751-8121/ad043a
10. Villazon, T., Chandran, A., Claeys, P.W.: Integrability and dark states in an anisotropic central spin model, Phys. Rev. Res. **2**, 032052(R) (2020) doi:10.1103/PhysRevResearch.2.032052

11. Villazon, T., Claeys, P.W., Pandey, M., Polkovnikov, A., Chandran, A.: Persistent dark states in anisotropic central spin models, Sci. Rep. **10**, 16080 (2020) doi:10.1038/s41598-020-73015-1
12. Wittmann W., K., Ymai, L.H., Tonel, A.P., Links, J., Foerster, A.: Control of tunneling in an atomtronic switching device, Commun. Phys. **1**, 91 (2018) doi:10.1038/s42005-018-0089-1
13. Wittmann W., K., Castro, E.R., Foerster, A., Santos, L.F.: Interacting bosons in a triple well: preface of many-body quantum chaos, Phys. Rev. E **105**, 034204 (2022) doi:10.1103/PhysRevE.105.034204
14. Wittmann W., K., Ymai, L.H., Barros, B.H.C., Links, J. Foerster, A.: Controlling entanglement in a triple-well system of dipolar atoms, Phys. Rev. A **108**, 033313 (2023) doi:10.1103/PhysRevA.108.033313
15. Ymai, L.H., Tonel A.P., Foerster, A., Links, J.: Quantum integrable multi-well tunneling models, J. Phys. A, Math. Theor. **50**, 264001 (2017) doi:10.1088/1751-8121/aa7227
16. Zhou, H.-Q., Links, J., McKenzie, R.H., Guan, X.-W.: Exact results for a tunnel-coupled pair of trapped Bose–Einstein condensates J. Phys. A: Math. Gen. **36**, L113 (2003) doi:10.1088/0305-4470/36/7/104

On the XXX spin-1/2 quantum chain with non-diagonal boundary fields

Andreas Klümper and Xin Zhang

Abstract The XXX spin-$\frac{1}{2}$ Heisenberg chain with non-diagonal boundary fields is a cornerstone model in the study of integrable systems with open boundaries. Despite its significance, solving this model exactly poses significant challenges due to the breaking of $U(1)$ symmetry. Building on the off-diagonal Bethe Ansatz (ODBA), we analyze a set of non-linear integral equations (NLIEs) that characterize the exact spectrum of the model.

For $U(1)$-symmetric chains, the NLIEs involve two functions, $a(x)$ and $\bar{a}(x)$, coupled by short-range integration kernels. The solutions exhibit characteristic features at a length scale that grows logarithmically with the system size, N.

In the absence of $U(1)$ symmetry, a novel third function, $c(x)$, emerges to account for the additional term in the T-Q relation. This introduces long-range kernel elements, leading to a winding phenomenon in $\log(1+a(x))$ and $\log(1+\bar{a}(x))$, where these functions undergo a steep $2\pi\mathrm{i}$ increase at a characteristic scale x_1. Other features are observed at the distinct scale $x_0 \sim \log N$ known from the NLIEs for the periodic boundary case. The ratio x_1/x_0 takes large/small values for short/long chains, for which we present explicit solutions derived through iterative numerical methods.

We also explore a scaling limit of the NLIEs, where short-range kernels simplify to delta functions while long-range Cauchy-like elements remain intact. This study establishes a framework for analyzing finite-size corrections and conformal properties of integrable spin chains with non-diagonal boundaries.

Andreas Klümper
Department of Physics, University of Wuppertal, Gaussstraße 20, 42119 Wuppertal, Germany
e-mail: kluemper@uni-wuppertal.de

Xin Zhang
Beijing National Laboratory for Condensed Matter Physics, Institute of Physics, Chinese Academy of Sciences, Beijing 100190, China
e-mail: xinzhang@iphy.ac.cn

D. R. Wood et al. (eds.), *2024 MATRIX Annals, Part II*, MATRIX Book Series 8,
https://doi.org/10.1007/978-3-032-16206-9_7

1 Introduction

The XXX spin-$\frac{1}{2}$ chain is a prototypical example of quantum integrable systems, renowned for its exact solvability under periodic boundary conditions through Bethe's Ansatz [1]. The foundation of integrability lies in the Yang-Baxter equation [2], a pivotal framework that has driven significant advances in quantum mechanics and statistical physics by enabling the discovery of numerous integrable models. However, the introduction of non-periodic boundary conditions adds complexity, requiring new theoretical tools and methods.

Research on quantum spin systems with non-trivial boundary conditions has a long history. It plays a crucial role in the study of single-impurity systems underlying the Kondo effect [3, 4], which has also inspired investigations into conformal field theories with boundaries [5, 6]. Beyond equilibrium systems with boundaries, spin chains coupled to dissipative boundary baths with controlled polarization have also attracted interest. In such systems, the dynamics become non-unitary, leading to relaxation into a non-equilibrium steady state. At strong dissipation, edge spin fluctuations are suppressed, effectively imposing Dirichlet boundary conditions on the quantum spin chain [7, 8]. Much of the recent research in this area has been driven by theoretical interest. Many studies focus on surface effects of a single impurity, yielding $\mathcal{O}(1)$ corrections to thermodynamic quantities in systems of size N. In this work, we take a step further by examining the mutual correlations of two boundaries at opposite ends of an open Heisenberg chain, where the boundary fields enclose an arbitrary angle. These correlations are of order $\mathcal{O}(1/N)$, revealing novel finite-size effects in boundary-driven quantum spin chains.

The XXX spin-$\frac{1}{2}$ chain with parallel boundary fields was successfully solved using both the coordinate and algebraic Bethe Ansatz [9, 10], and finite-size corrections were subsequently explored in [11]. For more general boundary conditions, Sklyanin's reflection algebra [10, 12] provided a powerful framework to address the factorization of scattering processes at the chain boundaries. This approach culminated in the proof of integrability for the Heisenberg spin chain with non-parallel boundary fields [13].

Non-parallel boundary fields, however, break the $U(1)$ symmetry of the system, presenting challenges for conventional Bethe Ansatz methods. To overcome these obstacles, advanced techniques such as $T-Q$ relations [14, 15], fusion hierarchies [16], and the off-diagonal Bethe Ansatz (ODBA) [17, 18] have been developed. These methods utilize inhomogeneous $T-Q$ relations and hierarchies of transfer matrices, with recent work demonstrating completeness for certain ODBA formulations [19].

Further progress has been achieved through alternative approaches, including the modified algebraic Bethe Ansatz, which utilizes chiral basis states [20–23], the chiral coordinate Bethe ansatz [24] and separation of variables techniques applied to increasingly complex boundary conditions [25–27]. These advancements have expanded the applicability of integrable systems to diverse boundary conditions and physical phenomena.

In [28] physical properties have been calculated that are independent of the angle between the boundary fields, such as the ground-state energy, low-lying excitations and surface terms. In this work, we build on some of the developments to analyze the isotropic Heisenberg chain with arbitrary boundary fields [29] which in turn among other things builds on the inhomogeneous $T-Q$ relation [17, 18]. We here discuss how to use the extended set of NLIEs of [29] to derive data beyond the expressions for the ground-state energy and surface energy. Notably for finite-size corrections, we provide a consistent perspective on deriving $\mathscr{O}(1/N)$ terms in a two step scaling limit.

The structure of the paper is as follows. In Section 2 we give a short review on the NLIEs for the periodic boundary case of the isotropic Heisenberg spin chain. We restrict this to the discussion of the ground state, but present numerical results for various system sizes. In Section 3, we summarize the ODBA approach [17, 18] for the open boundary case with non-parallel boundary fields. We identify an extended set of characteristic "auxiliary functions" that satisfy NLIEs. Section 4 indicates elements of the derivation of these NLIEs and discusses numerical solutions to these. There, qualitatively different features are observed for short and large system sizes. In Section 5, we present the multiplicative scaling limit of the NLIEs and some numerical studies for the simplified, but still non-trivial NLIEs. A conjectural approximate expression for the finite size data of the ground-state energy is given. Finally, conclusions are provided in Section 6.

2 Non-linear integral equations for periodic boundaries

The spin-1/2 Heisenberg model is a well studied quantum system, especially for periodic boundary conditions

$$H_p = \sum_{j=1}^{N} \vec{\sigma}_j \cdot \vec{\sigma}_{j+1}, \qquad \vec{\sigma}_{N+1} := \vec{\sigma}_1 \,. \tag{1}$$

In this paper we are interested in exact integral equations for eigenvalues and their (numerical) solutions. For periodic boundary conditions and arbitrary chain length these were derived in [30, 31] and read for the ground-state of the isotropic system

$$\begin{pmatrix} \log a \\ \log \overline{a} \end{pmatrix} = d + k * \begin{pmatrix} \log A \\ \log \overline{A} \end{pmatrix} . \tag{2}$$

These are non-linear integral equations as the logarithms of a, A, $\overline{a}$, and $\overline{A}$ enter where

$$A(x) := 1 + a(x), \quad \overline{A}(x) := 1 + \overline{a}(x) \,. \tag{3}$$

The functions $a(x)$ and $\overline{a}(x)$ are treated as independent, but "turn out" to be complex conjugate to each other. The integrals are of convolution type. In general the convolution $*$ of two functions f, g is defined by

$$f * g(x) = \frac{1}{2\pi} \int dy f(x-y) g(y). \tag{4}$$

The source (driving) term and the kernel matrix in (2) are

$$d(x) = N \log \tanh\left(\frac{\pi}{4}x\right) \cdot \begin{pmatrix} 1 \\ 1 \end{pmatrix}, \quad k(x) = \begin{bmatrix} \kappa(x,1) & -\kappa(x-\mathrm{i}2^-,1) \\ -\kappa(x+\mathrm{i}2^-,1) & \kappa(x,1) \end{bmatrix}, \tag{5}$$

where $2^- = 2 - 0^+$ and the function κ is related to the digamma function ψ

$$\kappa(x,r) := \frac{1}{4}\Big[\psi\left(\tfrac{1}{4}(r+3+\mathrm{i}x)\right) + \psi\left(\tfrac{1}{4}(r+3-\mathrm{i}x)\right) \\ -\psi\left(\tfrac{1}{4}(r+1+\mathrm{i}x)\right) - \psi\left(\tfrac{1}{4}(r+1-\mathrm{i}x)\right)\Big]. \tag{6}$$

The ground-state energy is given by a bulk term and a finite size contribution of order $\mathscr{O}(1/N)$ in terms of an integral involving the functions $\log A$ and $\log \overline{A}$

$$E = N(1 - 4\log 2) - 4\mathrm{i}\left[e' * (\log A + \log \overline{A})\big|_{-\mathrm{i}}\right], \qquad e(x) := \frac{\frac{\pi}{2}}{\cosh \frac{\pi}{2}x}, \tag{7}$$

here written in form of a convolution integral at a specific argument.

In Fig. 1 we show numerical results for chain lengths $N = 10^3, 10^6, 10^9$. The solution functions show a simple qualitative behaviour, $\log A$ is very close to 0 for arguments $|x| < \frac{2}{\pi}\log N$. For $|x| > \frac{2}{\pi}\log N$ the function takes non-zero values with algebraically fast convergence to $\log 2$ for $x \to \infty$. This behaviour is mainly dominated by the source term on the right hand side of (2). The contributions of the convolution integral are of course mostly of quantitative type, new qualitative features are a non vanishing imaginary part for arguments $|x| \simeq \frac{2}{\pi}\log N$ and algebraically decaying asymptotics. Note that similar NLIEs with different driving terms hold for the thermodynamics of the Heisenberg chain and related quantum systems [32–36].

3 Non-parallel boundary fields: eigenvalue equation and characteristic functions

For the spin-1/2 Heisenberg model with isotropic bulk interaction the Hamiltonian of the system with arbitrary boundary fields can be brought to the form

$$H_o = \sum_{j=1}^{N-1} \vec{\sigma}_j \cdot \vec{\sigma}_{j+1} + \frac{1}{p}\sigma_1^z + \frac{1}{q}\left(\sigma_N^z + \xi \sigma_N^x\right), \tag{8}$$

where N is the number of sites, and p, q, and ξ are boundary parameters, where ξ is equal to the tangent of the angle between the boundary fields.

The Hamiltonian is found as member of a family of commuting operators that are generated by an associated family of commuting transfer matrices. We study the

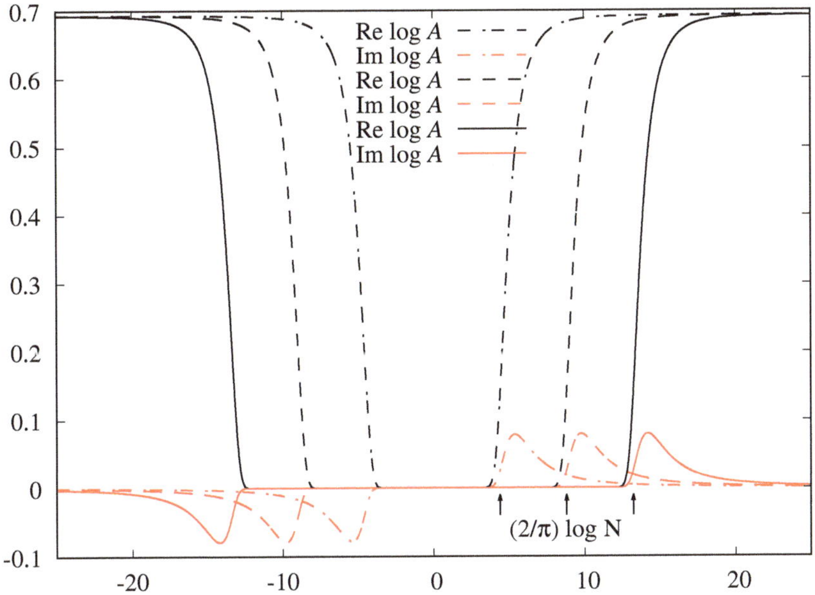

Fig. 1 Graphs of the functions $\log\Lambda$ resp. $\mathrm{Re}\log\Lambda$ and $\mathrm{Im}\log\Lambda$ for the ground-state eigenvalue of the spin-1/2 XXX chain with periodic boundary conditions and chain lengths $N = 10^3, 10^6, 10^9$ (dash-dotted, dashed, solid lines). There is one important length scale in the system, $\frac{2}{\pi}\log N$, where at arguments x of this order $\mathrm{Re}\log\Lambda(x)$ shows a transition from one flat inner region with values extremely close to 0 to the asymptotic behaviour $\log 2$, and where $\mathrm{Im}\log\Lambda(x)$ deviates noticeably from zero.

eigenvalue functions $\Lambda(x)$ for this family of transfer matrices for an even number of sites N. Using the inhomogeneous T-Q relation of [17, 18] with two Q-functions we have

$$\Lambda(x) = \lambda_1(x) + \lambda_2(x) + \lambda_3(x), \tag{9}$$

where we rescaled the function by dropping a constant factor $(\mathrm{i}/2)^{2N+2}$ and we also reparameterized the argument $u = \mathrm{i}x/2 - 1/2$ used in [18] by the new variable x. The summands in (9) are

$$\lambda_1(x) := \phi_1(x)\frac{q_1(x+2\mathrm{i})}{q_2(x)}, \quad \lambda_2(x) := \frac{\phi_2(x)}{q_1(x)q_2(x)}, \quad \lambda_3(x) := \phi_3(x)\frac{q_2(x-2\mathrm{i})}{q_1(x)}, \tag{10}$$

where $q_1(x)$ and $q_2(x)$ are polynomials of degree N with $q_2(x) = q_1(-x)$. The zeros of $q_1(x)$ are called Bethe roots.

The functions $\phi_1(x)$, $\phi_2(x)$, $\phi_3(x)$ are explicitly given by

$$\begin{aligned}
\phi_1(x) &:= \xi_1 \frac{\varphi(x)}{x}(x-\mathrm{i})^{2N+1}, \qquad \varphi(x) := (x+\mathrm{i}+\mathrm{i}p_1)(x+\mathrm{i}+\mathrm{i}p_2),\\
\phi_2(x) &:= 2(1-\xi_1)(x^2+1)^{2N+1},\\
\phi_3(x) &:= \xi_1 \frac{\overline{\varphi}(x)}{x}(x+\mathrm{i})^{2N+1}, \qquad \overline{\varphi}(x) := (x-\mathrm{i}-\mathrm{i}p_1)(x-\mathrm{i}-\mathrm{i}p_2),
\end{aligned} \tag{11}$$

and we used new combinations of the boundary parameters

$$\xi_1 := (1+\xi^2)^{1/2}, \qquad p_1 := -2p, \qquad p_2 := -2q/\xi_1. \tag{12}$$

In [18] the Bethe roots are calculated for relatively short chains by numerically solving for the Bethe equations. These equations are derived from the condition that all potential poles in (9) cancel, resulting in an analytic function $\Lambda(x)$. We do not write down these equations explicitly as we do not use them.

Here we adopt a different resp. opposite approach by using the analyticity of the eigenvalue function directly. For the details we have to refer to [29]. There, three combinations of functions are found to be fundamental to formulating a closed set of integral equations

$$a(x) := \frac{\lambda_2(x+\mathrm{i})+\lambda_3(x+\mathrm{i})}{\lambda_1(x+\mathrm{i})}, \quad A(x) := 1+a(x) = \frac{\Lambda(x+\mathrm{i})}{\lambda_1(x+\mathrm{i})}, \tag{13}$$

$$\overline{a}(x) := \frac{\lambda_1(x-\mathrm{i})+\lambda_2(x-\mathrm{i})}{\lambda_3(x-\mathrm{i})}, \quad \overline{A}(x) := 1+\overline{a}(x) = \frac{\Lambda(x-\mathrm{i})}{\lambda_3(x-\mathrm{i})}, \tag{14}$$

$$c(x) := \frac{\lambda_2(x)\Lambda(x)}{\lambda_1(x)\lambda_3(x)}, \qquad C(x) := 1+c(x) = \frac{[\lambda_1(x)+\lambda_2(x)][\lambda_2(x)+\lambda_3(x)]}{\lambda_1(x)\lambda_3(x)}, \tag{15}$$

with asymptotic behaviour

$$a(\infty) = \overline{a}(\infty) = 2/\xi_1 - 1, \qquad c(\infty) = 4(1/\xi_1^2 - 1/\xi_1). \tag{16}$$

A clarification of our terminology is in order. We introduce three functions, two of which share the same notation as those appearing in Sect. 2. However, despite their identical symbols, the functions $a(x)$ and $\bar{a}(x)$ are defined differently here and take different values. Nevertheless, strong similarities exist, the most significant being the (sub-)kernel matrix coupling these functions in the NLIEs (see (5) and (21)). For this reason, we have retained the same notation, trusting that the reader will recognize the distinction when interpreting plots. In particular, we distinguish cases where only $a(x)$ (or equivalently, $\log A$) is shown from those where both $a(x)$ and $c(x)$ (or $\log A$ and $\log C$) appear.

Any energy eigenvalue E of the Hamiltonian is obtained from the logarithmic derivative of the corresponding eigenvalue $\Lambda(x)$ of the transfer matrix. For the ground state eigenvalue with both p and q negative we have, ordered by bulk, surface and finite size terms

$$
\begin{aligned}
E =& N(1-4\log 2) && (17)\\
&+\pi-1-2\log 2+\frac{2}{p_1}+\psi\left(\frac{p_1}{4}\right)-\psi\left(\frac{p_1+2}{4}\right)+\frac{2}{p_2}+\psi\left(\frac{p_2}{4}\right)-\psi\left(\frac{p_2+2}{4}\right) && (18)\\
&+4e(x_0)-2\mathrm{i}\left[e'*(\log A+\log\overline{A})\big|_{-\mathrm{i}}\right]. && (19)
\end{aligned}
$$

The last line (19) is a $\mathscr{O}(1/N)$ contribution, because $x_0 \simeq \frac{2}{\pi}\log N$ and the function $\log A+\log\overline{A}$ takes non-negligeable values for arguments only outside the interval $[-x_0,x_0]$. Interestingly, only functions $\log A$ and $\log\overline{A}$ enter the energy formula, but do not satisfy a closed set of integral equations among themselves. Only the set of 3 auxiliary functions (13)-(15) allows for this.

Here, however, we have to skip the derivation of NLIEs for the above defined functions and refer to [29]. That work is considerably involved. Non-technical evidence of this is the qualitatively different behaviour of the functions in comparison to the periodic boundary case as shown in Fig. 2 for relatively short ($N=10^3$) and Fig. 3 for longeer ($N=10^6,10^9$) chains. Here, the functions $\log A$ and $\log\overline{A}$ show non-trivial asymptotics with increase resp. decrease of the imaginary parts by $2\cdot 2\pi\mathrm{i}$. In addition, the set of NLIEs comprises a new third function $C(x)=1+c(x)$ with no counter-part in the periodic boundary case.

4 Non-parallel boundary fields case: the non-linear integral equations

The derivation of NLIEs is done in [29] largely following [30,31,37,38] but facing a new challenge that the involved functions show non-vanishing asymptotics. For this, subtraction terms are introduced in the convolution part such that the involved functions show vanishing asymptotics. And for the compensation, counter terms in the driving (source) terms of the NLIEs are used. The subtractions and compensation terms can be chosen to be simply of rational function type with free parameters. The result is

$$
\begin{pmatrix}\log a\\ \log\overline{a}\\ \log c\end{pmatrix} = d+K*\begin{pmatrix}\log(A/A(\infty))-\log\left(\dfrac{x-x_{r+}}{x-x_{r-}}\cdot\dfrac{x-x_{l+}}{x-x_{l-}}\right)\\ \log(\overline{A}/\overline{A}(\infty))-\log\left(\dfrac{x-x_{r-}}{x-x_{r+}}\cdot\dfrac{x-x_{l-}}{x-x_{l+}}\right)\\ \log(C/C(\infty))\end{pmatrix}, \tag{20}
$$

with kernel matrix

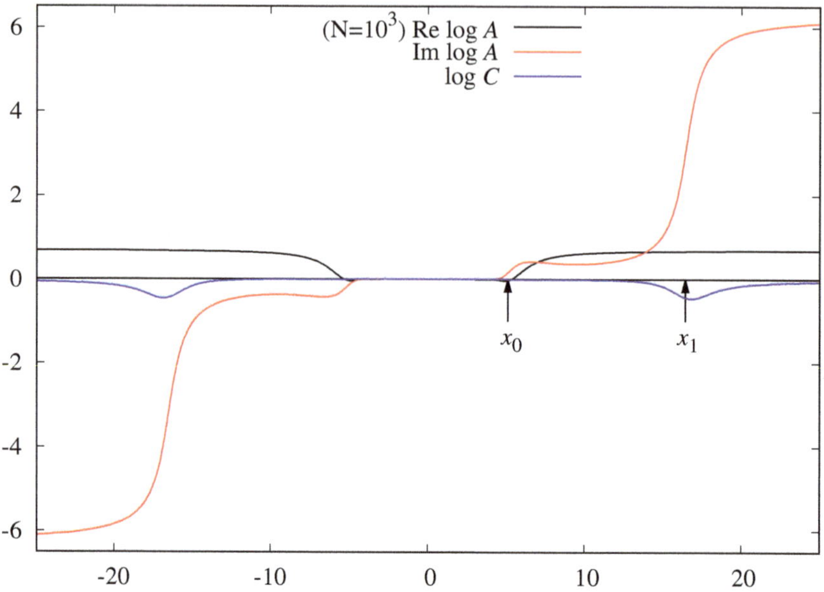

Fig. 2 Graphs of $\log A$ and $\log C$ for the ground-state eigenvalue of the spin-$1/2$ XXX chain with boundary parameters $p = -0.6$, $q = -0.3$, $\xi = 0.1$, and system size $N = 10^3$. The real (imaginary) part of $\log A$ is even (odd), while $\log C$ is real and even. Two key scales are visible: the zeros $\pm x_0$ of $\log C$, where $x_0 \simeq \frac{2}{\pi}\log N \approx 5.1$. In $[-x_0, x_0]$, $\log C$ remains small and positive, turning noticeably negative outside. Similarly, $\log A$ takes near-zero values in this interval, approaching non-zero asymptotics outside. At $x_1 \sim 16.5$ ($> x_0$) and at $-x_1$, $\log C$ has minima, while the imaginary part of $\log A$ shows a steady increase, ultimately reaching $+2\pi$ as $x \to +\infty$.

$$K(x) = \begin{bmatrix} \kappa(x,1) & -\kappa(x-\mathrm{i}2^-,1) & -\mathrm{i}/(x-\mathrm{i}) \\ -\kappa(x+\mathrm{i}2^-,1) & \kappa(x,1) & \mathrm{i}/(x+\mathrm{i}) \\ \mathrm{i}/(x+\mathrm{i}) & -\mathrm{i}/(x-\mathrm{i}) & 0 \end{bmatrix}, \tag{21}$$

where the function κ is the same as already given above in (6). Note that the logarithms of analytic functions as appearing above are analytic functions themselves without any $\pm 2\pi\mathrm{i}$ jumps.

Note that the careful prescription for the convolution integrals is necessary because of the slow asymptotics of K_{31} and K_{32} and the curious property that the functions $\log A$ and $\log \overline{A}$ show non-trivial windings, see Fig. 2 for a system with size $N = 10^3$, still to be considered small on grounds that become clearer shortly. The equations (20) are non-linear integral equations (NLIEs) for the functions a, $\overline{a}$, c, because $A = 1 + a$, $\overline{A} = 1 + \overline{a}$, $C = 1 + c$. Numerical solutions are obtained by iterative treatments and evaluations of the convolution integrals by the Fast Fourier Transform. In parallel, the zeros $\pm x_0$ of $\Lambda(x)$ are determined by solving $a(x_0 + \mathrm{i}) = -1$, evaluating the right-hand sides of the integral equations off the real axis. The value of x_0 is iteratively updated during the numerical treatment. This parameter also appears in the source term (24) but has no counterpart in the case of periodic boundary conditions, where the ground-state eigenvalue function – not discussed in Sect. 2 – has no real-axis zeros.

In the counter terms x_{r+} and x_{r-} are complex numbers with positive real part and positive resp. negative imaginary parts. And x_{l+} and x_{l-} are defined similarly with negative real parts. The introduced functions subtract the winding behaviour observed in the functions $\log A$ and $\log \overline{A}$. Now the inhomogeneity d is a tuple of three functions containing the counter terms

$$\begin{aligned} d_1(x) &= (2N+1)\log\tanh\left(\frac{\pi}{4}x\right) + \frac{\pi}{2}\mathrm{i} - \mathrm{i}\alpha(x-\mathrm{i},1) \\ &+ \mathrm{i}\alpha(x-\mathrm{i},p_1) + \mathrm{i}\alpha(x-\mathrm{i},p_2) - \mathrm{i}\alpha(x-x_0-\mathrm{i},1) - \mathrm{i}\alpha(x+x_0-\mathrm{i},1) \\ &+ \log\left(a(\infty)\cdot\frac{x-x_{r+}-2\mathrm{i}}{x-x_{r-}}\cdot\frac{x-x_{l+}-2\mathrm{i}}{x-x_{l-}}\right), \end{aligned} \tag{22}$$

$$\begin{aligned} d_2(x) &= (2N+1)\log\tanh\left(\frac{\pi}{4}x\right) - \frac{\pi}{2}\mathrm{i} + \mathrm{i}\alpha(x+\mathrm{i},1) \\ &- \mathrm{i}\alpha(x+\mathrm{i},p_1) - \mathrm{i}\alpha(x+\mathrm{i},p_2) + \mathrm{i}\alpha(x-x_0+\mathrm{i},1) + \mathrm{i}\alpha(x+x_0+\mathrm{i},1) \\ &+ \log\left(\overline{a}(\infty)\cdot\frac{x-x_{r-}+2\mathrm{i}}{x-x_{r+}}\cdot\frac{x-x_{l-}+2\mathrm{i}}{x-x_{l+}}\right), \end{aligned} \tag{23}$$

$$d_3(x) = \log\left(c(\infty)\cdot\frac{x^2(x^2-x_0^2)}{(x-x_{r-}+\mathrm{i})(x-x_{r+}-\mathrm{i})(x-x_{l-}+\mathrm{i})(x-x_{l+}-\mathrm{i})}\right). \tag{24}$$

Here $\alpha(x,r)$ is the integral of the function $\kappa(x,r)$ with respect to x

$$\alpha(x,r) := \mathrm{i}\log\frac{\Gamma\left(\frac{1}{4}(r+3-\mathrm{i}x)\right)\Gamma\left(\frac{1}{4}(r+1+\mathrm{i}x)\right)}{\Gamma\left(\frac{1}{4}(r+3+\mathrm{i}x)\right)\Gamma\left(\frac{1}{4}(r+1-\mathrm{i}x)\right)}. \tag{25}$$

Note that the functions $\kappa(x,r)$ and $\alpha(x,r)$ are real valued for real arguments x, r.

The concrete values of $x_{r\pm}$ and $x_{l\pm}$ drop out of the calculations and affect at best the accuracy of the calculations. For practical purposes we choose for these numbers

$$x_{r\pm} = \tilde{x}_1 \pm \mathrm{i}\,\delta, \qquad x_{l\pm} = -\tilde{x}_1 \pm \mathrm{i}\,\delta, \tag{26}$$

with some $\delta > 0$ and $\tilde{x}_1$ is an estimate of the location of the transition of the imaginary part of $\log A$ from small resp. practically zero values to $+2\pi$ as explained in Fig. 2. The *optimal* choice of the parameters $\tilde{x}_1$ and δ for the discussed counterterms ensures the fastest convergence of the iterative numerical treatment and provides the *best* estimate for the location x_1 of the transition in the imaginary part of $\log A$. Unlike the zero x_0 of the eigenvalue function $\Lambda(x)$, the parameter x_1 is not precisely defined. However, the transition in $\mathrm{Im}\log A$ arises from a well-defined complex conjugate pole-zero pair of $A(x)$. Here, we do not analyze these structures in detail, as their precise properties are irrelevant for calculating the energy which only depends on the real part of $\log A$ (19) which does not have any remarkable structure at x_1.

A practical approach for iterating the NLIEs is to use initial data resembling solutions of the Bethe ansatz equations, feasible for small system sizes ($N \sim 10$). For such systems, $a(x)$ remains near zero for x close to 0 but grows in magnitude for $|x|$ beyond $\frac{2}{\pi}\log N$, eventually with $a(x)$ encircling -1 counter-clockwise at some x_1, causing a sharp increase in $\mathrm{Im}\log A$, see Fig. 2. This behavior is critical for understanding the winding in $\log A$.

Convergence of the NLIEs was achieved for "short chains" (like $N = 10^3$ for $\xi = 0.1$), but increasing N "soon" results in the divergence of the iterations, independent of boundary parameters. We found that x_0, the zero of $c(x)$, always scales as $\frac{2}{\pi}\log N$, separating regions where $a(x)$ is near zero from those where it becomes order 1. For short chains, the position x_1 of the winding observed in $\log A$ satisfies $x_1 \gg x_0$, but increasing N leads to $x_1 \sim x_0$.

For large sizes, no initial data reliably captured the winding at $x_1 > x_0$. Surprisingly, the winding occurs in the "forbidden region" $[-x_0, x_0]$, where $a(x)$ is nearly zero. For large sizes, initial data with winding at $x_1 < x_0$ lead to convergence, see Fig. 3. For all systems, zeros of $q_1(x)$ and $q_2(x)$ form a pole-zero pair of $A(x)$, causing $A(x)$ taking a loop around 0. Tracking this pair as N increases shows it moving away from the origin initially but eventually stabilizing at a finite distance and moving back at very large N.

The condition for this to occur is that the separation between the zero and pole in the pair approaches zero exponentially fast. Under this condition, the aforementioned winding happens so rapidly that the convolution integrals generate contributions canceling the leading term, $(2N+1)\log\tanh\frac{\pi}{4}x$. Based on this reasoning, we successfully found solutions to the NLIEs for significantly larger system sizes, as illustrated in Fig. 3.

5 The two step scaling limit

In the periodic boundary case the finite size data of order $\mathcal{O}(1/N)$ are obtained in an *additive scaling limit* [31] of the NLIEs (2) for the $N \to \infty$ limits of the functions with shifted argument, e.g. for the "right movers"

$$a_r(x) := \lim_{N\to\infty} a\left(x + \tfrac{2}{\pi}\log N\right), \quad \text{and} \quad \overline{a}_r(x) := \lim_{N\to\infty} \overline{a}\left(x + \tfrac{2}{\pi}\log N\right), \tag{27}$$

and similar for functions obtained with a shift $x \to x - \frac{2}{\pi}\log N$ describing "left movers". These functions satisfy sets of NLIEs very similar to (2), however with source terms $-2\exp(-x)$. A similar scaling limit has to be considered here for the derivation of the finite size corrections. However, a direct application of the above strategy does not yield a closed set of NLIEs just among the "right movers" or among the "left movers". This happens because we have long ranged kernel elements that couple "right movers" and "left movers".

The scaling limit consists of two steps. In the first step we formulate a *multiplicative scaling limit* for all three functions (here we are allowed to use $x_0 = \frac{2}{\pi}\log N$ as

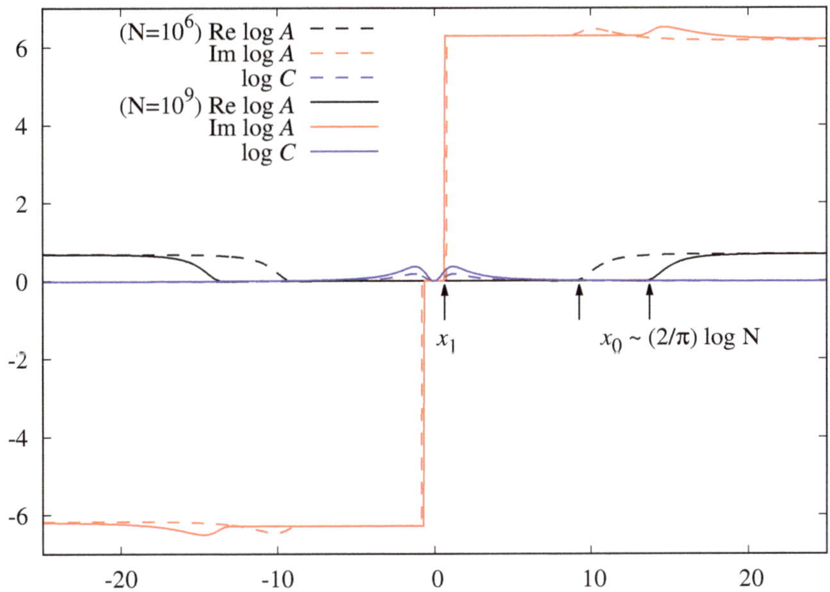

Fig. 3 Plot of $\log A$ and $\log C$ for the ground-state eigenvalue of the spin-$1/2$ XXX chain, with parameters as in Fig. 2, for $N = 10^6$ (dashed) and $N = 10^9$ (solid). The zeros of $\log C$ are at $\pm x_0$, with $x_0 = 9.27..\ (13.72..) \simeq \frac{2}{\pi}\log N$ for $N = 10^6$ (10^9). Unlike Fig. 2, the point x_1, where $\mathrm{Im}\log A$ jumps sharply from 0 to 2π, satisfies $x_1 < x_0$ and appears step-like. Here, $x_1 = 0.78..\ (0.64..)$ for $N = 10^6$ (10^9), showing that x_0 increases and x_1 decreases with N. Within $[-x_0, x_0]$, $\log C$ is positive and large, while outside it is flat and negative. Here *maxima* in $\log C$ align with the transitions of $\mathrm{Im}\log A$ at $\pm x_1$. Note that in the case of shorter chains, *minima* in $\log C$ appeared, see Fig. 2.

identity)

$$a_m(x) := \lim_{N\to\infty} a(x_0\cdot x), \quad \overline{a}_m(x) := \lim_{N\to\infty} \overline{a}(x_0\cdot x), \quad c_m(x) := \lim_{N\to\infty} c(x_0\cdot x), \tag{28}$$

which leads to a well-defined set of NLIEs for all three functions.

We impose that the kinks do not increase with system size, actually their values appear to shrink to the origin. Certainly in the multiplicative scaling limit the kink positions take the value 0 and many other functions in the source terms simplify drastically resulting in the following NLIEs

$$\log a_m(x) = \log a(\infty) + \frac{1}{2}\log\frac{A_m(x)}{\overline{A}_m(x)} - \frac{\mathrm{i}}{x-\mathrm{i}\varepsilon} * \log\frac{C_m}{C(\infty)}, \quad \text{for } x \notin [-1,1],$$
$$\text{else } a_m(x) = 0, \tag{29}$$

$$\log c_m(x) = \log\left(c(\infty)\cdot\frac{x^2-1}{x^2}\right) + \frac{\mathrm{i}}{x+\mathrm{i}\varepsilon} * \log\frac{A_m}{A(\infty)} - \frac{\mathrm{i}}{x-\mathrm{i}\varepsilon} * \log\frac{\overline{A}_m}{\overline{A}(\infty)}, \tag{30}$$

where $a(\infty)$ is the limit of the function $a(x)$ for $x \to \infty$, etc. In the derivation of (30) from (20) we used the limiting functions (28) with an additional simplification for their logarithms. The counter terms in (20) and (22,23), i.e. the rational functions with parameters $x_{r,\pm}$ and $x_{l,\pm}$, take the limits 1. The logarithms, however, of these functions are step functions that take values $\pm 2\pi\mathrm{i}$. Absorbing these constants into the functions $\log A_m$ and $\log \overline{A}_m$ leads to logarithms of A_m and $\overline{A}_m$ on the standard branch. The counter terms in (24) appearing in the denominator simplify to a x^4 such that the second equation in (30) results implying a pole of second order at $x = 0$ for the function $c_m(x)$. Numerical results for (29) and (30) are shown in Fig. 4.

In the second step, but not treated here, we consider the *additive scaling limit* of just the two functions a and $\overline{a}$ and take as input for the function c the data of c_m obtained in the *multiplicative scaling limit*.

Finally we give results for the finite size term of the ground-state energy for a system with angle ϕ between the boundary fields, i.e. $\xi = \tan\phi$ and v being the velocity of elementary excitations

$$E_N - Ne_0 - f_s = -\frac{\pi v}{24N}\left(1 - 6\left(1 - \frac{\phi}{\pi}\right)^2\right),$$

where e_0 is the ground state energy per site (17) and f_s the surface energy (18). This result is still of conjectural nature based on small ϕ data.

6 Conclusion

The XXX spin chain with non-diagonal boundary fields remains a central but challenging model in the study of quantum integrable systems with non-trivial boundaries [13–23, 25–28]. Despite its significance, finding an exact solution has been a longstanding challenge. Significant progress has been achieved through the development of inhomogeneous $T-Q$ relations and the off-diagonal Bethe Ansatz (ODBA) [17, 18] . These advances, along with subsequent extensions [19–23, 25–27], have paved the way for deeper investigations into the model's structure.

In this work, we explored exact non-linear integral equations (NLIEs) for three auxiliary functions derived in [29]. A major contribution is the introduction of the function $c(x)$, which incorporates the inhomogeneous term of the $T-Q$ relation, complementing the classical functions $a(x)$ and $\overline{a}(x)$. This framework captures essential phenomena such as the winding behavior, where $\log A(x)$ and $\log \overline{A}(x)$ exhibit steep increases by $2\pi\mathrm{i}$ at a characteristic scale x_1. Additionally, the function $\log C(x)$ becomes zero at a separate scale x_0. Importantly, our analysis revealed that x_1 and x_0 are independent, with their ratio x_1/x_0 being large for short chains and small for long chains. Numerical results validated these findings for both regimes.

The large-N results facilitated the formulation of a two-step scaling limit for the NLIEs. In the first step, we introduced a multiplicative scaling limit, in which the non-trivial long-range kernel elements are preserved. Unlike the periodic boundary

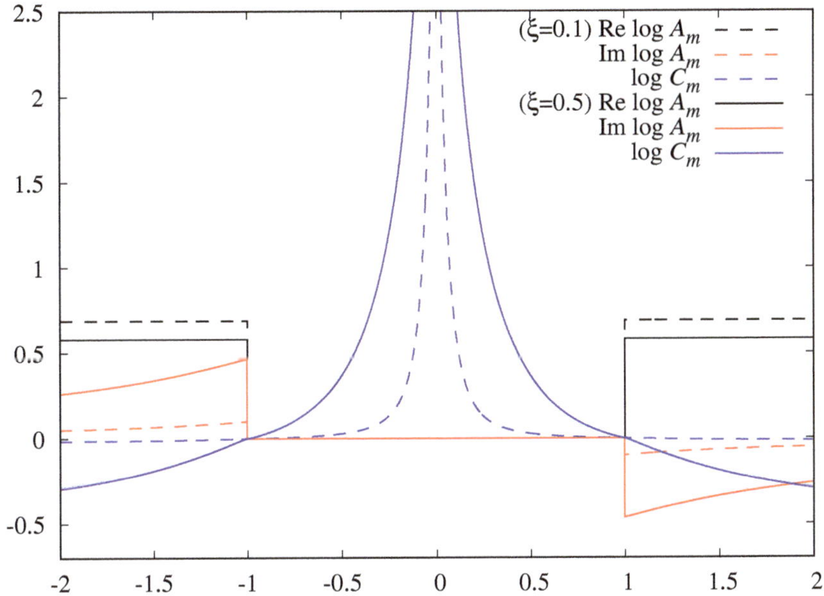

Fig. 4 Plot of the functions $\log A_m$ and $\log C_m$ defined in the multiplicative scaling limit and satisfying the NLIEs (30). The functions depend only on the parameter ξ. Here we show results for two cases, $\xi = 0.1$ and $\xi = 0.5$.

case, where solutions simplify to piecewise constant functions, the open boundary case retains a rich structure of non-trivial solutions. Based on these results, we conjectured a formula for the finite-size corrections to the ground-state energy. The second, additive scaling limit, which finalizes this analysis, will be detailed in future work.

Acknowledgements X.Z. acknowledges financial support from the Alexander von Humboldt Foundation. A.K. acknowledges financial support by Deutsche Forschungsgemeinschaft through FOR 2316 and hospitality by the Innovation Academy for Precision Measurement Science and Technology, Wuhan, A.K.'s work has been funded by the Chinese Academy of Sciences President's International Fellowship Initiative, Grant No. 2024PVA0036.

References

1. H. Bethe, *Zur Theorie der Metalle. I. Eigenwerte und Eigenfunktionen der linearen Atomkette.*, Z. Physik **71** (1931), no. 3/4, 205–26.
2. R. J. Baxter, *Exactly solved models in statistical mechanics*, Academic Press, 1982.
3. N. Andrei, *Diagonalization of the Kondo Hamiltonian*, Phys. Rev. Lett. 45.5 (1980): 379.

4. P. B. Wiegmann, *Exact solution of the sd exchange model (Kondo problem)*, J. Phys. C: Solid State Physics 14.10 (1981): 1463.
5. J. L. Cardy, *Conformal invariance and surface critical behavior*, Nucl. Phys. B 240.4 (1984): 514-532.
6. I. Affleck, and A. W. W. Ludwig, *Exact critical theory of the two-impurity Kondo model*, Phys. Rev. Lett. 68.7 (1992): 1046.
7. P. Zanardi and L. C. Venuti, *Coherent quantum dynamics in steady-state manifolds of strongly dissipative systems*, Phys. Rev. Lett. 113, 240406 (2014).
8. V. Popkov, S. Essink, C. Presilla, and G. M. Schütz, *Effective quantum Zeno dynamics in dissipative quantum systems*, Phys. Rev. A 98, 052110 (2018).
9. F. C. Alcaraz, M. N. Barber, M. T. Batchelor, R. Baxter, and G. Quispel, *Surface exponents of the quantum XXZ, Ashkin-Teller and Potts models*, J. Phys. A: Math. Gen. **20** (1987), 6397.
10. E. K. Sklyanin, *Boundary conditions for integrable quantum systems*, J. Phys. A: Math. Gen. **21** (1988), 2375.
11. H. Asakawa and M. Suzuki, *Finite-size corrections in the XXZ model and the Hubbard model with boundary fields*, Journal of Physics A: Mathematical and General **29** (1996), 225.
12. I. V. Cherednik, *Factorizing particles on a half-line and root systems*, Theor. Math. Phys. **61** (1984), 977–983.
13. H. De Vega and A. González-Ruiz, *Boundary K-matrices for the XYZ, XXZ and XXX spin chains*, J. Phys. A: Math. Gen. **27** (1994), 6129.
14. R.I. Nepomechie, *Solving the open XXZ spin chain with nondiagonal boundary terms at roots of unity*, Nucl. Phys. B **622** (2002), 615.
15. R.I. Nepomechie, *Functional relations and Bethe ansatz for the XXZ chain*, J. Stat. Phys. **111** (2003), 1363.
16. H. Frahm, J. H. Grelik, A. Seel, and T. Wirth, *Functional Bethe ansatz methods for the open XXX chain*, J. Phys. A **44** (2011), 015001.
17. J. Cao, W.-L. Yang, K. Shi, and Y. Wang, *Off-diagonal Bethe ansatz solutions of the anisotropic spin-1/2 chains with arbitrary boundary fields*, Nucl. Phys. B **877** (2013), 152–175.
18. Y. Wang, W.-L. Yang, J. Cao, and K. Shi, *Off-diagonal Bethe ansatz for exactly solvable models*, Springer, 2016.
19. R. I. Nepomechie, *An inhomogeneous T-Q equation for the open XXX chain with general boundary terms: completeness and arbitrary spin*, Journal of Physics A: Mathematical and Theoretical **46** (2013), no. 44, 442002.
20. S. Belliard, *Modified algebraic Bethe ansatz for XXZ chain on the segment – I: Triangular cases*, Nuclear Physics B **892** (2015), 1–20.
21. S. Belliard and R. A. Pimenta, *Modified algebraic Bethe ansatz for XXZ chain on the segment – II: General cases*, Nuclear Physics B **894** (2015), 527–552.
22. N. Crampe, *Algebraic Bethe ansatz for the XXZ Gaudin models with generic boundary*, SIGMA **13** (2017), 094.
23. J. Avan, S. Belliard, N. Grosjean, and R.A. Pimenta, *Modified algebraic Bethe ansatz for XXZ chain on the segment – III – proof*, Nuclear Physics B **899** (2015), 229–246.
24. X. Zhang, A. Klümper, and V. Popkov, *Chiral coordinate Bethe ansatz for phantom eigenstates in the open XXZ spin-1/2 chain*, Physical Review B **104** (2021), 195409.
25. H. Frahm, A. Seel, and T. Wirth, *Separation of Variables in the open XXX chain*, Nucl. Phys. B **802** (2008), 351–367.
26. G. Niccoli, *Non-diagonal open spin-1/2 XXZ quantum chains by separation of variables: complete spectrum and matrix elements of some quasi-local operators*, Journal of Statistical Mechanics: Theory and Experiment **2012** (2012), P10025.
27. S. Faldella, N. Kitanine, and G. Niccoli, *The complete spectrum and scalar products for the open spin-1/2 XXZ quantum chains with non-diagonal boundary terms*, Journal of Statistical Mechanics: Theory and Experiment **2014** (2014), P01011.
28. J.-S. Dong, P. Lu, P. Sun, Y. Qiao, J. Cao, K. Hao, and W.-L. Yang, *Exact surface energy and elementary excitations of the XXX spin-1/2 chain with arbitrary non-diagonal boundary fields*, Chinese Physics B **32** (2023), 017501.

29. Holger Frahm, Andreas Klümper, Dennis Wagner, and Xin Zhang, *Non-linear integral equations for the XXX spin-1/2 quantum chain with non-diagonal boundary fields*, arXiv:2502.07229
30. A. Klümper and M. T. Batchelor, *An analytic treatment of finite-size corrections of the spin-1 antiferromagnetic XXZ chain*, J. Phys. A **23** (1990), no. 5, L189–95.
31. Andreas Klümper, Murray T. Batchelor, and Paul A. Pearce, *Central charges of the 6- and 19-vertex models with twisted boundary conditions*, J. Phys. A **24** (1991), no. 13, 3111–33.
32. A. Klümper, *Free energy and correlation length of quantum chains related to restricted solid-on-solid lattice models*, Ann. Physik **1** (1992), 540.
33. C. Destri and H. J. de Vega, *New thermodynamic Bethe ansatz equations without strings*, Phys. Rev. Lett. **69** (1992), 2313.
34. A. Klümper, *Thermodynamics of the anisotropic spin-1/2 Heisenberg chain and related quantum chains*, Z. Phys. B **91** (1993), 507.
35. C. Destri and H. J. de Vega, *Unified approach to thermodynamic Bethe ansatz and finite size corrections for lattice models and field theories*, Nucl. Phys. B **438 [FS]** (1995), 413.
36. J. Suzuki, *Spinons in magnetic chains of arbitrary spins at finite temperatures*, J. Phys. A **32** (1999), 2341.
37. G. Jüttner and A. Klümper, *Exact calculation of thermodynamical quantities of the integrable $t-J$ model*, Europhys. Lett. **37** (1997), no. 5, 335–340.
38. G. Jüttner, A. Klümper, and J. Suzuki, *Exact thermodynamics and Luttinger liquid properties of the integrable $t-J$ model*, Nucl. Phys. B **487** (1997), no. 3, 650–674.

Ocneanu algebra of seams: Critical unitary E_6 RSOS lattice model

Paul A. Pearce and Jørgen Rasmussen

Abstract We consider the A series and exceptional E_6 Restricted Solid-On-Solid lattice models as prototypical examples of the critical Yang-Baxter integrable two-dimensional A-D-E lattice models. We focus on type I theories which are characterized by the existence of an extended chiral symmetry in the continuum scaling limit. Starting with the commuting family of column transfer matrices on the torus, we build matrix representations of the Ocneanu graph fusion algebra as integrable seams for arbitrary finite-size lattices. In the A cases, the Ocneanu algebra coincides with the Verlinde algebra. In the other cases, the seam algebra contains the fused adjacency and graph fusion algebras as subalgebras. Our matrix representation of the Ocneanu algebra encapsulates the quantum symmetry of the commuting family of transfer matrices. In the continuum scaling limit, the integrable seams realize the topological defects of the associated conformal field theory and the known toric matrices encode the twisted conformal partition functions of Petkova and Zuber.

1 Introduction

In this short article, we are concerned with two-dimensional critical Restricted Solid-On-Solid (RSOS) lattice models built on the Dynkin diagrams of simply-laced $sl(2)$ A-D-E Lie algebras as in Figure 1. These models are Yang-Baxter

Paul A. Pearce
School of Mathematics and Statistics, University of Melbourne, Parkville, VIC 3010, Australia; and
School of Mathematics and Physics, University of Queensland, St Lucia, QLD 4072, Australia
e-mail: papearce@unimelb.edu.au

Jørgen Rasmussen
School of Mathematics and Physics, University of Queensland, St Lucia, QLD 4072, Australia
e-mail: j.rasmussen@uq.edu.au

D. R. Wood et al. (eds.), *2024 MATRIX Annals, Part II*, MATRIX Book Series 8,
https://doi.org/10.1007/978-3-032-16206-9_8

integrable [1]. The first such models were built on the A-type Dynkin diagrams and solved off-criticality in 1984 by Andrews, Baxter and Forrester [2]. The full family of critical A-D-E lattice models was subsequently introduced and studied by Pasquier [3–5]. In the continuum scaling limit, the thermodynamic behaviour of these models is described by Conformal Field Theory (CFT) [6]. The A series is associated with the conformal unitary minimal models $\mathcal{M}(g-1,g)$ of Belavin, Polyakov and Zamolodchikov [7]. The unitary A-D-E CFTs are in fact (A,G) coset models [8, 6]

$$(A_{g-2},G) = \begin{cases} (A_{g-2},A_{g-1}), & g=4,5,6,\dots \\ (A_{g-2},D_{(g+2)/2}), & g=6,8,10,\dots \\ (A_{10},E_6), & g=12 \\ (A_{16},E_7), & g=18 \\ (A_{28},E_8), & g=30 \end{cases} \qquad c = 1-\frac{6}{g(g-1)} \tag{1.1}$$

where g is the Coxeter number of G and c is the central charge. These theories are rational [9], admitting a finite number of irreducible representations of the Virasoro algebra which close under fusion.

Much effort within CFT has been focussed on the study of boundary conditions. On the cylinder, the boundary conditions of unitary minimal CFTs and their conformal cylinder partition functions are understood [10, 11] in terms of nonnegative integer matrix representations (nimreps) of the Verlinde [12] and graph fusion [4] algebras. The irreducible representations are conjugate to conformal boundary conditions [13] labelled by the coset graph nodes $(r,a) \in (A,G)$. Much of this structure is common [14, 15] between the finite lattice models and the associated CFT. On the torus with periodic boundary conditions in both directions, the modular invariant partition functions [16] with $c<1$ are exhausted by the A-D-E classification of Cappelli, Itzykson and Zuber [17]. Incorporating the left and right chiral copies of the Virasoro algebra, the conformal partition functions are given as sesquilinear forms in Virasoro characters. Twisted boundary conditions are imposed by inserting topological defect seams into the bulk. The resulting twisted partition functions on the torus are understood in terms of the Ocneanu [18] fusion algebra which describes the quantum symmetry and how the left and right chiral halves of the theory are glued together. Indeed, working within CFT, Petkova and Zuber [19] have obtained explicit expressions for the toric matrices and the associated twisted conformal partition functions. The algebraic structure behind these results has been elucidated by Coquereaux and collaborators [20–23] who, following Ocneanu [18], identified the Ocneanu algebra with the quotient algebra $G\otimes_T G := (G\otimes G)/T$ where G denotes the graph algebra of type A,D,E and where T is the ambichiral subalgebra of G specified in [19]. In particular, the authors of [20, 21] give a detailed account of the prototypical type I example of E_6.

In parallel to the developments coming from CFT, twisted boundary conditions on the torus have also been studied from a purely lattice perspective [24–26] by inserting integrable seam segments into the commuting *row* transfer matrices.

These segments give rise to commuting integrable *vertical* seams on the square lattice which, in the continuum scaling limit, become the topological defects of the associated CFT. In this picture, the fusion of the integrable seams is simply implemented by matrix products of the integrable seams. This picture was implicitly confirmed in [26] by numerical studies of the conformal spectra of the commuting row transfer matrices.

Lie Algebra	Graph G	g	$\mathrm{Exp}(G)$	Type/H	Γ
A_L	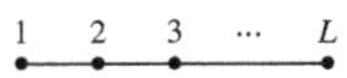	$L+1$	$1,2,\ldots,L$	I	$\mathbb{Z}_2$
$D_L\,(L$ even$)$	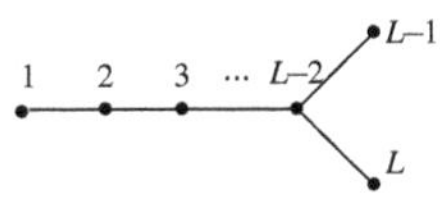	$2L-2$	$1,3,\ldots,2L-3,L-1$	I	$\mathbb{Z}_2/\mathbb{S}_3$
$D_L\,(L$ odd$)$	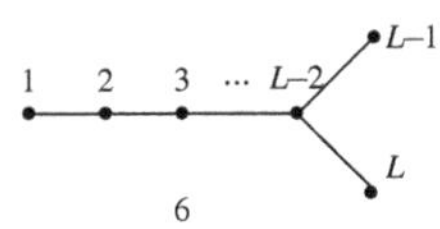	$2L-2$	$1,3,\ldots,2L-3,L-1$	II/A_{2L-3}	$\mathbb{Z}_2$
E_6	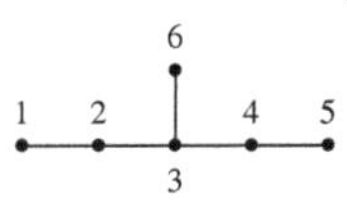	12	$1,4,5,7,8,11$	I	$\mathbb{Z}_2$
E_7	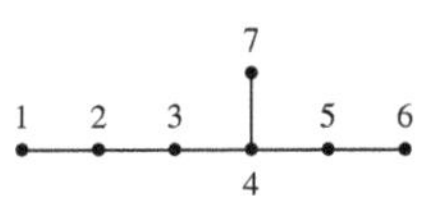	18	$1,5,7,9,11,13,17$	II/D_{10}	1
E_8	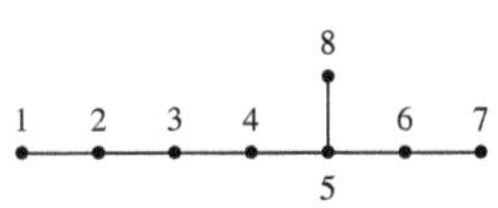	30	$1,7,11,13,17,19,23,29$	I	1

Fig. 1 Dynkin diagrams of the classical simply-laced $sl(2)$ *A-D-E* Lie algebras. The nodes associated with the identity and the fundamental are labelled 1 and 2 respectively. The fundamental is the unique neighbour of the identity. Also shown are the Coxeter numbers g, exponents $\mathrm{Exp}(G)$, the type I or II and the so-called parent graphs $H \neq G$. The diagram automorphism group Γ is generated by a single $\mathbb{Z}_2$ automorphism σ. The D_4 graph is an exception having the noncommutative automorphism group $\mathbb{S}_3$. The eigenvalues of G are $2\cos\frac{s\pi}{g}$ with $s \in \mathrm{Exp}(G)$. By abuse of notation, we use G to denote the graph, the set of its vertices with cardinality $|G|$ and its adjacency matrix so that $2I - G$ is the Cartan matrix. The meaning of G should be clear from context.

The aim of this article is to develop further the lattice approach to fusion of integrable seams on the lattice. As in [27, 28], we turn our attention to the "cross-channel", but here we use the commuting family of column transfer matrices to build the various types of integrable seams on the torus. Modifying the notation of [19, 26], the composite integrable seams are labelled by $x = (r, \kappa, a, b)$. The associated *column* transfer matrices are constructed by matrix products of elementary seam transfer matrices

$$\mathbf{T}_x(u) = \mathbf{T}_r(u,\xi)\,\sigma^\kappa\,\widehat{\mathbf{N}}_a\overline{\widehat{\mathbf{N}}}_b, \quad x=(r,\kappa,a,b), \quad r\in A_{g-2}, \quad \kappa=0,1, \quad a,b\in G \tag{1.2}$$

as shown in Figure 2. Here $\mathbf{T}_r(u,\xi)$ is the column transfer matrix for an r-type seam, ξ is a column inhomogeneity, $\widehat{\mathbf{N}}_a, \overline{\widehat{\mathbf{N}}}_b$ are graph fusion seams labelled by $a,b\in G$ (as defined in Section 4) and $\sigma^\kappa = \mathbf{I}$ or σ where the σ-parity $\kappa = 0,1$ labels respectively the identity $\mathbf{I}$ and the $\mathbb{Z}_2$ graph automorphism seam σ. The seam labels κ do not need to be explicitly added in x except for the D_L (L even) series of lattice models.

As pointed out in [26], due to the coset factorization structure [29], the r-type seams (with $r\in A_{g-2}$) play a somewhat trivial role. We will therefore usually set $r=1$ (no r-type seam) and focus on the so-called Wess-Zumino-Witten (WZW) factor associated with the graph G. For $x=(a,b)$, we assert that the integrable seams $\widehat{\mathbf{P}}_{a,b} = \widehat{\mathbf{N}}_a\overline{\widehat{\mathbf{N}}}_b$ yield matrix representations of the Ocneanu fusion algebra for arbitrary system sizes. In particular, for the critical A-D-E lattice models, the Ocneanu algebra encodes the quantum symmetries possessed by the commuting families of column transfer matrices on a finite lattice. Following [20], we focus here on the A series and the prototypical type I example of E_6. Our full findings for the D and other E critical RSOS lattice models will appear elsewhere.

2 Critical RSOS models on the torus

2.1 Critical A-D-E lattice models

Defining quantum dimensions by $S_a = [a]_x = \frac{x^a - x^{-a}}{x - x^{-1}}$ with $x = e^{i\lambda}$, the nondegenerate largest eigenvalue of the adjacency matrix G is $[2]_x = 2\cos\lambda$ and the associated (unnormalized) Perron-Frobenius eigenvector $\psi = (\psi_a)_{1\le a\le |G|}$ is

$$G\psi = [2]_x\,\psi, \qquad \psi = \begin{cases} \big([1]_x,[2]_x,\ldots,[L]_x\big), & G=A_L\\ \big([1]_x,[2]_x,\ldots,[\ell]_x,\frac{[\ell]_x}{[2]_x},\frac{[\ell]_x}{[2]_x}\big), & G=D_{\ell+2}\\ \big([1]_x,[2]_x,[3]_x,[2]_x,[1]_x,\frac{[3]_x}{[2]_x}\big), & G=E_6\\ \big([1]_x,[2]_x,[3]_x,[4]_x,\frac{[6]_x}{[2]_x},\frac{[4]_x}{[3]_x},\frac{[4]_x}{[2]_x}\big), & G=E_7\\ \big([1]_x,[2]_x,[3]_x,[4]_x,[5]_x,\frac{[7]_x}{[2]_x},\frac{[5]_x}{[3]_x},\frac{[5]_x}{[2]_x}\big), & G=E_8 \end{cases} \tag{2.1}$$

The allowed face weights of the critical A-D-E lattice models are then given by

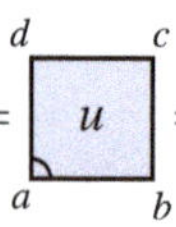

$$W\!\left(\begin{matrix}d & c\\ a & b\end{matrix}\,\middle|\,u\right) = \boxed{u} = \frac{\sin(\lambda-u)}{\sin\lambda}\,\delta_{ac} + \frac{\sin u}{\sin\lambda}\sqrt{\frac{\psi_a\psi_c}{\psi_b\psi_d}}\,\delta_{bd}, \quad \lambda=\frac{\pi}{g}, \quad 0\le u\le\lambda \tag{2.2}$$

where u is the spectral parameter, λ is the crossing parameter and the face orientation is marked by an arc in the bottom-left corner. It is understood that the

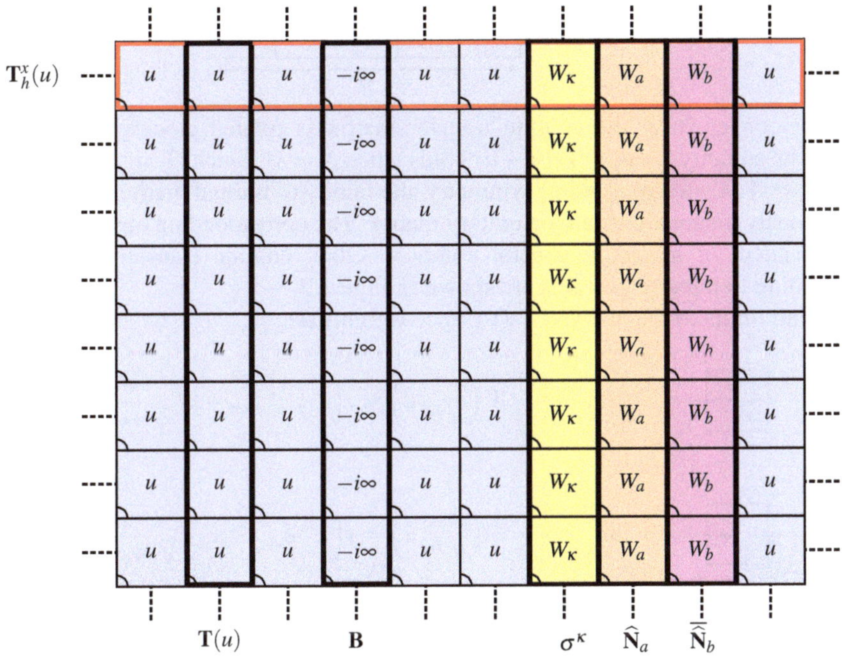

Fig. 2 An $N \times M$ periodic lattice with $(N,M) = (10,8)$ showing (i) the row transfer matrix $\mathbf{T}_h^x(u)$ with seam segment $x = (\kappa, a, b)$, (ii) the column/seam transfer matrices $\mathbf{T}(u)$, $\mathbf{B} = \mathbf{n}_2$, $\widehat{\mathbf{N}}_a$ and $\widehat{\widehat{\mathbf{N}}}_b$ as explained in Section 4 and (iii) the $\mathbb{Z}_2$ diagram automorphism seam σ^κ. The composite seam is the matrix product $\mathbf{T}_x = \sigma^\kappa \widehat{\mathbf{N}}_a \widehat{\widehat{\mathbf{N}}}_b$. The seam $\mathbf{T}_x$ has the same twisted partition function as the seam $\widehat{\mathbf{N}}_a \widehat{\widehat{\mathbf{N}}}_b \sigma^\kappa$. We assume $r = 1$ so there is no r-type seam shown. The labels W_κ, W_a and W_b indicate that special face weights are assigned to these faces.

face weights vanish if $G_{ab}G_{bc}G_{cd}G_{da} = 0$. The face weights are invariant under reflections in the two diagonals and satisfy the crossing symmetry

$$W\begin{pmatrix} d & c \\ a & b \end{pmatrix}\!\!\left. \right| u \Big) = \boxed{u}_{\,a\,b}^{\,d\,c} = \sqrt{\frac{\psi_a \psi_c}{\psi_b \psi_d}}\; \boxed{\lambda - u}_{\,a\,b}^{\,d\,c} = \sqrt{\frac{\psi_a \psi_c}{\psi_b \psi_d}}\; W\begin{pmatrix} c & b \\ d & a \end{pmatrix}\!\!\left. \right| \lambda - u \Big) \tag{2.3}$$

2.2 Transfer matrices, braid limit and Hamiltonians

The row and column families of commuting transfer matrices $\mathbf{T}_h(u), \mathbf{T}(u)$ of the critical A-D-E models on an $N \times M$ periodic lattice are defined by

$$\mathbf{T}_h(u) = \cdots \underbrace{\boxed{u}\,\boxed{u}\,\boxed{u}\,\boxed{\cdots}\,\boxed{u}}_{N} \cdots \tag{2.4}$$

$$\mathbf{T}(u) = \mathbf{T}_v(u) = \tag{2.5}$$

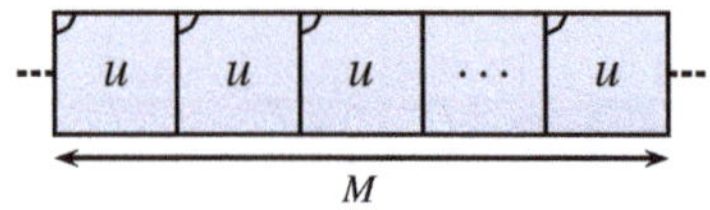

where, for convenience, the column transfer matrix is rotated clockwise by 90 degrees. The equality $\mathbf{T}_v(u) = \mathbf{T}_h(\lambda - u)$ holds when $N = M$. Since $[\mathbf{T}(u), \mathbf{T}(v)] = \mathbf{0}$ and $\mathbf{T}(u)^T = \mathbf{T}(\lambda - u)$ by crossing symmetry, the family of normal matrices $\mathbf{T}(u)$ is simultaneously diagonalizable by a unitary matrix. The corresponding eigenvectors are independent of u and in general complex. Other column transfer matrices corresponding to integrable seams are shown in Figure 2.

The braid limits of the allowed A-D-E face weights are

$$B\begin{pmatrix} d & c \\ a & b \end{pmatrix} =$$

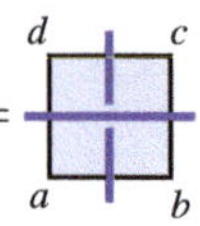

$$= -\lim_{u \to -i\infty} \frac{x^{\frac{1}{2}}}{\rho(u)} W\left(\begin{matrix} d & c \\ a & b \end{matrix}\middle| u\right) = i\left(-x^{-\frac{1}{2}}\,\delta_{ac} + x^{\frac{1}{2}}\sqrt{\frac{\psi_a\psi_c}{\psi_b\psi_d}}\,\delta_{bd}\right) \tag{2.6a}$$

$$\overline{B}\begin{pmatrix} d & c \\ a & b \end{pmatrix} =$$

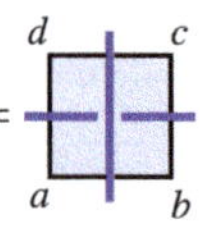

$$= -\lim_{u \to i\infty} \frac{x^{-\frac{1}{2}}}{\rho(u)} W\left(\begin{matrix} d & c \\ a & b \end{matrix}\middle| u\right) = i\left(x^{\frac{1}{2}}\,\delta_{ac} - x^{-\frac{1}{2}}\sqrt{\frac{\psi_a\psi_c}{\psi_b\psi_d}}\,\delta_{bd}\right) \tag{2.6b}$$

where $x = e^{i\lambda}$ and $\rho(u) = \sin(\lambda - u)\sin(\lambda + u)$. These braid limits relate to the left and right chiral halves of the theory and are related to each other by complex conjugation. Taking the action of the face transfer operators $X_j(u)$ to be diagonal and using the Temperley-Lieb generators e_j, these expressions become

$$b_j =$$

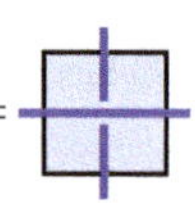

$$= -\lim_{u \to -i\infty} \frac{x^{\frac{1}{2}}}{\rho(u)} X_j(u) = i\left(-x^{-\frac{1}{2}} I + x^{\frac{1}{2}} e_j\right) \tag{2.7a}$$

$$\overline{b}_j =$$

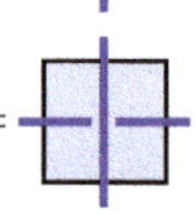

$$= -\lim_{u \to i\infty} \frac{x^{-\frac{1}{2}}}{\rho(u)} X_j(u) = i\left(x^{\frac{1}{2}} I - x^{-\frac{1}{2}} e_j\right) = b_j^{-1} \tag{2.7b}$$

The Hamiltonian limit of the horizontal transfer matrix $\mathbf{T}_h(u)$ (2.4) gives rise to periodic anyonic chains

$$\mathcal{H} = -\sin\lambda \,\frac{d}{du} \log \mathbf{T}_h(u)\Big|_{u=0} = -\sum_{j=1}^{N} e_j \tag{2.8}$$

The simplest case is the Hamiltonian limit of the tricritical Ising A_4 RSOS model [30] corresponding to the Fibonacci golden chain [31,32].

3 Ocneanu fusion algebra

Most of the content in this section is known [18–23]. We review some material to establish notation and to develop a more economical picture using new smaller extended toric matrices rather than the known larger (unextended) toric matrices. This extended picture holds for type I and also type II cases when the parent graph H is of type I.

3.1 Fusion matrices and structure constants

The fused adjacency matrices (intertwiners) n_i and graph fusion matrices $\widehat{N}_a$ are defined by the finitely truncated recursions

$$n_1 = I, \quad n_2 = G, \quad n_i n_2 = n_{i-1} + n_{i+1}, \quad n_g = 0, \qquad i \in A_{g-1} \tag{3.1}$$

$$\widehat{N}_1 = I, \quad \widehat{N}_2 = G, \quad G\widehat{N}_a = \sum_{b\in G} G_{ab}\widehat{N}_b, \qquad a \in G \tag{3.2}$$

For $G = A_{g-1}$, the fused adjacency matrices n_i are in fact the Verlinde [12] fusion matrices N_i. The fused adjacency and graph fusion matrices satisfy the commutative associative fusion algebras

$$n_i n_j = \sum_{k=1}^{g-1} N_{ij}{}^k n_k, \qquad \widehat{N}_a\widehat{N}_b = \sum_{c\in G} \widehat{N}_{ab}{}^c \widehat{N}_c, \qquad n_i\widehat{N}_a = \sum_{b\in G} n_{ia}{}^b \widehat{N}_b \tag{3.3}$$

where the structure constants are

$$N_{ij}{}^k = (N_i)_j{}^k \in \mathbb{N}, \;\; \widehat{N}_{ab}{}^c = (\widehat{N}_a)_b{}^c \in \begin{cases} \mathbb{N}, & \text{type I} \\ \mathbb{Z}, & \text{type II} \end{cases}, \quad n_{ia}{}^b = (n_i)_a{}^b \in \mathbb{N} \tag{3.4}$$

For type I theories, the N_i, $\widehat{N}_a$ and n_i form nimreps (nonnegative integer matrix representations) of these fusion algebras. The type II theories do not admit proper graph fusion algebras.

Denoting the complex unitary matrices of eigenvectors of N_2, $\widehat{N}_2$ by $S_i{}^\ell$ (modular matrix) and $\Psi_a{}^\ell$ with $S_1{}^\ell, \Psi_1{}^\ell > 0$, the corresponding Verlinde and Verlinde-type formulas are

$$N_{ij}{}^k = \sum_{\ell=1}^{g-1} \frac{S_i{}^\ell S_j{}^\ell \overline{S_k{}^\ell}}{S_1{}^\ell}, \quad \widehat{N}_{ab}{}^c = \sum_{\ell\in\text{Exp}(G)} \frac{\Psi_a{}^\ell \Psi_b{}^\ell \overline{\Psi_c{}^\ell}}{\Psi_1{}^\ell}, \quad n_{ia}{}^b = \sum_{\ell\in\text{Exp}(G)} \frac{S_i{}^\ell \Psi_a{}^\ell \overline{\Psi_b{}^\ell}}{S_1{}^\ell} \tag{3.5}$$

Here bars denote complex conjugation and the sum in the first expression is over $\ell \in \text{Exp}(A_{g-1})$. For structure constants larger than 1, it is necessary to introduce so-called bond variables (fusion graphs with multiple edges). Explicit fusion adjacency matrices for A_4 and E_6 are

$$N_1=\begin{pmatrix}1&0&0&0\\0&1&0&0\\0&0&1&0\\0&0&0&1\end{pmatrix},\quad N_2=\begin{pmatrix}0&1&0&0\\1&0&1&0\\0&1&0&1\\0&0&1&0\end{pmatrix},\quad N_3=\begin{pmatrix}0&0&1&0\\0&1&0&1\\1&0&1&0\\0&1&0&0\end{pmatrix},\quad N_4=\begin{pmatrix}0&0&0&1\\0&0&1&0\\0&1&0&0\\1&0&0&0\end{pmatrix} \tag{3.6}$$

$$\widehat{N}_1=\begin{pmatrix}1&0&0&0&0&0\\0&1&0&0&0&0\\0&0&1&0&0&0\\0&0&0&1&0&0\\0&0&0&0&1&0\\0&0&0&0&0&1\end{pmatrix},\quad \widehat{N}_2=\begin{pmatrix}0&1&0&0&0&0\\1&0&1&0&0&0\\0&1&0&1&0&1\\0&0&1&0&1&0\\0&0&0&1&0&0\\0&0&1&0&0&0\end{pmatrix},\quad \widehat{N}_3=\begin{pmatrix}0&0&1&0&0&0\\0&1&0&1&0&1\\1&0&2&0&1&0\\0&1&0&1&0&1\\0&0&1&0&0&0\\0&1&0&1&0&0\end{pmatrix}$$
$$\widehat{N}_4=\begin{pmatrix}0&0&0&1&0&0\\0&0&1&0&1&0\\0&1&0&1&0&1\\1&0&1&0&0&0\\0&1&0&0&0&0\\0&0&1&0&0&0\end{pmatrix},\quad \widehat{N}_5=\begin{pmatrix}0&0&0&0&1&0\\0&0&0&1&0&0\\0&0&1&0&0&0\\0&1&0&0&0&0\\1&0&0&0&0&0\\0&0&0&0&0&1\end{pmatrix},\quad \widehat{N}_6=\begin{pmatrix}0&0&0&0&0&1\\0&0&1&0&0&0\\0&1&0&1&0&0\\0&0&1&0&0&0\\0&0&0&0&0&1\\1&0&0&0&1&0\end{pmatrix} \tag{3.7}$$

where N_4 and $\widehat{N}_5$ encode the $\mathbb{Z}_2$ graph automorphism of A_4 and E_6. For type I theories, there exists an extended conformal chiral algebra with extended Virasoro characters encoded by the fundamental intertwiner $C=(n_{i1}{}^a)$. Accordingly, for the type I graphs, we have

$$n_i=\sum_{a\in G} n_{i1}{}^a\widehat{N}_a, \qquad \widehat{\chi}_{r,a}(q)=\sum_{s=1}^{g-1} n_{s1}{}^a\chi_{r,s}(q) \tag{3.8}$$

The first of these decompositions follows by specializing the compatibility condition in (3.3). For E_6:

$$C=(n_{i1}{}^a)=\begin{pmatrix}1&0&0&0&0&0\\0&1&0&0&0&0\\0&0&1&0&0&0\\0&0&0&1&0&1\\0&0&1&0&1&0\\0&1&0&1&0&0\\1&0&1&0&0&0\\0&1&0&0&0&1\\0&0&1&0&0&0\\0&0&0&1&0&0\\0&0&0&0&1&0\end{pmatrix};\qquad \begin{aligned}\widehat{\chi}_{r,1}(q)&=\chi_{r,1}(q)+\chi_{r,7}(q)\\ \widehat{\chi}_{r,2}(q)&=\chi_{r,2}(q)+\chi_{r,6}(q)+\chi_{r,8}(q)\\ \widehat{\chi}_{r,3}(q)&=\chi_{r,3}(q)+\chi_{r,5}(q)+\chi_{r,7}(q)+\chi_{r,9}(q)\\ \widehat{\chi}_{r,4}(q)&=\chi_{r,4}(q)+\chi_{r,6}(q)+\chi_{r,10}(q)\\ \widehat{\chi}_{r,5}(q)&=\chi_{r,5}(q)+\chi_{r,11}(q)\\ \widehat{\chi}_{r,6}(q)&=\chi_{r,4}(q)+\chi_{r,8}(q)\end{aligned} \tag{3.9}$$

3.2 Twisted conformal partition functions

The twisted partition functions of the *A*-*D*-*E* theories are given by Petkova and Zuber [19]. For the *A* series, with $i,j,k\in A_{g-1}$, the Ocneanu fusion matrices are simply the Verlinde fusion matrices N_i with conformal partition functions

$$Z_i(q)=\sum_{j,k\in A_{g-1}} N_{ij}{}^k\chi_j(q)\chi_k(\bar{q}) \tag{3.10}$$

More generally, for (A_{g-2}, G) cosets with G of type I, the twisted partition functions are given by the sesquilinear forms

$$Z^{(r,a,b,\kappa)}(q) = \sum_{(r',s'),(r'',s'')} N_{r,r'}{}^{r''} [P_{a,b}^{(\kappa)}]_{s's''} \chi_{r',s'}(q)\, \chi_{r'',s''}(\overline{q}) \tag{3.11a}$$

$$= \sum_{(r',a'),(r'',a'')} N_{r,r'}{}^{r''} [\widehat{P}_{a,b}^{(\kappa)}]_{a'a''} \widehat{\chi}_{r',a'}(q)\, \widehat{\chi}_{r'',a''}(\overline{q}) \tag{3.11b}$$

where the second form is new. Here $r, r', r'' \in A_{g-2}$; $s, s', s'' \in A_{g-1}$; $a, a', a'', b \in G$ and the *toric* and *extended toric* matrices are

$$[P_{a,b}^{(\kappa)}]_{s's''} = \sum_{c\in T_\kappa} n_{s'a}{}^{c}\, n_{s''b}{}^{c}, \qquad [\widehat{P}_{a,b}^{(\kappa)}]_{a'a''} = \sum_{c\in T_\kappa} \widehat{N}_{ac}{}^{a'}\, \widehat{N}_{bc}{}^{a''}, \qquad \kappa = 0,1 \tag{3.12}$$

The label $\kappa = 0,1$ is only needed for the D_L (L even) series and indicates the absence or presence of the $\mathbb{Z}_2$ automorphism. The *ambichiral vertices* T_0 (common to both left and right chiral subalgebras) and T_1 are

$$T_0 = \begin{cases} \{1,2,\ldots,L\}, & G = A_L \\ \{1,3,5,\ldots,2\ell-1,2\ell\}, & G = D_{2\ell} \\ \{1,5,6\}, & G = E_6 \\ \{1,7\}, & G = E_8 \end{cases} \qquad T_1 = \begin{cases} \{2,4,\ldots,2\ell-2\}, & G = D_{2\ell} \\ T_0, & \text{otherwise} \end{cases} \tag{3.13}$$

3.3 *Extended toric matrices and quantum symmetry*

In the following, we fix $r = 1$ and consider type 1 theories with $\kappa = 0$. Taking $x = (a,b)$, the extended toric matrices can be written as

$$\widehat{P}_{a,b} = (\widehat{N}_a)^T \widehat{N}_b, \qquad x = (a,b) \tag{3.14}$$

where the superscript T denotes transpose and, in accord with the tensor structure $G \otimes_{T_0} G$, the rows of the rectangular matrices $\widehat{N}_a$ are restricted to those labelled by the vertices of T_0. Alternatively, working with the full $|G| \times |G|$ matrices, the extended toric matrices are given by the bilinear map

$$\widehat{P}_{a,b} = (\widehat{N}_a, \widehat{N}_b) = \widehat{N}_a \widehat{P} \widehat{N}_b, \qquad (L,R) = L\widehat{P}R = L \otimes_{T_0} R, \qquad \widehat{P} = \widehat{P}_{1,1} = (I,I) \tag{3.15}$$

The projection $\widehat{P} = \widehat{P}_{1,1}$ joins the left and right chiral components through the ambichiral nodes T_0.

For E_6, the 6×6 array of extended toric matrices $\widehat{P}_{a,b}$ is shown in Figure 3. In accord with the quantum symmetry exhibited in Figure 3 and (3.17), the 36 extended toric matrices lie in a convex cone $\mathbb{K}$ consisting of nonnegative-integer linear (conical) combinations of the (ordered) Hilbert basis [33]

$$\{\widehat{P}_1,\widehat{P}_2,\ldots,\widehat{P}_{12}\}=\{\widehat{P}_{1,1},\widehat{P}_{2,1},\widehat{P}_{3,1},\widehat{P}_{4,1},\widehat{P}_{5,1},\widehat{P}_{6,1},\widehat{P}_{1,2},\widehat{P}_{2,2},\widehat{P}_{3,2},\widehat{P}_{4,2},\widehat{P}_{5,2},\widehat{P}_{6,2}\} \tag{3.16}$$

The 12 basis elements are in bijection with the vertices of the Ocneanu graph in Figure 4. They are conjugate to the 12 E_6 irreducible representations of $\mathrm{Vir}\otimes\overline{\mathrm{Vir}}$. The remaining 24 extended toric matrices are expressed as the conical linear combinations

$$\widehat{P}_{5,5}=\widehat{P}_{1,1},\qquad \widehat{P}_{4,5}=\widehat{P}_{2,1},\qquad \widehat{P}_{2,6}=\widehat{P}_{3,5}=\widehat{P}_{4,6}=\widehat{P}_{3,1} \tag{3.17a}$$

$$\widehat{P}_{2,5}=\widehat{P}_{4,1},\qquad \widehat{P}_{1,5}=\widehat{P}_{5,1},\qquad \widehat{P}_{1,6}=\widehat{P}_{5,6}=\widehat{P}_{6,5}=\widehat{P}_{6,1} \tag{3.17b}$$

$$\widehat{P}_{5,4}=\widehat{P}_{1,2},\qquad \widehat{P}_{4,4}=\widehat{P}_{2,2},\qquad \widehat{P}_{2,3}=\widehat{P}_{3,4}=\widehat{P}_{4,3}=\widehat{P}_{3,2} \tag{3.17c}$$

$$\widehat{P}_{2,4}=\widehat{P}_{4,2},\qquad \widehat{P}_{1,4}=\widehat{P}_{5,2},\qquad \widehat{P}_{1,3}=\widehat{P}_{5,3}=\widehat{P}_{6,4}=\widehat{P}_{6,2} \tag{3.17d}$$

$$\widehat{P}_{3,3}=\widehat{P}_{2,2}+\widehat{P}_{4,2},\quad \widehat{P}_{3,6}=\widehat{P}_{2,1}+\widehat{P}_{4,1},\quad \widehat{P}_{6,3}=\widehat{P}_{1,2}+\widehat{P}_{5,2},\quad \widehat{P}_{6,6}=\widehat{P}_{1,1}+\widehat{P}_{5,1} \tag{3.17e}$$

$$\begin{pmatrix}1&0&0&0&0&0\\0&0&0&0&0&0\\0&0&0&0&0&0\\0&0&0&0&0&0\\0&0&0&0&1&0\\0&0&0&0&0&1\end{pmatrix}\begin{pmatrix}0&1&0&0&0&0\\0&0&0&0&0&0\\0&0&0&0&0&0\\0&0&0&0&0&0\\0&0&0&1&0&0\\0&0&1&0&0&0\end{pmatrix}\begin{pmatrix}0&0&1&0&0&0\\0&0&0&0&0&0\\0&0&0&0&0&0\\0&0&0&0&0&0\\0&0&1&0&0&0\\0&1&0&1&0&0\end{pmatrix}\begin{pmatrix}0&0&0&1&0&0\\0&0&0&0&0&0\\0&0&0&0&0&0\\0&0&0&0&0&0\\0&1&0&0&0&0\\0&0&1&0&0&0\end{pmatrix}\begin{pmatrix}0&0&0&0&1&0\\0&0&0&0&0&0\\0&0&0&0&0&0\\0&0&0&0&0&0\\1&0&0&0&0&0\\0&0&0&0&0&1\end{pmatrix}\begin{pmatrix}0&0&0&0&0&1\\0&0&0&0&0&0\\0&0&0&0&0&0\\0&0&0&0&0&0\\0&0&0&0&0&1\\1&0&0&0&1&0\end{pmatrix}$$

$$\begin{pmatrix}0&0&0&0&0&0\\1&0&0&0&0&0\\0&0&0&0&0&1\\0&0&0&0&1&0\\0&0&0&0&0&0\\0&0&0&0&0&0\end{pmatrix}\begin{pmatrix}0&0&0&0&0&0\\0&1&0&0&0&0\\0&0&1&0&0&0\\0&0&0&1&0&0\\0&0&0&0&0&0\\0&0&0&0&0&0\end{pmatrix}\begin{pmatrix}0&0&0&0&0&0\\0&0&1&0&0&0\\0&1&0&1&0&0\\0&0&1&0&0&0\\0&0&0&0&0&0\\0&0&0&0&0&0\end{pmatrix}\begin{pmatrix}0&0&0&0&0&0\\0&0&0&1&0&0\\0&0&1&0&0&0\\0&1&0&0&0&0\\0&0&0&0&0&0\\0&0&0&0&0&0\end{pmatrix}\begin{pmatrix}0&0&0&0&0&0\\0&0&0&0&1&0\\0&0&0&0&0&1\\1&0&0&0&0&0\\0&0&0&0&0&0\\0&0&0&0&0&0\end{pmatrix}\begin{pmatrix}0&0&0&0&0&0\\0&0&0&0&0&1\\1&0&0&0&1&0\\0&0&0&0&0&1\\0&0&0&0&0&0\\0&0&0&0&0&0\end{pmatrix}$$

$$\begin{pmatrix}0&0&0&0&0&0\\0&0&0&0&0&1\\1&0&0&0&1&0\\0&0&0&0&0&1\\0&0&0&0&0&0\\0&0&0&0&0&0\end{pmatrix}\begin{pmatrix}0&0&0&0&0&0\\0&0&1&0&0&0\\0&1&0&1&0&0\\0&0&1&0&0&0\\0&0&0&0&0&0\\0&0&0&0&0&0\end{pmatrix}\begin{pmatrix}0&0&0&0&0&0\\0&1&0&1&0&0\\0&0&2&0&0&0\\0&1&0&1&0&0\\0&0&0&0&0&0\\0&0&0&0&0&0\end{pmatrix}\begin{pmatrix}0&0&0&0&0&0\\0&0&1&0&0&0\\0&1&0&1&0&0\\0&0&1&0&0&0\\0&0&0&0&0&0\\0&0&0&0&0&0\end{pmatrix}\begin{pmatrix}0&0&0&0&0&0\\0&0&0&0&0&1\\1&0&0&0&1&0\\0&0&0&0&0&1\\0&0&0&0&0&0\\0&0&0&0&0&0\end{pmatrix}\begin{pmatrix}0&0&0&0&0&0\\1&0&0&0&1&0\\0&0&0&0&0&2\\1&0&0&0&1&0\\0&0&0&0&0&0\\0&0&0&0&0&0\end{pmatrix}$$

$$\begin{pmatrix}0&0&0&0&0&0\\0&0&0&0&1&0\\0&0&0&0&0&1\\1&0&0&0&0&0\\0&0&0&0&0&0\\0&0&0&0&0&0\end{pmatrix}\begin{pmatrix}0&0&0&0&0&0\\0&0&0&1&0&0\\0&0&1&0&0&0\\0&1&0&0&0&0\\0&0&0&0&0&0\\0&0&0&0&0&0\end{pmatrix}\begin{pmatrix}0&0&0&0&0&0\\0&0&1&0&0&0\\0&1&0&1&0&0\\0&0&1&0&0&0\\0&0&0&0&0&0\\0&0&0&0&0&0\end{pmatrix}\begin{pmatrix}0&0&0&0&0&0\\0&1&0&0&0&0\\0&0&1&0&0&0\\0&0&0&1&0&0\\0&0&0&0&0&0\\0&0&0&0&0&0\end{pmatrix}\begin{pmatrix}0&0&0&0&0&0\\1&0&0&0&0&0\\0&0&0&0&0&1\\0&0&0&0&1&0\\0&0&0&0&0&0\\0&0&0&0&0&0\end{pmatrix}\begin{pmatrix}0&0&0&0&0&0\\0&0&0&0&0&1\\1&0&0&0&1&0\\0&0&0&0&0&1\\0&0&0&0&0&0\\0&0&0&0&0&0\end{pmatrix}$$

$$\begin{pmatrix}0&0&0&0&1&0\\0&0&0&0&0&0\\0&0&0&0&0&0\\0&0&0&0&0&0\\1&0&0&0&0&0\\0&0&0&0&0&1\end{pmatrix}\begin{pmatrix}0&0&0&1&0&0\\0&0&0&0&0&0\\0&0&0&0&0&0\\0&0&0&0&0&0\\0&1&0&0&0&0\\0&0&1&0&0&0\end{pmatrix}\begin{pmatrix}0&0&1&0&0&0\\0&0&0&0&0&0\\0&0&0&0&0&0\\0&0&0&0&0&0\\0&0&1&0&0&0\\0&1&0&1&0&0\end{pmatrix}\begin{pmatrix}0&1&0&0&0&0\\0&0&0&0&0&0\\0&0&0&0&0&0\\0&0&0&0&0&0\\0&0&0&1&0&0\\0&0&1&0&0&0\end{pmatrix}\begin{pmatrix}1&0&0&0&0&0\\0&0&0&0&0&0\\0&0&0&0&0&0\\0&0&0&0&0&0\\0&0&0&0&1&0\\0&0&0&0&0&1\end{pmatrix}\begin{pmatrix}0&0&0&0&0&1\\0&0&0&0&0&0\\0&0&0&0&0&0\\0&0&0&0&0&0\\0&0&0&0&0&1\\1&0&0&0&1&0\end{pmatrix}$$

$$\begin{pmatrix}0&0&0&0&0&1\\0&0&0&0&0&0\\0&0&0&0&0&0\\0&0&0&0&0&0\\0&0&0&0&0&1\\1&0&0&0&1&0\end{pmatrix}\begin{pmatrix}0&0&1&0&0&0\\0&0&0&0&0&0\\0&0&0&0&0&0\\0&0&0&0&0&0\\0&0&1&0&0&0\\0&1&0&1&0&0\end{pmatrix}\begin{pmatrix}0&1&0&1&0&0\\0&0&0&0&0&0\\0&0&0&0&0&0\\0&0&0&0&0&0\\0&1&0&1&0&0\\0&0&2&0&0&0\end{pmatrix}\begin{pmatrix}0&0&1&0&0&0\\0&0&0&0&0&0\\0&0&0&0&0&0\\0&0&0&0&0&0\\0&0&1&0&0&0\\0&1&0&1&0&0\end{pmatrix}\begin{pmatrix}0&0&0&0&0&1\\0&0&0&0&0&0\\0&0&0&0&0&0\\0&0&0&0&0&0\\0&0&0&0&0&1\\1&0&0&0&1&0\end{pmatrix}\begin{pmatrix}1&0&0&0&1&0\\0&0&0&0&0&0\\0&0&0&0&0&0\\0&0&0&0&0&0\\1&0&0&0&1&0\\0&0&0&0&0&2\end{pmatrix}$$

Fig. 3 The 6×6 array of the extended toric matrices $\widehat{P}_{a,b}$ of E_6. The 12 basis matrices (3.16) are linearly independent. The other 24 matrices are given by the quantum symmetry (3.17).

The Hilbert basis of extended toric matrices is unique up to these quantum symmetry equivalences.

It is useful to define the fusion product $\star$ of two extended toric matrices for type I theories by

$$\begin{aligned}\widehat{P}_{a,a'}\star\widehat{P}_{b,b'} &= (\widehat{N}_a,\widehat{N}_{a'})\star(\widehat{N}_b,\widehat{N}_{b'}) := (\widehat{N}_a\widehat{N}_b,\widehat{N}_{a'}\widehat{N}_{b'})\\ &= \sum_{c,c'\in G}\widehat{N}_{ab}{}^{c}\,\widehat{N}_{a'b'}{}^{c'}\,\widehat{P}_{c,c'}, \qquad a,a',b,b'\in G\end{aligned} \tag{3.18}$$

This defines the double graph fusion algebra. As we discuss in the next subsection, it reduces to the Ocneanu algebra when the right side is expressed in terms of the Hilbert basis using the quantum symmetry (3.17). In our notation, the left and right fundamentals are $\widehat{P}_{2,1}=(\widehat{N}_2,\widehat{N}_1)=(\widehat{N}_2,I)$ and $\widehat{P}_{1,2}=(\widehat{N}_1,\widehat{N}_2)=(I,\widehat{N}_2)$. For example, for E_6, straightforward calculation yields

$$\begin{aligned}\widehat{P}_{3,2}\star\widehat{P}_{2,1} &= (\widehat{N}_3,\widehat{N}_2)\star(\widehat{N}_2,\widehat{N}_1) = (\widehat{N}_3\widehat{N}_2,\widehat{N}_2\widehat{N}_1)\\ &= (\widehat{N}_2+\widehat{N}_4+\widehat{N}_6,\widehat{N}_2) = \widehat{P}_{2,2}+\widehat{P}_{4,2}+\widehat{P}_{6,2}\end{aligned} \tag{3.19}$$

$$\begin{aligned}\widehat{P}_{1,2}\star\widehat{P}_{3,3} &= (\widehat{N}_1,\widehat{N}_2)\star(\widehat{N}_3,\widehat{N}_3) = (\widehat{N}_1\widehat{N}_3,\widehat{N}_2\widehat{N}_3) = (\widehat{N}_3,\widehat{N}_2+\widehat{N}_4+\widehat{N}_6)\\ &= (\widehat{N}_3,\widehat{N}_2)+(\widehat{N}_3,\widehat{N}_4)+(\widehat{N}_3,\widehat{N}_6) = \widehat{P}_{3,2}+\widehat{P}_{3,4}+\widehat{P}_{3,6} = \widehat{P}_{2,1}+2\widehat{P}_{3,2}+\widehat{P}_{4,1}\end{aligned} \tag{3.20}$$

3.4 E_6 Ocneanu algebra and quantum symmetry

The Ocneanu graph $\widetilde{E}_6$ of E_6 is shown in Figure 4. The known twelve 12×12 E_6 Ocneanu graph fusion matrices (nimreps) $\widetilde{N}_\eta$ with $\eta=1,2,\ldots,12$ are shown in Figure 5. These fusion matrices satisfy the Ocneanu graph fusion algebra

$$\widetilde{N}_\eta\widetilde{N}_\mu = \sum_{\nu=1}^{12}\widetilde{N}_{\eta\mu}{}^{\nu}\,\widetilde{N}_\nu, \qquad \eta,\mu=1,2,\ldots,12 \tag{3.21}$$

and can be expressed compactly as

$$\widetilde{N}_\eta = \begin{cases}\widehat{N}_\eta\oplus\widehat{N}_\eta, & \eta=1,2,\ldots,6\\ \Sigma\,\widetilde{N}_{\bar\eta}\,\Sigma, & \eta=7\\ \widetilde{N}_7\widetilde{N}_{\eta-6}, & \eta=8,9,\ldots,12\end{cases} \qquad \Sigma = \begin{pmatrix}1&0&0&0&0&0&0&0&0&0&0&0\\0&0&0&0&0&0&1&0&0&0&0&0\\0&0&0&0&0&0&0&0&0&0&0&1\\0&0&0&0&0&0&0&0&0&0&1&0\\0&0&0&0&1&0&0&0&0&0&0&0\\0&0&0&0&0&1&0&0&0&0&0&0\\0&1&0&0&0&0&0&0&0&0&0&0\\0&0&0&0&0&0&0&1&0&0&0&0\\0&0&0&0&0&0&0&0&1&0&0&0\\0&0&0&0&0&0&0&0&0&1&0&0\\0&0&0&1&0&0&0&0&0&0&0&0\\0&0&1&0&0&0&0&0&0&0&0&0\end{pmatrix} \tag{3.22}$$

The $\mathbb{Z}_2$ chiral conjugation Σ implements the interchanges $\eta \leftrightarrow \bar{\eta}$ with $2 \leftrightarrow 7$, $3 \leftrightarrow 12$, $4 \leftrightarrow 11$ and $\bar{\eta} = \eta$ for $\eta = 1,5,6,8,9,10$.

It is readily verified that the basis toric matrices (3.16) satisfy the Ocneanu algebra

$$\widehat{P}_\eta \star \widehat{P}_\mu = \sum_{\nu=1}^{12} \widetilde{N}_{\eta\mu}{}^{\nu} \widehat{P}_\nu, \qquad \eta,\mu = 1,2,\ldots,12 \tag{3.23}$$

Due to the chiral action of the $\star$ fusion product, which multiplies left with left chiral components and right with right chiral components, this realization of the Ocneanu algebra does not yield simple matrix product representations of the algebra. If we now combine left and right chiral components we obtain

$$\widehat{P}_\eta \star \widehat{P}_{\bar{\mu}} = \sum_{\nu=1}^{12} \widetilde{N}_{\eta\bar{\mu}}{}^{\nu} \widehat{P}_\nu, \qquad \eta,\mu = 1,2,\ldots,12 \tag{3.24}$$

For $1 \le \eta,\mu \le 6$, $\widehat{P}_\eta \star \widehat{P}_{\bar{\mu}} = \widehat{P}_{\eta,\bar{\mu}}$ and restricting to $3 \le \mu \le 6$ in the right side above reproduces the quantum symmetry relations (3.17).

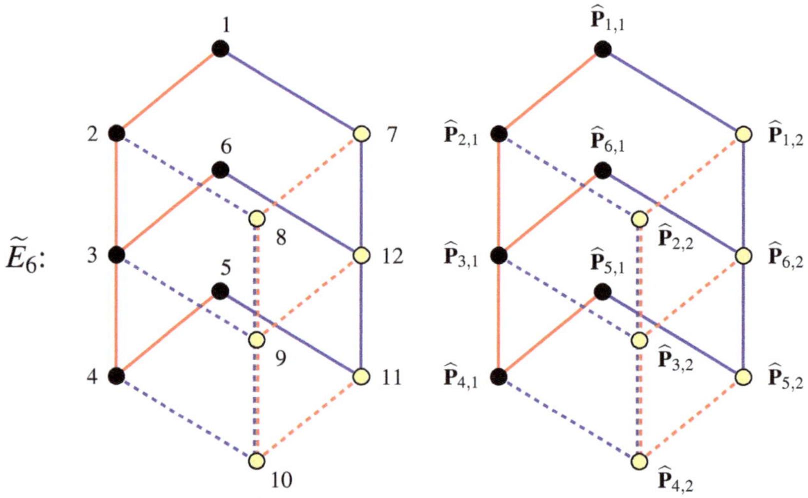

Fig. 4 Ocneanu fusion graph $\widetilde{E}_6$ of E_6 with 12 vertices and two alternative labellings of nodes. The 12 vertices are in bijection with (i) the Hilbert basis $\widehat{P}_\eta$ (3.16) with $\eta = 1,2,\ldots,12$ and (ii) the linearly independent integrable seams $\widehat{\mathbf{P}}_\eta$ with $\eta = 1,2,\ldots,12$ as in (4.10). The red and blue lines (solid or dashed) show the action of the fundamental seams $\widehat{\mathbf{P}}_2 = \widehat{\mathbf{P}}_{2,1}$ with $\widehat{\mathbf{P}}_7 = \widehat{\mathbf{P}}_{\bar{2}} = \widehat{\mathbf{P}}_{1,2}$ respectively. This action interlocks four copies of the E_6 graph. The seams satisfy $\overline{\widehat{\mathbf{P}}}_\eta = \widehat{\mathbf{P}}_{\bar{\eta}}$ where $\widehat{\mathbf{P}}_a = \widehat{\mathbf{P}}_{a,1}$ and $\widehat{\mathbf{P}}_{a+6} = \widehat{\mathbf{P}}_{a,2}$ with $a = 1,2,\ldots,6$.

$$
\begin{pmatrix}
1&0&0&0&0&0&0&0&0&0&0&0\\
0&1&0&0&0&0&0&0&0&0&0&0\\
0&0&1&0&0&0&0&0&0&0&0&0\\
0&0&0&1&0&0&0&0&0&0&0&0\\
0&0&0&0&1&0&0&0&0&0&0&0\\
0&0&0&0&0&1&0&0&0&0&0&0\\
0&0&0&0&0&0&1&0&0&0&0&0\\
0&0&0&0&0&0&0&1&0&0&0&0\\
0&0&0&0&0&0&0&0&1&0&0&0\\
0&0&0&0&0&0&0&0&0&1&0&0\\
0&0&0&0&0&0&0&0&0&0&1&0\\
0&0&0&0&0&0&0&0&0&0&0&1
\end{pmatrix}
\begin{pmatrix}
0&1&0&0&0&0&0&0&0&0&0&0\\
1&0&1&0&0&0&0&0&0&0&0&0\\
0&1&0&1&0&1&0&0&0&0&0&0\\
0&0&1&0&1&0&0&0&0&0&0&0\\
0&0&0&1&0&0&0&0&0&0&0&0\\
0&0&1&0&0&0&0&0&0&0&0&0\\
0&0&0&0&0&0&0&1&0&0&0&0\\
0&0&0&0&0&0&1&0&1&0&0&0\\
0&0&0&0&0&0&0&1&0&1&0&1\\
0&0&0&0&0&0&0&0&1&0&1&0\\
0&0&0&0&0&0&0&0&0&1&0&0\\
0&0&0&0&0&0&0&0&1&0&0&0
\end{pmatrix}
\begin{pmatrix}
0&0&1&0&0&0&0&0&0&0&0&0\\
0&1&0&1&0&1&0&0&0&0&0&0\\
1&0&2&0&1&0&0&0&0&0&0&0\\
0&1&0&1&0&1&0&0&0&0&0&0\\
0&0&1&0&0&0&0&0&0&0&0&0\\
0&1&0&1&0&0&0&0&0&0&0&0\\
0&0&0&0&0&0&0&0&1&0&0&0\\
0&0&0&0&0&0&0&1&0&1&0&1\\
0&0&0&0&0&0&1&0&2&0&1&0\\
0&0&0&0&0&0&0&1&0&1&0&1\\
0&0&0&0&0&0&0&0&1&0&0&0\\
0&0&0&0&0&0&0&1&0&1&0&0
\end{pmatrix}
\begin{pmatrix}
0&0&0&1&0&0&0&0&0&0&0&0\\
0&0&1&0&1&0&0&0&0&0&0&0\\
0&1&0&1&0&1&0&0&0&0&0&0\\
1&0&1&0&0&0&0&0&0&0&0&0\\
0&1&0&0&0&0&0&0&0&0&0&0\\
0&0&1&0&0&0&0&0&0&0&0&0\\
0&0&0&0&0&0&0&0&0&1&0&0\\
0&0&0&0&0&0&0&0&1&0&1&0\\
0&0&0&0&0&0&0&1&0&1&0&1\\
0&0&0&0&0&0&1&0&1&0&0&0\\
0&0&0&0&0&0&0&1&0&0&0&0\\
0&0&0&0&0&0&0&0&1&0&0&0
\end{pmatrix}
$$

$$
\begin{pmatrix}
0&0&0&0&1&0&0&0&0&0&0&0\\
0&0&0&1&0&0&0&0&0&0&0&0\\
0&0&1&0&0&0&0&0&0&0&0&0\\
0&1&0&0&0&0&0&0&0&0&0&0\\
1&0&0&0&0&0&0&0&0&0&0&0\\
0&0&0&0&0&1&0&0&0&0&0&0\\
0&0&0&0&0&0&0&0&0&0&1&0\\
0&0&0&0&0&0&0&0&0&1&0&0\\
0&0&0&0&0&0&0&0&1&0&0&0\\
0&0&0&0&0&0&0&1&0&0&0&0\\
0&0&0&0&0&0&1&0&0&0&0&0\\
0&0&0&0&0&0&0&0&0&0&0&1
\end{pmatrix}
\begin{pmatrix}
0&0&0&0&0&1&0&0&0&0&0&0\\
0&0&1&0&0&0&0&0&0&0&0&0\\
0&1&0&1&0&0&0&0&0&0&0&0\\
0&0&1&0&0&0&0&0&0&0&0&0\\
0&0&0&0&0&1&0&0&0&0&0&0\\
1&0&0&0&1&0&0&0&0&0&0&0\\
0&0&0&0&0&0&0&0&0&0&0&1\\
0&0&0&0&0&0&0&0&1&0&0&0\\
0&0&0&0&0&0&0&1&0&1&0&0\\
0&0&0&0&0&0&0&0&1&0&0&0\\
0&0&0&0&0&0&0&0&0&0&0&1\\
0&0&0&0&0&0&1&0&0&0&1&0
\end{pmatrix}
\begin{pmatrix}
0&0&0&0&0&0&1&0&0&0&0&0\\
0&0&0&0&0&0&0&1&0&0&0&0\\
0&0&0&0&0&0&0&0&1&0&0&0\\
0&0&0&0&0&0&0&0&0&1&0&0\\
0&0&0&0&0&0&0&0&0&0&1&0\\
0&0&0&0&0&0&0&0&0&0&0&1\\
1&0&0&0&0&0&0&0&0&0&0&1\\
0&1&0&0&0&0&0&0&1&0&0&0\\
0&0&1&0&0&0&0&1&0&1&0&0\\
0&0&0&1&0&0&0&0&1&0&0&0\\
0&0&0&0&1&0&0&0&0&0&0&1\\
0&0&0&0&0&1&1&0&0&0&1&0
\end{pmatrix}
\begin{pmatrix}
0&0&0&0&0&0&0&1&0&0&0&0\\
0&0&0&0&0&0&1&0&1&0&0&0\\
0&0&0&0&0&0&0&1&0&1&0&1\\
0&0&0&0&0&0&0&0&1&0&1&0\\
0&0&0&0&0&0&0&0&0&1&0&0\\
0&0&0&0&0&0&0&0&1&0&0&0\\
0&1&0&0&0&0&0&0&1&0&0&0\\
1&0&1&0&0&0&0&1&0&1&0&1\\
0&1&0&1&0&1&1&0&2&0&1&0\\
0&0&1&0&1&0&0&1&0&1&0&1\\
0&0&0&1&0&0&0&0&1&0&0&0\\
0&0&1&0&0&0&0&1&0&1&0&0
\end{pmatrix}
$$

$$
\begin{pmatrix}
0&0&0&0&0&0&0&0&1&0&0&0\\
0&0&0&0&0&0&0&1&0&1&0&1\\
0&0&0&0&0&0&1&0&2&0&1&0\\
0&0&0&0&0&0&0&1&0&1&0&1\\
0&0&0&0&0&0&0&0&1&0&0&0\\
0&0&0&0&0&0&0&1&0&1&0&0\\
0&0&1&0&0&0&0&1&0&1&0&0\\
0&1&0&1&0&1&1&0&2&0&1&0\\
1&0&2&0&1&0&0&2&0&2&0&2\\
0&1&0&1&0&1&1&0&2&0&1&0\\
0&0&1&0&0&0&0&1&0&1&0&0\\
0&1&0&1&0&0&0&0&2&0&0&0
\end{pmatrix}
\begin{pmatrix}
0&0&0&0&0&0&0&0&0&1&0&0\\
0&0&0&0&0&0&0&0&1&0&1&0\\
0&0&0&0&0&0&0&1&0&1&0&1\\
0&0&0&0&0&0&1&0&1&0&0&0\\
0&0&0&0&0&0&0&1&0&0&0&0\\
0&0&0&0&0&0&0&0&1&0&0&0\\
0&0&0&1&0&0&0&0&1&0&0&0\\
0&0&1&0&1&0&0&1&0&1&0&1\\
0&1&0&1&0&1&1&0&2&0&1&0\\
1&0&1&0&0&0&0&1&0&1&0&1\\
0&1&0&0&0&0&0&0&1&0&0&0\\
0&0&1&0&0&0&0&1&0&1&0&0
\end{pmatrix}
\begin{pmatrix}
0&0&0&0&0&0&0&0&0&0&1&0\\
0&0&0&0&0&0&0&0&0&1&0&0\\
0&0&0&0&0&0&0&0&1&0&0&0\\
0&0&0&0&0&0&0&1&0&0&0&0\\
0&0&0&0&0&0&1&0&0&0&0&0\\
0&0&0&0&0&0&0&0&0&0&0&1\\
0&0&0&0&1&0&0&0&0&0&0&1\\
0&0&0&1&0&0&0&0&1&0&0&0\\
0&0&1&0&0&0&0&1&0&1&0&0\\
0&1&0&0&0&0&0&0&1&0&0&0\\
1&0&0&0&0&0&0&0&0&0&0&1\\
0&0&0&0&0&1&1&0&0&0&1&0
\end{pmatrix}
\begin{pmatrix}
0&0&0&0&0&0&0&0&0&0&0&1\\
0&0&0&0&0&0&0&0&1&0&0&0\\
0&0&0&0&0&0&0&1&0&1&0&0\\
0&0&0&0&0&0&0&0&1&0&0&0\\
0&0&0&0&0&0&0&0&0&0&0&1\\
0&0&0&0&0&0&1&0&0&0&1&0\\
0&0&0&0&0&1&1&0&0&0&1&0\\
0&0&1&0&0&0&0&1&0&1&0&0\\
0&1&0&1&0&0&0&0&2&0&0&0\\
0&0&1&0&0&0&0&1&0&1&0&0\\
0&0&0&0&0&1&1&0&0&0&1&0\\
1&0&0&0&1&0&0&0&0&0&0&2
\end{pmatrix}
$$

Fig. 5 The twelve 12×12 E_6 Ocneanu graph fusion matrices (nimreps) $\widetilde{N}_\eta$ with $\eta = 1, 2, \ldots, 12$.

4 *A-D-E* integrable seams

In this section we use bold face symbols $\mathbf{n}_s, \widehat{\mathbf{N}}_a, \widehat{\overline{\mathbf{N}}}_b$ to denote integrable seams with system size M (see Figure 2). This distinguishes them from their associated nimreps such as $n_s, \widehat{N}_a, \widehat{N}_b$. The nimreps give the adjacency conditions on the horizontal bonds of the vertical integrable seams $\mathbf{n}_s, \widehat{\mathbf{N}}_a, \widehat{\overline{\mathbf{N}}}_b$.

4.1 Construction of *s*-type seams

The fundamental braid seams $\mathbf{B}, \overline{\mathbf{B}}$ are built analogously to the transfer matrix $\mathbf{T}(u)$ in (2.5) but with the face weights replaced by the complex weights (2.6a) and (2.6b) respectively. These are just the braid limits of the commuting family $\mathbf{T}(u)$, suitably normalized, so $\mathbf{B}$ and $\overline{\mathbf{B}}$ commute with $\mathbf{T}(u)$ and with each other. Since $\overline{\mathbf{B}} = \mathbf{B}^T$, $\mathbf{B}$ is

Hermitian. It also has the same real eigenvalues as G. The fused s-type braid seams $\mathbf{n}_s, \bar{\mathbf{n}}_s$ are defined recursively by the analog of (3.1)

$$\mathbf{n}_0 = 0, \qquad \mathbf{n}_1 = I, \qquad \mathbf{n}_2 = \mathbf{B}, \qquad \mathbf{n}_s\mathbf{n}_2 = \mathbf{n}_{s-1} + \mathbf{n}_{s+1} \tag{4.1a}$$

$$\bar{\mathbf{n}}_0 = 0, \qquad \bar{\mathbf{n}}_1 = I, \qquad \bar{\mathbf{n}}_2 = \overline{\mathbf{B}}, \qquad \bar{\mathbf{n}}_s\bar{\mathbf{n}}_2 = \bar{\mathbf{n}}_{s-1} + \bar{\mathbf{n}}_{s+1} \tag{4.1b}$$

Alternatively, these braid seams are constructed [34] using Wenzl-Jones projectors to build the fused transfer matrices $\mathbf{T}^{1,s}(u)$ and then taking the braid limits. The fused braid matrices $\mathbf{n}_s, \bar{\mathbf{n}}_s$ share the same real eigenvalues which are a subset of the eigenvalues of the Verlinde matrices N_s. The $\mathbf{n}_s, \bar{\mathbf{n}}_s$ integrable seams are called s-type seams because they are associated with the s label in the (r,s) Kac labelling of the irreps of Vir.

If $G = A_{g-1}$ then, for arbitrary system size M, $\mathbf{N}_s = \overline{\mathbf{N}}_s = \mathbf{n}_s$ give matrix representations of the Verlinde algebra

$$\mathbf{N}_i\mathbf{N}_j = \sum_{k=1}^{g-1} N_{ij}{}^k\mathbf{N}_k \tag{4.2}$$

In these cases, the Ocneanu algebra reduces to the Verlinde algebra.

4.2 Construction of a-type and b-type seams

For D and E theories, we now extend the decomposition (3.8) to relate fused seams of types s, a and b

$$\mathbf{n}_s = \sum_{a\in G} n_{s1}{}^a\widehat{\mathbf{N}}_a, \qquad \bar{\mathbf{n}}_s = \sum_{b\in G} n_{s1}{}^b\widehat{\overline{\mathbf{N}}}_b, \qquad s = 1,2,\ldots,g-1 \tag{4.3}$$

The face weights of the fused seams $\widehat{\mathbf{N}}_a, \widehat{\overline{\mathbf{N}}}_b$ can be calculated systematically as explained in [26]. However, the form of these face weights is not unique. Indeed, they depend on a choice of gauge. Here we will work in a different gauge to [26]. For arbitrary system size M, the fused seam matrices constructed here provide matrix representations of the same fused adjacency and graph fusion algebras as in (3.3) with the same structure constants:

$$\mathbf{n}_i\mathbf{n}_j = \sum_{k=1}^{g-1} N_{ij}{}^k\mathbf{n}_k, \qquad \widehat{\mathbf{N}}_a\widehat{\mathbf{N}}_b = \sum_{c\in G}\widehat{N}_{ab}{}^c\widehat{\mathbf{N}}_c, \qquad \mathbf{n}_i\widehat{\mathbf{N}}_a = \sum_{b\in G} n_{ia}{}^b\widehat{\mathbf{N}}_b \tag{4.4a}$$

$$\bar{\mathbf{n}}_i\bar{\mathbf{n}}_j = \sum_{k=1}^{g-1} N_{ij}{}^k\bar{\mathbf{n}}_k, \qquad \widehat{\overline{\mathbf{N}}}_a\widehat{\overline{\mathbf{N}}}_b = \sum_{c\in G}\widehat{N}_{ab}{}^c\widehat{\overline{\mathbf{N}}}_c, \qquad \bar{\mathbf{n}}_i\widehat{\overline{\mathbf{N}}}_b = \sum_{c\in G} n_{ib}{}^c\widehat{\overline{\mathbf{N}}}_c \tag{4.4b}$$

For E_6, the rectangular intertwiner $C = (n_{s1}{}^b)$ in (3.9) admits a generalized left inverse C_{left}^{-1} such that $C_{\text{left}}^{-1}C = I$. So (4.3) can be inverted to give the graph fusion seam matrices as

$$\widehat{\mathbf{N}}_a = \mathbf{n}_a,\ a = 1,2,3; \quad \widehat{\mathbf{N}}_4 = \mathbf{n}_6 - \mathbf{n}_2, \quad \widehat{\mathbf{N}}_5 = \mathbf{n}_5 - \mathbf{n}_3, \quad \widehat{\mathbf{N}}_6 = \mathbf{n}_2 + \mathbf{n}_4 - \mathbf{n}_6 \quad (4.5a)$$

$$\overline{\widehat{\mathbf{N}}}_b = \bar{\mathbf{n}}_b,\ b = 1,2,3; \quad \overline{\widehat{\mathbf{N}}}_4 = \bar{\mathbf{n}}_6 - \bar{\mathbf{n}}_2, \quad \overline{\widehat{\mathbf{N}}}_5 = \bar{\mathbf{n}}_5 - \bar{\mathbf{n}}_3, \quad \overline{\widehat{\mathbf{N}}}_6 = \bar{\mathbf{n}}_2 + \bar{\mathbf{n}}_4 - \bar{\mathbf{n}}_6 \quad (4.5b)$$

With these, it is readily verified using Mathematica [35] that, for even system sizes up to $M = 12$, all of the relations (4.4) are satisfied and that all of these matrices commute. In fact these relations hold for all M by construction.

The local face weights of $\widehat{\mathbf{N}}_a = \mathbf{n}_a$ with $a = 1,2,3$ follow from [34]. For $a = 3$, these face weights involve bond variables. For $\widehat{\mathbf{N}}_a$ with $a = 4,5,6$, we find that the allowed face weights are given by

$\widehat{\mathbf{N}}_4$:

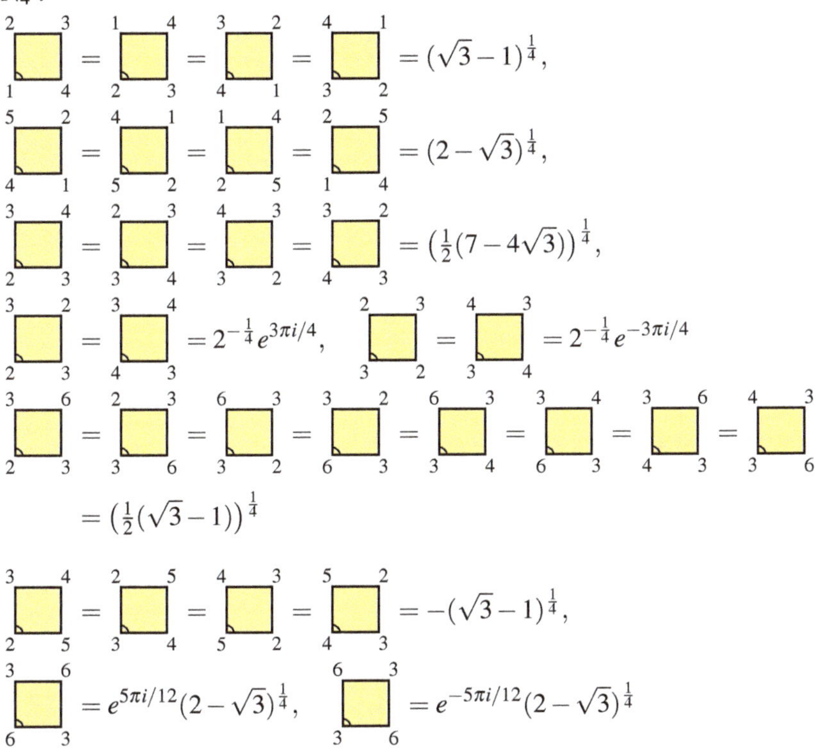

(4.6)

$\widehat{\mathbf{N}}_5$:

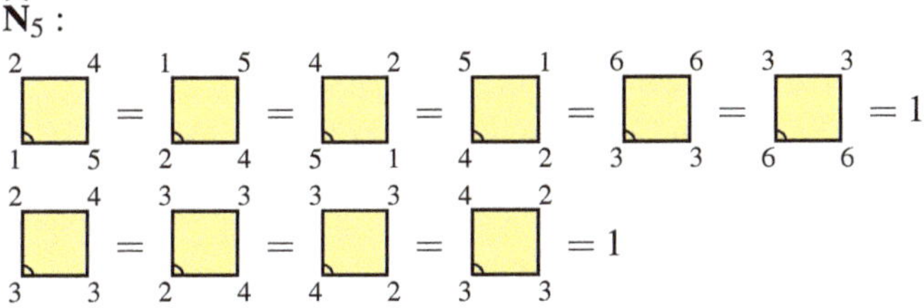

(4.7)

$\widehat{\mathbf{N}}_6$:

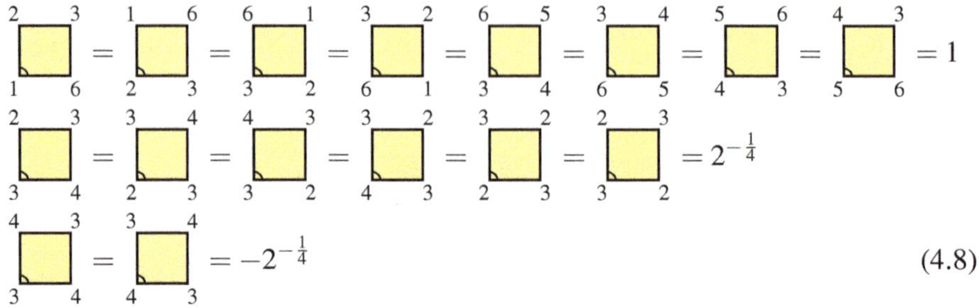

(4.8)

The face weights not explicitly listed vanish. Here $\widehat{\mathbf{N}}_5 = \sigma$ is the $\mathbb{Z}_2$ E_6 graph automorphism seam. For $\overline{\widehat{\mathbf{N}}}_a$ with $a = 4,5,6$, the face weights are obtained by complex conjugation. Reflection in the horizontal implements complex conjugation. We note that $\widehat{\mathbf{N}}_a = \overline{\widehat{\mathbf{N}}}_a$ is real for the ambichiral vertices $a = 1,5,6$.

4.3 E_6 Ocneanu algebra of seams and quantum symmetry

For even system sizes up to $M = 12$, it is now readily confirmed that the E_6 integrable seams satisfy the double graph fusion algebra

$$\widehat{\mathbf{P}}_{a,b} = \widetilde{\mathbf{N}}_{a,b} = \widehat{\mathbf{N}}_a \overline{\widehat{\mathbf{N}}}_b, \qquad \widetilde{\mathbf{N}}_{ab} \widetilde{\mathbf{N}}_{a'b'} = \sum_{a'',b''=1}^{6} \widehat{N}_{aa'}{}^{a''} \widehat{N}_{bb'}{}^{b''} \widetilde{\mathbf{N}}_{a''b''}, \tag{4.9}$$

with $\overline{\widehat{\mathbf{P}}}_{a,b} = \widehat{\mathbf{P}}_{b,a}$. Alternatively, using ordered sets, we can define

$$\{\widehat{\mathbf{P}}_1, \widehat{\mathbf{P}}_2, \ldots, \widehat{\mathbf{P}}_{12}\} = \{\widehat{\mathbf{P}}_{1,1}, \widehat{\mathbf{P}}_{2,1}, \ldots, \widehat{\mathbf{P}}_{6,1}, \widehat{\mathbf{P}}_{1,2}, \widehat{\mathbf{P}}_{2,2}, \ldots, \widehat{\mathbf{P}}_{6,2}\} \tag{4.10}$$

with $\overline{\widehat{\mathbf{P}}}_\eta = \widehat{\mathbf{P}}_{\bar{\eta}}$. For $M \le 12$, it is verified that these integrable seams satisfy the Ocneanu algebra

$$\widehat{\mathbf{P}}_\eta \widehat{\mathbf{P}}_\mu = \sum_{\nu=1}^{12} \widetilde{N}_{\eta\mu}{}^{\nu} \widehat{\mathbf{P}}_\nu, \qquad \eta, \mu = 1, 2, \ldots, 12 \tag{4.11}$$

In this realization involving fusion of seams, the fusion product $\star$ is replaced with the usual matrix product. In a sense, the decomposition of fusion products of seams on the lattice is an analog of the local operator product expansion in CFT.

In accord with the quantum symmetry (3.17) exhibited by the 36 "extended toric seams" (see Figure 3), the Hilbert basis of composite seams (4.10) is in bijection with the 12 nodes of the Ocneanu graph in Figure 4. The remaining 24 "extended toric seams" are given by the conical linear combinations

$$\widehat{\mathbf{P}}_{5,5} = \widehat{\mathbf{P}}_{1,1}, \qquad \widehat{\mathbf{P}}_{4,5} = \widehat{\mathbf{P}}_{2,1}, \qquad \widehat{\mathbf{P}}_{2,6} = \widehat{\mathbf{P}}_{3,5} = \widehat{\mathbf{P}}_{4,6} = \widehat{\mathbf{P}}_{3,1} \tag{4.12a}$$

$$\widehat{\mathbf{P}}_{2,5} = \widehat{\mathbf{P}}_{4,1}, \qquad \widehat{\mathbf{P}}_{1,5} = \widehat{\mathbf{P}}_{5,1}, \qquad \widehat{\mathbf{P}}_{1,6} = \widehat{\mathbf{P}}_{5,6} = \widehat{\mathbf{P}}_{6,5} = \widehat{\mathbf{P}}_{6,1} \tag{4.12b}$$

$$\widehat{\mathbf{P}}_{5,4} = \widehat{\mathbf{P}}_{1,2}, \qquad \widehat{\mathbf{P}}_{4,4} = \widehat{\mathbf{P}}_{2,2}, \qquad \widehat{\mathbf{P}}_{2,3} = \widehat{\mathbf{P}}_{3,4} = \widehat{\mathbf{P}}_{4,3} = \widehat{\mathbf{P}}_{3,2} \tag{4.12c}$$

$$\widehat{\mathbf{P}}_{2,4} = \widehat{\mathbf{P}}_{4,2}, \qquad \widehat{\mathbf{P}}_{1,4} = \widehat{\mathbf{P}}_{5,2}, \qquad \widehat{\mathbf{P}}_{1,3} = \widehat{\mathbf{P}}_{5,3} = \widehat{\mathbf{P}}_{6,4} = \widehat{\mathbf{P}}_{6,2} \tag{4.12d}$$

$$\widehat{\mathbf{P}}_{3,3} = \widehat{\mathbf{P}}_{2,2} + \widehat{\mathbf{P}}_{4,2}, \quad \widehat{\mathbf{P}}_{3,6} = \widehat{\mathbf{P}}_{2,1} + \widehat{\mathbf{P}}_{4,1}, \quad \widehat{\mathbf{P}}_{6,3} = \widehat{\mathbf{P}}_{1,2} + \widehat{\mathbf{P}}_{5,2}, \quad \widehat{\mathbf{P}}_{6,6} = \widehat{\mathbf{P}}_{1,1} + \widehat{\mathbf{P}}_{5,1} \tag{4.12e}$$

where again this was checked in Mathematica out to even system sizes $M = 12$. Replacing μ with $\bar{\mu}$ in (4.11) gives

$$\widehat{\mathbf{P}}_{\eta,\mu} - \widehat{\mathbf{N}}_\eta \overline{\widehat{\mathbf{N}}_\mu} - \widehat{\mathbf{N}}_\eta \widehat{\mathbf{N}}_{\bar{\mu}} = \sum_{\nu=1}^{12} \widetilde{N}_{\eta\bar{\mu}}{}^{\nu} \widehat{\mathbf{P}}_\nu, \qquad \eta,\mu = 1,2,\ldots,12 \tag{4.13}$$

For $1 \le \eta \le 6$ and $3 \le \mu \le 6$, this reproduces the quantum symmetry (4.12). Using this to express the right side of the double graph fusion algebra (4.9) in the basis seams (4.10) reproduces the Ocneanu algebra (4.11).

In general, the representations may not be faithful for small values of M. For E_6, the representations are not faithful for $M < 6$. The quantum symmetry encodes the rules for fusing left and right chiral seams. The E_6 Cayley table of the Ocneanu algebra of seams is shown in Figure 6.

	1	2	3	4	5	6	7	8	9	10	11	12
1	1	2	3	4	5	6	7	8	9	10	11	12
2	2	1+3	2+4+6	3+5	4	3	8	7+9	8+10+12	9+11	10	9
3	3	2+4+6	1+2(3)+5	2+4+6	3	2+4	9	8+10+12	7+2(9)+11	8+10+12	9	8+10
4	4	3+5	2+4+6	1+3	2	3	10	9+11	8+10+12	7+9	8	9
5	5	4	3	2	1	6	11	10	9	8	7	12
6	6	3	2+4	3	6	1+5	12	9	8+10	9	12	7+11
7	7	8	9	10	11	12	1+12	2+9	3+8+10	4+9	5+12	6+7+11
8	8	7+9	8+10+12	9+11	10	9	2+9	1+3+8 +10+12	2+4+6+7 +2(9)+11	3+5+8 +10+12	4+9	3+8+10
9	9	8+10+12	7+2(9)+11	8+10+12	9	8+10	3+8+10	2+4+6+7 +2(9)+11	1+2(3)+5+ 2(8) +2(10)+2(12)	2+4+6+7 +2(9)+11	3+8+10	2+4+2(9)
10	10	9+11	8+10+12	7+9	8	9	4+9	3+5+8 +10+12	2+4+6+7 +2(9)+11	1+3+8 +10+12	2+9	3+8+10
11	11	10	9	8	7	12	5+12	4+9	3+8+10	2+9	1+12	6+7+11
12	12	9	8+10	9	12	7+11	6+7+11	3+8+10	2+4+2(9)	3+8+10	6+7+11	1+5+2(12)

Fig. 6 Cayley table of the (commutative) E_6 Ocneanu algebra of seams with the compact notation $\eta = \widehat{\mathbf{P}}_\eta$ for $\eta = 1,2,\ldots,12$. The fusion of these seams reproduces the known E_6 Ocneanu Cayley table [22]. The top-left quadrant is the Cayley table of the E_6 graph fusion algebra.

4.4 E_6 Ocneanu algebra: quotient ring and quantum dimensions

There is an injective ring homomorphism

$$\phi : \langle \widehat{\mathbf{P}}_{a,1}, \widehat{\mathbf{P}}_{a,2} \,|\, a = 1, \ldots, 6 \rangle \to \mathbb{Z}[x,y]/\langle p_1(x), p_2(x,y) \rangle \tag{4.14}$$

where

$$\begin{aligned} p_1(x) &= \prod_{\ell \in \mathrm{Exp}(E_6)} (x - 2\cos\tfrac{\ell\pi}{g}) = x^6 - 5x^4 + 5x^2 - 1 \\ p_2(x,y) &= y^2 + (x^5 - 5x^3 + 4x)y - 1, \qquad p_2(y,y) = p_1(y) \end{aligned} \tag{4.15}$$

and

$$\widehat{\mathbf{P}}_{1,1} \mapsto 1, \quad \widehat{\mathbf{P}}_{2,1} \mapsto x, \quad \widehat{\mathbf{P}}_{3,1} \mapsto x^2 - 1, \quad \widehat{\mathbf{P}}_{4,1} \mapsto x^5 - 4x^3 + 2x \tag{4.16a}$$

$$\widehat{\mathbf{P}}_{5,1} \mapsto x^4 - 4x^2 + 2, \quad \widehat{\mathbf{P}}_{6,1} \mapsto -x^5 + 5x^3 - 4x, \quad \widehat{\mathbf{P}}_{a,2} \mapsto \phi(\widehat{\mathbf{P}}_{a,1})\,y, \quad a = 1, 2, \ldots, 6 \tag{4.16b}$$

The E_6 integrable seams all mutually commute and are simultaneously diagonalizable. The eigenvalues therefore yield a 1-dimensional representation of the E_6 Ocneanu algebra in terms of $s\ell(2)$ quantum dimensions on $\mathrm{Vir} \otimes \overline{\mathrm{Vir}}$. For the spin-$\frac{1}{2}$ fundamentals, let us set $x = q^s + q^{-s}$, $y = q^{\bar{s}} + q^{-\bar{s}}$ with $q = e^{\pi i/12}$ to ensure the fusion rules truncate correctly. We then need to solve

$$p_1(x) = 0, \qquad p_2(x,y) = 0 \tag{4.17}$$

These equations admit $6 \times 2 = 12$ solutions for $x = q^s + q^{-s} = 2\cos\frac{s\pi}{12}$ and $y = q^{\bar{s}} + q^{-\bar{s}} = 2\cos\frac{\bar{s}\pi}{12}$ where, explicitly,

$$\begin{aligned} (s,\bar{s}) \in \mathscr{S} = \{&(1,1),(4,4),(5,5),(7,7),(8,8),(11,11), \\ &(1,7),(7,1),(4,8),(8,4),(5,11),(11,5)\} \end{aligned} \tag{4.18}$$

These indices coincide with the operator content of the E_6 modular invariant partition function $Z(q) = \sum_{r=1}^{g-2} \sum_{(s,\bar{s}) \in \mathscr{S}} \chi_{r,s}(q) \chi_{r,\bar{s}}(\bar{q})$ for which the conformal weights satisfy $\Delta_{r,s} - \Delta_{r,\bar{s}} \in \mathbb{Z}$. As required, the set $\mathscr{S}$ is symmetric under interchange of s and $\bar{s}$. In the $(s,\bar{s})$ sector, the corresponding quantum dimensions are

$$\begin{array}{ll}
(1,1): & 1 \\
(2,1): & q^s+q^{-s} \\
(3,1): & q^{2s}+1+q^{-2s} \\
(4,1): & q^{5s}+q^{3s}+q^{-3s}+q^{-5s} \\
(5,1): & q^{4s}+q^{-4s} \\
(6,1): & -q^{5s}+q^s+q^{-s}-q^{-5s} \\
(1,2): & q^{\bar{s}}+q^{-\bar{s}} \\
(2,2): & q^{s+\bar{s}}+q^{s-\bar{s}}+q^{-s+\bar{s}}+q^{-s-\bar{s}} \\
(3,2): & q^{2s+\bar{s}}+q^{2s-\bar{s}}+q^{\bar{s}}+q^{-\bar{s}}+q^{-2s+\bar{s}}+q^{-2s-\bar{s}} \\
(4,2): & q^{5s+\bar{s}}+q^{5s-\bar{s}}+q^{3s+\bar{s}}+q^{3s-\bar{s}}+q^{-3s+\bar{s}}+q^{-3s-\bar{s}}+q^{-5s+\bar{s}}+q^{-5s-\bar{s}} \\
(5,2): & q^{4s+\bar{s}}+q^{4s-\bar{s}}+q^{-4s+\bar{s}}+q^{-4s-\bar{s}} \\
(6,2): & -q^{5s+\bar{s}}-q^{5s-\bar{s}}+q^{s+\bar{s}}+q^{s-\bar{s}}+q^{-s+\bar{s}}+q^{-s-\bar{s}}-q^{-5s+\bar{s}}-q^{-5s-\bar{s}}
\end{array} \tag{4.19}$$

Taking the union over the 12 sectors (4.18), we thus find that the distinct eigenvalues of the integrable seams are given by the roots of the minimal polynomials

$$\begin{array}{llll}
(1,1): & \lambda-1 & (1,2): & p_1(\lambda) \\
(2,1): & p_1(\lambda) & (2,2): & \lambda^4-4\lambda^3+4\lambda-1 \\
(3,1): & \lambda^3-2\lambda^2-2\lambda & (3,2): & \lambda^7-30\lambda^5+60\lambda^3-8\lambda \\
(4,1): & p_1(\lambda) & (4,2): & \lambda^4-4\lambda^3+4\lambda-1 \\
(5,1): & \lambda^2-1 & (5,2): & p_1(\lambda) \\
(6,1): & \lambda^3-2\lambda & (6,2): & \lambda^3-2\lambda^2-2\lambda
\end{array} \tag{4.20}$$

The toric seam matrices satisfy the polynomial equations obtained by setting the associated minimal polynomial to zero. We observe that some integrable seams have a zero eigenvalue and so are not invertible. Strictly speaking, in this sense, these seams are not symmetries — they are "non-invertible symmetries" [36].

4.5 E_6 Ocneanu algebra of seams and structure constants

For E_6, it follows from (4.15) and Cayley-Hamilton that

$$p_1(\mathbf{B}) = p_1(\overline{\mathbf{B}}) = \mathbf{0} \tag{4.21}$$

Additionally, for system sizes $M \le 12$, we have verified in Mathematica that

$$p_2(\mathbf{B},\overline{\mathbf{B}}) = \overline{\mathbf{B}}^2 - \widehat{\mathbf{P}}_{6,1}\overline{\mathbf{B}} - \mathbf{I} = \mathbf{0} \tag{4.22}$$

From (4.15) and (4.16b), $\mathbf{B}$ only enters this polynomial equation via the combination

$$\widehat{\mathbf{P}}_{6,1} = \widehat{\mathbf{N}}_6 = -\mathbf{B}^5 + 5\mathbf{B}^3 - 4\mathbf{B} \tag{4.23}$$

so (4.22) can be rewritten as

$$\overline{\mathbf{N}}_2^2 = \mathbf{I} + \widehat{\mathbf{N}}_6\overline{\mathbf{N}}_2 \tag{4.24}$$

This is in accord with the product $7*7=1+12$ in the Cayley table of Figure 6.

Let $a,b,c=1,2,\ldots,6$ and $a',b',c'=1,2$. Armed with the reduction (4.24), the product of any two seams $\widehat{\mathbf{P}}_{a,a'}$ and $\widehat{\mathbf{P}}_{b,b'}$ can now be obtained using commutativity and associativity. We can do this separately for the four quadrants (a',b') of the Cayley table Figure 6 with $a',b'=1,2$. But, since by commutativity the $(1,2)$ and $(2,1)$ quadrants agree, we only need to consider the three cases $a'b'=1,2,4$:

$$a'b'=1: \quad \widehat{\mathbf{N}}_a\widehat{\mathbf{N}}_b=\sum_c \widehat{N}_{ab}{}^c\,\widehat{\mathbf{N}}_c=\sum_c \widehat{N}_{ab}{}^c\,\widehat{\mathbf{P}}_{c,1} \tag{4.25}$$

$$a'b'=2: \quad (\widehat{\mathbf{N}}_a\overline{\widehat{\mathbf{N}}}_2)\widehat{\mathbf{N}}_b=\widehat{\mathbf{N}}_a(\widehat{\mathbf{N}}_b\overline{\widehat{\mathbf{N}}}_2)=\sum_c \widehat{N}_{ab}{}^c(\widehat{\mathbf{N}}_c\overline{\widehat{\mathbf{N}}}_2)=\sum_c \widehat{N}_{ab}{}^c\,\widehat{\mathbf{P}}_{c,2} \tag{4.26}$$

$$\begin{aligned} a'b'=4: \quad (\widehat{\mathbf{N}}_a\overline{\widehat{\mathbf{N}}}_2)(\widehat{\mathbf{N}}_b\overline{\widehat{\mathbf{N}}}_2)&=\sum_c \widehat{N}_{ab}{}^c\,\widehat{\mathbf{N}}_c\overline{\widehat{\mathbf{N}}}_2^{\,2}=\sum_c \widehat{N}_{ab}{}^c\,\widehat{\mathbf{N}}_c\,(\mathbf{I}+\widehat{\mathbf{N}}_6\overline{\widehat{\mathbf{N}}}_2) \qquad (4.27)\\ &=\sum_c \widehat{N}_{ab}{}^c\,\widehat{\mathbf{N}}_c+\sum_{c,d}\widehat{N}_{ab}{}^c\,\widehat{N}_{c6}{}^d(\widehat{\mathbf{N}}_d\overline{\widehat{\mathbf{N}}}_2)=\sum_c \widehat{N}_{ab}{}^c\,\widehat{\mathbf{P}}_{c,1}+\sum_{c,d}\widehat{N}_{ab}{}^c\,\widehat{N}_{c6}{}^d\widehat{\mathbf{P}}_{d,2} \end{aligned}$$

This observation gives explicitly the structure constants $\widetilde{N}_\eta$ of the E_6 Ocneanu algebra, as in Figure 5, in terms of the graph fusion and double graph fusion structure constants

$$\widetilde{N}_{a,a';b,b'}{}^{c,c'}=(\delta_{a',b'}\delta_{c',1}+\delta_{a'b',2}\delta_{c',2})\widehat{N}_{ab}{}^c+\delta_{a'b',4}\delta_{c',2}\sum_{d=1}^{6}\widehat{N}_{ab}{}^d\widehat{N}_{d6}{}^c \tag{4.28}$$

We deduce that the integrable seams $\widehat{\mathbf{P}}_{a,b}$ in (3.16) satisfy the E_6 Ocneanu algebra for arbitrary system sizes M once the seam relation $p_2(\mathbf{B},\overline{\mathbf{B}})=\mathbf{0}$ is established. This proof follows from (4.19). For a given eigenvector of $\mathbf{T}(u)$, the eigenvalues of $\mathbf{B},\overline{\mathbf{B}}$ are $2\cos\frac{s\pi}{12}, 2\cos\frac{\bar{s}\pi}{12}$ with $(s,\bar{s})\in\mathscr{S}$ as in (4.18). But again, it is readily verified that

$$p_2(2\cos\tfrac{s\pi}{12},2\cos\tfrac{\bar{s}\pi}{12})=0, \qquad (s,\bar{s})\in\mathscr{S} \tag{4.29}$$

Essentially, the proof that the integrable seams satisfy the E_6 Ocneanu algebra for arbitrary system sizes M follows by simultaneous diagonalization since the seam eigenvalues, in the form of the quantum dimensions, satisfy the E_6 Ocneanu algebra.

5 Conclusion

In this short article, we have developed further the lattice approach to fusion in critical unitary RSOS lattice models. Using Yang-Baxter integrability, it is now possible to construct column transfer matrices (seams) for topological defects that appear in the associated CFT in the continuum scaling limit. Like CFT topological defects, the lattice seams commute and propagate freely along the lattice. Remarkably, the lattice seams satisfy the Verlinde algebra, for A theories, and the graph fusion and Ocneanu algebras for D and E theories. In addition, the CFT quantum symmetry is already exhibited by the finite size integrable seams on the lattice. Importantly, fusion is implemented by simple matrix multiplication of square matrices associated with the integrable seams. Solving the boundary inversion relations for $\mathbf{T}_h(u)$ as in [26], it is seen that the boundary free energies of the s, a and b-type seams vanish whereas this is not the case for the general r-type seams. It would be of interest to find the boundary entropies of the seams as well as to find projectors onto the $(s,\bar{s})$ and (a,b) sectors of $\mathbf{T}(u)$.

Here we only considered type I theories and focussed on the A series and the exceptional E_6 theory of the critical unitary RSOS models on the torus. Crucially, we have identified the E_6 Ocneanu algebra as a quotient polynomial ring in two indeterminates (4.14). We will publish details for all the critical A-D-E RSOS models elsewhere.

Our new methods provide powerful tools to investigate the Ocneanu algebra of seams for other critical lattice models including models of higher spin or higher rank as well as models defined on the cylinder. Indeed, it would be of interest to apply the new methods to critical nonunitary A-D-E RSOS models [37], fused A-D-E RSOS models [38] such as the superconformal series [39, 40], as well as the critical dilute A-D-E RSOS lattice models [41–43] including the critical (A,G) series in branch 2 and tricritical (G,A) series in branch 1 [44].

Acknowledgements The Australian Research Council (ARC) is acknowledged for support under Discovery Project DP200102316. We thank Jean-Bernard Zuber for comments and Robert Coquereaux for detailed comments on the manuscript. Work on this project was carried out as part of the MATRIX workshop Mathematics and Physics of Integrability (MPI2024):
https://www.matrix-inst.org.au/events/mathematics-and-physics-of-integrability

References

1. R.J. Baxter, *Exactly Solved Models in Statistical Mechanics*, Academic Press (1982).
2. G.E. Andrews, R.J. Baxter, P.J. Forrester, *Eight-vertex SOS model and generalized Rogers-Ramanujan-type identities*, J. Stat. Phys. **35** (1984) 193–266.
3. V. Pasquier, *Two-dimensional critical systems labelled by Dynkin diagrams*, Nucl. Phys. **B285** (1987) 162–172.
4. V. Pasquier, *Operator content of the ADE lattice models*, J. Phys. A: Math. Gen. **20** (1987) 5707–5717.
5. V. Pasquier, *Lattice derivation of modular invariant partition functions on the torus*, J. Phys. A: Math. Gen. **20** (1987) L1229–L1237.
6. P. Di Francesco, P. Mathieu, D. Sénéchal, *Conformal Field Theory*, Springer (1997).
7. A.A. Belavin, A.M. Polyakov, A.B. Zamolodchikov, *Infinite conformal symmetry in two-dimensional quantum field theory*, Nucl. Phys. **B241** (1984) 333–380.
8. P. Goddard, A. Kent, D. Olive, *Virasoro algebras and coset space models*, Phys. Lett. **B152** (1985) 88–92.
9. G. Moore, N. Seiberg, *Classical and quantum conformal field theory*, Commun. Math. Phys. **123** (1989) 177–254.
10. R.E. Behrend, P.A. Pearce, V.B. Petkova, J.-B. Zuber, *On the classification of bulk and boundary conformal field theories*, Phys. Lett. **B444** (1998) 163–166.
11. R.E. Behrend, P.A. Pearce, V.B. Petkova, J.-B. Zuber, *Boundary conditions in rational conformal field theories*, Nucl. Phys. **B579** (2000) 707–773.

12. E. Verlinde, *Fusion rules and modular transformations in 2D conformal field theory*, Nucl. Phys. **B300** (1988) 360–376.
13. R.E. Behrend, P.A. Pearce, *Integrable and conformal boundary conditions for* $\widehat{s\ell}(2)$ *A-D-E lattice models and unitary minimal conformal field theories*, J. Stat. Phys. **102** (2001) 577–640.
14. V. Pasquier and H. Saleur, *Common Structures Between Finite Systems and Conformal Field Theories Through Quantum Groups* Nucl. Phys. **B330** (1990) 523–556.
15. R.E. Behrend, P.A. Pearce, J.-B. Zuber, *Integrable boundaries, conformal boundary conditions and A-D-E fusion rules*, J. Phys. A: Math. Gen. **31** (1998) L763–770.
16. J.L. Cardy, *Operator content of two-dimensional conformally invariant theories*, Nucl. Phys. **B270** (1986) 186–204.
17. A. Cappelli, C. Itzykson, J. B. Zuber. *The A-D-E classification of minimal and* $A_1^{(1)}$ *conformal invariant theories*, Commun. Math. Phys. **113** (1987) 1–26.
18. A. Ocneanu, *Paths on Coxeter Diagrams: From Platonic Solids and Singularities to Minimal Models and Subfactors*, in Lectures on Operator Theory, Fields Institute, Waterloo, Ontario, April 26–30, 1995, (Notes taken by S. Goto) Fields Institute Monographs, AMS 1999, Rajarama Bhat et al, eds;
Quantum symmetries for $SU(3)$ *CFT Models*, Lectures at Bariloche Summer School, Argentina, Jan 2000, AMS Contemporary Mathematics, R. Coquereaux, A. Garcia and R. Trinchero, eds;
The classification of subgroups of quantum $SU(N)$, Bariloche, 2000:
https://www.cpt.univ-mrs.fr/~coque/Bariloche2000/Bariloche2000/Bariloche2000.html
19. V.B. Petkova, J.-B. Zuber, *Generalised twisted partition functions*, Physics Letters **B504** (2001) 157–164; *Conformal field theories, graphs and quantum algebras*, MathPhys Odyssey 2001: Integrable Models and Beyond, 415–435.
20. R. Coquereaux, *Notes on the quantum tetrahedron*, Moscow Mathematical Journal **2** (2002) 41–80, https://arxiv.org/abs/math-ph/0011006.
21. R. Coquereaux, G. Schieber, *Twisted partition functions for ADE boundary conformal field theories and Ocneanu algebras of quantum symmetries*, J. Geom. Phys. **42** (2002) 216–258.
22. R. Coquereaux, M. Huerta, *Torus structure on graphs and twisted partition functions for minimal and affine models*, J. Geom. Phys. **48** (2003) 580–634.
23. R. Coquereaux, R. Trinchero, *On quantum symmetries of ADE graphs*, Adv. Theor. Math. Phys. **8** (2004) 189–216.
24. C.H. Otto Chui, C. Mercat, W.P. Orrick, P.A. Pearce, *Integrable lattice realizations of conformal twisted boundary conditions*, Phys. Lett. **B517** (2001) 429–435.
25. C.H. Otto Chui, C. Mercat, P.A. Pearce, *Integrable boundaries and universal TBA functional equations*, MathPhys Odyssey 2001: Integrable Models and Beyond, 391–413.
26. C.H. Otto Chui, C. Mercat, P.A. Pearce, *Integrable and conformal twisted boundary conditions for* $sl(2)$ *A-D-E lattice models*, J. Phys. A: Math. Gen. **36** (2003) 2623–2662.
27. J. Belletête, A.M. Gainutdinov, J.L Jacobsen, H. Saleur, T.S. Tavares, *Topological defects in periodic RSOS models and anyonic chains*, https://arxiv.org/abs/2003.11293;
Topological defects in lattice models and affine Temperley-Lieb algebra, Commun. Math. Phys. **400** (2023) 1203–1254.
28. T.S. Tavares, M. Sinha, L. Grans-Samuelsson, A. Roy, H. Saleur, *Integrable RG flows on topological defect lines in 2D conformal field theories*, https://arxiv.org/pdf/2408.08241.
29. P. Goddard, A. Kent, and D. Olive, *Virasoro algebras and coset space models*, Phys. Lett. **B152** (1985) 88–92.
30. A. Klümper, P.A. Pearce, *Analytic calculation of scaling dimensions: Tricritical hard squares and critical hard hexagons*, J. Stst. Phys. **64** (1991) 13–76.
31. A. Feiguin, S. Trebst, A.W.W. Ludwig, M. Troyer, A. Kitaev, Z. Wang, and M.H. Freedman, *Interacting anyons in topological quantum liquids: The golden chain*, Phys. Rev. Lett. **98** (2007) 160409.
32. J. Heymann, T. Quella, *Revisiting the symmetry-resolved entanglement for non-invertible symmetries in 1+1d conformal field theories*, https://arxiv.org/pdf/2409.02315.

33. Wikipedia: The Free Encyclopedia, Wikimedia Foundation Inc. https://en.wikipedia.org/wiki/Hilbert_basis_(linear_programming)
34. P.A. Pearce, Y-K. Zhou, *Fusion of A-D-E lattice models*, Int.J.Mod. Phys. **8** (1994) 3531–3577.
35. Wolfram Research, *Mathematica Edition: Version 13*, Wolfram Research Inc., Champaign, Illinois (2024).
36. S. Schäfer-Nameki, *ICTP lectures on (non-)invertible generalized symmetries*, Physics Reports **1063** (2024) 1–55.
37. P.J. Forrester, R.J. Baxter, *Further exact solutions of the eight-vertex SOS model and generalizations of the Rogers-Ramanujan identities*, J. Stat. Phys. **38** (1985) 435–472.
38. E. Date, M. Jimbo, A. Kuniba, T. Miwa, M. Okado, *Exactly solvable SOS models: I. Local height probabilities and theta function identities*, Nucl. Phys. **B290** [FS20] (1987) 231–273; *Exactly solvable SOS models: II. Proof of star-triangle relation and combinatorial identities*, Adv. Stud. Pure Math. **16** (1988) 17–122
39. P. Goddard, A. Kent, D. Olive, *Unitary representations of the Virasoro and super-Virasoro algebras*, Commun. Math. Phys. **103** (1986) 105–119.
40. C. Richard, P.A. Pearce, *Integrable lattice realizations of $N = 1$ superconformal boundary conditions*, Nucl. Phys. **B631** (2002) 447–470.
41. S.O. Warnaar, B. Nienhuis, K.A. Seaton, *New construction of solvable lattice models including an Ising model in a magnetic field*, Phys. Rev. Lett. **69** (1992) 710–712.
42. P. Roche, *On the construction of integrable dilute ADE lattice models*, Phys. Lett. **B285** (1992) 49–53.
43. S.O. Warnaar, P.A. Pearce, K.A. Seaton, B. Nienhuis, *Order parameters of the dilute A models*, J. Stat. Phys. **74** (1994) 469–531.
44. D.L. O'Brien, P.A. Pearce, *Lattice realizations of unitary minimal modular invariant partition functions*, J. Phys. A Math. Gen. **28** (1995) 4891–4905.

Chapter 2

MATRIX Program: Multivariate Dependence Modelling: Theory and Applications

A high-dimensional Bayesian skew-Gaussian spatial model for analyzing rainfall trends

Arghadeep Basu and Arnab Hazra

Abstract Analyzing the effect of climate change on rainfall is crucial from the perspective of agriculture, water resources management, and disaster management. Motivated by the key features present in a gridded rainfall dataset published by the India Meteorological Department (IMD) for the years 1951–2022 at a spatial resolution of $0.25^\circ \times 0.25^\circ$ across India, we propose a high-dimensional Bayesian skew-Gaussian conditional autoregressive model to analyze spatially-varying temporal patterns in Indian monsoon rainfall. We model the mean component in a semiparametric regression framework with spatially-varying coefficients and assume the skewness component to be also spatially-varying. We briefly discuss some theoretical properties of the proposed model. We draw inferences using Gibbs sampling by exploiting the hierarchical definition of a skew-Gaussian distribution as a random location-mixture of a Gaussian distribution. We study the posterior mean and the corresponding T-statistics of the overall change in rainfall amounts during the observational period at each grid cell. Our results indicate a significant negative trend in various parts of the Gangetic River basin, where agriculture is predominantly dependent on the availability of rainwater.

Key words: High-dimensional Bayesian model; Rainfall trends estimation; Skew-Gaussian process

---

Arghadeep Basu
Department of Statistics, University of Georgia, Athens, Georgia 30602, United States
e-mail: arghadeep.basu@uga.edu

Arnab Hazra
Department of Mathematics and Statistics, Indian Institute of Technology Kanpur, Kanpur 208016, India
e-mail: ahazra@iitk.ac.in

D. R. Wood et al. (eds.), *2024 MATRIX Annals, Part II*, MATRIX Book Series 8,
https://doi.org/10.1007/978-3-032-16206-9_9

1 Introduction

Rainfed agriculture refers to a type of farming where rainfall is the primary source of water for crops. In India, rainfed agriculture covers 67% of the total sown areas, the world's largest extent of such farming venkateswarlu2011rainfed. However, future climate assessments suggest that dry regions are gradually becoming drier, which could significantly impact the productivity of rainfed agriculture and exacerbate food security concerns [3]. To address these issues, the Government of India established the National Rainfed Area Authority (NRAA) in 2007 and implemented the Rainfed Area Development Programme (RADP) in 22 states of India during 2012–2013 [11]. Later, in 2015, the government consolidated various irrigation schemes into the Pradhan Mantri Krishi Sinchayee Yojana (PMKSY) to enhance irrigation facilities. However, according to the Economic Survey Report 2017–2018, 52% of the total cropped area still lacks irrigation, relying solely on rainfall. Consequently, the monsoon months (June, July, August, and September; JJAS, henceforth) are critical for farming. Moreover, rainfall anomalies lead to floods and droughts, causing severe crop damage. Accurate assessment of the availability of rainwater is essential to avoid crop failure, and statistical modeling of rainfall data plays a crucial role in agro-meteorology in this context.

We analyze a gridded rainfall dataset for the years 1951–2022 across India. Total rainfall amounts during monsoon months are usually nonzero throughout India, and we ignore a tiny fraction of cases (0.62% in our case) where observed rainfall is zero. We usually model rainfall data for each year using a spatial stochastic process with continuous responses [4, 14]. Gaussian processes (GPs) are prevalent models in spatial geostatistics because of their several favorable theoretical and computational characteristics [6]. However, there are three main issues with fitting a GP model for our case study: (i) because of the large spatial dimension of the dataset, the inference involves computing the determinant and inverse of the large covariance matrix, which is computationally demanding, (ii) a GP is a valid model for point-referenced spatial data, while due to the gridded spatial structure, it is more appropriate to choose areal models like the conditional autoregressive (CAR) model, and (iii) the data exhibit spatially-varying temporal trend and skewness in the marginal and joint distributions, indicating the need for a skewness component to be added to a GP.

The CAR models were first proposed by [2]. They are based on Markovian properties such that the conditional distribution of a component of the random vector (observed data or parameter vector in a Bayesian setting) depends only on a set of neighbors and are particular cases of Markov random fields [17]. They are appropriate models for areal data and allow a sparse inverse covariance matrix; hence, issues (i) and (ii) can be addressed using a CAR model for the spatial fields. The long-term spatially-varying trend can be modeled flexibly using a semiparametric regression framework [7, 8], and the skewness can be modeled using a skew-GP [19, 13]. However, as of our knowledge, no study has considered these aspects simultaneously, which is required to estimate the effects of climate change in a high-dimensional skewed spatial GP.

We propose a Bayesian skew-CAR model to examine the spatially-varying temporal patterns within Indian monsoon rainfall in a high-resolution setting. The study assumes spatially-varying coefficients to model the mean component in a semiparametric regression framework and introduces a spatially-varying skewness component. The paper also outlines the model's theoretical properties. To draw inferences, we utilize Gibbs sampling [5] by leveraging the hierarchical nature of a skew-GP in terms of a random location-mixture of a GP. The study investigates the posterior mean and corresponding T-statistics to analyze the overall changes in rainfall at each pixel during the study period.

The paper is organized as follows. In Section 2, we provide an overview of the Indian monsoon rainfall dataset and present some exploratory analyses. The statistical methodology is described in Section 3. The Bayesian computation for the proposed model is discussed in Section 4. Section 5 contains a discussion of the results, while the paper concludes with some final remarks in Section 6.

2 The Indian Monsoon Rainfall Data

2.1 Data Description and Pre-processing

We obtain daily gridded rainfall data from the India Meteorological Department (IMD) at a resolution of $0.25^\circ \times 0.25^\circ$ across India for 1951–2022. This dataset is prepared by [16] using daily rainfall records from 6995 rain gauge stations in India. Focusing on the analysis of total rainwater availability during the monsoon months, we obtain total rainfall across the months JJAS for each year. A total of 88 grid cells near the northeastern regions are removed due to the inconsistency of the gridded data with the available literature. Finally, data are available at 4876 spatial grid cells over 72 years. The observed data for the years 1951 and 2022 are presented in Figure 1. While higher monsoon rainfall is observable near the Western Ghats and the northeastern Himalayan regions, low rainfall is observed throughout western India except the Western Ghats region for both years. Different spatial patterns are observable near Madhya Pradesh, Uttarakhand, and Arunachal Pradesh for these years.

2.2 Exploratory Data Analysis

We first explore the temporal trend by fitting a simple linear regression model with year as a covariate separately at each grid cell and observe that such a model does not fit the data well (not shown). Besides, assuming a linear trend across 72 years is questionable. Hence, we further replace the simple linear regression model with a semiparametric regression model, where the expected rainfall amounts are assumed to be a smooth function of the years. We rewrite the smooth function as a linear combination of cubic B-splines with equidistant knots, and based on 10-fold

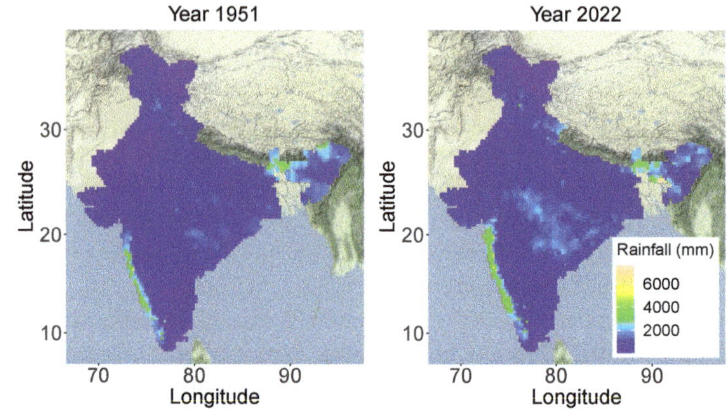

Fig. 1 Gridded monsoon rainfall for the years 1951 (left) and 2022 (right). Both panels share the same legend.

cross-validation and the minimum prediction mean square error criterion, we set the number of splines to be $K = 5$ (not shown). Based on this semiparametric model, the estimated annual rate of change (ROC) for each grid cell is presented in the left panel of Figure 2. Further, after obtaining the residuals, we calculate their marginal empirical skewness and present them in the right panel of Figure 2. While this spatial profile appears to be more jittery than that for ROC, both profiles indicate the need for spatially-varying spline coefficients and skewness. We then explore the presence of multivariate skewness using the multivariate normality test of [12] for a large number of pairs of sites. However, the tests suggest the presence of multivariate skewness in the residuals for most cases. Further, we calculate the empirical semivariance of the regression residuals, the empirical spline coefficients, and the residual skewness and present in Figure 3 after appropriate averaging. All three panels indicate the need for spatially dependent likelihood and priors for the spatial parameters.

3 Methodology

Suppose we denote the monsoon rainfall at a grid cell $\boldsymbol{s}_i$ for the t-th year by $Y_{i,t}$, where $i = 1,\ldots,N$ and $t = 1,\ldots,T$. Further, we denote the response vector for year t by $\boldsymbol{Y}_t = [Y_{1,t},\ldots,Y_{N,t}]'$, the k-th spline by $B_k(\cdot), k = 1,\ldots,K$, the vector of splines evaluated at t by $\boldsymbol{B}_t = [B_1(t),\ldots,B_K(t)]'$, and the spline coefficients for the i-th grid cell by $\boldsymbol{\beta}_i = [\beta_{i,1},\ldots,\beta_{i,K}]'$. We model $\boldsymbol{Y}_t$ hierarchically as

$$
\begin{aligned}
\boldsymbol{Y}_t | X_t &\overset{\text{Indep}}{\sim} \text{MVN}_N\left(\boldsymbol{\beta}^* \boldsymbol{B}_t + \boldsymbol{\lambda} X_t, \sigma^2 [\boldsymbol{M} - \rho \boldsymbol{A}]^{-1}\right), \\
X_t &\overset{\text{IID}}{\sim} \text{Half-Normal}(0, \sigma^2), \;\; t = 1,\ldots,T, \qquad (1)
\end{aligned}
$$

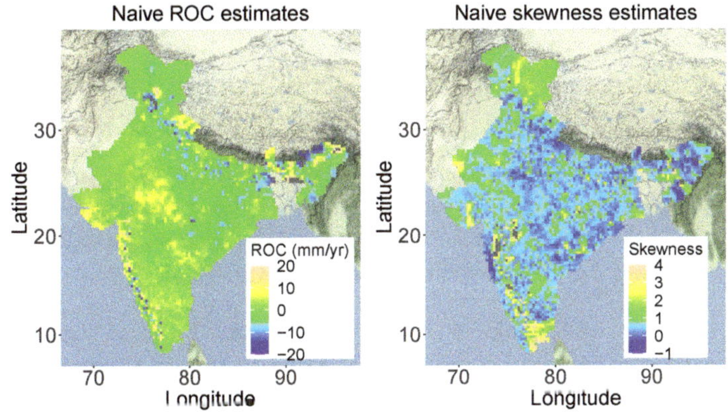

Fig. 2 Naive estimates of the annual rate of change (ROC) in monsoon rainfall (left) and naive estimates of marginal residual skewness (right).

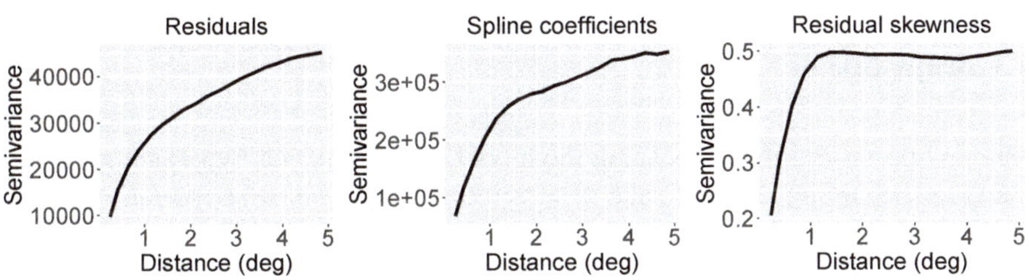

Fig. 3 Average (across years) empirical semivariance for the residuals after fitting a semiparametric mean component (left), average (across splines) empirical semivariance of the naive estimates of the spline coefficients (middle), and the empirical semivariance of the naive estimates of the marginal residual skewness (right).

where MVN_N denotes a N-variate normal distribution, $\boldsymbol{\beta}^*$ is a $(N \times K)$-dimensional matrix with its i-th row being $\boldsymbol{\beta}_i$, $\boldsymbol{\lambda} = [\lambda_1, \ldots, \lambda_N]'$ denotes a skewness-related parameter vector, σ^2 denotes a global variance-related parameter, $\boldsymbol{A}$ denotes a $(N \times N)$-dimensional matrix with its (i,i')-th entry being 1 if $\boldsymbol{s}_i$ and $\boldsymbol{s}'_i$ are neighbors and zero otherwise, $\boldsymbol{M}$ is a diagonal matrix with its diagonal elements being the row-wise sums of $\boldsymbol{A}$, and ρ denotes the spatial autocorrelation parameter. Here, the precision-related matrix $\boldsymbol{M} - \rho\boldsymbol{A}$ is a standard choice for gridded spatial data based on a conditional autoregressive Gaussian graphical model and obtained via Brook's lemma [2]. Thus, from (1), the conditional joint distribution of $\boldsymbol{Y} = [\boldsymbol{Y}'_1, \ldots, \boldsymbol{Y}'_T]'$ given $\boldsymbol{X} = [X_1, \ldots, X_T]'$ can be written as

$$\boldsymbol{Y}|\boldsymbol{X} \sim \text{MVN}_{NT}\left([\boldsymbol{B} \otimes \boldsymbol{I}_N]\boldsymbol{\beta} + [\boldsymbol{X} \otimes \boldsymbol{\lambda}], \sigma^2 \boldsymbol{I}_T \otimes [\boldsymbol{M} - \rho\boldsymbol{A}]^{-1}\right), \tag{2}$$

where $\boldsymbol{B}$ is a $(T \times K)$-dimensional matrix with its t-th row given by $\boldsymbol{B}_t$ and $\boldsymbol{\beta} = [\boldsymbol{\beta}^{*'}_{\cdot,1}, \ldots, \boldsymbol{\beta}^{*'}_{\cdot,K}]'$ with $\boldsymbol{\beta}^*_{\cdot,k}$ denoting the k-th column of $\boldsymbol{\beta}^*$.

After marginalizing over X_t, matching the notation of [1], we obtain

$$\boldsymbol{Y}_t \overset{\text{Indep}}{\sim} \text{MSN}_N\left(\boldsymbol{\beta}^*\boldsymbol{B}_t, \sigma^2([\boldsymbol{M}-\rho\boldsymbol{A}]^{-1}+\boldsymbol{\lambda}\boldsymbol{\lambda}'), [\boldsymbol{M}-\rho\boldsymbol{A}]\boldsymbol{\lambda}\right), \quad (3)$$

where MSN_N denotes an N-variate skew-normal distribution. When λ_i are equal, say $\lambda_i = \lambda_0$ for each i, (2) reduces to $\boldsymbol{Y}_t \overset{\text{Indep}}{\sim} \text{MSN}_N(\boldsymbol{\beta}^*\boldsymbol{B}_t, \sigma^2([\boldsymbol{M}-\rho\boldsymbol{A}]^{-1} + \lambda_0\boldsymbol{1}\boldsymbol{1}'), \lambda_0(1-\rho)\boldsymbol{M}\boldsymbol{1})$ and we have $\boldsymbol{Y}_t \overset{\text{Indep}}{\sim} \text{MVN}_N(\boldsymbol{\beta}^*\boldsymbol{B}_t, \sigma^2[\boldsymbol{M}-\rho\boldsymbol{A}]^{-1})$ if $\lambda_0 = 0$. Thus, the proposed model is a more general case of the spatial skew-Gaussian process with a fixed skewness-related component and the spatial Gaussian process. The mean vector and the covariance matrix of $\boldsymbol{Y}_t$ are given by

$$E[\boldsymbol{Y}_t] = \boldsymbol{\beta}^*\boldsymbol{B}_t + \sqrt{2/\pi}\sigma\boldsymbol{\lambda}, \;\; V[\boldsymbol{Y}_t] = \sigma^2([\boldsymbol{M}-\rho\boldsymbol{A}]^{-1} + (1-2/\pi)\boldsymbol{\lambda}\boldsymbol{\lambda}'). \quad (4)$$

Further, we assume the priors for the spatially -varying parameter vectors to be $\boldsymbol{\beta} \sim \text{MVN}_{NK}(\boldsymbol{0}, \sigma_\beta^2[\boldsymbol{M}-\rho_\beta\boldsymbol{A}]^{-1} \otimes \boldsymbol{I}_K)$, $\boldsymbol{\lambda} \sim \text{MVN}_N(\boldsymbol{0}, \sigma_\lambda^2[\boldsymbol{M}-\rho_\lambda\boldsymbol{A}]^{-1})$, and for the hyperparameters, we assume $\sigma^2, \sigma_\beta^2, \sigma_\lambda^2 \overset{\text{IID}}{\sim} \text{Inverse-Gamma}(0.01, 0.01)$.

4 Computation

We draw posterior inferences about the model parameters and hyperparameters using Gibbs sampling. For the sake of feasible computation, it is crucial to assume ρ, ρ_β, and ρ_λ to be known. While drawing inferences for a different combination of choices of these parameters and then comparing the cases based on the deviance information criterion (DIC) or Watanabe Akaike information criterion (WAIC) is recommended [10], for the sake of simplicity, we choose $\rho = \rho_\beta = \rho_\lambda = 0.5$ as suggested by fitting CAR models to the residuals, estimated spline coefficients, and the marginally estimated skewness parameters and obtaining maximum likelihood estimates as a part of exploratory data analysis. While none of such estimates are exactly equal to 0.5, given the high uncertainty in such estimates, we fix them to 0.5, a middle point between spatial independence ($\rho = \rho_\beta = \rho_\lambda = 0$) and intrinsic CAR model scenario ($\rho = \rho_\beta = \rho_\lambda = 1$). The full conditional posterior distributions for the elements of $\Theta = \{\beta, \lambda, \mathbf{X_0}, \sigma^2, \sigma_\beta^2, \sigma_\lambda^2\}$ are as follows. Here, *rest* denotes $\boldsymbol{Y}$ and all the elements of Θ except the one(s) being updated.

- $\pi(\boldsymbol{\beta}|rest) \sim \text{MVN}_{NK}(\boldsymbol{\mu}_\beta, \boldsymbol{\Sigma}_\beta)$, where $\boldsymbol{\Sigma}_\beta^{-1} = [\boldsymbol{M}-\rho\boldsymbol{A}] \otimes [\sigma^{-2}\boldsymbol{B}'\boldsymbol{B} + \sigma_\beta^{-2}\boldsymbol{I}_K]$ and $\boldsymbol{\mu}_\beta = \sigma^{-2}\boldsymbol{\Sigma}_\beta([\boldsymbol{B}\otimes\boldsymbol{I}_N]'[\boldsymbol{I}_T \otimes (\boldsymbol{M}-\rho\boldsymbol{A})][\boldsymbol{Y} - \boldsymbol{X}\otimes\boldsymbol{\lambda}])$
- $\pi(\boldsymbol{\lambda}|rest) \sim \text{MVN}_N(\boldsymbol{\mu}_\lambda, \boldsymbol{\Sigma}_\lambda)$, where $\boldsymbol{\Sigma}_\lambda^{-1} = \sigma^{-2}(\sum_{t=1}^T X_t^2)[\boldsymbol{M}-\rho\boldsymbol{A}] + \sigma_\lambda^{-2}[\boldsymbol{M} - \rho_\lambda\boldsymbol{A}]$ and $\boldsymbol{\mu}_\lambda = \sigma^{-2}\boldsymbol{\Sigma}_\lambda([\boldsymbol{M}-\rho\boldsymbol{A}][\sum_{t=1}^T X_t(\boldsymbol{Y}_t - \boldsymbol{\beta}^*\boldsymbol{B}_t)])$
- $\pi(X_t|rest) \overset{\text{Indep}}{\sim} \text{Truncated-N}_{(0,\infty)}\left(\frac{[\boldsymbol{Y}_t - \boldsymbol{\beta}^*\boldsymbol{B}_t]'[\boldsymbol{M}-\rho\boldsymbol{A}]\boldsymbol{\lambda}}{1+\boldsymbol{\lambda}'[\boldsymbol{M}-\rho\boldsymbol{A}]\boldsymbol{\lambda}}, \frac{\sigma^2}{1+\boldsymbol{\lambda}'[\boldsymbol{M}-\rho\boldsymbol{A}]\boldsymbol{\lambda}}\right)$
- $\pi(\sigma^2|\text{rest}) \sim \text{Inverse-Gamma}(0.01 + (N+1)T/2, 0.01 + 0.5\boldsymbol{\varepsilon}'(\boldsymbol{I}_T \otimes [\boldsymbol{M}-\rho_\lambda\boldsymbol{A}])\boldsymbol{\varepsilon} + 0.5\sum_{t=1}^T X_t^2)$, where $\boldsymbol{\varepsilon} = \boldsymbol{Y} - [\boldsymbol{B}\otimes\boldsymbol{I}_N]\boldsymbol{\beta} - [\boldsymbol{X}\otimes\boldsymbol{\lambda}]$.
- $\pi(\sigma_\beta^2|\text{rest}) \sim \text{Inverse-Gamma}(0.01 + NK/2, 0.01 + \boldsymbol{\beta}'([\boldsymbol{M}-\rho\boldsymbol{A}]\otimes\boldsymbol{I}_k)\boldsymbol{\beta})$
- $\pi(\sigma_\lambda^2|\text{rest}) \sim \text{Inverse-Gamma}\left(0.01 + N/2, 0.01 + \boldsymbol{\lambda}'[\boldsymbol{M}-\rho_\lambda\boldsymbol{A}]\boldsymbol{\lambda}\right)$

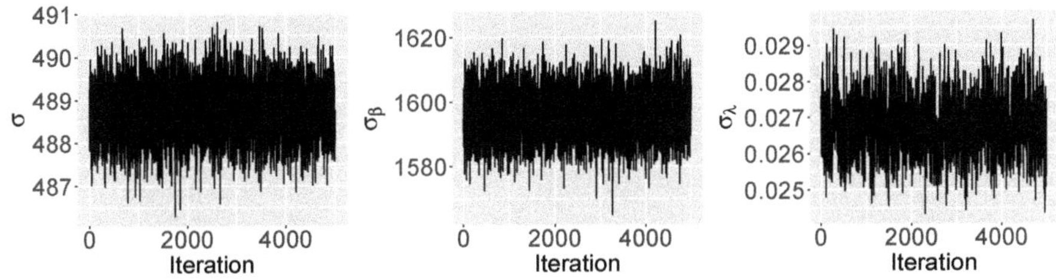

Fig. 4 Trace plots of the variance-related parameters and hyperparameters.

We run the Gibbs sampling for 10,000 iterations, and discard first 5,000 iterations as burn-in. The chains are monitored through trace plots; the post-burn-in samples for σ^2, σ_β^2 and σ_λ^2 are presented in Figure 4. All the panels show good convergence and mixing of the Gibbs sampler. The computing time is 402 minutes on a desktop with AMD Ryzen 9 5900x processor and 64GB RAM.

5 Results and Discussions

We calculate the posterior mean and standard deviation (SD) profiles of the annual ROC in monsoon rainfall based on the Markov chain Monte Carlo (MCMC) samples obtained in Section 4. We calculate the T-statistic, i.e., the ratio of posterior mean ROC and the corresponding posterior SD, for every grid cell; an absolute T-statistic value larger than two indicates significant changes and its sign indicates whether the change is positive or negative. We present the spatial maps of the posterior means and the T-statistics for ROC in Figure 5. For a total of 410 grid cells, the T-statistic turns out to be significantly negative, while it is significantly positive for 457 grid cells. Comparing with the empirical ROC estimates presented in the left panel of Figure 2, we observe that the posterior mean and the empirical estimates are close to each other, indicating a good model fit. From the plot for T-statistics, we observe that the estimated ROC is significantly negative for parts of Haryana and Punjab, some parts around the lower Gangetic River basin as well as in Arunachal Pradesh and Assam. This is concerning as a significant portion of these regions lies among the most fertile and rainfed regions of India. Besides, the estimated ROC is significantly positive, mainly for parts of Madhya Pradesh, Gujarat, Uttarakhand, and Meghalaya.

Further, the spatial maps of the posterior mean and T-statistic of the skewness-related parameter $\boldsymbol{\lambda}$ (not shown) show significant positive values near the Western Ghats. The posterior mean and SD of σ are 488.72 mm and 0.63 mm, respectively. For σ_β, they are 1596.44 mm and 7.32 mm, respectively, while for σ_λ, they are 0.03 mm and 7×10^{-4} mm, respectively. The posterior median, 0.025^{th} and 0.975^{th} quantiles of the random effects X_t are presented in Figure 6.

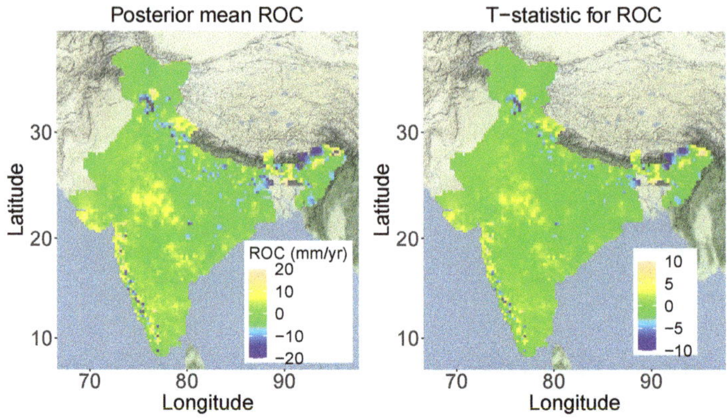

Fig. 5 Posterior mean profile the annual rate of change (ROC) in monsoon rainfall (left) and the corresponding T-statistics values across the grid cells (right).

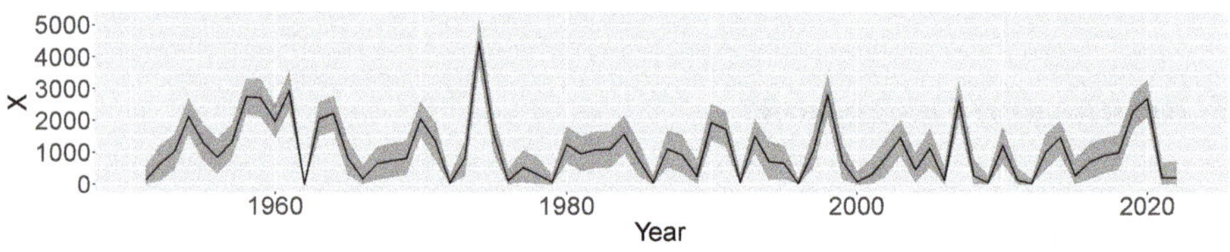

Fig. 6 Posterior median (solid line), and the 0.025-th and 0.975-th quantiles (grey ribbon) of the random effects X_t. A high value (during 1974, for example) indicates an event of country-wide higher rainfall on average.

6 Conclusions

We present a novel Bayesian model that analyzes spatially-varying temporal patterns using a high-dimensional skew-Gaussian CAR model. Our model, initially motivated by Indian monsoon rainfall features, is also applicable for analyzing long-term trends in other climatological variables. Traditional linear regression models with only year as a covariate are unsuitable for meteorological parameters due to multiple linear or nonlinear, micro-scale or macro-scale factors (e.g., El Nino); however, data for all such potential covariates are often unavailable. To address this, we use B-splines in a semiparametric Bayesian regression framework to model the mean component, which enables the model to explain local temporal behaviors in a data-driven approach. Additionally, our model assumes spatially-varying spline and skewness coefficients. The model allows posterior inference using Gibbs sampling. However, our model requires spatial autocorrelation-related parameters and hyperparameters to be pre-fixed due to computational constraints. For an exact Bayesian inference, one possible remedy is discretizing the parameter space of the spatial autocorrelation-related parameters, pre-calculating the necessary quadratic forms and determinants of the final covariance matrices over the parameter space, and using the stored information within MCMC; such an idea is explored by [15]. Future endeavors would be to extend our model to fit winter rainfall data with a large proportion of nil rainfall cases and explore faster computational schemes like Max-and-Smooth [9].

From an application perspective, analyzing the impact of climate change on rainfall is crucial for agriculture, water resource management, and disaster management. Our results highlight a significant negative trend in various parts of the Gangetic River basin, where agriculture relies heavily on rainwater availability.

Acknowledgements The authors would like to thank Laleh Tafakori from RMIT University, Australia, and an anonymous reviewer for their several important suggestions, which improved the flow and the content of the paper. The research of the second author is partially supported by the Startup Research Grant, Science and Engineering Research Board, India, with Award No. SERB-MATH-2023632.

References

1. Azzalini, A.: The skew-normal and related families. Cambridge University Press, Cambridge, UK (2014)
2. Besag, J.: Spatial interaction and the statistical analysis of lattice systems. J. R. Stat. Soc. Ser. B Methodol. **36**, 192–225 (1974)
3. Chauhan, B. S., Mahajan, G., and Randhawa, R. K., Singh, H., Kang, M. S.: Global warming and its possible impact on agriculture in India. Adv. Agron. **123**, 65–121 (2014)
4. Christakos, G.: Random field models in earth sciences. Academic Press, San Diego, USA (1992)
5. Gelfand, A. E.: Gibbs sampling. J. Am. Stat. Assoc. **95**, 1300–1304 (2000)
6. Gelfand, A. E., Schliep, E. M.: Spatial statistics and Gaussian processes: A beautiful marriage. Spat. Stat. **18**, 86–104 (2016)
7. Hazra, A., Reich, B. J., Staicu, A-M: A multivariate spatial skew-t process for joint modeling of extreme precipitation indexes. Environmetrics. **31**, e2602 (2020)
8. Hazra, A., Huser, R.: Estimating high-resolution Red Sea surface temperature hotspots, using a low-rank semiparametric spatial model. Ann. Appl. Stat. **15**, 572–596 (2021)
9. Hazra, A., Huser, R., Jóhannesson, A. V.: Bayesian latent Gaussian models for high-dimensional spatial extremes. In: Statistical Modeling Using Bayesian Latent Gaussian Models: With Applications in Geophysics and Environmental Sciences. (2023)
10. Hazra, A., Huser, R., Bolin, D.: Efficient modeling of spatial extremes over large geographical domains. J. Comput. Graph. Stat. (2024) doi: 10.1080/10618600.2024.2409784
11. Katyaini, S., Barua, A.: Assessment of interstate virtual water flows embedded in agriculture to mitigate water scarcity in India (1996–2014). Water Resour. Res. **53**, 7382–7400 (2017)
12. Mardia, K. V.: Measures of multivariate skewness and kurtosis with applications. Biometrika. **57**, 519–530 (1970)
13. Mastrantonio, G., Gelfand, A. E., Jona Lasinio, G.: The wrapped skew Gaussian process for analyzing spatio-temporal data. Stoch. Environ. Res. Risk Assess. **30**, 2231–2242 (2016)
14. Olea, R. A.: Geostatistics for engineers and earth scientists. Springer, New York, USA (1999)
15. Pakrashi, A., Hazra, A., Raveendran, S. M., Balakrishnan, K.: Approximate Bayesian inference for high-resolution spatial disaggregation using alternative data sources. J. Agric. Biol. Environ. Stat. (2025) doi: 10.1007/s13253-025-00695-5
16. Pai, D. S., Rajeevan, M., Sreejith, O. P., Mukhopadhyay, B., Satbha, N. S.: Development of a new high spatial resolution (0.25$\times$ 0.25) long period (1901–2010) daily gridded rainfall data

set over India and its comparison with existing data sets over the region. Mausam. **65**, 1–18 (2014)
17. Rue, H., Held, L.: Gaussian Markov random fields: theory and applications. CRC Press, New York, USA (2005)
18. Venkateswarlu, B.: Rainfed agriculture in India: issues in technology development and transfer. In: Model training course on "impact of climate change in rainfed agriculture and adaptation strategies" (2011) via ICAR-Central Research Institute for Dryland Agriculture.
19. Zhang, H., El-Shaarawi, A.: On spatial skew-Gaussian processes and applications. Environmetrics. **21**, 33–47 (2010)

Maximum likelihood estimation under inequality constraints in univariate linear models

Katarzyna Filipiak, Dietrich von Rosen, Martin Singull

Abstract This paper determines maximum likelihood estimates under general univariate linear models with a priori information related to maximum effects in the models. The negative log-likelihood functions and the constraints are convex functions, so convex optimization theory can be utilized to obtain relevant estimates. In particular, the complementary slackness condition, common in convex optimization, implies two alternative types of solutions, strongly dependent on the data and the restriction.

1 Introduction

Estimation techniques for estimating regression parameters are often based on penalized fitting functions with the ultimate goal to find estimators with desirable properties and also sometimes simultaneously include appropriate covariables. Several methods have been proposed; for example, [3] proposed bridge regression, [6] introduced LASSO, [2] came up with SCAD, [8] proposed the elastic net and [7] introduced the adaptive LASSO. However, if the variable selection is not of main concern, the ridge regression method [5] is also commonly used (see also Tikhonov regularization) and can be viewed as a penalized estimation method. In fact, the

Katarzyna Filipiak
Poznań University of Technology, Institute of Mathematics, Piotrowo 3A, PL-60965 Poznań, Poland
e-mail: katarzyna.filipiak@put.poznan.pl

Dietrich von Rosen
Linköping University, Department of Mathematics, SE-581 83 Linköping, Sweden
e-mail: dietrich.von.rosen@liu.se

Martin Singull
Linköping University, Department of Mathematics, SE-581 83 Linköping, Sweden
e-mail: martin.singull@liu.se

D. R. Wood et al. (eds.), *2024 MATRIX Annals, Part II*, MATRIX Book Series 8,
https://doi.org/10.1007/978-3-032-16206-9_10

above given references connect to ridge regression which appeared earlier. All the approaches modify the estimation function and the usual scenario is that the knowledge between data and the statistical model is vague.

In this work it will be utilized that there is specific knowledge (a priori information) about the maximum effects in a model which will be studied; for example, in physical and chemical processes one often knows the "bounds" of the processes. This knowledge is supposed to be quantifiable and then it can be built into the models via inequality constraints leading to a model closer to reality.

So far these ideas have not been implemented in statistics. In this article it will be shown how it can be carried out in linear models. We will stand on convex optimization theory. This means that the estimation functions (least squares function and negative log-likelihood function) should be convex functions with respect to the parameters in the model, which sometimes requires a model reparametrization. Moreover, the function which quantifies the prior information should also be convex. The approach in the present work also demands the functions to be differentiable which except ridge regression, usually is not the case with the above-mentioned variable selection methods.

In the article bold lower cases denote vectors whereas bold upper cases denote matrices. Other notations will be defined when they are introduced in the main body of the text.

2 Safety belt estimates and the general univariate linear model

Maximum likelihood estimation will be applied, mainly because estimation of the mean parameters and estimation of the variance can be handled simultaneously.

Let $\mathbf{y}$: $n \times 1$, be the response vector. We will not distinguish between the observation vector $\mathbf{y}$, which is thought of being a realization of a random vector $\mathbf{y}$, and the random vector $\mathbf{y}$ itself. This should however not lead to any confusion. Moreover, let $\mathbf{X}$: $n \times k$, represent the usual known design matrix in a linear model (usually of full column rank, with the first column consisting of vector of ones) with corresponding unknown mean effects β: $k \times 1$, and let ε: $n \times 1$, be the vector of unobservable independently normally distributed variables with mean 0 and common variance σ^2. These assumptions can be summarized as

$$\mathbf{y} \sim \mathcal{N}_n(\mathbf{X}\beta, \sigma^2\mathbf{I}_n), \tag{1}$$

where $\mathbf{I}_n$ stands for the $n \times n$ identity matrix. A crucial assumption is that there exist a priori information about β so that it makes sense to assume that $\beta^\top \beta \leq t$ for some given $t > 0$, "$^\top$" denotes the transpose, or a bit more general $\beta^\top \mathbf{H}\beta \leq t$, for some positive semi-definite matrix $\mathbf{H}$. The matrix $\mathbf{H}$ can be used to select components from β, for example via $\mathbf{H} = \begin{pmatrix} \mathbf{0} & \mathbf{0} \\ \mathbf{0} & \mathbf{I} \end{pmatrix}$, where $\mathbf{0}$ stands for a matrix with elements equal to 0. Obviously $\mathbf{H} = \mathbf{I}$ is another choice. However it can also be written that for $\mathbf{H} = (h_{ij})$ and $\beta = (\beta_i)$, $\beta^\top \mathbf{H}\beta = \sum_{ij} h_{ij}\beta_i\beta_j$ indicating the effect of choice of elements in $\mathbf{H}$. The use of an $\mathbf{H}$-matrix in ridge regression analysis has been considered in the literature (e.g., see [4, p.113]) and for example by choosing $\mathbf{H} = \mathbf{X}^\top\mathbf{X}$ one obtains the James-Stein estimator of β.

Since the a priori information is quantified via inequality constrains on the parameter space (squares of the parameters), we do not deal with linear subspaces and therefore cannot refer to the usual linear models theory. Moreover, in many applications $\beta^\top\mathbf{H}\beta \leq t$ makes sense. For example, if using linear regression for the study of a human body height over time we usually know pretty precisely what the maximum growth intensity will be. In fact, in most planned experiments one has the knowledge about maximum effects. In particular, measurement instruments have to be calibrated within certain ranges. However, in observational studies with many interacting variables, the upper bound t can be more difficult to set.

A general rule in statistics is to include a priori information when building a model as long as it is mathematical motivated and one does not lose interpretability. The next proposition balances inclusion of information and interpretability.

The proposition presents maximum likelihood estimates under inequality restrictions on $\beta^\top\mathbf{H}\beta$, where $\mathbf{H}$ is supposed to be non-random known positive semi-definite matrix. Moreover, it is supposed that the design matrix $\mathbf{X}$ is of full rank.

Proposition 1. *In model* (1) *with a priori information* $\beta^\top\mathbf{H}\beta \leq t$, *where* $t > 0$ *is a known constant and* $\mathbf{H}$ *is a known non-random positive semi-definite matrix, maximum likelihood estimates* $\widehat{\beta}$ *and* $\widehat{\sigma}^2$ *of* β *and* σ^2, *respectively, are presented:*

(i) *if* $\mathbf{y}^\top\mathbf{X}(\mathbf{X}^\top\mathbf{X})^{-1}\mathbf{H}(\mathbf{X}^\top\mathbf{X})^{-1}\mathbf{X}^\top\mathbf{y} \leq t$ *then*

$$\widehat{\beta} = (\mathbf{X}^\top\mathbf{X})^{-1}\mathbf{X}^\top\mathbf{y}, \qquad n\widehat{\sigma}^2 = (\mathbf{y}-\mathbf{X}\widehat{\beta})^\top(\mathbf{y}-\mathbf{X}\widehat{\beta});$$

(ii) *if* $\mathbf{y}^\top\mathbf{X}(\mathbf{X}^\top\mathbf{X})^{-1}\mathbf{H}(\mathbf{X}^\top\mathbf{X})^{-1}\mathbf{X}^\top\mathbf{y} > t$ *then*

$$\widehat{\beta} = (\mathbf{X}^\top\mathbf{X}+\widehat{\lambda}\mathbf{H})^{-1}\mathbf{X}^\top\mathbf{y}, \qquad n\widehat{\sigma}^2 = (\mathbf{y}-\mathbf{X}\widehat{\beta})^\top(\mathbf{y}-\mathbf{X}\widehat{\beta}), \tag{2}$$

where $\widehat{\lambda}$ *is the solution to the non-linear equation in* λ,

$$\mathbf{y}^\top\mathbf{X}(\mathbf{X}^\top\mathbf{X}+\lambda\mathbf{H})^{-1}\mathbf{H}(\mathbf{X}^\top\mathbf{X}+\lambda\mathbf{H})^{-1}\mathbf{X}^\top\mathbf{y} = t.$$

Proof. First, we find an upper bound of a likelihood with respect to σ^2 *and thereafter convex optimization with respect to* β *takes place. Thus, we put* $\theta = \sigma^2$ *and then the negative log-likelihood function equals*

$$f(\beta,\theta) = \tfrac{n}{2}\log(2\pi) + \tfrac{n}{2}\log(\theta) + \tfrac{1}{2}\theta^{-1}(\mathbf{y}-\mathbf{X}\beta)^\top(\mathbf{y}-\mathbf{X}\beta).$$

Note, that the above function is convex with respect to θ^{-1} *with minimum for* $\theta^{-1} = n/((\mathbf{y}-\mathbf{X}\beta)^\top(\mathbf{y}-\mathbf{X}\beta))$. *Therefore*

$$\begin{aligned} f(\beta,\theta) &\geq f(\beta,(\mathbf{y}-\mathbf{X}\beta)^\top(\mathbf{y}-\mathbf{X}\beta)/n) \\ &= \tfrac{n}{2}\log(2\pi) + \tfrac{n}{2}\log((\mathbf{y}-\mathbf{X}\beta)^\top(\mathbf{y}-\mathbf{X}\beta)/n) + \tfrac{n}{2} \end{aligned} \tag{3}$$

and

$$\min_{\beta^\top \mathbf{H}\beta \le t, \theta} f(\beta, \theta) = \min_{\beta^\top \mathbf{H}\beta \le t} f(\beta, (\mathbf{y} - \mathbf{X}\beta)^\top (\mathbf{y} - \mathbf{X}\beta)/n).$$

Finding a minimizer of (3) *under the a priori information $\beta^\top \mathbf{H}\beta \le t$ is equivalent to finding a minimizer of $(\mathbf{y} - \mathbf{X}\beta)^\top (\mathbf{y} - \mathbf{X}\beta)$ under the condition $\beta^\top \mathbf{H}\beta \le t$.*

The mathematical problem can be introduced by estimating β, under a priori information on β, using a linear model and applying the least squares approach, i.e.,

$$\min_{\beta} \quad (\mathbf{y} - \mathbf{X}\beta)^\top (\mathbf{y} - \mathbf{X}\beta), \quad \beta^\top \mathbf{H}\beta \le t,$$

where $t > 0$ is given. To study this problem a well known approach is to set up the Lagrangian function:

$$L(\beta, \lambda) = (\mathbf{y} - \mathbf{X}\beta)^\top (\mathbf{y} - \mathbf{X}\beta) + \lambda(\beta^\top \mathbf{H}\beta - t), \tag{4}$$

where $\lambda > 0$ is the Lagrangian multiplier.

Note that $(\mathbf{y} - \mathbf{X}\beta)^\top (\mathbf{y} - \mathbf{X}\beta)$ and $\beta^\top \mathbf{H}\beta$ are convex functions in β which follows by differentiating the functions twice. Therefore the Karush-Kuhn-Tucker (KKT) conditions give necessary and sufficient conditions for solving (4) (e.g., see [1]). The KKT-conditions equal

$$\frac{d\, L(\beta, \lambda)}{d\, \beta} = \mathbf{0}, \quad \beta^\top \mathbf{H}\beta - t \le 0, \quad \lambda \ge 0, \quad \lambda(\beta^\top \mathbf{H}\beta - t) = 0. \tag{5}$$

The derivative in (5) is a vector derivative.

Spelling out the derivative implies that the following system of equations has to be solved:

$$-2\mathbf{X}^\top (\mathbf{y} - \mathbf{X}\beta) + 2\lambda \mathbf{H}\beta = \mathbf{0}, \tag{6}$$

$$\beta^\top \mathbf{H}\beta - t \le 0, \qquad \lambda \ge 0, \tag{7}$$

$$\lambda(\beta^\top \mathbf{H}\beta - t) = 0. \tag{8}$$

Condition (8) is usually called the complementary slackness condition and is very important in our approach. The condition implies that either $\lambda = 0$ or $\beta^\top \mathbf{H}\beta - t = 0$. If $\lambda = 0$ then from (6) it appears that

$$\widehat{\beta} = (\mathbf{X}^\top \mathbf{X})^{-1} \mathbf{X}^\top \mathbf{y},$$

which is the usual least squares estimator. However if $\lambda > 0$ then $\beta^\top \mathbf{H}\beta - t = 0$ and then the task is to choose β and λ so that (6) – (8) are true. Usually this problem has to be solved numerically. Hence the two results for $\widehat{\beta}$ of the proposition have been established. Coming to $\theta = (\mathbf{y} - \mathbf{X}\beta)^\top (\mathbf{y} - \mathbf{X}\beta)/n$ and inequality (3) it follows that $\widehat{\sigma}^2$ is obtained.

In our approach, which is mathematically optimal, if the least squares estimates are too far from what can be assumed through the a priori information, the estimates are shrunk via the Lagrangian multiplier. Thus, on top of the usual estimation approach, a "safety belt" has been put so that in particular a "bad" $\mathbf{X}$ matrix can be handled, i.e., when the eigenvalues of $(\mathbf{X}^\top\mathbf{X})^{-1}$ become large, or if there exist influential observations. In these cases $\widehat{\beta}$ and $\widehat{\sigma}^2$ in (2) can be used.

The next proposition, which is presented, with sketch of a proof, includes a different type of restriction. It is the kind of restriction which will be possible to handle in a straightforward manner when finding maximum likelihood estimates in the general multivariate linear model (MANOVA model). In the univariate case this restriction is of the form $(\sigma^2)^{-1}\beta^\top\mathbf{H}\beta \leq t$, where as before t is a positive known constant and $\mathbf{H}$ is a known positive semi-definite matrix. This kind of restriction has been discussed in, e.g., [4, Corollary 3.2.2], when comparing ridge regression estimates with least squares estimates. However, our motivation for considering $(\sigma^2)^{-1}\beta^\top\mathbf{H}\beta \leq t$ is different. In this article one of the main ideas is to rely on convex optimization when finding estimators. The approach in this paper is inspired by the fact that the normal distribution (univariate and multivariate) belongs to an exponential family.

Proposition 2. *In model* (1) *with a priori information* $(\sigma^2)^{-1}\beta^\top\mathbf{H}\beta \leq t$, *where* $t > 0$ *is a known constant and* $\mathbf{H}$ *is a known non-random positive semi-definite matrix, maximum likelihood estimates* $\widehat{\beta}$ *and* $\widehat{\sigma}^2$ *of* β *and* σ^2, *respectively, are presented. Put*

$$\widetilde{\beta} = (\mathbf{X}^\top\mathbf{X})^{-1}\mathbf{X}^\top\mathbf{y}, \qquad n\widetilde{\sigma}^2 = (\mathbf{y}-\mathbf{X}\widetilde{\beta})^\top(\mathbf{y}-\mathbf{X}\widetilde{\beta}).$$

(i) *If* $\widetilde{\beta}^\top\mathbf{H}\widetilde{\beta} \leq \widetilde{\sigma}^2 t$, *then maximum likelihood estimates are given by* $\widehat{\beta} = \widetilde{\beta}$ *and* $\widehat{\sigma}^2 = \widetilde{\sigma}^2$.

(ii) *If* $\widetilde{\beta}^\top\mathbf{H}\widetilde{\beta} > \widetilde{\sigma}^2 t$ *then maximum likelihood estimates of* β *and* σ^2 *satisfy*

$$\widehat{\beta} = (\mathbf{X}^\top\mathbf{X} + \widehat{\lambda}\mathbf{H})^{-1}\mathbf{X}^\top\mathbf{y}, \qquad n\widehat{\sigma}^2 = (\mathbf{y}-\mathbf{X}\widehat{\beta})^\top(\mathbf{y}-\mathbf{X}\widehat{\beta}) + \widehat{\lambda}\widehat{\beta}^\top\mathbf{H}\widehat{\beta},$$

where $\widehat{\lambda}$ *is the solution to* $\mathbf{y}^\top\mathbf{X}(\mathbf{X}^\top\mathbf{X}+\lambda\mathbf{H})^{-1}\mathbf{H}(\mathbf{X}^\top\mathbf{X}+\lambda\mathbf{H})^{-1}\mathbf{X}^\top\mathbf{y} = t\widehat{\sigma}^2$.

Proof. (sketch of the proof) Let us introduce the canonical parameters $\theta_1 = (\sigma^2)^{-1}\beta$ *and* $\theta_2 = (\sigma^2)^{-1}$. *Then, the negative log-likelihood function can be expressed as*

$$f(\theta_1,\theta_2) = \tfrac{n}{2}\log(2\pi) - \tfrac{n}{2}\log\theta_2 + \tfrac{1}{2}\theta_2\mathbf{y}^\top\mathbf{y} - \mathbf{y}^\top\mathbf{X}\theta_1 + \tfrac{1}{2}\theta_2^{-1}\theta_1^\top\mathbf{X}^\top\mathbf{X}\theta_1,$$

which is a convex function of θ_1 *and* θ_2. *Moreover, since the condition* $g(\beta,\sigma^2) = (\sigma^2)^{-1}\beta^\top\mathbf{H}\beta - t$ *can be also expressed in terms of canonical variables, i.e.,* $g(\theta_1,\theta_2) = \theta_2^{-1}\theta_1^\top\mathbf{H}\theta_1 - t$, *it can be checked that this is also a convex function of* θ_1 *and* θ_2. *Therefore applying the KKT conditions for solving*

$$L(\theta_1,\theta_2) = f(\theta_1,\theta_2) + \lambda g(\theta_1,\theta_2),$$

where $\lambda > 0$ *is the Lagrangian multiplier, we obtain*

$$\widehat{\theta_2^{-1}\theta_1} = (\mathbf{X}^\top\mathbf{X}+\lambda\mathbf{H})^{-1}\mathbf{X}^\top\mathbf{y}$$

and

$$n\widehat{\theta_2^{-1}} = \mathbf{y}^\top\mathbf{y} - \widehat{\theta_2^{-1}\theta_1}^\top\mathbf{X}^\top\mathbf{X}\widehat{\theta_2^{-1}\theta_1} - \lambda\widehat{\theta_2^{-1}\theta_1}^\top\mathbf{H}\widehat{\theta_2^{-1}\theta_1}.$$

Coming back to original notation we obtain the results.

Acknowledgements This work was supported by the Poznań University of Technology under Grant no. 0213/SBAD/0118 (K. Filipiak).

References

1. *Boyd, S., Vandenberghe, L., Convex Optimization, Cambridge University Press, 2004.*
2. *Fan, J., Li, R., Variable selection via nonconcave penalized likelihood and its oracle properties, Journal of the American Statistical Association, 96, 1348–1360, 2001.*
3. *Frank, I.E. and Friedman, J.H., A statistical view of some chemometrics regression tools, Technometrics, 35, 109–135, 1993.*
4. *Gruber, M., Improving Efficiency by Shrinkage: The James-Stein and Ridge Regression Estimators, CRC Press, New York, 1998.*
5. *Hoerl, A.E., Kennard, R.W., Ridge regression: biased estimation for nonorthogonal problems, Technometrics, 12, 55–67, 1970.*
6. *Tibshirani, R., Regression shrinkage and selection via the Lasso, Journal of the Royal Statistical Society, Series B, 58, 267–288, 1996.*
7. *Zou, H., The adaptive Lasso and its oracle properties, Journal of the American Statistical Association, 101, 1418–1429, 2006.*
8. *Zou, H., Hastie, T., Regularization and variable selection via the elastic net, Journal of the Royal Statistical Society, Series B, 67, 301–320, 2005.*

Estimation of latent variables in high-dimensional factor models with weak residual dependence

Xinyao Fan and Harry Joe

Abstract For D-dimensional Gaussian 1-factor, bi-factor, oblique factor models and their copula-based counterparts, proxies have previously been defined to estimate the latent variables for large D. It is shown that the previously-defined proxies are asymptotically consistent and hence robust to some interpretable assumptions of weak residual conditional dependence of observed variables given the latent variables. Proofs make use of techniques of likelihoods with mis-specified models. Some simulation results provide concrete behavior of the effect of weak residual dependence and increasing D.

1 Introduction

Factor models are useful tools for modeling the dependence of large sets of variables when the dependence between the observed variables can be explained by several latent variables. When the variables can be divided into groups, structured factor Gaussian and copula models in [9] and [10] are often good first-order models to explain the dependence of variables. The models assume the conditional independence of observed variables given latent variables. As the dimension increases (the number of observed variables linked to each latent variable increases), it may be reasonable to assume a slight departure from the conditional independence assumption, such as weak conditional dependence given the latent variables. In this article, we refer to this as weak residual dependence.

Xinyao Fan
University of British Columbia, Vancouver, Canada
e-mail: xinyao.fan@stat.ubc.ca

Harry Joe
University of British Columbia, Vancouver, Canada
e-mail: Harry.Joe@ubc.ca

D. R. Wood et al. (eds.), *2024 MATRIX Annals, Part II*, MATRIX Book Series 8,
https://doi.org/10.1007/978-3-032-16206-9_11

For multivariate dependence modeling, extending Gaussian classical factor models to have (a) flexible non-Gaussian tail behavior via factor copulas and (b) weak conditional dependence given latent variables is important for high-dimensional applications.

For a large dimension D for the total number of variables, proxy variables consist of estimates for latent variables based on conditional expectations given the observed variables, for Gaussian factor models or their copula counterparts. For Gaussian factor models, the proxy variables are the factor scores as given in [8]. Consistency results for these proxy variables are proved in [4], under some non-stringent conditions.

The main purpose of this article is to show that proxy variables, defined based on the conditional independence assumption, are robust to weak residual dependence. Detailed proofs using non-standard asymptotic methods are given for 1-factor and bi-factor Gaussian models, and 1-factor copula models, based on interpretable sufficient conditions for weak residual dependence. By interpreting the sufficient conditions, we can intuit sufficient weak residual dependence conditions that might hold more generally. The adequacy of the proxy estimation method is further demonstrated for 1-factor and bi-factor copula models through simulation studies.

Section 2 provides the definitions and notation for Gaussian and copula-based factor models with residual dependence. Section 3 summarizes the conditional expectation form of proxy variables for the Gaussian and copula models. Section 4 provides some interpretable sufficient conditions on weak residual dependence for the consistency of proxy variables for the case of a known factor model. Section 5 has sufficient conditions on residual dependence for the consistency of the proxies in 1-factor models when parameters are estimated. Section 6 presents a sequential estimation method for practical use when linking copula families are to be estimated from data. Section 7 includes simulation studies to provide additional support for the usefulness of the proxy variables in factor copula models when there is weak residual dependence. Section 8 has the conclusions and explains how results in this article can be applied. The proofs of the main results are mostly lengthy, because of the need to keep track of asymptotic order of many terms, and they are given in Section 9.

Abbreviations that are used include cdf for cumulative distribution function and MLE for maximum likelihood estimator or estimation. Copulas which are cumulative distribution functions with $U(0,1)$ univariate margins are denoted with C and subscripts. The copulas are assumed to be absolutely continuous, and the corresponding densities are denoted with c.

2 Factor models with residual dependence

This section defines the Gaussian and copula factor models with their notation to prepare for the analysis of estimation of latent variables in the subsequent sections. Factor models have parsimonious dependence and there is more chance to depart from the dependence structure as the number of variables increases. As discussed in [7], adding conditional dependence given latent variables may possibly lead to a more interpretable model than adding additional latent variables.

2.1 Gaussian factor models

When there are many variables that are related, the factor models assuming conditional independence might be good first approximations but it is possible that the conditional (residual) dependence exists after accounting for latent variables. In this section, we assume that the observed variables in the Gaussian factor models have been standardized to be standard normal $\mathcal{N}(0,1)$.

Let $\boldsymbol{W} \sim \mathcal{N}(\boldsymbol{0},\boldsymbol{I}_p)$ be a p-vector of independent latent variables and let $\boldsymbol{Z}_D = (Z_j)$ be a D-vector of observed variables that are each $\mathcal{N}(0,1)$. Let $\boldsymbol{\varepsilon}_D \sim \mathcal{N}(\boldsymbol{0},\boldsymbol{\Omega}_D)$ be a random vector of idiosyncratic noise or specific factors that are independent of $\boldsymbol{W}$, where $\boldsymbol{\Omega}_D$ is a correlation matrix.

The stochastic representation for the Gaussian p-factor model with residual dependence is:

$$\boldsymbol{Z}_D = \boldsymbol{A}_D\boldsymbol{W} + \boldsymbol{\Psi}_D\boldsymbol{\varepsilon}_D \tag{1}$$

where $\boldsymbol{A}_D = (\alpha_{jk})$ is a $D \times p$ loading matrix, $\boldsymbol{\Psi}_D = \mathrm{diag}(\psi_j : j = 1,\ldots,D)$ and the loading parameters $\{\alpha_{jk}\}$ are such that $\psi_j^2 = 1 - \sum_{k=1}^{p} \alpha_{jk}^2 > 0$ for all j. The correlation matrix of $\boldsymbol{Z}_D$ is $\mathrm{Cor}(\boldsymbol{Z}_D) = \boldsymbol{A}_D\boldsymbol{A}_D^\top + \boldsymbol{\Gamma}_D$, where $\boldsymbol{\Gamma}_D = \boldsymbol{\Psi}_D\boldsymbol{\Omega}_D\boldsymbol{\Psi}_D$. Note that $\boldsymbol{\Gamma}_D = \mathrm{Cov}(\boldsymbol{Z}_D|\boldsymbol{W})$ is the covariance matrix of $\boldsymbol{Z}$ given $\boldsymbol{W}$. If $\boldsymbol{\Omega}_D = \boldsymbol{I}_D$, then the components Z_j of $\boldsymbol{Z}$ are conditionally independent given $\boldsymbol{W}$ (for the usual p-factor model).

In the remainder of this article, because of non-identifiability to rotation of the loading matrix for $p \geq 2$, we focus on the 1-factor model with residual dependence if the variables are not grouped. If the 1-factor model is not an adequate fit for data, the 1-factor model with residual dependence is an alternative ([11]) to adding additional factors and it might be more interpretable in some applications. For high-dimensional $\boldsymbol{Z}_D$, estimation of latent variables is studied in [4] for the 1-factor model but not the p-factor for $p \geq 2$.

For variables that have been divided into G non-overlapping groups, we will consider structured bi-factor models and then the structured loading matrix (with many 0s in it) is based on $p = G + 1$ latent variables with one global latent variable and G group-based latent variables. We will also consider the oblique factor model for G non-overlapping groups where the structured loading matrix $\boldsymbol{A}$ is $D \times G$ and $\boldsymbol{W}$ consists of dependent latent variables.

Definition 1. (1-factor Gaussian with residual dependence). The stochastic representation, with $\boldsymbol{A}_D = \boldsymbol{\alpha}_D = (\alpha_1,\ldots,\alpha_D)^\top$, is

$$Z_j = \alpha_j W + \psi_j \varepsilon_j, \quad j = 1,\ldots,D, \tag{2}$$

where $W \sim \mathcal{N}(0,1)$, $\boldsymbol{\varepsilon}_D = (\varepsilon_1,\ldots,\varepsilon_D)^\top \sim \mathcal{N}(\mathbf{0},\boldsymbol{\Omega}_D)$ is independent of W, and $\alpha_j \in (-1,1)$ and $\psi_j^2 = 1-\alpha_j^2$ for all j. Note that $\mathrm{Cov}(\mathbf{Z}_D|W) = \boldsymbol{\Gamma}_D = \boldsymbol{\Psi}_D\boldsymbol{\Omega}_D\boldsymbol{\Psi}_D$ with $\boldsymbol{\Psi}_D = \mathrm{diag}(\psi_1,\ldots,\psi_D)$.

Definition 2. (Bi-factor Gaussian with residual dependence).
With G groups, let $D = \sum_{g=1}^G d_g$, with d_g variables from group g. Write the loading matrix as $\boldsymbol{A}_D = [\boldsymbol{a}_{0,D}, \mathrm{diag}(\boldsymbol{b}_{d_1 1},\ldots,\boldsymbol{b}_{d_G G})]$ where $\boldsymbol{a}_{0,D} = (\boldsymbol{b}_{0,d_1 1}^\top,\ldots,\boldsymbol{b}_{0,d_G G}^\top)^\top$ and for $g \in \{1,\ldots,G\}$, $\boldsymbol{b}_{d_g g} = (\alpha_{1g},\ldots,\alpha_{d_g g})^\top$ and $\boldsymbol{b}_{0,d_g g} = (\alpha_{0,1g},\ldots,\alpha_{0,d_g g})^\top$. The stochastic representation is

$$Z_{jg} = \alpha_{0,jg}W_0 + \alpha_{jg}W_g + \psi_{jg}\varepsilon_{jg}, \quad j \in \{1,\ldots,d_g\}, \quad g \in \{1,\ldots,G\}, \tag{3}$$

$\boldsymbol{\varepsilon}_g^\top = (\varepsilon_{1g},\ldots,\varepsilon_{jg},\ldots,\varepsilon_{d_g g})$, $\boldsymbol{\varepsilon}_D^\top = (\boldsymbol{\varepsilon}_1^\top,\ldots,\boldsymbol{\varepsilon}_G^\top)$ with $\boldsymbol{\varepsilon}_D \sim \mathcal{N}(\mathbf{0},\boldsymbol{\Omega}_D)$, W_0 and $\{W_g\}$ are mutually independent latent $\mathcal{N}(0,1)$ random variables independent of $\boldsymbol{\varepsilon}$, and $\alpha_{0,jg}, \alpha_{jg} \in (-1,1)$ such that $\psi_{jg}^2 = 1-\alpha_{0,jg}^2-\alpha_{jg}^2 < 1$ for all (j,g).

Definition 3. (Gaussian oblique factor with residual dependence). With G groups, let $D = \sum_{g=1}^G d_g$, with d_g variables from group g. The stochastic representation is

$$Z_{jg} = \alpha_{jg}W_g + \psi_{jg}\varepsilon_{jg}, \quad j \in \{1,\ldots,d_g\}, \quad g \in \{1,\ldots,G\}, \tag{4}$$

where $\boldsymbol{\varepsilon}_g^\top = (\varepsilon_{1g},\ldots,\varepsilon_{jg},\ldots,\varepsilon_{d_g g})$, $\boldsymbol{\varepsilon}_D^\top = (\boldsymbol{\varepsilon}_1^\top,\ldots,\boldsymbol{\varepsilon}_G^\top)$ with $\boldsymbol{\varepsilon}_D \sim \mathcal{N}(\mathbf{0},\boldsymbol{\Omega}_D)$, $\boldsymbol{W} = (W_1,\ldots,W_G)^\top \sim \mathcal{N}(\mathbf{0},\boldsymbol{\Sigma}_W)$ independent of $\boldsymbol{\varepsilon}$, and $\alpha_{jg} \in (-1,1)$ such that $\psi_{jg}^2 = 1-\alpha_{jg}^2$ for all (j,g). Here, $\boldsymbol{\Sigma}_W$ is a correlation matrix.

2.2 Factor copula models

In this section, the copula counterparts of the Gaussian factor models are given. Loading parameters for the main factor are replaced with bivariate linking copulas of the observed and corresponding latent variable, and loading parameters of other latent variables (e.g., p-factor and bi-factor) are transformed to partial correlations given previous latent variables and then replaced with bivariate linking copulas to represent conditional dependence.

With copulas, we assume that observed and latent variables have been transformed to be uniform random variables in the interval $(0,1)$. Generalizing the Gaussian factor models, the copula densities of (structured) factor copula models with weak residual dependence are summarized below based on [11].

Definition 4. (1-factor copula model with residual dependence). Let $\boldsymbol{U}_D = (U_1, U_2, \ldots, U_d)$ be a vector of dependent $U(0,1)$ random variables and let $V \sim U(0,1)$ be the latent variable. For $j \in \{1,\ldots,D\}$, let C_{j0} be the absolutely continuous bivariate copula cdf for (U_j,V) with density c_{j0} and conditional cdf $C_{j|0}(u|v) = \Pr(U_j \le u|V=v) = \partial C_{j0}(u,v)/\partial v$. Assume the conditional distribution of $\boldsymbol{U}_D$ given V has copula density c_{res}. The joint density of $\boldsymbol{U}_D$ is:

$$c_{U_D}(\boldsymbol{u}_D) = \int_0^1 \underbrace{c_{\text{res}}\big(C_{1|0}(u_1|v),\ldots,C_{D|0}(u_D|v)\big)}_{\text{part 1}} \times \underbrace{\Big\{\prod_j c_{j0}(u_j,v)\Big\}}_{\text{part 2}} \,\mathrm{d}v. \tag{5}$$

Part 1 involves the conditional dependence of $[U_1|V],\ldots,[U_D|V]$. If $c_{\text{res}} \equiv 1$ on $(0,1)^D$ (independence copula) to represent conditional independence given V, then the integrand only involves part 2, and c_{U_D} becomes the density for the 1-factor copula model.

Definition 5. (Bi-factor copula model with residual dependence). With G groups, let $D = \sum_{g=1}^G d_g$, with d_g variables from group g. Let $\boldsymbol{D} = (d_1,\ldots,d_G)$. Let $\boldsymbol{V} = (V_0,V_1,\ldots,V_G)$ be the vector of mutually independent latent variables. Let $\boldsymbol{U}_g^\top = (U_{1g},\ldots,U_{d_g g})$, $g \in \{1,\ldots,G\}$, and $\boldsymbol{U}_{\boldsymbol{D}}^\top = (\boldsymbol{U}_1^\top,\ldots,\boldsymbol{U}_G^\top)$. Let

$$\widetilde{\boldsymbol{u}}_g = \big(C_{U_{1g}|V_0,V_g}(u_{1g}|v_0,v_g),\ldots,C_{U_{d_g g}|V_0,V_g}(u_{d_g g}|v_0,v_g)\big).$$

Assuming the conditional distribution of $\boldsymbol{U}_D$ given $\boldsymbol{V} = \boldsymbol{v}$ has copula density $c_{\text{res}}(\widetilde{\boldsymbol{u}}_1,\ldots,\widetilde{\boldsymbol{u}}_G)$, the joint density of $\boldsymbol{U}_D$ is:

$$c_{\boldsymbol{U_D}}(\boldsymbol{u_D}) = \int_0^1 \cdots \int_0^1 \underbrace{\Big\{\prod_{g=1}^G \prod_{j=1}^{d_g} c_{U_{jg}V_0}(u_{jg},v_0)\cdot c_{U_{jg}V_g;V_0}\big(C_{U_{jg}|V_0}(u_{jg}|v_0),v_g\big)\Big\}}_{\text{part1}} \times \underbrace{c_{\text{res}}\big(\widetilde{\boldsymbol{u}}_1,\ldots,\widetilde{\boldsymbol{u}}_G\big)}_{\text{part2}} \,\mathrm{d}v_0 \cdots \mathrm{d}v_g. \tag{6}$$

For notation, $c_{U_{jg}V_g;V_0}$ is the copula density to represent conditional dependence of U_{jg},V_g given V_0. Part 1 in (6) has the joint density of observed and latent variables assuming conditional independence, while part 2 accounts for conditional dependence of the U_{jg} given the latent variables.

Definition 6. (Oblique factor copula model with residual dependence). With G groups, let $D = \sum_{g=1}^G d_g$, with d_g variables from group g. Let $\boldsymbol{D} = (d_1,\ldots,d_G)$. Let $\boldsymbol{V} = (V_1,\ldots,V_G)$. Let $\boldsymbol{U}_g^\top = (U_{1g},\ldots,U_{d_g})$, $g \in \{1,\ldots,G\}$, $\boldsymbol{U}_{\boldsymbol{D}}^\top = (\boldsymbol{U}_1^\top,\ldots,\boldsymbol{U}_G^\top)$. Let $\widetilde{\boldsymbol{u}}_g = (C_{U_{1g}|\boldsymbol{V}}(u_{1g}|\boldsymbol{v}),\ldots,C_{U_{d_g g}|\boldsymbol{V}}(u_{d_g g}|\boldsymbol{v}))$. Assume the conditional distribution of $\boldsymbol{U}_D$ given $\boldsymbol{V} = \boldsymbol{v}$ has copula density $c_{\text{res}}(\widetilde{\boldsymbol{u}}_1,\ldots,\widetilde{\boldsymbol{u}}_G)$. Let $c_{\boldsymbol{V}}$ be the joint copula density of the latent variables. The joint density of $\boldsymbol{U}_D$ is:

$$c_{\boldsymbol{U_D}}(\boldsymbol{u_D}) = \int_0^1 \cdots \int_0^1 \underbrace{\Big\{\prod_{g=1}^G \prod_{j=1}^{d_g} c_{U_{jg},V_g}(u_{jg},v_g)\Big\} c_{\boldsymbol{V}}(\boldsymbol{v})}_{\text{part1}} \times \underbrace{c_{\text{res}}\big(\widetilde{\boldsymbol{u}}_1,\ldots,\widetilde{\boldsymbol{u}}_G\big)}_{\text{part2}} \,\mathrm{d}v_1 \cdots \mathrm{d}v_G. \tag{7}$$

Part 1 in (7) has the joint density of observed and latent variables assuming conditional independence, while part 2 accounts for conditional dependence of the U_{jg} given the latent variables.

3 Proxies for the latent variables

Proxy estimates are defined in [4] for the factor models in Section 2 assuming $\boldsymbol{\Omega}$ is an identity matrix or c_{res} is an independence copula, that is, assuming conditional independence of observed variables given the latent variables. In high dimensions, some weak residual or conditional dependence given the latent variables is not unreasonable and one would hope that the proxy variables are robust. The next Section 4 has some sufficient conditions to explain and interpret weak residual dependence.

3.1 Proxies in Gaussian factor models

Consider a $D \times 1$ random vector z_D from the p-factor Gaussian model with weak residual dependence as discussed in Section 2.1. The correlation matrix of z_D has a factor structure $\boldsymbol{\Sigma}_D = \boldsymbol{A}_D\boldsymbol{A}_D^\top + \boldsymbol{\Gamma}_D$, where $\boldsymbol{A}_D$ is a $D \times p$ matrix of factor loadings, with p being the number of factors. Here, $\boldsymbol{\Gamma}_D$ represents the covariance of $(\psi_j \varepsilon_j : 1 \le j \le D)$. The matrix $\boldsymbol{\Gamma}_D$ is the conditional covariance matrix of the observed variables given the latent variables, denoted as $\boldsymbol{\Gamma}_D = \text{Cor}(\boldsymbol{Z}_D, \boldsymbol{Z}_D|\boldsymbol{W})$; it is not diagonal under the assumption of residual dependence. Define $\boldsymbol{\Psi}_D^2 = \text{diag}(\psi_1^2, \dots, \psi_D^2)$, where the off-diagonal elements of $\boldsymbol{\Gamma}_D$ have been set to 0. The proxies are defined as the conditional expectation of the latent variables given the observed variables, assuming conditional independence.

In Gaussian factor models with residual dependence in Section 2.1, the conditional expectation proxies are defined with the loading matrix $\boldsymbol{A}_D$ and $\boldsymbol{\Psi}_D$ following the same expressions as presented in [4].

1. For the 1-factor model in Definition 1, let (w^0, z_D) be a realization of $(W, \boldsymbol{Z}_D)$. The proxy for W (or estimate of w^0) given z_D, based on $\boldsymbol{\Omega}_D = \boldsymbol{I}_D$, is:

$$\begin{aligned}\widetilde{w}_D &= \boldsymbol{A}_D{}^\top(\boldsymbol{A}_D\boldsymbol{A}_D^\top + \boldsymbol{\Psi}_D^2)^{-1} z_D \\ &= (1 + \boldsymbol{A}_D{}^\top \boldsymbol{\Psi}_D^{-2} \boldsymbol{A}_D)^{-1} \boldsymbol{A}_D^\top \boldsymbol{\Psi}_D^{-2} z_D,\end{aligned} \tag{8}$$

   if $\boldsymbol{A}_D\boldsymbol{A}_D^\top + \boldsymbol{\Psi}_D^2$ is non-singular and $\boldsymbol{\Psi}_D$ has no zeros on its diagonal.
2. For the bi-factor model in Definition 2, let $(w_0^0, w_1^0, \dots, w_G^0, z_D)$ be a realization of $(W_0, W_1, \dots, W_G, \boldsymbol{Z}_D)$. The proxies (estimates of $w_0^0, w_1^0, \dots, w_G^0$) given z_D, based on $\boldsymbol{\Omega}_D = \boldsymbol{I}_D$, are:

$$\widetilde{w}_0 = (\boldsymbol{a}_0)^\top (\boldsymbol{A}_D\boldsymbol{A}_D^\top + \boldsymbol{\Psi}_D^2)^{-1} \boldsymbol{z}_D, \tag{9}$$

$$\widetilde{w}_g(\widetilde{w}_0) = (\boldsymbol{b}_g^\top, 0)\boldsymbol{\Sigma}_g^{-1}(\boldsymbol{z}_g^\top, \widetilde{w}_0)^\top, \quad g \in \{1,\dots,G\}, \tag{10}$$

where $\boldsymbol{a}_0 = \boldsymbol{a}_{0,D}$ is the first column of the loading matrix, $\boldsymbol{z}_g = (z_{1g},\dots,z_{d_g g})^\top$, $\boldsymbol{b}_{0g} = \boldsymbol{b}_{0,d_g g}$ and $\boldsymbol{b}_g = \boldsymbol{b}_{d_g g}$ are the $d_g \times 1$ global and local loading vectors for group g. Let $\boldsymbol{B}_g = [\boldsymbol{b}_{0g}, \boldsymbol{b}_g]$ (matrix of size $d_g \times 2$). Let $\boldsymbol{\Sigma}_g$ be the correlation matrix of $(\boldsymbol{Z}_g^\top, W_0)$. Then $\boldsymbol{\Sigma}_g = \begin{bmatrix} \boldsymbol{B}_g\boldsymbol{B}_g^\top + \boldsymbol{\Psi}_g^2 & \boldsymbol{b}_{0g} \\ \boldsymbol{b}_{0g}^\top & 1 \end{bmatrix}$ for $g \in \{1,\dots,G\}$ with $\boldsymbol{\Psi}_D^2 = \mathrm{diag}(\boldsymbol{\Psi}_1,\dots,\boldsymbol{\Psi}_G)$.

3. For the oblique factor model in Definition 3, the proxies (estimates of $w_1^0,\dots,w_G^0$) given $\boldsymbol{z}_D$, based on $\boldsymbol{\Omega}_D - \boldsymbol{I}_D$, are based on separate 1-factor models for each group.

3.2 Proxies in factor copula models

In this section, the proxies, based on the assumption that c_{res} comes from the independence copula, are summarized.

1. For the 1-factor copula model in Definition 4, let $(v^0, \boldsymbol{u}_D)$ be a realization of $(V, \boldsymbol{U}_D)$. The proxy estimate of v^0 is:

$$\widetilde{v}_D = \widetilde{v}_D(\boldsymbol{u}_D) = \frac{\int_0^1 v \prod_{j=1}^D c_{j0}(u_j, v)\,\mathrm{d}v}{\int_0^1 \prod_{j=1}^D c_{j0}(u_j, v)\,\mathrm{d}v}. \tag{11}$$

2. For the bi-factor copula model in Definition 5, let $(v_0^0, v_1^0, \dots, v_G^0, \boldsymbol{u}_D)$ be a realization of $(V_0, V_1, \dots, V_G, \boldsymbol{U}_D)$ with $\boldsymbol{u}_D = (\boldsymbol{u}_1, \dots, \boldsymbol{u}_G)$. Based on conditional expectations, the proxy estimates of v_0^0 and v_g^0 are given below as $\widetilde{v}_{0\boldsymbol{D}}(\boldsymbol{u}_D)$ and $\widetilde{v}_{g\boldsymbol{D}}(\boldsymbol{u}_g, \widetilde{v}_{0\boldsymbol{D}})$. Let

$$f_g(\boldsymbol{u}_g, v_0) = \int_0^1 \prod_{j=1}^{d_g} c_{U_{jg},V_g;V_0}(C_{U_{jg}|V_0}(u_{jg}|v_0), v_g)\,\mathrm{d}v_g, \quad g \in \{1,\dots,G\}.$$

With $\boldsymbol{D} = (d_1, \dots, d_g)$,

$$\widetilde{v}_{0\boldsymbol{D}}(\boldsymbol{u}_D) = \frac{\int_0^1 v_0 \prod_{g=1}^G \{\prod_{j=1}^{d_g} c_{U_{jg},V_0}(u_{jg}, v_0) \cdot f_g(\boldsymbol{u}_g, v_0)\}\,\mathrm{d}v_0}{\int_0^1 \prod_{g=1}^G \{\prod_{j=1}^{d_g} c_{U_{jg},V_0}(u_{jg}, v_0) \cdot f_g(\boldsymbol{u}_g, v_0)\}\,\mathrm{d}v_0}. \tag{12}$$

For $g \in \{1,\dots,G\}$,

$$\widetilde{v}_{g\boldsymbol{D}}(\boldsymbol{u}_g, \widetilde{v}_{0\boldsymbol{D}}) = \frac{\int_0^1 v_g \prod_{j=1}^{d_g} c_{U_{jg},V_0}(u_{jg}, \widetilde{v}_{0\boldsymbol{D}}) \cdot c_{U_{jg},V_g;V_0}(C_{U_{jg}|V_0}(u_{jg}|\widetilde{v}_{0\boldsymbol{D}}), v_g)\, \mathrm{d}v_g}{\int_0^1 \prod_{j=1}^{d_g} c_{U_{jg},V_0}(u_{jg}, \widetilde{v}_{0\boldsymbol{D}}) \cdot c_{U_{jg},V_g;V_0}(C_{U_{jg}|V_0}(u_{jg}|\widetilde{v}_{0\boldsymbol{D}}), v_g)\, \mathrm{d}v_g}. \tag{13}$$

3. For the oblique factor copula in Definition 6, let $(v_1^0, \ldots, v_G^0, \boldsymbol{u}_1, \ldots, \boldsymbol{u}_G)$ be a realization of $(V_1, \ldots, V_G, \boldsymbol{U}_D)$. The proxies are given below as $\widetilde{v}_{g\boldsymbol{D}}(\boldsymbol{u}_g)$ for $g \in \{1, \ldots, G\}$:

$$\widetilde{v}_{g\boldsymbol{D}}(\boldsymbol{u}_g) = \frac{\int_0^1 v_g \prod_{j=1}^{d_g} c_{U_{jg},V_g}(u_{jg}, v_g)\, \mathrm{d}v_g}{\int_0^1 \prod_{j=1}^{d_g} c_{U_{jg},V_g}(u_{jg}, v_g)\, \mathrm{d}v_g}. \tag{14}$$

4 Consistency of proxies: model known

General proofs for factor copula models are not possible or would take too much space because of the generality for c_{res}. For Gaussian factor models with parameters that are loading and correlation matrices, proofs are possible with interpretable conditions for weak residual dependence. These provide insights into conditions for the consistency of the proxies in the factor copula counterparts, and these conditions provide more concise definitions of weak residual dependence.

The theorems are established assuming all parameters in the factor model and residual dependence are known. We have to know that proxy estimates are consistent in this case in order to discuss a method for determining parametric bivariate linking copula families for factor copula models and estimating the parameters.

With proxy estimates of latent variables in high dimensions, existing methods for vine copulas can be used to choose bivariate linking copula families in Definitions 4 to 6.

To consider factor copula models, the variables are monotonically related to each other and to the latent variables. Without this assumption, there could be identifiability issues. For data, assuming the variables are monotonically related and the correlation matrix $\boldsymbol{R}_{data}$ of normal scores summarizes the dependence well, the difference with $\boldsymbol{R}_{model}$ based on a 1-factor or structured factor model can be computed to assess which variables have residual dependence; see more details in Section 6.

In this section, sufficient conditions are obtained on the residual dependence so that the proxies defined in Section 3 for Gaussian and copula factor models with weak residual dependence remain consistent as the dimension $D \to \infty$. The interpretation of the sufficient conditions is that the total dependence of any variable with the other variables, conditioned on the latent variables, is bounded as $D \to \infty$.

Section 4.1 discusses results for the consistency of factor scores in Gaussian factor models, while Section 4.2 presents results for the consistency of proxy variables in factor copula models. The assumptions regarding the strength of the residual dependence are similar for both Gaussian factor models and factor copula models.

4.1 Consistency in Gaussian factor models

For the 1-factor Gaussian model with weak residual dependence in Definition 1, suppose the maximum eigenvalue of $\boldsymbol{\Gamma}_D$ is bounded as $D \to \infty$, the model is an approximate factor model from the definition in [1]. This assumption is sufficient for the proxy in (8) to be asymptotically consistent. An assumption that is more straightforward and can be extended to the copula case is provided below.

Assumption 1 Let $\boldsymbol{\Gamma}_D = (\gamma_{st})_{1\le s,t\le D}$ be the conditional covariance matrix of observed variables given the latent variables in Definition 1. Let $S_D = \sum_{j=1}^{D} \psi_j \varepsilon_j$ and $\bar{\varepsilon} = S_D/D$, then $\mathrm{E}(S_D^2) = \sum_{s=1}^{D}\sum_{t=1}^{D} \gamma_{st}$. Assume

$$0 < \liminf_{D\to\infty} D^{-1}\mathrm{E}(S_D^2) < \limsup_{D\to\infty} D^{-1}\mathrm{E}(S_D^2) < M,$$

where M is a positive constant.

Remark 1. Let $\gamma_{s+}^{(D)} = \sum_{t=1}^{D} \gamma_{st}$ for $s \in \{1,\dots,D\}$. Below are examples where Assumption 1 is satisfied and not satisfied.

- (Satisfied). If $\gamma_{s+}^{(D)}$ is $O(1)$ as $D \to \infty$ for all s, then Assumption 1 is satisfied; for example, ε's are indexed to have ante-dependence of order 1: $0 < r_1 < \gamma_{j,j+1} < r_2 < 1$ for all j, and $\gamma_{jk} = \prod_{i=j}^{k-1} \gamma_{i,i+1}$ for $k - j \ge 2$.
- (Not satisfied). If $\gamma_{s+}^{(D)}$ is $O(D)$ as $D \to \infty$ for at least one s, then Assumption 1 is not satisfied; for example, ε_1 is dominating: $0 < r_1 < \gamma_{1j} < r_2 < 1$ for all j, and $\gamma_{jk} = \gamma_{1j} \cdot \gamma_{1k}$.

Theorem 1. *(Asymptotic properties of proxies in 1-factor Gaussian with residual dependence model). For the 1-factor Gaussian model with residual dependence in Definition 1, with given sequence $\{\alpha_1, \alpha_2, \dots\}$, suppose there is a realized infinite sequence of observed variables $z_1, z_2, \dots$ and a realized infinite sequence of disturbance terms $e_1, e_2, \dots$ with realized latent variable w^0 (independent of dimension D). For the truncated sequence to the first D variables, let $\boldsymbol{z}_D = (z_1, \dots, z_D)^\top$, $\boldsymbol{e}_D = (e_1, \dots, e_D)^\top$, and let the loading matrix be the vector $(\alpha_1, \dots, \alpha_D)^\top$. Assume*

$$-1 < \liminf_{j\to\infty} \alpha_j < \limsup_{j\to\infty} \alpha_j < 1, \quad \lim_{D\to\infty} D^{-1} \sum_{j=1}^{D} |\alpha_j| \to const, \quad const \neq 0.$$

With Assumption 1 on residual dependence, for the proxy defined in equation (8), *then $\widetilde{w}_D - w^0 = O_p(D^{-1/2})$ as $D \to \infty$.*

Proof. The proof is in Section 9.1.

Remark 2. Assumption 1 implies $\mathrm{Var}(\bar{\varepsilon}) = O(D^{-1})$, the same order as the case of independent and identically distributed. The convergence rate at $\sqrt{D}$ for the proxy variable is the same with the assumption of conditional independence of observed variables given the latent variable.

The consistency of proxies in the oblique factor model is a straightforward corollary of Theorem 1, with an assumption of residual dependence similar to Assumption 1. The observed variables $\mathbf{Z}_g$, $g \in \{1,\ldots,G\}$ in one group g are linked to the same (realized) latent variable w_g^0 and these variables satisfy a 1-factor model with weak residual dependence. The details of the theorems are omitted.

The conditions of the bi-factor Gaussian model for consistency of the proxy variables are a bit stronger than that required for the 1-factor model.

Assumption 2 *For the bi-factor model with weak residual dependence and a fixed number of G groups in* (3), $D = \sum_{g=1}^{G} d_g$. *Let* $\boldsymbol{\Gamma_D} = (\gamma_{st})_{1\le s,t\le D}$ *be the conditional covariance matrix of disturbances* $\{\varepsilon_{jg}\}$. *Assume*

$$0 < \liminf_{D\to\infty} D^{-1} \sum_{s=1}^{D}\sum_{t=1}^{D} |\gamma_{st}| < \limsup_{D\to\infty} D^{-1} \sum_{s=1}^{D}\sum_{t=1}^{D} |\gamma_{st}| < M,$$

where M is a positive constant.

Theorem 2. *(Asymptotic properties of factor scores in bi-factor Gaussian model with residual dependence model). For the bi-factor Gaussian factor model with fixed G groups defined in Definition 2, with given sequences* $\{\alpha_{0,1g},\alpha_{0,2g},\ldots\}$ *and* $\{\alpha_{1g},\alpha_{2g},\ldots\}$ *for* $g \in \{1,\ldots,G\}$, *suppose there are realized infinite sequences of observed variables* $\mathbf{z}_1^\top = (z_{1,1},z_{2,1},\ldots)$, $\ldots,\mathbf{z}_G^\top = (z_{1,G},z_{2,G},\ldots)$ *and realized sequences of disturbance terms* $\boldsymbol{e}_1^\top = (e_{1,1},e_{2,1},\ldots)$, $\ldots,\boldsymbol{e}_G^\top = (e_{1,G},e_{2,G},\ldots)$, *with latent variables* $\mathbf{w}^0 = (w_0^0,w_1^0,\ldots,w_G^0)$. *Truncate the sequences to the first* d_g *variables* $\mathbf{z}_{d_g,g}$ *for* $g \in \{1,\ldots,G\}$, *with no* d_g *dominating others. Let* $\mathbf{z_D} = (\mathbf{z}_{d_1,1}^\top,\ldots,\mathbf{z}_{d_G,G}^\top)^\top$, $\boldsymbol{e_D} = (\boldsymbol{e}_{d_1,1}^\top,\ldots,\boldsymbol{e}_{d_G,G}^\top)$ *and assume that the loading matrices*

$$\boldsymbol{A_D} = [\boldsymbol{a}_{0,D},\mathrm{diag}(\boldsymbol{b}_{d_1 1},\ldots,\boldsymbol{b}_{d_G G})]$$

are of full rank, with bounded condition number over $d_g \to \infty$ *for all g. With* $\mathbf{b}_{d_g g} = (\alpha_{1g},\ldots,\alpha_{d_g,g})^\top$, *assume* $D^{-1}\|\boldsymbol{a}_{0,D}\|_1 \not\to 0$, $d_g^{-1}\|\boldsymbol{b}_{d_g g}\|_1 \not\to 0$ *for* $g \in \{1,\ldots,G\}$, *and* $\liminf_{j,g} \psi_{jg}^2 > 0$. *Let* $\widetilde{w}_{\boldsymbol{D}} = (\widetilde{w}_0,\ldots,\widetilde{w}_G)$ *be the factor scores defined in* (9) *and* (10). *With Assumption 2 on the residual dependence, then*

$$\widetilde{w}_g - w_g^0 = O_p(D^{-1/2}), \quad g \in \{0,1,2,\ldots,G\}.$$

Proof. See Section 9.2 for the proof.

The assumptions $D^{-1}\|\boldsymbol{a}_{0,D}\|_1 \not\to 0$, $d_g^{-1}\|\boldsymbol{b}_{d_g g}\|_1 \not\to 0$ for $g \in \{1,\ldots,G\}$ are similar to an assumption in Theorem 1. The average absolute loading in any column of $\boldsymbol{A_D}$ does not go to 0 as $D \to \infty$ (otherwise, the latent variables do not explain enough dependence in the observed variables).

4.2 Consistency in factor copula models

In this section, we present the conditions on the residual dependence for the consistency of the proxy variables in the 1-factor copula models with residual dependence defined in Section 2.2. We assume that the linking copulas between the observed variables and the latent variables, as well as the copula for the residual dependence, are known. (This step is needed before the case where parametric families of linking copulas are estimated.) For tractability of a proof, we also assume that residual dependence conditioned on the latent variables, is summarized by a multivariate Gaussian copula. The results under this simplified assumption provide insights into (a) the 1-factor copula model when the residual dependence is modeled by other well-behaved parametric copulas instead of Gaussian copulas, and (b) the bi-factor copula model with weak residual dependence.

The regularity conditions on the linking copulas stated in Assumption 1 in [4] are also necessary. These conditions require the linking copulas from the observed variables to the latent variables to be well-behaved, and the dependence between the observed and latent variables to be strong enough and bounded away from comonotonicity and countermonotonicity. (If the dependence can approach comonotonicity or countermonotonicity, then one of the variables can be taken as the proxy.) Additionally, further assumptions on the residual dependence are required, analogous to Section 4.1. It is required that for any variable, the total dependence of the variable with the other variables, conditioned on the latent variables, is bounded as $D \to \infty$. In the following remark, we provide an outline of the proof for the consistency of proxy variables in the 1-factor copula model.

Remark 3. Consider a realization $(v^0, \boldsymbol{u}_D)$ of $(V, \boldsymbol{U}_D)$ where v^0 is to be estimated based on $\boldsymbol{u}_D$. Let $\overline{L}_D(v;\boldsymbol{u}_D)$ be the average negative log-likelihood treating v as a parameter, that is,

$$\overline{L}_D(v;\boldsymbol{u}_D) := \underbrace{-D^{-1}\sum_{j=1}^{D}\log c_{j0}(u_j,v)}_{:=\overline{qL}_D(v,\boldsymbol{u}_D)} \underbrace{-\frac{1}{D}\log c_{\mathrm{res}}(C_{1|0}(u_1|v),\ldots,C_{D|0}(u_D|v))}_{:=\overline{L}_D^{(r)}(v,\boldsymbol{u}_D)}$$

so that $\overline{L}_D(v;\boldsymbol{u}_D) = \overline{qL}_D(v;\boldsymbol{u}_D) + \overline{L}_D^{(r)}(v;\boldsymbol{u}_D)$, where $\overline{L}_D^{(r)}$ comes from the residual dependence given the latent variable.

From the proof of Theorem 5 in [4], the proxy $\widetilde{v}_D$ and maximum likelihood estimate v_D^* with $c_{\mathrm{res}} \equiv 1$ are asymptotically equivalent. With residual dependence, the optimum v_D^* estimated from $\overline{qL}_D$ is now a quasi-MLE. Therefore, to show the consistency of proxy variables, we need to show that the quasi-MLE v_D^* converges to the realized latent variable v^0 under some conditions on the weak residual dependence as $D \to \infty$.

The gradient of the negative log-likelihood for v is

$$\frac{\partial \overline{L}_D}{\partial v} = \frac{\partial \overline{qL}_D}{\partial v} + \frac{\partial \overline{L}_D^{(r)}}{\partial v} \tag{15}$$

The proof mainly consists of two steps.
A. Show that the limiting inference (score) function or derivative of $\overline{L}_D$ with respect to v has a unique solution at v^0.
B. Show that the gradient of the $\overline{L}_D^{(r)}$ with respect to v at v^0 goes to zero as $D \to \infty$.

To prove $v_D^* \to v^0$, for Step B, it is sufficient to justify the Fisher consistency, by showing

$$\left.\frac{\partial \overline{L}_D^{(r)}}{\partial v}\right|_{v=v^0} \to_p 0, \quad D \to \infty. \tag{16}$$

This implies $\lim_{D\to\infty} \partial \overline{qL}_D(v;\boldsymbol{u}_D)/\partial v \mid_{v=v^0} = 0$ since, from step A (under usual regularity conditions for the likelihood), $\lim_{D\to\infty} \partial \overline{L}_D(v;\boldsymbol{u}_D)/\partial v|_{v=v^0} = 0$.

The Fisher consistency can be derived by a first-order Taylor expansion. Expanding $\partial \overline{qL}_D(v^0;\boldsymbol{u}_D)/\partial v$ around the quasi-MLE v_D^* leads to

$$\frac{\partial \overline{qL}_D(v_D^*;\boldsymbol{u}_D)}{\partial v} + \frac{\partial^2 \overline{qL}_D(v_D^*;\boldsymbol{u}_D)}{\partial v^2}(v^0 - v_D^*) .$$

Since $\partial \overline{qL}_D(v_D^*;\boldsymbol{u}_D)/\partial v = 0$ and assuming $\lim_{D\to\infty} |\partial^2 \overline{qL}_D(v_D^*;\boldsymbol{u}_D)/\partial v^2|$ is bounded, then $v_D^* - v^0 \to 0$ when $D \to \infty$.

The Fisher consistency in (16) is justified in Theorem 3 for the 1-factor copula model with Assumption 3 and Condition 1 below.

Assumption 3 Consider the 1-factor model with residual dependence in Definition 4.
(i) Assume the bivariate linking copulas $\{C_{j0}(u,v)\}$ are such that the derivatives $\{\partial C_{j|0}(u|v)/\partial v\}$ are uniformly bounded for all j and the limits of the derivatives go to 0 when $u \to 1$ or $u \to 0$ for $v_A \le v \le v_B$, where v_A is close to 0 and v_B is close to 1.
(ii) Assume the bivariate linking copulas $\{C_{j0}(u,v)\}$ are such that the ratio of derivatives $[\partial^2 C_{j|0}(u|v)/\partial v \partial u]/[\partial C_{j|0}(u|v)/\partial u]$ are finite and uniformly bounded over j when $u \to 1$ or $u \to 0$ and for $v_A \le v \le v_B$, where v_A is close to 0 and v_B is close to 1.
(iii) Let $\widetilde{U}_j^0 = C_{j|0}(U_j|V = v^0)$, and $E_j^0 = \Phi^{-1}(\widetilde{U}_j^0)$ for all j. Assume that $\sum_{j=1}^D \sum_{k=1}^D |\mathrm{Cov}(\widetilde{U}_j^0, \widetilde{U}_k^0)| = O(D)$ or $\sum_{j=1}^D \sum_{k=1}^D |\mathrm{Cov}(E_j^0, E_k^0)| = O(D)$.

Condition 1 *For a Gaussian copula for residual dependence, with correlation matrix parameter $\boldsymbol{\Gamma}_D = (\gamma_{jk})$, with its (j,k) entry given by $\mathrm{Cov}(E_j^0, E_k^0)$. Let $\boldsymbol{\Gamma}_D^{-1} = (\gamma^{jk})_{1\le j,k\le D}$. Assume $\sum_{j=1}^D |\gamma^{jk}| = O(1)$ and $\sum_{j=1}^D |\gamma_{jk}| = O(1)$ for all k.*

Remark 4. One example where Condition 1 is satisfied is for an m-truncated Gaussian partial correlation vine. In [6], a positive $D \times D$ correlation matrix can be re-parametrized into D correlations and $D(D-1)/2$ partial correlations which are algebraically independent in $(-1,1)$. The re-parametrization is not unique and can be represented by vine trees. Consider the vine trees truncated to m levels, from Proposition 2 in [7], there are $(D-m)(D-m-1)/2$ positions of the inverse correlation matrix $\boldsymbol{\Gamma}_D^{-1}$ that are zero and thus Condition 1 is satisfied.

Theorem 3. *(Consistency of proxy in 1-factor copula model with weak residual dependence when linking copulas are known). For the 1-factor model with weak residual dependence in Definition 4, suppose there is a realized infinite sequence* $u_1, u_2, \ldots,$ *with latent variable* v^0 *(independent of dimension D). Truncate the sequence to the first D variables and let* $\boldsymbol{u}_D = (u_1, \ldots, u_D)^\top$*. Define the average negative quasi-log-likelihood in v as*

$$\overline{qL}_D(v) = -D^{-1} \sum_{j=1}^{D} \log c_{j0}(u_j, v). \tag{17}$$

Assume $\lim_{D\to\infty} \overline{qL}_D$ *has a global minimum for all* $D \geq d$ *(for some large* $d > 0$*) and the second derivative* $\overline{qL}_D''$ *at the global minimum has a limit infimum that is strictly positive. Also, assume the likelihood function in* (17) *satisfies the usual regularity conditions in Assumption 1 in [4]. With Assumption 3, Condition 1, stochastic increasingness* ($1 - C_{j|0}(\cdot|v)$ *increasing in v for all j), and* $v_A < v^0 < v_B$*, the proxy* $\widetilde{v}_D$ *defined in* (11) *satisfies* $\widetilde{v}_D - v^0 = o_p(1)$ *as* $D \to \infty$.

Proof. The proof is in Section 9.3.

Remark 5. Assumption 3 requires the linking copulas from the observed variables to the latent variable are well-behaved such that the limits of some partial derivatives or ratios of some partial derivatives satisfy certain conditions when the observed variables reach the boundary. If the linking copulas for modeling the residual dependence are all bivariate Gaussian copulas, these conditions are easy to check and are satisfied. This assumption is mild, as it holds for commonly used parametric copula families such as the Gumbel, BB1 and Frank families; this can be verified through analytical and numerical checks. The conditions in Assumption 3(iii) on the residual correlation matrix are parallel to the Gaussian copula case. These assumptions on the linking copulas and on the residual dependence provide insights for the bi-factor and oblique-factor models.

5 Consistency of proxy variables with estimated parameters

In this section, rather than completely specified factor models, we assume there are loading parameters for the Gaussian factor models or parameters for bivariate linking copulas in the factor copula models that have to be estimated. For factor models with residual dependence, the estimation based on the log-likelihood assuming $\boldsymbol{\Omega}_D = \boldsymbol{I}_D$ or c_{res} being an independence copula implies estimation with a mis-specified likelihood, and the resulting estimator from this mis-specified log-likelihood is a quasi-MLE.

In this section, weak sufficient conditions for residual dependence are given so that it can be proved for the 1-factor Gaussian models with residual dependence that the quasi-MLEs of loading parameters are consistent. Note that parameters for the residual dependence are not involved in the estimation of the proxies. The log-likelihoods in this section are functions of the loading parameters and not the latent variables (as in Section 4.1).

Let $\boldsymbol{Y}_{1,D},\ldots,\boldsymbol{Y}_{N,D},\ldots$ be an (infinite) random sample from a density g_D and let the realizations be $\boldsymbol{y}_{1,D},\ldots,\boldsymbol{y}_{N,D},\ldots$. ($Y=Z$ for Gaussian factor models and $Y=U$ for factor copula models.) The ith observation vector is $\boldsymbol{y}_{i,D}=(y_{i1},\ldots,y_{iD})^\top$. Suppose $g_D(\boldsymbol{y}_D)$ is the density with residual dependence and $f_D(\boldsymbol{y}_D,\boldsymbol{\theta})$ is the parametric density of the corresponding factor model assuming conditional independence (given latent variables). The quasi-MLE maximizes the quasi-log-likelihood

$$\hat{\boldsymbol{\theta}}_D=\hat{\boldsymbol{\theta}}_{D,N}=\hat{\boldsymbol{\theta}}_D(\boldsymbol{y}_{1,D},\ldots,\boldsymbol{y}_{N,D})=\operatorname{argmax}\sum_{i=1}^{N}\log f_D(\boldsymbol{y}_{i,D};\boldsymbol{\theta}).$$

Under some regularity conditions, such as in [13], including the MLE asymptotically being in the interior of the parameter space, then $\hat{\boldsymbol{\theta}}_D\to_p\boldsymbol{\theta}_D^*$ as $N\to\infty$ (with fixed D), where $\boldsymbol{\theta}_D^*$ minimizes the Kullback-Leibler (KL) divergence

$$\int g_D(\boldsymbol{y}_D)\log\Big[\frac{g_D(\boldsymbol{y}_D)}{f_D(\boldsymbol{y}_D;\boldsymbol{\theta})}\Big]\,\mathrm{d}\boldsymbol{y}_D. \tag{18}$$

For $j=1,2,\ldots$, let $\boldsymbol{\theta}_j$ be the dependence parameter vector for observation j linking to the latent variables. The parameter vectors $\boldsymbol{\theta}_j$ are specified in the Gaussian factor models with weak residual dependence as below.

- Gaussian 1-factor: θ_j is a loading parameter or correlation parameter with the latent variable, that is $\boldsymbol{\theta}_j=\alpha_j$.
- Gaussian bi-factor: $\boldsymbol{\theta}_j=(\alpha_{0,jg},\alpha_{jg})$ if observation j is in group g.
- Gaussian oblique factor: $\boldsymbol{\theta}_j=\alpha_{jg}$ if observation j is in group g. For Gaussian oblique factor models, there are additional parameters $\{\rho_{ij}:1\le i<j\le G\}$ for the correlations of the latent variables.

For factor copula models, the parameters are in the copulas in Definitions 4 to 6.

For fixed D, let $\boldsymbol{\theta}_D^*=\{\theta_{j,D}^*:j\in\{1,\ldots,D\}\}$ be the vector minimizing the KL divergence of f_D relative to g_D in (18).

A theorem is given only for the case of the Gaussian 1-factor model (see Theorem 4) and the proof is long. The details are much more complicated than in [4]. The theorem is stated in Section 5.2 with its proof in Section 9.4. The conditions for the theorem can be intuitively extrapolated to 1-factor and group-based structured factor copula models. Section 7 has simulation studies showing the effect of weak residual dependence on parameter and latent variable estimation in factor copula models. For a specific simulation scenario with weak residual dependence, the key properties of the score vector and Hessian matrix in the proof of Theorem 4 and its lemmas can be verified numerically.

5.1 Estimation of parameters in blocks

To avoid the rate of convergence with both N and D increasing to infinity, we use the property of closure under margins for factor models so that parameters can be estimated in blocks as D gets large. This is valid with a super-population assumption for the variables (and corresponding parameters linking to latent variables). This is the methodology of proofs used in [4] for parameter estimation.

With residual dependence, for block-estimation to be valid, there is an additional assumption. The variables are assumed to be organized in blocks such that the conditional correlations between the blocks conditioned on the latent variables is weaker than the conditional correlations for the observed variables within blocks.

With the assumption that the residual dependence is sampled from some super-populations, such that for any variable, the total dependence of the variable with other variables given the latent variables is bounded as $D \to \infty$, the model maintains the "closed-under-margin" property by blocks. Therefore, the block-estimation method discussed in Section 5.3 in [4] can be utilized to estimate the increasing parameters in the loading matrix for Gaussian factor models or on the linking copulas in factor copula models in sequential blocks. The block size d_{block} is similar across blocks, and the number of blocks increases with the growing number of observed variables.

5.2 Differences between quasi-MLEs and true parameters

In this section, we mainly show that with the super-population assumption on the residual dependence, the differences of the quasi-MLEs estimated in sequential blocks and the true parameters are of order $O(N^{-1/2}) + O(d_{\text{block}}^{-1})$. The remark below discusses an outline of this result.

Remark 6. Consider a random sample of size N with D variables. The variables are split into K blocks for parameter estimation. Let the kth block be $\mathscr{B}_k$ and the block size $d_k \approx d_{\text{block}}$, for $k = 1, \dots, K$. The differences between the quasi-MLEs and true parameters in each block come from two aspects.

Firstly, based on analysis of Kullback-Leibler divergence between f_D and g_D, under conditions in Assumption 4 on the weak residual dependence,

$$\theta^*_{j,d_{\text{block}}} - \theta_j = O(d_{\text{block}}^{-1}) \quad \text{for } j \in \mathscr{B}_k. \tag{19}$$

Secondly, consider the sampling variability. The quasi-MLEs for a sample of size N converge to the minimizer of the KL divergence at rate $O(1/\sqrt{N})$. Let the quasi-MLEs be $\hat{\theta}_{j,N,d_{\text{block}}}$ in one block $\mathscr{B}_k$. Under the general regularity conditions for quasi-maximum likelihood estimation [13],

$$\hat{\theta}_{j,N,d_{\text{block}}} - \theta^*_{j,\text{block}} = O_p(N^{-1/2}), \quad j \in \mathscr{B}_k.$$

Therefore, the differences between the quasi-MLEs and the true parameters are of order $O(N^{-1/2}) + O(d_{\text{block}}^{-1})$. If the block size is sufficiently large, the differences between the quasi-MLEs and the true parameters in each block will be dominated by sampling variability.

The assumptions on residual dependence, ensuring (19) is satisfied, are outlined in Assumption 4. The statements for the 1-factor Gaussian model are presented in Theorem 4. The conditions in Assumption 4 can be extended to other factor (copula) models, such as the 1-factor copula model and bi-factor Gaussian (copula) models with residual dependence. The mathematical proof for the 1-factor Gaussian model is tractable but lengthy. The details of the other models are omitted due to the tedious calculations.

Assumption 4 *(Super-population): (1) In the first-level super-population, the observed variables (or their correlations, partial correlations, or linking copulas) are sampled from appropriate super-populations. (2) In the second-level super-population, the conditional dependencies of any pair of variables are assumed to be sampled from some super-populations such that for any observed variable j, the total dependence of this variable with other variables with indices in* $\{1,\ldots,j-1,j+1,\ldots\}$*, given the latent variables, is bounded* $O(1)$ *as* $D \to \infty$*.*

In the 1-factor Gaussian factor model with weak residual dependence in (2)*, for a fixed dimension, let* $\boldsymbol{\Gamma}_D = (\gamma_{st})_{1\le s,t\le D}$ *be the conditional covariance matrix of observed variables given the latent variables. Assume that*

$$\sum_{\ell=1}^{D} |\gamma_{\ell k}| = \sum_{\ell=1}^{D} |\gamma_{k\ell}| = O(1).$$

uniformly holds for all k.

The super-population assumption in the first level essentially avoids the cases where additional variables, indexed by j as $j \to \infty$, have dependence with the latent variable that converges to independence or perfect dependence. The super-population assumption in the second level means that the pattern of conditional dependence of a variable j with the remaining variables is similar over index j.

The conditions specified in the second paragraph of Assumption 4 are stricter than those in Assumption 1, because the condition in Assumption 4 serves as a sufficient condition to establish the condition in Assumption 1. The examples provided in Remark 1 show cases where the assumption could fail or be satisfied.

Theorem 4. *(Difference between KL minimizer and vector of true loadings in 1-factor Gaussian model with residual dependence). For the model in Definition 1 with loadings* α_j *for* $1 \le j \le D$*, let the correlation matrix of* g_D *be* $\boldsymbol{R}_D^0 = \boldsymbol{\alpha}_D\boldsymbol{\alpha}_D^\top + \boldsymbol{\Gamma}_D$*, where* $\boldsymbol{\Gamma}_D = (\gamma_{ij})_{1\le i,j\le D}$*. The correlation matrix of* f_D *is* $\boldsymbol{\Sigma}_D = \boldsymbol{\beta}_D\boldsymbol{\beta}_D^\top + \boldsymbol{\Psi}_D^2$*, with* $\boldsymbol{\Psi}_D^2 = \mathrm{diag}(1-\beta_j^2; 1 \le j \le D)$*.*

Let $\boldsymbol{\beta}_D^*$ *be the minimizer of the KL divergence between true* g_D *and* f_D*. With the super-population assumption and conditions in Assumption 4, as well as the loading parameters being bounded away from* ± 1*, then* $\boldsymbol{\beta}_D^* - \boldsymbol{\alpha}_D = O(D^{-1})$*.*

Proof. See Section 9.4 for the proof.

In [4], when $\boldsymbol{\Omega}_D = \boldsymbol{I}_D$ or c_{res} is the independence copula, under the super-population assumption, the differences between the estimated loading matrix for the Gaussian factor models or the estimated parameters on the linking copulas for the factor copula model and the true parameters are of order $O(N^{-1/2})$ uniformly in each block when the block size is large enough. The proxies are defined the same way as in [4], involving only the links from the observed variables to the latent variables. Therefore, Lipschitz continuity of the proxy variables presented in Section 5.1 of [4] still holds. Following the same logic, the consistency also holds when the parameters are estimated as both N and D increase to infinity.

6 Sequential estimation of parameters and latent variables

This section summarizes how estimation for factor copula models can proceed with weak residual dependence. The estimation method is used in Section 7 for some simulation examples. It is also used in the applications in [3].

We suppose the context of the data suggests that a factor model should be suitable. The steps in using proxies and assessing weak residual dependence are the following, assuming there is a random sample $\{(y_{i1},\dots,y_{id}) : i = 1,\dots,N\}$, from a multivariate cdf $F_{\boldsymbol{Y}}$.

1. Fit d univariate distributions and apply probability integral transforms, or apply rank transforms to get $\{\boldsymbol{u}_i = (u_{i1},\dots,u_{id}) : i = 1,\dots,N\}$.
2. Apply the standard normal quantile function Φ^{-1} to each u_{ij} to get vectors of normal scores $\{(z_{i1},\dots,z_{id}) : i = 1,\dots,n\}$.
3. Let $\boldsymbol{R}_{data}$ be the $d \times d$ correlation matrix of normal scores.
4. If there is no group structure, fit a 1-factor Gaussian model to $\boldsymbol{R}_{data}$. If there is a group structure with $G \geq 2$ groups, fit bi-factor and oblique-factor Gaussian models to $\boldsymbol{R}_{data}$ and keep the better fitting model. Let $\boldsymbol{R}_{model}$ be the model-based correlation matrix computed as a function of the loading parameters.
5. Compare $\boldsymbol{R}_{data}$ and $\boldsymbol{R}_{model}$ to note the (j,k) entries where $|R_{data,jk} - R_{model,jk}|$ exceeds a threshold and might indicate weak residual dependence.
6. Assuming few instances of exceeding the threshold, use the residual dependence pattern to find a parsimonious truncated vine structure (e.g., [7]) for the weak residual dependence.
7. If normal scores plots suggest departures from a Gaussian copula for the multivariate dependence, apply the proxy estimation theory in previous Sections of this article and in [4] for estimation of a parametric factor copula model.

The steps for estimating latent variables with proxies are as follows. Simple proxies (as in [4]) are used in the first stage to help in selecting parametric linking copula families from the observed to their corresponding latent variables and obtaining the initial estimates of the copula parameters. Then, "conditional expectation" proxies in Section 3.2 are constructed and used to estimate the parameters by optimizing the approximate (complete) log-likelihood with the latent variables replaced by the proxy variables.

Some details are repeated from [4] for the 1-factor copula model; they are similar for the bi-factor model except the simple proxy is based on a bi-factor Gaussian approximation.

For the parametric 1-factor copula with weak residual dependence in Definition 4, suppose there is realized random sample of size N, with ith observation vector $\boldsymbol{u}_i = (u_{i1}, \dots, u_{iD})$ as an independent realization of $\boldsymbol{U} = (U_1, \dots, U_D)$. If the latent variables $(v_1, \dots, v_N)$ are assumed observed, the complete log-likelihood, with parametric bivariate linking copula $C_{j0}(\cdot; \boldsymbol{\theta}_j)$ for U_j $(1 \le j \le D)$ and copula $C_{\text{res}}(\cdot; \boldsymbol{\theta}_{res})$ for residual dependence, is

$$\sum_{i=1}^{N} \log c_{\boldsymbol{U},V}(u_{i1}, \dots, u_{iD}, v_i; \boldsymbol{\theta}_D) = \sum_{i=1}^{N} \sum_{j=1}^{D} \log c_{j0}(u_{ij}, v_i; \boldsymbol{\theta}_j)$$
$$+ \sum_{i=1}^{N} \log c_{\text{res}}(C_{1|0}(u_{i1}|v_i; \boldsymbol{\theta}_1), \dots, C_{D|0}(u_{iD}|v_i; \boldsymbol{\theta}_D); \boldsymbol{\theta}_{\text{res}}). \tag{20}$$

There are two stages in the proxy estimation.

- Stage 1: Define the simple proxy random variable as

$$U_0 = P_D^U \Big(D^{-1} \sum_{j=1}^{D} U_j \Big), \tag{21}$$

  where P_D^U is the cdf of $\bar{U}_D := D^{-1} \sum_{j=1}^{D} U_j$. (21) has the version of proxy in [11]. For observation i, let $\bar{u}_i = D^{-1} \sum_{j=1}^{D} u_{ij}$ and let the sample proxy be $u_{i,0} = [\text{rank}(\bar{u}_i) - 0.5]/N$; $\text{rank}(\bar{u}_i)$ is defined as the rank of $\bar{u}_i$ based on $\bar{u}_1, \dots, \bar{u}_N$. Combining the bivariate scatterplots of proxy variables versus observed variables, empirical tail-weighted dependence measures and Akaike information criterion (AIC) values of a few candidate parametric copula families, one can decide on copula families C_{j0}. Substitute $v_i = u_{i,0}$ in the log-likelihood (20), and obtain the first-stage estimates of the parameters $\boldsymbol{\theta}$ from this as an approximate log-likelihood. Also, for a simple parametric model for weak residual dependence (such as for a Markov tree structure), $\boldsymbol{\theta}_{res}$ can be estimated.
- Stage 2: Construct the conditional expectation proxies based on (11) with the first-stage estimated parameters of $\boldsymbol{\theta}_D = \{\boldsymbol{\theta}_j, j = 1, \dots, D\}$. One-dimensional Gauss-Legendre quadrature can be used. Denote the improved proxies as $\widetilde{u}_{i,0}$ for $i \in \{1, \dots, N\}$. Substitute $v_i = \widetilde{u}_{i,0}$ in the log-likelihood (20) to obtain the second-stage estimates of the parameters.

7 Simulation studies

The theorems in previous sections provide intuitive conditions for weak residual dependence in factor copula models. Proofs emulating Theorems 3 and 4 would take too much space, so instead this section summarizes some representative simulation results for factor copulas to support and explain ideas in previous sections, and show the effects of weak residual dependence.

For the simulations, to concentrate on the behavior under weak residual dependence, we assume univariate margins are known or have been well estimated so that the probability integral transforms lead to a sample from a factor copula. In all the following settings, the parameters on the linking copulas are designed to be generated uniformly from a bounded subset of the parameter space; these are examples of sampling from some super-populations. A few different scenarios were assessed and some summaries are given in two subsections for the considered factor copula models.

7.1 1-factor copula with weak residual dependence

The sample size is N and there are D variables with bivariate linking copulas $C_{jV}(\cdot,\cdot;\boldsymbol{\theta}_j)$, $j \in \{1,\cdots,D\}$, to the latent variable. The parameters $\{\boldsymbol{\theta}_1,\ldots,\boldsymbol{\theta}_D\}$ are uniformly sampled from $U(\theta_L,\theta_U)$, where θ_L and θ_U are chosen so that the Kendall's taus of the bivariate copulas range between 0.4 and 0.8 for moderate to strong dependence. The parameters change for different generated random samples in different simulation runs. For simplicity, the weak residual dependence in simulated data sets assume the variable indices have been permuted so that a 1-truncated or 2-truncated D-vine structure can be used – for 1-truncated, there is weak conditional dependence for variables with neighboring indices, and for 2-truncated, there is additional weak dependence for variables indexed 2-apart. The parameters for weak residual dependence have Kendall tau values in the interval $(0.1,0.3)$.

The details of the three simulation settings are shown in Table 1. Setting 1 in Table 1 is designed to compare with the results presented in [4] in the 1-factor copula model. To further explore the influence of the weak residual dependence on parameter estimation and proxy estimation, additional settings are designed within Setting 2. The linking copulas in the truncated vine for residual dependence include Gaussian or Frank copulas. In Setting 3, the linking copulas from the observed to the latent variables comes from different families.

The simulation results for Settings 1, 2, and 3 are presented in Tables 2, 3 and 4, respectively. In Settings 1 and 2, the family linking copulas are assumed known and only copula parameters are estimated. In Setting 3, the parametric linking copula families are chosen from among a few candidates based on AIC. Similar to [4], the main summary is the Mean Absolute Error (MAE) or Root Mean Squared Error (RMSE) of the estimated parameters for the D linking copulas from the observed to the latent variables for two different methods, denoted as $m \in \{1,2\}$ in the equation

(23). Here, $m = 1$ indicates the simple proxy approach (based on (21)) and $m = 2$ represents the sequential proxy approach.

Table 1 Three simulation settings for the 1-factor copula model; in setting 3, the number of linking copulas in different families is approximately $D/3$, and the ν parameter of t copulas is fixed at 5. The parameters on the linking copulas from the observed to the latent variables are chosen with Kendall's tau in [0.4,0.8] while the parameters on the linking copulas in the residual dependence are chosen such that Kendall's tau is in $[0.1, 0.3]$. For Setting 2.2, the residual dependence is modeled by a 2-truncated D-vine with Kendall's tau in the first level of the D-vine in $[0.2, 0.3]$ and that in the second level of the D-vine in $[0.1, 0.2]$.

	Linking Families	Residual Dependence Structure
setting 1	Frank	1-truncated D-vine (all Gaussian copulas)
setting 2.1	Gumbel	1-truncated D-vine (all Gaussian copulas)
setting 2.2	Gumbel	2-truncated D-vine (all Gaussian copulas)
setting 2.3	Gumbel	1-truncated D-vine (all Frank copulas)
setting 3	Gumbel, t, BB1	1-truncated D-vine (all Gaussian copulas)

	$[\boldsymbol{\theta}_L, \boldsymbol{\theta}_U]$ (Kendall's τ)	$[\boldsymbol{\theta}_L^{\text{res}}, \boldsymbol{\theta}_U^{\text{res}}]$ (Kendall's τ)
setting 1	$[4.2, 18.5]$ $(0.4, 0.8)$	$[0.15, 0.45](0.1, 0.3)$
setting 2.1	$[1.67, 5]$ $(0.4, 0.8)$	$[0.15, 0.45](0.1, 0.3)$
setting 2.2	$[1.67, 5]$ $(0.4, 0.8)$	$U(0.30, 0.45)(0.2, 0.3)$ $U(0.15, 0.30)(0.1, 0.2)$
setting 2.3	$[1.67, 5]$ $(0.4, 0.8)$	$[0.95, 2.95](0.1, 0.3)$
setting 3	$[1.67, 5]$, $[0.59, 0.95]$, $[0.5, 1.30] \times [1.35, 3.0]$ $(0.4, 0.8)$	$[0.15, 0.45](0.1, 0.3)$

Summaries from the simulation settings are the following based on N^* random samples of size N being generated.

a. The RMSEs of the proxies of the two methods are:

$$\hat{v}_{\text{RMSE}}^m = \left\{ \frac{1}{N^* N} \sum_{s=1}^{N^*} \sum_{i=1}^{N} \left(\hat{v}_{si}^m - v_{si} \right)^2 \right\}^{1/2}, \quad m \in \{1, 2\}. \tag{22}$$

b. The MAE of the differences between the estimates obtained from the proxy methods and the true parameters is:

$$\hat{\theta}_{\text{MAE}}^m = \frac{1}{N^* D} \sum_{s=1}^{N^*} \sum_{j=1}^{D} |\hat{\theta}_{sj}^m - \theta_{sj}|,$$

where θ_{sj} is the parameter of C_{j0} generated at the sth simulation and $\hat{\theta}_{sj}^m$ is the corresponding estimate using method m.

c. The differences of Kendall's taus (functions of the estimated bivariate linking copula) of the estimates obtained from the proxy methods and the maximum likelihood with exact likelihood:

$$\hat{\theta}^m_{|\text{diff}|} = \frac{1}{N^*D}\sum_{s=1}^{N^*}\sum_{j=1}^{D}|\hat{\theta}^m_{sj} - \hat{\theta}^0_{sj}|,\ \hat{\tau}^m_{|\text{diff}|} = \frac{1}{N^*D}\sum_{s=1}^{N^*}\sum_{j=1}^{D}|\hat{\tau}^m_{sj} - \hat{\tau}^0_{sj}|,\ m \in \{1,2\}. \quad (23)$$

d. In Setting 3, the averaged differences of some dependence measures between the true and fitted models over the D bivariate linking copulas:

$$[\hat{M}^{\text{diff}}]^m_{\text{mean}} = \frac{1}{N^*D}\sum_{s=1}^{N^*}\sum_{j=1}^{D}|[\widehat{M}^{\text{model}}]_j - [M^{\text{true}}]_j|, \quad m \in \{1,2\}, \qquad (24)$$

In Setting 3, summaries include the averaged differences of the dependence measures between the true and fitted models over the D bivariate linking copulas in (24). The three dependence measures are Kendall's tau and tail-weighted upper/lower tail ζ_α (with $\alpha = 20$), as defined in [12], to assess central dependence and tail properties.

From the results in Table 2 for Setting 1, the simple and sequential proxy approaches can provide accurate parameter estimates when D is large, and the sequential procedure performs better compared to the simple proxy, as it has more accurate estimates of the parameters and the realized latent variables. In the 1-factor model with weak residual dependence, the differences in the estimated parameters and those obtained from the exact likelihood are in a similar scale as those in the 1-factor copula model (without residual dependence) presented in [4]. However, the RMSEs of the estimated proxies are larger compared to those results in [4] with the assumption the conditional independence given the latent variables.

Table 2 1-factor copula models with all Frank linking copulas (Setting 1); simulation size $N^* = 1000$, sample size $N = 500$, $\boldsymbol{\theta}$ uniform in $(\boldsymbol{\theta}_L, \boldsymbol{\theta}_U)$ as specified in Table 1. Summaries from (23), (24), and (22) for 3 approaches — superscript $m = 0$: exact likelihood; superscript $m = 1$: simple proxy; superscript $m = 2$: stage 2 proxy.

D	$\hat{\theta}^0_{\text{MAE}}$	$\hat{\theta}^1_{\text{MAE}}$	$\hat{\theta}^2_{\text{MAE}}$	$\hat{\theta}^1_{\lvert\text{diff}\rvert}$	$\hat{\theta}^2_{\lvert\text{diff}\rvert}$	$\lvert\hat{\tau}^{\text{diff}}\rvert^1_{\text{mean}}$	$\lvert\hat{\tau}^{\text{diff}}\rvert^2_{\text{mean}}$	$\hat{v}^{m=1}_{\text{RMSE}}$	$\hat{v}^{m=2}_{\text{RMSE}}$
20	0.456	0.598	0.646	0.477	0.469	0.012	0.009	0.051	0.037
40	0.437	0.483	0.488	0.298	0.217	0.007	0.004	0.039	0.027
60	0.429	0.461	0.458	0.234	0.163	0.005	0.003	0.033	0.023
80	0.418	0.443	0.437	0.203	0.140	0.004	0.003	0.030	0.020

From the results in Table 3 for Setting 2, comparing Settings 2.1 and 2.2, when the weak residual dependence becomes more complex (going from 1-truncated to 2-truncated for modeling the residual dependence), the RMSE of the estimated parameters (of linking copulas from the observed variables to the latent variable) of the two approaches is slightly larger, as well as the RMSE of the proxies for the two approaches. However, as the dimension increases, the sequential approach can still provide accurate parameter estimates and proxy estimates, even in cases where the conditional independence is farther from holding in Setting 2.2. Comparing Settings 2.3 and 2.1, the results are similar. This indicates that the sequential approach can be useful when the linking copulas for modeling the residual dependence are not Gaussian but belong to some other well-behaved copula families.

Table 3 Results for Setting 2: 1-factor copula models with all Gumbel linking copulas with weak residual dependence; simulation size $N^* = 1000$, sample size $N = 500$. The three stacked panels show the results for Settings 2.1, 2.2, and 2.3, respectively, from top to bottom. Summaries from (23), (24), and (22) for 2 approaches —superscript $m = 1$: simple proxy; superscript $m = 2$: stage 2 proxy. The differences in (23), (24) are between the estimated parameters and true parameters (instead of estimated parameters from the exact approach). The implementation of the exact maximum likelihood approach is limited to scenarios where the residual dependence is modeled using a 1-truncated vine.

D	$\hat{\theta}^1_{\text{RMSE}}$	$\hat{\theta}^2_{\text{RMSE}}$	$\hat{\theta}^1_{\lvert\text{diff}\rvert}$	$\hat{\theta}^2_{\lvert\text{diff}\rvert}$	$\lvert\hat{\tau}^{\text{diff}}\rvert^1_{\text{mean}}$	$\lvert\hat{\tau}^{\text{diff}}\rvert^2_{\text{mean}}$	$\hat{v}^{m=1}_{\text{RMSE}}$	$\hat{v}^{m=2}_{\text{RMSE}}$
20	0.225	0.218	0.167	0.165	0.015	0.014	0.054	0.042
40	0.203	0.149	0.148	0.112	0.013	0.011	0.041	0.030
60	0.192	0.136	0.139	0.103	0.012	0.010	0.035	0.024
80	0.188	0.135	0.138	0.103	0.012	0.010	0.032	0.022
20	0.237	0.308	0.187	0.246	0.019	0.021	0.067	0.053
40	0.194	0.176	0.143	0.134	0.013	0.012	0.051	0.039
60	0.190	0.146	0.138	0.111	0.012	0.011	0.043	0.032
80	0.182	0.137	0.133	0.104	0.012	0.010	0.039	0.028
20	0.235	0.233	0.175	0.177	0.017	0.015	0.057	0.044
40	0.201	0.152	0.146	0.114	0.013	0.011	0.043	0.032
60	0.194	0.138	0.141	0.105	0.012	0.010	0.036	0.026
80	0.192	0.135	0.139	0.102	0.012	0.010	0.033	0.023

Table 4 Results for Setting3: 1-factor copula model with linking copulas from Gumbel, t, and BB1 families; Simulation size $N^* = 1000$, sample size $N = 500$. In each simulation $\boldsymbol{\theta}$ is uniform in $(\boldsymbol{\theta}_L, \boldsymbol{\theta}_U)$. The summaries are for (24) and (22) and are shown in order $m = 1/m = 2$ respectively in columns 2 to 4. With linking copula families to be decided, $[\tau^{\text{diff}}]_{\text{mean}}$, $[\zeta^{\text{diff}}_{\alpha,U}]_{\text{mean}}$, $[\zeta^{\text{diff}}_{\alpha,L}]_{\text{mean}}$, are the averaged differences in the dependence measures between true and fitted models over D linking copulas. $\alpha = 20$ in $\zeta_{\alpha,U}$ and $\zeta_{\alpha,L}$ defined in [12] to summarize the dependence in the upper and lower joint tails.

D	$[\hat{\tau}^{\text{diff}}]^m_{\text{mean}}$	$[\hat{\zeta}^{\text{diff}}_{\alpha,U}]^m_{\text{mean}}$	$[\hat{\zeta}^{\text{diff}}_{\alpha,L}]^m_{\text{mean}}$	$\hat{v}^{m=1}_{\text{RMSE}}$	$\hat{v}^{m=2}_{\text{RMSE}}$
30	0.016/0.016	0.027/0.022	0.033/0.029	0.051	0.038
45	0.014/0.014	0.024/0.020	0.029/0.025	0.042	0.031
60	0.014/0.013	0.023/0.019	0.027/0.024	0.038	0.027
90	0.013/0.012	0.022/0.019	0.025/0.022	0.033	0.022

Setting 3 indicates that the sequential approach can identify linking copula families with similar tail behaviors to the correct ones in most cases, based on the small differences in upper/lower tail-weighted dependence measures and Kendall's tau when the dimension is large. Comparing the results in Table 3 to those results in [4], when the conditional independence assumption is slightly violated, the RMSE of the proxies is slightly larger, while the differences in the dependence measures are on a similar scale.

7.2 Bi-factor copula with weak residual dependence

A simulation setting is summarized in Table 5 to illustrate the sequential approach in bi-factor copula models. The sample size is N and there are D variables and $2D$ linking copulas. The number of groups $G = 3$ and the size of each group is $D/3$. The parameters of the D copulas linking the observed variables and the global latent variable are generated uniformly in (θ_L, θ_U) so that there is a wide range for the dependence between the observed variables and the global latent variable. As for the D bivariate copulas for conditional dependence, the parameters are generated uniformly from $(\theta_{LL}, \theta_{UU})$ so that the within-group dependence is strong. The residual dependence is assumed to be a 1-truncated D-vine structure, and the parameters on the linking copulas are uniformly from $(\boldsymbol{\theta}_L^{\text{res}}, \boldsymbol{\theta}_U^{\text{res}})$ such that the residual dependence is weak. The simulation size is $N^* = 1000$.

Table 5 The simulation setting for the bi-factor copula model with weak residual dependence; the linking families are for global and group; weak residual dependence is modeled by a 1-truncated D-vine. The range of Kendall's tau corresponding to the range of parameters is included after the parameter interval.

	linking families	$[\boldsymbol{\theta}_L, \boldsymbol{\theta}_U](\tau)$	$[\boldsymbol{\theta}_{LL}, \boldsymbol{\theta}_{UU}](\tau)$	$[\boldsymbol{\theta}_L^{\text{res}}, \boldsymbol{\theta}_U^{\text{res}}](\tau)$
setting 1	Frank/Frank/Gaussian	[1.87,8] (0.2,0.6)	[4.2,11.5] (0.4,0.7)	[0.1,0.3]

Summaries are given in Table 6. When the bi-factor models slightly deviate from the conditional independence assumption, the MAEs of the estimated parameters on the global linking copulas are slightly larger, but those and MAEs of the estimated parameters on the local linking copulas are on the same scale as the previous results in bi-factor models (with conditional independence given latent variables) shown in [4]. With proxies, the RMSEs for global latent variables are slightly smaller, and the RMSEs for local latent variables are larger than those in bi-factor models with conditional independence. The estimation improves as D increases.

Table 6 Bi-factor copula model with all linking copulas in the Frank family; Simulation size $N^* = 1000$, sample size $N = 1200$, $d_g \in \{10,20,30\}$ for $g \in \{1,2,3\}$. The parameters are generated as specified in Table 5. For D global/local linking copulas, summaries of $\hat{\theta}_{\text{MAE}}, \hat{\theta}_{\text{diff}}, \hat{\tau}_{\text{diff}}$ for approaches superscript $m = 0$: exact likelihood; superscript $m = 2$: stage 2 proxy are shown; For MAE, the results are shown for m=0/m=2 respectively.

D	Global linking copulas			Local linking copulas			RMSE$_{\text{proxy}}$
Frank	$\hat{\theta}^{m=0/m=2}_{\text{glob:MAE}}$	$\hat{\theta}^{m=2}_{\text{glob:}\lvert\text{diff}\rvert}$	$\hat{\tau}^{m=2}_{\text{glob:}\lvert\text{diff}\rvert}$	$\hat{\theta}^{m=0/m=2}_{\text{loc:MAE}}$	$\hat{\theta}^{m=2}_{\text{loc:}\lvert\text{diff}\rvert}$	$\hat{\tau}^{m=2}_{\text{loc:}\lvert\text{diff}\rvert}$	$\hat{v}_0/\hat{v}_g$
30	0.240/0.458	0.375	0.013	0.333/1.027	0.812	0.038	0.057/0.151
60	0.215/0.351	0.277	0.009	0.236/0.447	0.320	0.017	0.044/0.111
90	0.210/0.325	0.257	0.008	0.212/0.302	0.200	0.012	0.038/0.096

8 Conclusion

The proxy variable estimates of latent variables in factor models in [4] are defined based on the assumption of conditional independence given latent variables. In this article, some interpretable sufficient conditions on weak residual (conditional) dependence are obtained so that the proxies are still consistent when the conditional independence assumption is slightly violated.

For high-dimensional factor copula models with a large sample size, simulation studies show that the sequential estimation approach is robust when there is weak residual dependence. It is possible to efficiently estimate the latent variables, select the families of parametric linking copulas, and estimate the copula parameters. As shown in [4], a latent variable can be well estimated when the cumulative dependence of observed variables to a latent variable is strong enough; more variables would be needed if the individual dependencies with the latent variable are weaker. Small amounts of mis-specification are acceptable either in the copula families linking to the latent variables or weak residual dependence given the latent variables.

Results on consistency of factor scores for 1-factor and bi-factor Gaussian models with weak residual dependence are useful results on their own, but our motivation is to get an idea of weak residual dependence conditions that could apply for high-dimensional factor copula models.

For a large number of variables that can be divided into G groups with stronger within-group dependence, several parsimonious dependence structures given below can be considered that make use of the correlation matrix of normal scores of the observed variables and proxy variables. These are applied in [3] for data sets with daily stock returns, changes in daily currency exchange rates and gene expressions, and will be used in subsequent research.

- An oblique factor model with weak residual dependence, where there is a 1-factor model with weak residual dependence for each of G groups, and the G latent variables are dependent to explain the between-group dependence.

- A bi-factor structure with weak residual dependence, where there is global latent variable and G group-based latent variables that are mutually independent, and there is weak residual dependence of variables in each group conditioned on the global latent variable and the corresponding group-based latent variable.
- A parsimonious dependence structure constructed from applying truncated vine algorithms to the combination of observed and proxy variables, with one latent variable for each group.

Each of the above leads to a truncated vine structure (rooted at latent or proxy variables). With initially estimated proxy variables based on the Gaussian copula then one can check if there can be improvements to bivariate linking Gaussian copulas between each observed variable and the corresponding proxy variables.

Finally, the techniques of proofs in Section 9 are useful for studying asymptotic properties of estimators with slightly mis-specified models.

9 Proofs

9.1 Proof of Theorem 1

The short proof is modified from [4] in order to highlight the steps for the longer proof of Theorem 2.

Proof. Let $\boldsymbol{e}_D = (e_1, \ldots, e_D)$ be one realization of $\boldsymbol{\varepsilon}_D$. Let $q_D = \boldsymbol{A}_D^\top \boldsymbol{\Psi}_D^{-2} \boldsymbol{A}_D > 0$. Then, from (8),

$$\widetilde{w}_D - w^0 = (1 + q_D)^{-1} \boldsymbol{A}_D^\top \boldsymbol{\Psi}_D^{-1} \boldsymbol{e}_D + (1 + q_D^{-1})^{-1} w^0 - w^0.$$

Note that $Y = D^{-1/2} \boldsymbol{A}_D^\top \boldsymbol{\Psi}_D^{-1} \boldsymbol{e}_D$ is a realization of $D^{-1/2} \boldsymbol{A}_D^\top \boldsymbol{\Psi}_D^{-1} \boldsymbol{\varepsilon}_D$ which can be written as $D^{-1/2} \sum_{j=1}^D \alpha_j \varepsilon_j / \psi_j$. The latter random quantity has a variance $D^{-1} \sum_{j=1}^D \sum_{k=1}^D \alpha_j \alpha_k \gamma_{jk} / [\psi_j \psi_k]$. By Assumption 1 and with loadings that are bounded away from ± 1, this variance is $O(1)$ so that Y can be considered as $O_p(1)$. With the boundedness assumptions for the α_j's, $\bar{q}_D = D^{-1} q_D = O(1)$. Then $\widetilde{w}_D - w^0 = D^{-1/2} (D^{-1} + \bar{q}_D)^{-1} (D^{-1/2} \boldsymbol{A}_D^\top \boldsymbol{\Psi}_D^{-1} \boldsymbol{\varepsilon}_D) + O(D^{-1})$ is asymptotically $O_p(D^{-1/2})$. □

9.2 Proof of Theorem 2

Proof. Let $\boldsymbol{Q}_D = \boldsymbol{A}_D^\top \boldsymbol{\Psi}_D^{-2} \boldsymbol{A}_D$. With $p = G + 1$, (8) extends to

$$\widetilde{\boldsymbol{w}}_D - \boldsymbol{w}^0 = (\boldsymbol{I}_p + \boldsymbol{Q}_D)^{-1} \boldsymbol{A}_D^\top \boldsymbol{\Psi}_D^{-1} \boldsymbol{e}_D + \left[(\boldsymbol{I}_p + \boldsymbol{Q}_D^{-1})^{-1} \boldsymbol{w}^0 - \boldsymbol{w}^0 \right]. \tag{25}$$

The proof is outlined in two steps. For **step1**, the second term in the right side of (25) is shown to be $O(D^{-1})$. For **step2**, the first term in the right side of (25) is shown to be $O_p(D^{-1/2})$.

step1: Since $\boldsymbol{A_D}$ is of full rank and the entries of $\boldsymbol{\Psi_D}$ are uniformly bounded away from 0. Then $\overline{\boldsymbol{Q}}_{\boldsymbol{D}} = D^{-1}\boldsymbol{A}_{\boldsymbol{D}}^{\top}\boldsymbol{\Psi}_{\boldsymbol{D}}^{-2}\boldsymbol{A_D}$ is positive definite for any fixed D and $\overline{\boldsymbol{Q}}_D = O(1)$. For the second term in the right side of (25), by first-order Taylor expansion, $(\boldsymbol{I}_p + \boldsymbol{Q}_{\boldsymbol{D}}^{-1})^{-1} - \boldsymbol{I}_p = (\boldsymbol{I}_p + D^{-1}\overline{\boldsymbol{Q}}_D^{-1})^{-1} - \boldsymbol{I}_p = \boldsymbol{I}_p - D^{-1}\overline{\boldsymbol{Q}}_D + O(D^{-2}) - \boldsymbol{I}_p = O(D^{-1})$.

step2: The first term in the right side of (25) can be rewritten as

$$D^{-1/2}(D^{-1}\boldsymbol{I}_p + \overline{\boldsymbol{Q}}_{\boldsymbol{D}})^{-1}D^{-1/2}\boldsymbol{A}_{\boldsymbol{D}}^{\top}\boldsymbol{\Psi}_{\boldsymbol{D}}^{-1}\boldsymbol{e_D}. \tag{26}$$

$D^{-1/2}\boldsymbol{A}_{\boldsymbol{D}}^{\top}\boldsymbol{\Psi}_{\boldsymbol{D}}^{-1}\boldsymbol{e}_D$ is a realization of $D^{-1/2}\boldsymbol{A}_{\boldsymbol{D}}^{\top}\boldsymbol{\Psi}_{\boldsymbol{D}}^{-1}\boldsymbol{\varepsilon}_D$. The latter quantity will be shown to be $O_p(1)$. It is sufficient to justify that the covariance matrix of this quantity is bounded. Since $\boldsymbol{\varepsilon_D} \sim \mathcal{N}(\boldsymbol{0}, \boldsymbol{\Omega_D})$, the covariance matrix of $D^{-1/2}\boldsymbol{A}_{\boldsymbol{D}}^{\top}\boldsymbol{\Psi}_{\boldsymbol{D}}^{-1}\boldsymbol{\varepsilon_D}$ is

$$D^{-1}\boldsymbol{A}_{\boldsymbol{D}}^{\top}\boldsymbol{\Psi}_{\boldsymbol{D}}^{-1}\boldsymbol{\Omega_D}\boldsymbol{\Psi}_{\boldsymbol{D}}^{-1}\boldsymbol{A_D} := D^{-1}\boldsymbol{B},$$

where $\boldsymbol{B}$ is $p \times p$. Let $\mathbf{1}$ be the $p \times 1$ vector with all elements equal to 1. Then, the sum of the elements in the $p \times p$ covariance matrix can be written as

$$\begin{aligned}\text{trace}[D^{-1}\mathbf{1}^{\top}\boldsymbol{B}\mathbf{1}] &= \text{trace}[D^{-1}\mathbf{1}^{\top}\boldsymbol{A}_{\boldsymbol{D}}^{\top}\boldsymbol{\Psi}_{\boldsymbol{D}}^{-1}\boldsymbol{\Omega_D}\boldsymbol{\Psi}_{\boldsymbol{D}}^{-1}\boldsymbol{A_D}\mathbf{1}] \\ &= \text{trace}[D^{-1}\boldsymbol{\Omega_D}\boldsymbol{\Psi}_{\boldsymbol{D}}^{-1}\boldsymbol{A_D}\mathbf{1}\mathbf{1}^{\top}\boldsymbol{A}_{\boldsymbol{D}}^{\top}\boldsymbol{\Psi}_{\boldsymbol{D}}^{-1}] = \text{trace}[D^{-1}\boldsymbol{\Omega_D}\boldsymbol{K_D}],\end{aligned}$$

where $\boldsymbol{K}_D = \boldsymbol{\Psi}_{\boldsymbol{D}}^{-1}\boldsymbol{A_D}\mathbf{1}\mathbf{1}^{\top}\boldsymbol{A}_{\boldsymbol{D}}^{\top}\boldsymbol{\Psi}_{\boldsymbol{D}}^{-1}$. Let $\boldsymbol{\alpha}_i^{\top}$ to represent the ith row of $\boldsymbol{A}_D$ so that $\boldsymbol{A}_D = [\boldsymbol{\alpha}_1^{\top}, \ldots, \boldsymbol{\alpha}_D^{\top}]^{\top}$. Then $[\boldsymbol{K}_D]_{ij} = \psi_i^{-1}\psi_j^{-1}\boldsymbol{\alpha}_i^{\top}\mathbf{1}\mathbf{1}^{\top}\boldsymbol{\alpha}_j$.

With the assumptions on the loading parameters, there are constants $0 < M_1 < M_2 < 1$ such that $M_1 < \psi_j < M_2$ and $\|\boldsymbol{\alpha}_j\|_2 < 1$ for $j = 1, \ldots, D$. For the absolute value of the (i,j) entry (i can equal j) in $\boldsymbol{K_D}$, $|k_{ij}| = |\psi_i^{-1}\psi_j^{-1}\boldsymbol{\alpha}_i^{\top}\mathbf{1}\mathbf{1}^{\top}\boldsymbol{\alpha}_j| \le \psi_i^{-1}\psi_j^{-1}\|\boldsymbol{\alpha}_i\|_1\|\boldsymbol{\alpha}_j\|_1 \le \psi_i^{-1}\psi_j^{-1} \times p\|\boldsymbol{\alpha}_i\|_2\|\boldsymbol{\alpha}_j\|_2 < pM_1^{-2}$, by the Cauchy-Schwarz inequality. That is, the elements in $\boldsymbol{K_D}$ are uniformly bounded.

The summation of absolute values of all elements in the covariance matrix $D^{-1}\boldsymbol{B}$ is:

$$\left|\text{trace}[D^{-1}\boldsymbol{\Omega}_D\boldsymbol{K}_D]\right| = D^{-1}\sum_{s=1}^{D}\sum_{t=1}^{D}|\omega_{st}k_{st}| \le pM_1^{-2}D^{-1}\sum_{s=1}^{D}\sum_{t=1}^{D}|\omega_{st}|.$$

Based on the Assumption 2 on the order of residual dependence, the summation of absolute values of entries in the covariance matrix $D^{-1}\boldsymbol{B}_{p\times p}$ is of $O(1)$. Therefore, $D^{-1/2}\boldsymbol{A}_{\boldsymbol{D}}^{\top}\boldsymbol{\Psi}_{\boldsymbol{D}}^{-1}\boldsymbol{\varepsilon}_D$ is $O_p(1)$. Since $D^{-1/2}\boldsymbol{A}_{\boldsymbol{D}}^{\top}\boldsymbol{\Psi}_{\boldsymbol{D}}^{-1}\boldsymbol{e}_D$ is a realization of $D^{-1/2}\boldsymbol{A}_{\boldsymbol{D}}^{\top}\boldsymbol{\Psi}_{\boldsymbol{D}}^{-1}\boldsymbol{\varepsilon}_D$, so it is also $O_p(1)$. Since $(D^{-1}\boldsymbol{I}_p + \overline{\boldsymbol{Q}}_D)$ in (26) is $O(1)$ as $D \to \infty$, then $\widetilde{w}_D - \boldsymbol{w}^0 = O_p(D^{-1/2}) + O(D^{-1}) = O_p(D^{-1/2})$. □

9.3 Proof of Theorem 3

Proof. For the model defined in equation (5), write the averaged negative log-likelihood as the summation of two parts:

$$\overline{L}_D(v,\boldsymbol{u}_D) := \underbrace{-\frac{1}{D}\sum_{j=1}^{D}\log c_{j0}(u_j,v)}_{:=\overline{qL}_D(v,\boldsymbol{u}_D)} \underbrace{-\frac{1}{D}\log c_{\mathrm{res}}(C_{1|0}(u_1|v),\ldots,C_{D|0}(\boldsymbol{u}_D|v))}_{:=\overline{L}_D^{(r)}(v,\boldsymbol{u}_D)},$$

where $\overline{L}_D^{(r)}$ comes from the residual dependence given the latent variables and $\overline{qL}_D$ is the averaged negative quasi-log-likelihood based on conditional independence given the latent variables.

Step A: As mentioned in Remark 3, consider v as the parameter in the log-likelihood function $\overline{L}_D$. According to the general theory of MLE under standard regularity conditions in [13], and mixing conditions such as [2], the negative log-likelihood $\overline{L}_D$ has a minimizer located at the true parameters v^0 as $D \to \infty$.

Step B in Remark 3 is verified under the Assumptions 3 and Condition 1. It will be shown that with the assumptions that

$$\left.\frac{\partial \overline{L}_D^{(r)}}{\partial v}\right|_{v=v^0} \to_p 0, \quad D \to \infty.$$

Define random variables $\widetilde{U}_j(v) = C_{j|0}(U_j|v)$, and $\widetilde{U}_j^0 = C_{j|0}(U_j|v^0)$ with realized value v^0. Then $\widetilde{U}_j^0 \sim U(0,1)$. Let $\widetilde{u}_j(v) = C_{j|0}(u_j|v)$ and $\widetilde{u}_j^0 = \widetilde{u}_j(v^0)$ be realized values of $\widetilde{U}_j(v)$ and $\widetilde{U}_j^0$ respectively. Let $E_j^0 := \Phi^{-1}(\widetilde{U}_j^0) \sim \mathcal{N}(0,1)$ and let one realization be $e_j^0 = \Phi^{-1}(\widetilde{u}_j^0)$.

Denote the derivative of $\Phi^{-1}(\widetilde{u}_j(v))$ with respect to v evaluated at v^0 as m_j^0, so that

$$m_j^0 = \left.\frac{\partial \Phi^{-1}(C_{j|0}(u_j|v))}{\partial v}\right|_{v=v^0} = \frac{1}{\phi(\Phi^{-1}(\widetilde{u}_j^0))}\left.\frac{\partial C_{j|0}(u_j|v)}{\partial v}\right|_{v=v^0}$$
$$= \frac{1}{\phi(e_j^0)}\left.\frac{\partial \widetilde{u}_j(v)}{\partial v}\right|_{v=v^0}.$$

Let $m_j(u,v) = \partial\Phi^{-1}(C_{j|0}(u_j|v))/\partial v$. With Assumption 3 and also Assumption 1 in [4], $\lim_{u\to 1} m_j(u,v)$ and $\lim_{u\to 0} m_j(u,v)$ are finite for $v_A < v < v_B$, and $|m_j(u,v)|$ is uniformly bounded over j, $v_A \le v \le v_B$ and $0 \le u \le 1$. Furthermore, from the stochastic increasing condition with $1 - C_{j|0}(\cdot|v)$ increasing in v for all j, then $\partial C_{j|0}(u_j|v)/\partial v \le 0$ and $m_j^0 \le 0$ for all j. Let B_m be an upper bound on $-m_j(u,v)$ over all j, all $0 < u < 1$, and $v_A < v < v_B$.

Write the derivative of $\overline{L}^{(r)}(v,\boldsymbol{u}_D)$ with respect to v evaluated at v^0 as

$$\frac{\partial \overline{L}_D^{(r)}(v,\boldsymbol{u}_D)}{\partial v}\Big|_{v=v^0} = D^{-1}\frac{\partial \log c_{\text{res}}(\widetilde{u}_1(v),\ldots,\widetilde{u}_D(v))}{\partial v}\Big|_{v=v^0}. \tag{27}$$

From Condition 1, then

$$c_{\text{res}}(\widetilde{u}_1(v),\ldots,\widetilde{u}_D(v)) = \frac{\phi_D\big(\Phi^{-1}(\widetilde{u}_1(v)),\ldots,\Phi^{-1}(\widetilde{u}_D(v));\boldsymbol{\Gamma}_D\big)}{\prod_{j=1}^D \phi(\Phi^{-1}(\widetilde{u}_j(v))},$$

where ϕ_D is the D-variate Gaussian density, and ϕ, Φ are respectively the univariate standard normal densiy and cdf. Therefore,

$$\begin{aligned}
&D^{-1}\partial \log c_{\text{res}}(\widetilde{u}_1(v),\ldots,\widetilde{u}_D(v);\boldsymbol{\theta}^{(r)})\ /\ \partial v\,\big|_{v=v^0} \\
&= D^{-1}\partial \log \phi_d\big(\Phi^{-1}(\widetilde{u}_1(v)),\ldots,\Phi^{-1}(\widetilde{u}_D(v));\boldsymbol{\Gamma}_D\big)\ /\ \partial v\,\big|_{v=v^0} - \\
&D^{-1}\sum_{j=1}^D \partial \log \phi\big(\Phi^{-1}(\widetilde{u}_j(v))\big)\ /\ \partial v\,\big|_{v=v^0} \\
&= D^{-1}(\boldsymbol{m}_D^0)^\top \boldsymbol{\Gamma}_D^{-1}\boldsymbol{e}_D^0 - D^{-1}\sum_{j=1}^D \frac{\phi'(e_j^0)}{\phi(e_j^0)} m_j^0 \qquad (28) \\
&= D^{-1}(\boldsymbol{m}_D^0)^\top (\boldsymbol{\Gamma}_D^{-1} - \boldsymbol{I}_D)\boldsymbol{e}_D^0, \qquad (29)
\end{aligned}$$

where $\boldsymbol{m}_D^0 = (m_1^0,\ldots,m_D^0)^\top$ and $\boldsymbol{e}_D^0 = (e_1^0,\ldots,e_D^0)^\top$. In the last term of (28), $\phi'(x) = -x\phi(x)$, so that it simplifies to $-D^{-1}\sum_{j=1}^D e_j^0 m_j^0 = -D^{-1}(\boldsymbol{m}_D^0)^\top \boldsymbol{I}_D \boldsymbol{e}^0$.

It will proved in Lemma 1 below that $\{D^{-1}(\boldsymbol{m}_D^0)^\top(\boldsymbol{\Gamma}_D^{-1} - \boldsymbol{I}_D)\boldsymbol{e}_D^0\}$ is a realization of a random sequence whose variance converges to 0 as $D \to \infty$. Then, combining (27) and (29), $\partial \overline{L}_D^{(r)}(v,\boldsymbol{u}_D)/\partial v\,\big|_{v=v^0} \to 0$ as $D \to \infty$. Hence Step B is justified.

As noted in Remark 3 after step B, the quasi-MLE v_D^* which is the root $\partial \overline{qL}_D(v,u_D)/\partial v = 0$ is such that $v_D^* - v^0 = o_p(1)$ as $D \to \infty$ with a Taylor expansion. Also, due to the asymptotic equivalence of the proxy $\widetilde{v}_D$ and quasi-MLE v_D^* based on the Laplace approximation in [4], $\widetilde{v}_D - v_D^* = O(D^{-1})$. The consistency of the proxy variable is still satisfied since $\widetilde{v}_D - v^0 = (\widetilde{v}_D - v_D^*) + (v_D^* - v^0) = O(D^{-1}) + o_p(1) = o_p(1)$ as $D \to \infty$. □

Lemma 1. *$\{D^{-1}(\boldsymbol{m}_D^0)^\top(\boldsymbol{\Gamma}_D^{-1} - \boldsymbol{I}_D)\boldsymbol{e}_D^0\}$ is a realization of a random sequence whose variance converges to 0 as $D \to \infty$.*

Proof. The subscript D is suppressed for $\boldsymbol{\Gamma}, \boldsymbol{m}^0, \boldsymbol{e}^0$ etc. Let $\boldsymbol{M} = (M_j)$ and $\boldsymbol{E}^0 = (E_j^0)$ be the random counterparts of $\boldsymbol{m}^0$ and $\boldsymbol{e}^0$ in (29). The component e_j^0 is the realization of $E_j^0 \sim \mathcal{N}(0,1)$ and that $\boldsymbol{E}^0 \sim \mathcal{N}_D(\boldsymbol{0},\boldsymbol{\Gamma})$. M_j is a function of a random variable with distribution $C_{j|0}(\cdot|v^0)$, and is correlated with E_j^0. With $v_A < v_0 < v_B$, and the assumptions of Theorem 3, there is a positive constant B_m such that $0 \le -M_j \le B_m$.

Let $\boldsymbol{Y} := D^{-1}\boldsymbol{M}^\top(\boldsymbol{\Gamma}^{-1} - \boldsymbol{I})\boldsymbol{E}^0$ be the random version of $D^{-1}(\boldsymbol{m}^0)^\top(\boldsymbol{\Gamma}^{-1} - \boldsymbol{I})\boldsymbol{e}^0$. Let $\boldsymbol{X}^\top = \boldsymbol{M}^\top(\boldsymbol{\Gamma}^{-1} - \boldsymbol{I})$ be a random variable so that $\boldsymbol{X} \in \mathbb{R}^D$, then $\boldsymbol{Y} = D^{-1}\boldsymbol{X}^\top\boldsymbol{E}^0$. It will be shown that $\text{Var}(\boldsymbol{Y}) = D^{-2}\,\text{Var}(\boldsymbol{X}^\top\boldsymbol{E}^0) \to 0$ when $D \to \infty$. Write the variance of $\boldsymbol{Y}$ as

$$D^{-2}\operatorname{Var}(\boldsymbol{X}^\top\boldsymbol{E}^0) = D^{-2}\mathrm{E}([\boldsymbol{X}^\top\boldsymbol{E}^0]^2) - [D^{-1}\mathrm{E}(\boldsymbol{X}^\top\boldsymbol{E}^0)]^2. \tag{30}$$

In (30), the expectation in the second term can be rewritten as

$$\begin{aligned} D^{-1}\mathrm{E}(\boldsymbol{X}^\top\boldsymbol{E}^0) &= D^{-1}\mathrm{E}[\boldsymbol{M}^\top(\boldsymbol{\Gamma}^{-1}-\boldsymbol{I})\boldsymbol{E}^0] \\ &= D^{-1}\mathrm{E}\big[\mathrm{E}(\boldsymbol{M}^\top|\boldsymbol{E}^0)(\boldsymbol{\Gamma}^{-1}-\boldsymbol{I})\boldsymbol{E}^0\big]. \end{aligned} \tag{31}$$

Let $\boldsymbol{K} = \mathrm{E}[\boldsymbol{M}^\top|\boldsymbol{E}^0](\boldsymbol{\Gamma}^{-1}-\boldsymbol{I}) \in \mathbb{R}^D$, so that the ith entry of the vector is $K_i = \sum_{\ell=1}^D \mathrm{E}[M_\ell|\boldsymbol{E}^0]\gamma^{k\ell} - \mathrm{E}[M_i|\boldsymbol{E}^0] = O_p(1)$ with Condition 1, Assumption 3 on weak residual dependence and uniformly bounded $\{M_\ell\}$. Thus, $\boldsymbol{K}$ is a realization of an $O_p(1)$ random vector with uniformly bounded entries. Then, based on the law of large numbers,

$$D^{-1}\sum_{j=1}^D K_j E_j^0 \to_p 0. \tag{32}$$

Based on the dominating convergence theorem, the limit in (31) is 0.

Without D^{-2}, the first term in the right side of (30) can be rewritten as

$$\begin{aligned} \mathrm{E}[(\boldsymbol{X}^\top\boldsymbol{E}^0)^2] &= \mathrm{E}\big[\boldsymbol{M}^\top(\boldsymbol{\Gamma}^{-1}-\boldsymbol{I})\boldsymbol{E}^0(\boldsymbol{E}^0)^\top(\boldsymbol{\Gamma}^{-1}-\boldsymbol{I})\boldsymbol{M}\big] \\ &= \mathrm{E}\big[\operatorname{trace}\{(\boldsymbol{\Gamma}^{-1}-\boldsymbol{I})\boldsymbol{E}^0(\boldsymbol{E}^0)^\top(\boldsymbol{\Gamma}^{-1}-\boldsymbol{I})\boldsymbol{M}\boldsymbol{M}^\top\}\big] \\ &= \mathrm{E}\big[\operatorname{trace}\{(\boldsymbol{\Gamma}^{-1}-\boldsymbol{I})\boldsymbol{E}^0(\boldsymbol{E}^0)^\top(\boldsymbol{\Gamma}^{-1}-\boldsymbol{I})\,\mathrm{E}(\boldsymbol{M}\boldsymbol{M}^\top|\boldsymbol{E}^0)\}\big]. \end{aligned} \tag{33}$$

The entries of $\mathrm{E}(\boldsymbol{M}\boldsymbol{M}^\top|\boldsymbol{E}^0)$ are positive and uniformly bounded by B_m^2. Let $\boldsymbol{A} = (\boldsymbol{\Gamma}^{-1}-\boldsymbol{I})\boldsymbol{E}^0(\boldsymbol{E}^0)^\top(\boldsymbol{\Gamma}^{-1}-\boldsymbol{I}) = (A_{ks})$. Then from (33),

$$D^{-2}\operatorname{trace}\{(\boldsymbol{\Gamma}^{-1}-\boldsymbol{I})\boldsymbol{E}^0(\boldsymbol{E}^0)^\top(\boldsymbol{\Gamma}^{-1}-\boldsymbol{I})\,\mathrm{E}(\boldsymbol{M}\boldsymbol{M}^\top|\boldsymbol{E}^0) \le D^{-2}B_m^2\sum_{k=1}^D\sum_{s=1}^D |A_{ks}|.$$

Let $\boldsymbol{\Xi} = (\boldsymbol{\Gamma}^{-1}-\boldsymbol{I})\boldsymbol{\Gamma}(\boldsymbol{\Gamma}^{-1}-\boldsymbol{I}) = \boldsymbol{\Gamma}^{-1}+\boldsymbol{\Gamma}-2\boldsymbol{I} = (\xi_{ks})$. Then

$$\begin{aligned} D^{-2}\boldsymbol{1}^\top\boldsymbol{\Xi}\boldsymbol{1} &= D^{-2}\sum_{k=1}^D\sum_{s=1}^D \xi_{ks} \\ &\le D^{-2}\sum_{k=1}^D\sum_{s=1}^D \{|\gamma^{ks}+\gamma_{ks}-2I(k=s)|\} = O(D^{-1}), \end{aligned}$$

with Condition (1). Note that $D^{-2}\boldsymbol{A}$ is a random matrix with expectation $D^{-2}\boldsymbol{\Xi}$. Since $\boldsymbol{E}^0$ is a multivariate normal random vector, as $D\to\infty$,

$$D^{-2}\{(\boldsymbol{\Gamma}^{-1}-\boldsymbol{I})\boldsymbol{E}^0(\boldsymbol{E}^0)^\top(\boldsymbol{\Gamma}^{-1}-\boldsymbol{I})\} - D^{-2}\boldsymbol{\Xi} \to_p 0. \tag{34}$$

Hence,

$$\lim_{D\to\infty} D^{-2}\mathrm{E}([\boldsymbol{X}^\top \boldsymbol{E}^0]^2) \to 0. \tag{35}$$

Combining (32) and (35),

$$D^{-2}\,\mathrm{Var}(\boldsymbol{X}^\top \boldsymbol{E}^0) = D^{-2}\mathrm{E}([\boldsymbol{X}^\top \boldsymbol{E}^0]^2) - [D^{-1}\mathrm{E}(\boldsymbol{X}^\top \boldsymbol{E}^0)]^2 \to 0.$$

□

9.4 Proof of Theorem 4

Proof. The objective KL divergence function is defined as

$$\begin{aligned}
D_{\mathrm{KL}}(f_D, g_D) &\overset{Z_D\sim g_D}{=} \int \log[g_D(z)/f_D(z)]\cdot g_D(z)\,\mathrm{d}z \\
&= \mathrm{E}\Big\{-\tfrac{1}{2}\log|\boldsymbol{R}_D^0| - \tfrac{1}{2}\mathbf{Z}_D^\top \boldsymbol{R}_D^{-1}\mathbf{Z}_D + \tfrac{1}{2}\log|\boldsymbol{\Sigma}_D(\boldsymbol{\beta}_D)| + \tfrac{1}{2}\mathbf{Z}_D^\top \boldsymbol{\Sigma}_D^{-1}(\boldsymbol{\beta}_D)\mathbf{Z}_D\Big\} \\
&= -\tfrac{1}{2}\log|\boldsymbol{R}_D^0| + \tfrac{1}{2}\log|\boldsymbol{\Sigma}_D(\boldsymbol{\beta}_D)| - \tfrac{1}{2}D + \tfrac{1}{2}\,\mathrm{trace}[\boldsymbol{\Sigma}_D^{-1}(\boldsymbol{\beta}_D)\boldsymbol{R}_D^0] =: D_{\mathrm{KL}}(\boldsymbol{\beta}_D),
\end{aligned}$$

and the minimizer is defined as

$$\boldsymbol{\beta}_D^* = \arg\min_{\boldsymbol{\beta}} D_{\mathrm{KL}}(\boldsymbol{\beta}_D).$$

Hence $\boldsymbol{\beta}_D^*$ minimizes the objective function

$$\log|\boldsymbol{\Sigma}_D(\boldsymbol{\beta}_D)| + \mathrm{trace}[\boldsymbol{\Sigma}^{-1}(\boldsymbol{\beta}_D)\boldsymbol{R}_D^0]. \tag{36}$$

Let the score vector (gradient of objective function (36) with respect to $\boldsymbol{\beta}_D$ be $\boldsymbol{s}_D(\boldsymbol{\beta}_D)$ and let the Hessian matrix be $\boldsymbol{H}_D(\boldsymbol{\beta}_D)$. Let the jth element of $\boldsymbol{s}_D(\boldsymbol{\beta}_D)$ be $s_j(\boldsymbol{\beta}_D)$. Based on the mean-value theorem,

$$s_D(\boldsymbol{\beta}_D^*) = s_D(\boldsymbol{\alpha}_D) + \boldsymbol{H}_D(\bar{\boldsymbol{\alpha}}_D)(\boldsymbol{\beta}_D^* - \boldsymbol{\alpha}_D), \tag{37}$$

where $\bar{\boldsymbol{\alpha}}_D$ is a vector between $\boldsymbol{\alpha}_D$ and $\boldsymbol{\beta}_D^*$. Since the estimator $\boldsymbol{\beta}_D^*$ is the solution of a system of inference (or score) functions, $\boldsymbol{s}_D(\boldsymbol{\beta}_D^*) = \mathbf{0}$. Thus,

$$\boldsymbol{\beta}_D^* - \boldsymbol{\alpha}_D = -\boldsymbol{H}_D^{-1}(\bar{\boldsymbol{\alpha}}_D)\boldsymbol{s}_D(\boldsymbol{\alpha}_D). \tag{38}$$

All rates below are as $D \to \infty$ and are under the assumption that the α_j's are uniformly bounded away from -1 and 1. It will be proved in Lemma 3 that (I) the elements of the score vector $\boldsymbol{s}(\boldsymbol{\beta}_D)$ evaluated at the true parameters $\boldsymbol{\alpha}_D$ are uniformly $O(D^{-1})$, and in Lemma 4 that (II) the diagonal entries in the Hessian matrix $\boldsymbol{H}_D(\boldsymbol{\beta}_D)$ are uniformly $O(1)$ and the off-diagonal entries are uniformly $O(D^{-1})$.

Next, from Lemma 2, $\boldsymbol{H}_D^{-1}(\bar{\boldsymbol{\alpha}}_D)$ has uniformly bounded diagonal elements and off-diagonal elements that are uniformly $O(D^{-1})$. Let $\beta_{jD}^* - \alpha_{jD}$ be an element of the left side of (38). The properties of the score vector and inverse Hessian imply that $\boldsymbol{\beta}^*_{jD} - \boldsymbol{\alpha}_{jD} = O(D^{-1}) + D \cdot O(D^{-2}) = O(D^{-1})$ uniformly over j. □

Lemma 2. *Let $\boldsymbol{H}_D$ be a non-singular Hessian matrix with diagonal entries that are uniformly bounded and off-diagonal entries that are $O(D^{-1})$. Then the inverse Hessian matrix has the same properties and is diagonally dominant.*

Proof. Write $\boldsymbol{H}_D = \boldsymbol{H}_{0D} + D^{-1}\boldsymbol{\delta}_D$ where $\boldsymbol{H}_{0D} = \mathrm{diag}(\boldsymbol{H}_D)$ and $\boldsymbol{\delta}_D$ is a $D \times D$ matrix that is uniformly $O(1)$. Then

$$\begin{aligned}\boldsymbol{H}_D^{-1} &= (\boldsymbol{H}_{0D} + D^{-1}\boldsymbol{\delta}_D)^{-1} = [(\boldsymbol{I}_D + D^{-1}\boldsymbol{\delta}_D\boldsymbol{H}_{0D}^{-1})\boldsymbol{H}_{0D}]^{-1} \\ &- \boldsymbol{H}_{0D}^{-1}(\boldsymbol{I}_D + D^{-1}\boldsymbol{\delta}_D\boldsymbol{H}_{0D}^{-1})^{-1} \\ &= \boldsymbol{H}_{0D}^{-1}\{\boldsymbol{I}_D - D^{-1}\boldsymbol{\delta}_D\boldsymbol{H}_{0D}^{-1} + D^{-2}[\boldsymbol{\delta}_D\boldsymbol{H}_{0D}^{-1}]^2 + \ldots\}.\end{aligned}$$

Since $\boldsymbol{H}_{0D}^{-1}$ is diagonal and has all entries are $O(1)$, then $\boldsymbol{\delta}_D^{(-j)}\boldsymbol{H}_{0D}^{-1}$ is uniformly bounded. Hence, in the Taylor expansion, the second-order term is $O(D^{-1})$ and the third order is $O(D^{-2})$, etc. This proves that the inverse Hessian matrix is diagonally dominant and the off-diagonal elements are uniformly $O(D^{-1})$. □

Below, we sequentially show the results (I) and (II) in the proof of Theorem 4. The ideas in the proofs are straightforward, but there are many calculations of partial derivatives and determining orders of different terms as $D \to \infty$. All the presented orders in D are under the assumption that α_j's are uniformly bounded away from ± 1.

Lemma 3. , *The elements of the score vector $\boldsymbol{s}_D(\boldsymbol{\beta}_D)$ in* (37)*, evaluated at the true parameters $\boldsymbol{\alpha}_D$, are uniformly $O(D^{-1})$.*

Proof. For simpler notation, the subscript D will be suppressed for most variables.

An outline is summarized below in a few steps and substeps:

Step A: compute the expression of score vector below shown in (42);

Step B: evaluate the order of score vector at the true parameters $\boldsymbol{\alpha}_D$.

Step A:

Consider the jth element of the gradient of (36).

From Section 15.8 of [5], for a covariance matrix $\boldsymbol{\Sigma}$ parametrized by $\boldsymbol{\beta}$,

$$\frac{\log|\boldsymbol{\Sigma}|}{\partial\beta_j} = \mathrm{trace}\left(\boldsymbol{\Sigma}^{-1}\frac{\partial\boldsymbol{\Sigma}}{\partial\beta_j}\right), \quad \frac{\partial\boldsymbol{\Sigma}^{-1}}{\partial\beta} = -\boldsymbol{\Sigma}^{-1}\frac{\partial\boldsymbol{\Sigma}}{\partial\beta_j}\boldsymbol{\Sigma}^{-1}.$$

Thus, the gradient of (36) with respect to $\boldsymbol{\beta}_j$ is

$$\mathrm{trace}\left[\boldsymbol{\Sigma}^{-1}\frac{\partial\boldsymbol{\Sigma}}{\partial\beta_j}\right] - \mathrm{trace}\left[\boldsymbol{\Sigma}^{-1}\frac{\partial\boldsymbol{\Sigma}}{\partial\beta_j}\boldsymbol{\Sigma}^{-1}\boldsymbol{R}_D^0\right]. \tag{39}$$

Next, we simplify each term in (39). For the first term, the derivative of $\partial\boldsymbol{\Sigma}(\boldsymbol{\beta})/\partial\beta_j$ has only jth row and jth column being non-zero; specifically:

$$\frac{\partial \boldsymbol{\Sigma}}{\partial \beta_j} = \begin{bmatrix} 0 & 0 & \dots & 0 & \beta_1 & 0 & \dots & 0 \\ 0 & 0 & \dots & 0 & \beta_2 & 0 & \dots & 0 \\ 0 & \vdots & \vdots & \vdots & \vdots & \vdots & \vdots & 0 \\ \beta_1 & \beta_2 & \dots & \beta_{j-1} & 0 & \beta_{j+1} & \dots & \beta_D \\ 0 & \vdots & \vdots & \vdots & \vdots & \vdots & \vdots & 0 \\ 0 & 0 & \dots & 0 & \beta_D & 0 & \dots & 0 \end{bmatrix}.$$

Let $\boldsymbol{\Sigma}^{-1}(\boldsymbol{\beta}) = (\sigma^{ij})_{1\le i,j\le D}$. After some algebraic computations, the first term in (39) becomes

$$\text{trace}\left[\boldsymbol{\Sigma}^{-1}\frac{\partial \boldsymbol{\Sigma}}{\partial \boldsymbol{\beta}_j}\right] = 2\sum_{1\le \ell\le D, \ell\ne j}^{D} \beta_\ell \sigma^{j\ell} =: 2b_j^{(-j)}(\boldsymbol{\beta}_D). \tag{40}$$

For $k \ne j$, $k \in \{1,\dots,j-1,j+1,D\}$, let

$$b_k^{(-j)}(\boldsymbol{\beta}_D) = \sum_{1\le \ell\le D, \ell\ne j}^{D} \beta_\ell \sigma^{kl}. \tag{41}$$

For a fixed j, the second term of (39) becomes

$$\boldsymbol{\Sigma}^{-1}\frac{\partial \boldsymbol{\Sigma}}{\partial \beta_j} = \begin{bmatrix} \beta_1\sigma^{1j} & \beta_2\sigma^{1j} & \dots & \beta_{j-1}\sigma^{1j} & b_1^{(-j)} & \beta_{j+1}\sigma^{1j} & \dots & \beta_D\sigma^{1j} \\ \beta_1\sigma^{2j} & \beta_2\sigma^{2j} & \dots & \beta_{j-1}\sigma^{2j} & b_2^{(-j)} & \beta_{j+1}\sigma^{2j} & \dots & \beta_D\sigma^{2j} \\ \vdots & \vdots & \vdots & \vdots & \vdots & \vdots & & \\ \beta_1\sigma^{jj} & \beta_2\sigma^{jj} & \dots & \beta_{j-1}\sigma^{jj} & b_j^{(-j)} & \beta_{j+1}\sigma^{jj} & \dots & \beta_D\sigma^{jj} \\ \vdots & \vdots & \vdots & \vdots & \vdots & \vdots & & \\ \beta_1\sigma^{Dj} & \beta_2\sigma^{Dj} & \dots & \beta_{j-1}\sigma^{Dj} & b_D^{(-j)} & \beta_{j+1}\sigma^{Dj} & \dots & \beta_D\sigma^{Dj} \end{bmatrix}.$$

By symmetry of $\boldsymbol{\Sigma}^{-1}$,

$$\boldsymbol{\Sigma}^{-1}\frac{\partial \boldsymbol{\Sigma}}{\partial \beta_j}\boldsymbol{\Sigma}^{-1} = \begin{bmatrix} 2b_1^{(-j)}\sigma^{1j} & 2b_1^{(-j)}\sigma^{2j} & \dots & 2b_1^{(-j)}\sigma^{Dj} \\ 2b_2^{(-j)}\sigma^{1j} & 2b_2^{(-j)}\sigma^{2j} & \dots & 2b_2^{(-j)}\sigma^{Dj} \\ \vdots & \vdots & \vdots & \vdots \\ 2b_D^{(-j)}\sigma^{1j} & 2b_D^{(-j)}\sigma^{2j} & \dots & 2b_D^{(-j)}\sigma^{Dj} \end{bmatrix}.$$

Letting $\boldsymbol{R}_D^0 = (r_{ij})_{1\le i,j\le D}$, the second term of the (39) is

$$2\sum_{k=1}^{D}\sum_{m=1}^{D} b_k^{(-j)}\sigma^{mj} r_{mk}.$$

Overall, the gradient in (39) at the fixed index j is

$$s_j(\boldsymbol{\beta}_D) = 2\Big(\underbrace{b_j^{(-j)}(\boldsymbol{\beta}_D)}_{\text{term1}} - \sum_{k=1}^{D}\sum_{m=1}^{D} b_k^{(-j)}\sigma^{mj} r_{mk}\Big). \tag{42}$$

This concludes Step A.

Step B: Consider $s_j(\boldsymbol{\alpha}_D)$ on the right side of (37). We will show that $s_j(\boldsymbol{\alpha}_D) \to 0$ as $D \to \infty$. More specifically, $s_j(\boldsymbol{\alpha}_D) = O(D^{-1})$ with Assumption 1. We need to evaluate the order of two terms in (42). The second term in (42) can be written as

$$\sum_{k=1}^{D}\sum_{m=1}^{D} b_k^{(-j)}\sigma^{mj} r_{mk} = \underbrace{\sum_{m=1}^{D} b_j^{(-j)}\sigma^{mj} r_{mj}}_{\text{term2}} + \underbrace{\sum_{k=1,k\neq j}^{D}\sum_{m=1}^{D} b_k^{(-j)}\sigma^{mj} r_{mk}}_{\text{term3}}. \tag{43}$$

Step B can be carried out in four sub-steps:

- **Step B.0**: Some simplified expression are provided.
- **Step B.1**: The term1 in (42) with $\beta_D = \alpha_D$ is simplified and evaluated.
- **Step B.2**: The term2 in (43) with $\beta_D = \alpha_D$ is simplified and evaluated.
- **Step B.3**: The term3 in (43) with $\beta_D = \alpha_D$ is simplified and evaluated.

Step B.0 The simplified expressions of (1) σ^{mi}, (2) $b_j^{(-j)}(\boldsymbol{\alpha}_D)$, (3) $b_k^{(-j)}(\boldsymbol{\alpha}_D)$, (4) $\sum_{m=1}^{D}\sigma^{mj}\alpha_m\alpha_k$, (5) $\sum_{m=1}^{D}\sigma^{mj}\gamma_{mk}$, and (6) $\sum_{m=1}^{D}\sigma^{mj} r_{mk}$ are provided for later use.

(1) For the inverse of $\boldsymbol{\Sigma}$ with 1-factor structure, from special case of $\boldsymbol{\Sigma}^{-1}$, page 136 in [6],

$$\sigma^{mi} = -(1+q_D)^{-1}\psi_m^{-2}\alpha_m\alpha_i\psi_i^{-2} \qquad \text{if } m \neq i, \tag{44a}$$
$$\sigma^{mi} = \psi_m^{-2} - (1+q_D)^{-1}\psi_m^{-2}\alpha_m\alpha_m\psi_m^{-2} \qquad \text{if } m = i, \tag{44b}$$

where

$$q_D = \sum_{\ell=1}^{D} \alpha_\ell^2/\psi_\ell^2 \tag{45}$$

is $O(D)$. Therefore, σ^{mi} for $m \neq i$ is $O(D^{-1})$, and $\sigma^{mm} = \psi_m^{-2} + O(D^{-1})$. The inverse matrix of $\boldsymbol{\Sigma}$ will approach a diagonal matrix as $D \to \infty$.

(2) Define $q_D^{(-j)} = q_D - \alpha_j^2\psi_j^{-2}$. Let $\beta_\ell = \alpha_\ell$ in (40). Then

$$\begin{aligned} b_j^{(-j)}(\boldsymbol{\alpha}_D) &= \sum_{\ell=1,\ell\neq j}^{D} \alpha_\ell\sigma^{j\ell} = -\sum_{1\le\ell\le D,\ell\neq j} \alpha_\ell(1+q_D)^{-1}\psi_j^{-2}\alpha_j\alpha_\ell\psi_\ell^{-2} \\ &= -\psi_j^{-2}\alpha_j(1+q_D)^{-1}\sum_{1\le\ell\le D,\ell\neq j}\alpha_\ell[\alpha_\ell\psi_\ell^{-2}] = -\psi_j^{-2}\alpha_j(1+q_D)^{-1}q_D^{(-j)}. \end{aligned}$$

(3) Similarly, let $\beta_\ell = \alpha_\ell$ in (41). With $k \neq j$,

$$b_k^{(-j)}(\boldsymbol{\alpha}_D) = \sum_{\ell=1,\ell\neq j}^{D} \alpha_\ell \sigma^{k\ell} = \alpha_k \sigma^{kk} + \sum_{\ell=1,\ell\neq j,\ell\neq k}^{D} \alpha_\ell \sigma^{\ell k}$$
$$= \alpha_k \big[\psi_k^{-2} - (1+q_D)^{-1}\psi_k^{-2}\alpha_k\alpha_k\psi_k^{-2}\big] - \sum_{\ell=1,\ell\neq j,\ell\neq k}^{D} \alpha_\ell (1+q_D)^{-1}\psi_\ell^{-2}\alpha_\ell\alpha_k\psi_k^{-2}$$
$$= \alpha_k\psi_k^{-2} - \sum_{\ell=1,\ell\neq j}^{D} \alpha_\ell(1+q_D)^{-1}\psi_l^{-2}\alpha_\ell\alpha_k\psi_k^{-2}$$
$$= \alpha_k\psi_k^{-2} - (1+q_D)^{-1}\alpha_k\psi_k^{-2}\sum_{\ell=1,\ell\neq j}^{D}\alpha_\ell^2\psi_\ell^{-2}$$
$$= \alpha_k\psi_k^{-2}\Big[1-(1+q_D)^{-1}\sum_{\ell=1,\ell\neq j}^{D}\alpha_\ell^2\psi_\ell^{-2}\Big] = \alpha_k\psi_k^{-2}\big[1-(1+q_D)^{-1}q_D^{(-j)}\big].$$

Let $q_D^{(-j)}/(1+q_D) = \eta_D^{(-j)}$. In summary,

$$b_j^{(-j)}(\boldsymbol{\alpha}_D) = -\psi_j^{-2}\alpha_j\eta_D^{(-j)} =: \delta_D^{(-j)}, \tag{46}$$
$$b_k^{(-j)}(\boldsymbol{\alpha}_D) = \alpha_k\psi_k^{-2} - \alpha_k\psi_k^{-2}\eta_D^{(-j)} = (1-\eta_D^{(-j)})\alpha_k\psi_k^{-2}, \quad k\neq j. \tag{47}$$

Also, from the definition of $\eta_D^{(-j)}$, algebraic calculations lead to:

$$\eta_D^{(-j)} + (1+q_D)^{-1} = 1-(1+q_D)^{-1}\alpha_j^2\psi_j^{-2} = O(1), \tag{48}$$
$$\eta_D^{(-j)} = 1-\alpha_j^2\psi_j^{-2}(1+q_D)^{-1} - (1+q_D)^{-1} = O(1). \tag{49}$$

(4) By substituting σ^{mj} in (44b), $\sum_{m=1}^{D}\sigma^{mj}\alpha_m\alpha_k$ can be simplified. For $m\neq j$,

$$\sum_{m=1,m\neq j}^{D}\sigma^{mj}\alpha_m\alpha_k = -\alpha_k\sum_{m=1,m\neq j}^{D}(1+q_D)^{-1}\psi_m^{-2}\alpha_m^2\alpha_j\psi_j^{-2}$$
$$= -(1+q_D)^{-1}\alpha_k\alpha_j\psi_j^{-2}\sum_{m=1,m\neq j}^{D}\psi_m^{-2}\alpha_m^2 = -\eta_D^{(-j)}\alpha_k\alpha_j\psi_j^{-2}.$$

For $m=j$, with (48),

$$\sigma^{jj}\alpha_j\alpha_k = \alpha_k\alpha_j\psi_j^{-2}\big[1-(1+q_D)^{-1}\alpha_j^2\psi_j^{-2}\big] = \alpha_k\alpha_j\psi_j^{-2}\big[\eta_D^{(-j)}+(1+q_D)^{-1}\big].$$

Therefore, from summing the two parts,

$$\sum_{m=1}^{D}\sigma^{mj}\alpha_m\alpha_k = \sum_{m=1,m\neq j}^{D}\sigma^{mj}\alpha_m\alpha_k + \sigma^{jj}\alpha_j\alpha_k = (1+q_D)^{-1}\alpha_k\alpha_j\psi_j^{-2}. \tag{50}$$

(5) Substituting (44a), (44b) and (48) leads to

$$\sum_{m=1}^{D}\sigma^{mj}\gamma_{mk}=\sum_{m=1,m\neq j}^{D}\sigma^{mj}\gamma_{mk}+\sigma^{jj}\gamma_{jk}$$

$$=-\sum_{m=1,m\neq j}^{D}(1+q_D)^{-1}\psi_m^{-2}\alpha_m\alpha_j\psi_j^{-2}\gamma_{mk}+\psi_j^{-2}\left[1-(1+q_D)^{-1}\alpha_j^2\psi_j^{-2}\right]\gamma_{jk}$$

$$=-(1+q_D)^{-1}\alpha_j\psi_j^{-2}\sum_{m=1,m\neq j}^{D}\psi_m^{-2}\alpha_m\gamma_{mk}+\psi_j^{-2}\eta_D^{(-j)}\gamma_{jk}+\psi_j^{-2}(1+q_D)^{-1}\gamma_{jk}. \tag{51}$$

(6) With $r_{mk}=\alpha_m\alpha_k+\gamma_{mk}$, combining the results in (4) and (5) leads to

$$\sum_{m=1}^{D}\sigma^{mj}r_{mk}=\sum_{m=1}^{D}\sigma^{mj}\alpha_m\alpha_k+\sum_{m=1}^{D}\sigma^{mj}\gamma_{mk}$$

$$=(1+q_D)^{-1}\alpha_k\alpha_j\psi_j^{-2}-(1+q_D)^{-1}\alpha_j\psi_j^{-2}\sum_{m=1,m\neq j}^{D}\psi_m^{-2}\alpha_m\gamma_{mk}$$

$$+\psi_j^{-2}\eta_D^{(-j)}\gamma_{jk}+\psi_j^{-2}(1+q_D)^{-1}\gamma_{jk}. \tag{52}$$

Step B.1: By (46), the term 1 in (42) is $\delta_D^{(-j)}$.

Step B.2: For term2 in (43), substituting (46) and (52) with $k=j$ leads to:

$$\sum_{m=1}^{D}b_j^{(-j)}\sigma^{mj}r_{mj}$$

$$=\delta_D^{(-j)}\Big[(1+q_D)^{-1}\alpha_j^2\psi_j^{-2}-(1+q_D)^{-1}\alpha_j\psi_j^{-2}\sum_{m=1,m\neq j}^{D}\psi_m^{-2}\alpha_m\gamma_{mj}$$

$$+\psi_j^{-2}\eta_D^{(-j)}\gamma_{jj}+\psi_j^{-2}(1+q_D)^{-1}\gamma_{jj}\Big]. \tag{53}$$

Using (49), $\delta_D^{(-j)}=O(1)$ in (53) assuming loading parameters bounded away from ±1. As for the four terms inside the brackets in (53), the order of the first and the fourth term is $O(D^{-1})$ because $(1+q_D)^{-1}=O(D^{-1})$. With Assumption 4 for the $\{\gamma_{mj}\}$, the second term is also $O(D^{-1})$ with $\sum_{m=1,m\neq j}\psi_m^{-2}\alpha_m\gamma_{mk}=O(1)$. Furthermore, the third term is $\eta_D^{(-j)}=O(1)$ by noticing $\gamma_{jj}=\psi_j^2$. Overall, the order of term2 in (43) is $\eta_D^{(-j)}\delta_D^{(-j)}+O(D^{-1})$.

Step B.3: For term3 in (43), the substitution (52) and (47) leads to,

$$
\begin{aligned}
&\sum_{k=1,k\neq j}^{D}\sum_{m=1}^{D} b_k^{(-j)}\sigma^{mj} r_{mk} \\
&= (1-\eta_D^{(-j)})\Big[(1+q_D)^{-1}\alpha_j\psi_j^{-2}\sum_{k=1,k\neq j}^{D}\alpha_k^2\psi_k^{-2} \\
&-(1+q_D)^{-1}\alpha_j\psi_j^{-2}\sum_{k=1,k\neq j}^{D}\alpha_k\psi_k^{-2}\sum_{m=1,m\neq j}^{D}\psi_m^{-2}\alpha_m\gamma_{mk} \\
&+\psi_j^{-2}\eta_D^{(-j)}\sum_{k=1,k\neq j}^{D}\alpha_k\psi_k^{-2}\gamma_{jk}+\psi_j^{-2}(1+q_D)^{-1}\sum_{k=1,k\neq j}^{D}\alpha_k\psi_k^{-2}\gamma_{jk}\Big].
\end{aligned}
\tag{54}
$$

In (54), the scalar $(1-\eta_D^{(-j)})$ is $O(D^{-1})$; the first term, second term, and third term inside the bracket are all $O(1)$ while the fourth term is $O(D^{-1})$. Thus, term3 in (43) is $O(D^{-1})$.

Combining the results in **B.1,B.2,B.3**, term1 in (42) is $\delta_D^{(-j)}$, term2 in (43) is $\delta_D^{(-j)}\eta_D^{(-j)}+O(D^{-1})$, and term 3 in (43) is $O(D^{-1})$. The score vector in (42) equals to 2[term1-(term2+term3)] $=2[\delta_D^{(-j)}(1-\eta_D^{(-j)})+O(D^{-1})]$. Since $\delta_D^{(-j)}(1-\eta_D^{(-j)})=O(D^{-1})$ by (49), $s_j(\boldsymbol{\beta}_D)$ in (42) is of order $O(D^{-1})$. With the super-population assumption, the indices are "exchangeable" and the $\{s_j(\boldsymbol{\beta}_D)\}$ are uniformly $O(D^{-1})$. □

Lemma 4. *The diagonal entries in the Hessian matrix* $\boldsymbol{H}_D(\boldsymbol{\beta}_D)$ *in* (37) *are uniformly* $O(1)$ *and the off-diagonal entries are uniformly* $O(D^{-1})$.

Proof. For simpler notation, the subscript D will be suppressed for most variables.

Following similar techniques to the proof of the preceding lemma, the order of off-diagonal and diagonal elements in the Hessian matrix will be shown respectively in **Step C** and **Step D**. Based on the super-population assumption, considering the non-diagonal $(1,2)$ element and diagonal (1,1) element in the Hessian matrix are sufficient, and this will simplify the notation.

Step C. The order of off-diagonal entries in the Hessian matrix.

Take the derivatives of the score vector in (42), the off-diagonal (1,2) entry in the Hessian matrix can be expressed as

$$
\begin{aligned}
\frac{1}{2}\frac{\partial s_1(\boldsymbol{\beta}_D)}{\partial\boldsymbol{\beta}_2} &= \underbrace{\partial\Big[b_1^{(-1)}(\boldsymbol{\beta}_D)\Big(1-\sum_{m=1}^{D}\sigma^{m1}r_{m1}\Big)\Big]\Big/\partial\beta_2}_{\text{term1}} \\
&-\underbrace{\sum_{k=2}^{D}\sum_{m=1}^{D}\frac{\partial(b_k^{(-1)}\sigma^{m1}r_{mk})}{\partial\beta_2}}_{\text{term2}}.
\end{aligned}
\tag{55}
$$

The term2 in (55) can be rewritten as

$$\underbrace{\sum_{k=2}^{D}\sum_{m=1}^{D}\frac{\partial b_k^{(-1)}}{\partial\beta_2}\sigma^{m1}r_{mk}}_{\text{term21}}+\underbrace{\sum_{k=2}^{D}\sum_{m=1}^{D}b_k^{(-1)}\frac{\partial\sigma^{m1}}{\partial\beta_2}r_{mk}}_{\text{term22}}$$
$$+\underbrace{\sum_{k=2}^{D}\sum_{m=1}^{D}b_k^{(-1)}\sigma^{m1}\frac{\partial r_{mk}}{\partial\beta_2}}_{\text{term23}}. \quad (56)$$

For notational simplicity, let

$$\begin{aligned}\Delta_1 &= \frac{\partial(1-\alpha_2^2)^{-1}}{\partial\alpha_2}=2\alpha_2/(1-\alpha_2^2)^2,\\ \Delta_2 &= \frac{\partial\alpha_2(1-\alpha_2^2)^{-1}}{\partial\alpha_2}=(1+\alpha_2^2)/(1-\alpha_2^2)^2,\text{ and}\\ \Delta_3 &= \frac{\partial q_D}{\partial\alpha_2}=\frac{\partial[\alpha_2^2/(1-\alpha_2^2)]}{\partial\alpha_2}=2\alpha_2/(1-\alpha_2^2)^2\text{ from (45).}\end{aligned}$$

The evaluation of the order of off-diagonal entries is done in four sub-steps.

- **Step C.0**: Some simplified expressions are provided for later use.
- **Step C.1**: The term1 in (55) is simplified and evaluated at $\boldsymbol{\beta}_D=\boldsymbol{\alpha}_D$.
- **Step C.2**: The term21 in the expression (56) is simplified and evaluated at $\boldsymbol{\beta}_D=\boldsymbol{\alpha}_D$.
- **Step C.3**: The term22 in the expression (56) is simplified and evaluated at $\boldsymbol{\beta}_D=\boldsymbol{\alpha}_D$.
- **Step C.4**: The term23 in the expression (56) is simplified and evaluated.

Step C.0: With some algebraic calculations, the derivatives of $\sigma^{k\ell}$ in (44a) and (44b) with respect to β_2 when $k\neq 2$ is

$$\frac{\partial\sigma^{k\ell}(\boldsymbol{\beta}_D)}{\partial\beta_2}\bigg|_{\boldsymbol{\beta}_D=\boldsymbol{\alpha}_D}=\begin{cases}(1+q_D)^{-2}\Delta_3\psi_k^{-2}\alpha_k\alpha_\ell\psi_\ell^{-2}=O(D^{-2}) & \ell\neq 2,\\ -(1+q_D)^{-1}\alpha_k\psi_k^{-2}\Delta_2+O(D^{-2})=O(D^{-1}) & \ell=2.\end{cases} \quad (57)$$

When $k=2$, the derivative of $\sigma^{2\ell}$ with respect to β_2 evaluated is

$$\frac{\partial\sigma^{2\ell}(\boldsymbol{\beta}_D)}{\partial\beta_2}\bigg|_{\boldsymbol{\beta}_D=\boldsymbol{\alpha}_D}=\begin{cases}-\frac{[(1+q_D)\alpha_\ell\psi_\ell^{-2}\Delta_2-\psi_2^{-2}\alpha_2\alpha_\ell\psi_\ell^{-2}\Delta_3]}{(1+q_D)^2}\\ \quad =O(D^{-1}) & \ell\neq 2,\\ \Delta_1+O(D^{-1}) & \ell=2.\end{cases} \quad (58)$$

For $k\neq 2$, the derivative of $b_k^{(-1)}(\boldsymbol{\beta}_D)=\sum_{\ell=2}^{D}\beta_\ell\sigma^{k\ell}$ in (41) with respect to β_2 evaluated at $\boldsymbol{\beta}_D=\boldsymbol{\alpha}_D$ is simplified below by substituting the order of expressions in (44a) and (57):

$$\frac{\partial b_k^{(-1)}(\boldsymbol{\beta}_D)}{\partial \beta_2}\Big|_{\boldsymbol{\beta}_D=\boldsymbol{\alpha}_D} = \sigma^{k2}(\boldsymbol{\alpha}_D) + \beta_2\Big(\frac{\partial \sigma^{k2}}{\partial \beta_2}\Big|_{\boldsymbol{\alpha}_D}\Big) + \sum_{\ell=3}^{D}\beta_\ell\Big(\frac{\partial \sigma^{k\ell}}{\partial \beta_2}\Big|_{\boldsymbol{\alpha}_D}\Big)$$
$$= O(D^{-1}) + O(D^{-1}) + O(D^{-1}) = O(D^{-1}). \tag{59}$$

In the case of $k=2$, the derivative of $b_2^{(-1)}$ with respect with to β_2 evaluated at $\boldsymbol{\beta}_D = \boldsymbol{\alpha}_D$, using (58) and q_D in (45), is

$$\frac{\partial b_2^{(-1)}(\boldsymbol{\beta}_D)}{\partial \beta_2}\Big|_{\boldsymbol{\beta}_D=\boldsymbol{\alpha}_D} = \Big[\sigma^{22} + \beta_2\frac{\partial \sigma^{22}}{\partial \beta_2} + \sum_{\ell=3}^{D}\beta_\ell \frac{\partial \sigma^{2\ell}}{\partial \beta_2}\Big]\Big|_{\boldsymbol{\beta}_D=\boldsymbol{\alpha}_D}$$
$$= (1-\alpha_2^2)^{-1} + 2\alpha_2^2(1-\alpha_2^2)^{-2} - \sum_{\ell=3}^{D}(1+q_D)^{-1}\alpha_\ell^2\psi_\ell^{-2}\Delta_2 + O(D^{-1})$$
$$= \Delta_2 - (1+q_D)^{-1}\sum_{\ell=3}^{D}\Delta_2\alpha_\ell^2\psi_\ell^{-2} + O(D^{-1}) = O(D^{-1}).$$

Step C.1. From (52) with Assumption 4, $\sum_{m=1}^{D}\sigma^{m1}r_{m1} = \psi_1^{-2}\eta_D^{(-1)}\gamma_{11} + O(D^{-1}) = \eta_D^{(-1)} + O(D^{-1})$ as $\gamma_{11} = \psi_1^2$, and thus from (49),

$$1 - \sum_{m=1}^{D}\sigma^{m1}r_{m1} = O(D^{-1}). \tag{60}$$

Next,

$$\sum_{m=1}^{D}\frac{\partial(\sigma^{m1}r_{m1})}{\partial \beta_2} = \sum_{m=1}^{D}\frac{\partial \sigma^{m1}}{\partial \beta_2}r_{m1} + \sum_{m=1}^{D}\frac{\partial r_{m1}}{\partial \beta_2}\sigma^{m1}$$
$$= \frac{\partial \sigma^{21}}{\partial \beta_2}r_{21} + \sum_{m=1,m\neq 2}^{D}\frac{\partial \sigma^{m1}}{\partial \beta_2}r_{m1} + \sigma^{21}\beta_1 \tag{61}$$
$$= O(D^{-1}) + O(D^{-1}) + O(D^{-1}) = O(D^{-1}), \tag{62}$$

using $r_{m1} = \beta_m\beta_1 + \gamma_{m1}$, and $r_{11} = 1$. The orders in (62) are based on the orders computed in (57), (58), and (44a). The derivative of $\partial\sigma^{k\ell}/\partial\beta_2 = O(D^{-1})$ when $k=2, \ell\neq 2$ or $k\neq 2, \ell=2$ and is $O(D^{-2})$ when $k\neq 2, \ell\neq 2$.

With the above calculation and the term evaluated from step C.0 in (59), the order of term1 in (55) evaluated at $\boldsymbol{\beta}_D = \boldsymbol{\alpha}_D$ is

$$\frac{\partial b_1^{(-1)}(\boldsymbol{\alpha}_D)}{\partial \beta_2} \times \Big(1 - \sum_{m=1}^{D}\sigma^{m1}r_{m1}\Big) - b_1^{(-1)}(\boldsymbol{\alpha}_D)\sum_{m=1}^{D}\frac{\partial(\sigma^{m1}r_{m1})}{\partial \beta_2} \tag{63}$$
$$= O(D^{-2}) + O(D^{-1}) = O(D^{-1}).$$

The first term in (63) is based on (59) and (60) while the second term in (63) is based on (62) and $b_1^{(-1)}(\boldsymbol{\alpha}_D) = O(1)$ in (46). Therefore, the term1 in (55) is $O(D^{-1})$.

Step C.2. From equation (52), $\sum_{m=1}^{D}\sigma^{m1}r_{mk}=\psi_1^{-2}\eta_D^{(-1)}\gamma_{1k}+O(D^{-1})$. Then, the term21 in (56) evaluated at $\boldsymbol{\beta}_D=\boldsymbol{\alpha}_D$ equals to

$$\sum_{k=2}^{D}\frac{\partial b_k^{(-1)}}{\partial\beta_2}\big(\psi_1^{-2}\eta_D^{(-1)}\gamma_{1k}+O(D^{-1})\big). \tag{64}$$

From (59) in Step C.0, $\partial b_k^{(-1)}/\partial\beta_2$ is uniformly $O(D^{-1})$ for $k=2,\dots,D$. With the assumption on the weak residual dependence $\sum_{k=1}^{D}|\gamma_{1k}|=O(1)$ and loadings are bounded away from ±1, and hence the order of term in (64) is $O(D^{-1})$. Thus, the term21 in (56) is $O(D^{-1})$.

Step C.3. Substituting the expression (47) in the term22 in (56), and evaluating at $\boldsymbol{\beta}_D=\boldsymbol{\alpha}_D$ leads to

$$(1-\eta_D^{(-1)})\sum_{k=2}^{D}\alpha_k\psi_k^{-2}\sum_{m=1}^{D}\frac{\partial\sigma^{m1}}{\partial\beta_2}\times(\alpha_m\alpha_k+\gamma_{mk}). \tag{65}$$

For the double summation term in (65), substitute (57) leads to

$$\begin{aligned}
&\sum_{k=2}^{D}\alpha_k\psi_k^{-2}\sum_{m=1}^{D}\frac{\partial\sigma^{m1}}{\partial\beta_2}(\alpha_m\alpha_k+\gamma_{mk})\\
&=\sum_{k=2}^{D}\alpha_k\psi_k^{-2}\sum_{m\neq2}\frac{\partial\sigma^{m1}}{\partial\beta_2}(\alpha_m\alpha_k+\gamma_{mk})+\sum_{k=2}^{D}\alpha_k\psi_k^{-2}\frac{\partial\sigma^{21}}{\partial\beta_2}(\alpha_2\alpha_k+\gamma_{2k})\\
&=\sum_{k=2}^{D}\alpha_k\psi_k^{-2}\Big\{\sum_{m\neq2}(1+q_D)^{-2}\Delta_3\psi_m^{-2}\alpha_m\alpha_1\psi_1^{-2})(\alpha_m\alpha_k+\gamma_{mk})\\
&+\big[-(1+q_D)^{-1}\alpha_1\psi_1^{-2}\Delta_2+O(D^{-2})\big](\alpha_2\alpha_k+\gamma_{2k})\Big\}\\
&=\sum_{k=2}^{D}\alpha_k\psi_k^{-2}\Big\{(1+q_D)^{-2}\Delta_3\alpha_1\psi_1^{-2}\alpha_k\sum_{m\neq2}\alpha_m^2\psi_m^{-2}\\
&+(1+q_D)^{-2}\Delta_3\alpha_1\psi_1^{-2}\sum_{m\neq2}\alpha_m\psi_m^{-2}\gamma_{mk}-(1+q_D)^{-1}\alpha_1\alpha_2\psi_1^{-2}\Delta_2\alpha_k\\
&-(1+q_D)^{-1}\alpha_1\psi_1^{-2}\Delta_2\gamma_{2k}+O(D^{-2})\Big\}\\
&=\sum_{k=2}^{D}\alpha_k\psi_k^{-2}\Big\{\Delta_3(1+q_D)^{-2}q_D^{(-2)}\alpha_1\alpha_k\psi_1^{-2}\\
&+\Delta_3(1+q_D)^{-2}\alpha_1\psi_1^{-2}\sum_{m\neq2}^{D}\alpha_m\psi_m^{-2}\gamma_{mk}\\
&+O(D^{-2})-(1+q_D)^{-1}\alpha_1\psi_1^{-2}\Delta_2\alpha_2\alpha_k-(1+q_D)^{-1}\alpha_1\psi_1^{-2}\Delta_2\gamma_{2k}\Big\}.
\end{aligned}$$

Further algebraic simplications lead to:

$$
\begin{aligned}
&= \Delta_3(1+q_D)^{-1}\eta_D^{(-2)}\alpha_1\psi_1^{-2}\sum_{k=2}^{D}\alpha_k^2\psi_k^{-2} \\
&+ \Delta_3(1+q_D)^{-2}\alpha_1\psi_1^{-2}\sum_{k=2}^{D}\alpha_k^2\psi_k^{-2}\sum_{m\neq 2}^{D}\alpha_m\psi_m^{-2}\gamma_{mk} + O(D^{-1}) \\
&-(1+q_D)^{-1}\alpha_1\psi_1^{-2}\Delta_2\alpha_2\sum_{k=2}^{D}\alpha_k^2\psi_k^{-2} - (1+q_D)^{-1}\alpha_1\psi_1^{-2}\Delta_2\sum_{k=2}^{D}\alpha_k\psi_k^{-2}\gamma_{2k} \qquad (66) \\
&= O(1)+O(D^{-1})+O(D^{-1})+O(1)+O(D^{-1}) = O(1).
\end{aligned}
$$

$\eta^{(-1)}$ and $\eta^{(-2)}$ are in (48) with $j=1$ and $j=2$ respectively. By noticing $\sum_{k=2}^{D}\alpha_k^2\psi_k^{-2} = O(D)$ and $\sum_{m\neq 2}\alpha_m\psi_m^{-2}\gamma_{mk} = O(1)$, the summation in (66) is $O(1)$. Since $1-\eta_D^{(-1)} = O(D^{-1})$, the order of term in (65) is $O(D^{-1})$. Thus, the term22 in (56) is $O(D^{-1})$.

Step C.4. Since $\partial r_{mk}/\partial\beta_2$ is non-zero only when $m=2$ or $k=2$. term23 in (56) can be simplified, using (44a) and (47), to

$$
\begin{aligned}
&b_2^{(-1)}\sum_{m\neq 2}^{D}\sigma^{m1}\alpha_m + \sigma^{12}\sum_{k=3}^{D}\alpha_k b_k^{(-1)} \\
&= -b_2^{(-1)}\alpha_1\psi_1^{-2}(1+q_D)^{-1}\sum_{m\neq 2}^{D}\alpha_m^2\psi_m^{-2} + \sigma^{12}(1-\eta_D^{(-1)})\sum_{k=3}^{D}\alpha_k^2\psi_k^{-2} \\
&= O(D^{-1}) + O(D^{-1}) = O(D^{-1}).
\end{aligned}
$$

Recall $b_2^{(-1)}$ in (47), σ^{12} in (44a) and $1-\eta_D^{(-1)}$ are all $O(D^{-1})$. The term23 in (56) is also $O(D^{-1})$.

Combining Steps C.0, C.1, C.2, and C.3 and using the super-population assumption, the off-diagonal entries are uniformly of order $O(D^{-1})$.

Step D. The order of diagonal entries in the Hessian matrix.

For the diagonal entries, consider

$$
\frac{1}{2}\frac{\partial s_1(\boldsymbol{\beta}_D)}{\partial\beta_1} = \underbrace{\partial\left[b_1^{(-1)}(\boldsymbol{\beta}_D)\Big(1-\sum_{m=1}^{D}\sigma^{m1}r_{m1}\Big)\right]\Big/\partial\beta_1}_{\text{term1}} \underbrace{-\sum_{k=2}^{D}\sum_{m=1}^{D}\frac{\partial(b_k^{(-1)}\sigma^{m1}r_{mk})}{\partial\beta_1}}_{\text{term2}}. \qquad (67)
$$

Similar to the analysis of the off-diagonal entries, we evaluate the order of term 1 and term2 in (67) at $\boldsymbol{\beta}_D = \boldsymbol{\alpha}_D$ respectively.

Step D.1: Evaluate term1 in (67)

(1) Using the expressions in (40) and (44a),

$$\frac{\partial b_1^{(-1)}(\boldsymbol{\beta}_D)}{\partial \beta_1}\bigg|_{\boldsymbol{\beta}_D=\boldsymbol{\alpha}_D} = \sum_{\ell=2}^{D} \beta_\ell \frac{\partial \sigma^{\ell 1}}{\partial \beta_1}\bigg|_{\boldsymbol{\beta}_D=\boldsymbol{\alpha}_D} = O(1). \tag{68}$$

(2) Using the expressions in (44a) and (44b),

$$\begin{aligned}
\sum_{m=1}^{D}\left(\frac{\partial \sigma^{1m}}{\partial \beta_1}\bigg|_{\boldsymbol{\beta}_D=\boldsymbol{\alpha}_D}\right) r_{m1} &= \frac{\partial \sigma^{11}}{\partial \beta_1} + \sum_{m=2}^{D} \frac{\partial \sigma^{1m}}{\partial \beta_1}(\alpha_m \alpha_1 + \gamma_{1m}) \\
&= 2\alpha_1 \psi_1^{-4} + O(D^{-1}) - (1+q_D)^{-1}(1+\alpha_1^2)\psi_1^{-4}\alpha_1 \sum_{m\neq 2}^{D} \alpha_m^2 \psi_m^{-2} \\
&\quad (1+q_D)^{-1}(1+\alpha_1^2)\psi_1^{-4} \sum_{m=2}^{D} \alpha_m \psi_m^{-2} \gamma_{1m} + O(D^{-1}) \qquad (69) \\
&= O(1) + O(D^{-1}) + O(1) + O(D^{-1}) = O(1).
\end{aligned}$$

(3) Using $r_{m1} = \beta_m \beta_1 + \gamma_{1m}$, when $m \neq 1$, $\partial r_{m1}/\partial \beta_1 = \beta_m$ and when $m = 1$, $\partial r_{m1}/\partial \beta_1 = 0$. Therefore using (44b),

$$\sum_{m=1}^{D} \frac{\partial r_{m1}}{\partial \beta_1}\bigg|_{\boldsymbol{\beta}_D=\boldsymbol{\alpha}_D} \sigma^{m1} = \sum_{m=2}^{D} \alpha_m \sigma^{m1} = O(1). \tag{70}$$

Based on the order in (68), (69), (70), and (60), for the derivative of the term1 in (67) evaluated at $\boldsymbol{\beta}_D = \boldsymbol{\alpha}_D$,

$$\begin{aligned}
&\frac{\partial b_1^{(-1)}(\boldsymbol{\alpha}_D)}{\partial \beta_1} \times \left(1 - \sum_{m=1}^{D} \sigma^{m1} r_{m1}\right) - \sum_{m=1}^{D} \frac{\partial(\sigma^{m1} r_{m1})}{\partial \beta_1} \times b_1^{(-1)}(\boldsymbol{\alpha}_D) \\
&= O(D^{-1}) - b_1^{(-1)}(\boldsymbol{\alpha}_D)\left[\sum_{m=1}^{D} \frac{\partial \sigma^{1m}}{\partial \beta_1}\bigg|_{\boldsymbol{\beta}_D=\boldsymbol{\alpha}_D} r_{m1} + \sum_{m=1}^{D} \frac{\partial r_{m1}}{\partial \beta_1}\bigg|_{\boldsymbol{\beta}_D=\boldsymbol{\alpha}_D} \sigma^{m1}\right] \\
&= O(D^{-1}) + O(1) = O(1). \qquad (71)
\end{aligned}$$

Step D.2: Evaluate term2 in (67).

For $k = 2, \ldots, D$, from (41),

$$\frac{\partial b_k^{(-1)}(\boldsymbol{\beta}_D)}{\partial \beta_1}\bigg|_{\boldsymbol{\beta}_D=\boldsymbol{\alpha}_D} = \sum_{\ell=2}^{D} \beta_\ell \frac{\partial \sigma^{\ell k}}{\partial \beta_1}\bigg|_{\boldsymbol{\beta}_D=\boldsymbol{\alpha}_D} = O(D^{-1}). \tag{72}$$

The term2 in (67) can be rewritten as

$$\underbrace{\sum_{k=2}^{D}\sum_{m=1}^{D} \frac{\partial b_k^{(-1)}}{\partial \beta_1} \sigma^{m1} r_{mk}}_{\text{term21}} + \underbrace{\sum_{k=2}^{D}\sum_{m=1}^{D} b_k^{(-1)} \frac{\partial \sigma^{m1}}{\partial \beta_1} r_{mk}}_{\text{term22}} + \underbrace{\sum_{k=2}^{D}\sum_{m=1}^{D} b_k^{(-1)} \sigma^{m1} \frac{\partial r_{mk}}{\partial \beta_1}}_{\text{term23}}. \tag{73}$$

Since the technique applied to evaluate the order of each term in (73) is quite similar to Step C, we just briefly discuss the order of terms in (73).

The term21 in (73) is $O(1)$ because $\partial b_k^{(-1)}/\partial\beta_1 = O(D^{-1})$ uniformly for $k = 2,\ldots,D$, $\sum_{m=1}^{D}\sigma^{m1}r_{mk} = \psi_1^{-2}\eta_D^{(-1)}\gamma_{1k} + O(D^{-1})$ uniformly, and $\sum_{k=1}^{D}\gamma_{1k} = O(1)$ with Assumption 4.

The term22 in (73) is $O(1)$. Substitute the expression (47) in the term22 in (73), and evaluate at $\boldsymbol{\beta}_D = \boldsymbol{\alpha}_D$ leads to

$$(1-\eta_D^{(-1)})\sum_{k=2}^{D}\alpha_k\psi_k^{-2}\sum_{m=1}^{D}\frac{\partial\sigma^{m1}}{\partial\beta_1}(\alpha_m\alpha_k+\gamma_{mk}). \tag{74}$$

Similar to the derivation of (65), excluding the scalar $1-\eta_D^{(-1)}$, for the summation term in (74), let $\Delta_{11} = 2\alpha_1\psi_1^{-4}$ and $\Delta_{22} = (1+\alpha_1^2)\psi_1^{-4}$. Substitute $\partial\sigma^{m1}/\partial\beta_1$ computed using the expression in (44b) leads to

$$\begin{aligned}
&\sum_{k=2}^{D}\alpha_k\psi_k^{-2}\Big(\Delta_{11}(\alpha_1\alpha_k+\gamma_{1k})\\
&-\sum_{m=2}^{D}\Big[(1+q_D)^{-1}\alpha_m\psi_m^{-2}\Delta_{22}+O(D^{-2})\Big](\alpha_m\alpha_k+\gamma_{mk})\Big)\\
&=\Delta_{11}\alpha_1\sum_{k=2}^{D}\alpha_k^2\psi_k^{-2}+\Delta_{11}\sum_{k=2}^{D}\alpha_k\psi_k^{-2}\gamma_{1k}\\
&-(1+q_D)^{-1}\Delta_{22}\sum_{k=2}^{D}\alpha_k^2\psi_k^{-2}\sum_{m=2}^{D}\alpha_m^2\psi_m^{-2}\\
&-(1+q_D)^{-1}\Delta_{22}\sum_{k=2}^{D}\alpha_k\psi_k^{-2}\sum_{m=2}^{D}\alpha_m\psi_m^{-2}\gamma_{mk}\\
&=O(D)+O(1)+O(D)+O(1)=O(D).
\end{aligned} \tag{75}$$

As $1-\eta_D^{(-1)} = O(D^{-1})$, the term in (74) is $O(1)$.

As for the term23 in (73), $\partial r_{mk}/\partial\beta_1$ is not zero only when $m=1$ or $k=1$. Similar with step C.4, the term23 in (73) can be simplified into

$$b_1^{(-1)}\sum_{m=2}^{D}\sigma^{m1}\alpha_m+\sigma^{11}\sum_{k=2}^{D}\alpha_k b_k^{(-1)} = O(1)+O(1)=O(1) \tag{76}$$

by noticing that $\{\alpha_m\}$ are uniformly bounded, $b_1^{(-1)} = O(1)$ in (46), $\sum_{m=2}^{D}|\sigma^{m1}| = O(1)$ from (44a), $\sigma^{11} = O(1)$ in (44b), and $b_k^{(-1)} = O(D^{-1})$ in (47) for $k \neq 1$.

Combining all these results and under the super-population assumption, the diagonal terms in the Hessian matrix are uniformly $O(1)$. □

Acknowledgements This research has been supported by NSERC Discovery Grant GR010293. Thanks to the referees for their feedback and careful reading.

References

1. Chamberlain, G., Rothschild, M.: Arbitrage, factor structure, and mean-variance analysis on large asset markets. Econometrica **51**(5), 1281–1304 (1983)
2. Doukhan, P., Louhichi, S.: A new weak dependence condition and applications to moment inequalities. Stochastic Processes and their Applications **84**(2), 313–342 (1999)
3. Fan, X.: Dependence modeling in high dimensions with latent variables. Ph.D. thesis, University of British Columbia (2024)
4. Fan, X., Joe, H.: High-dimensional factor copula models with estimation of latent variables. Journal of Multivariate Analysis **201**(105263) (2024)
5. Harville, D.A.: Matrix Algebra From a Statistician's Perspective. Springer, New York (1997)
6. Joe, H.: Dependence Modeling with Copulas. Chapman & Hall/CRC, Boca Raton, FL (2014)
7. Joe, H.: Parsimonious graphical dependence models constructed from vines. Canadian Journal of Statistics **46**(4), 532–555 (2018)
8. Johnson, R.A., Wichern, D.W.: Applied Multivariate Statistical Analysis, 5th edn. Prentice Hall, Englewood Cliffs, NJ (2002)
9. Krupskii, P., Joe, H.: Factor copula models for multivariate data. Journal of Multivariate Analysis **120**, 85–101 (2013)
10. Krupskii, P., Joe, H.: Structured factor copula models: Theory, inference and computation. Journal of Multivariate Analysis **138**, 53–73 (2015)
11. Krupskii, P., Joe, H.: Approximate likelihood with proxy variables for parameter estimation in high-dimensional factor copula models. Statistical Papers **63**(2), 543–569 (2022)
12. Lee, D., Joe, H., Krupskii, P.: Tail-weighted dependence measures with limit being the tail dependence coefficient. Journal of Nonparametric Statistics **30**(2), 262–290 (2018)
13. White, H.: Maximum likelihood estimation of misspecified models. Econometrica **50**(1), 1–25 (1982)

Chapter 3

MATRIX Program: Mathematical Models for Lipids and Cells in Atherosclerotic Plaques

Quantifying bead accumulation inside a cannibalistic macrophage population

Ryan J. Murphy, Adelle C.F. Coster, Jennifer A. Flegg, Joseph P. Ndenda, Mary R. Myerscough, and Helen M. Byrne

Abstract Macrophages are a key cell type in the immune system that can promote healing and disease progression. In this paper, we use statistical modelling to revisit the mathematical modelling and *in vitro* experiments of Ford et al. [1]. We focus on experiments that quantify the accumulation of latex beads in macrophages. This combined mathematical, statistical, and experimental approach provides a powerful framework to interpret and quantify fundamental macrophage processes, such as cell division, cell death, and the cannibalistic ingestion and removal of dead cells. We find that the previously reported estimate of a key mathematical model parameter obtained by visual inspection is not contained within the corresponding confidence interval obtained using our objective statistical modelling. We also explore how uncertainty in parameter estimates influences key predictions, such as the distribution of the number of beads per cell. This analysis provides a platform for future studies exploring substance accumulation in macrophage populations.

Ryan J. Murphy
UniSA STEM, The University of South Australia, Mawson Lakes, SA 5095, Australia
e-mail: ryan.murphy@unisa.edu.au

Jennifer A. Flegg
School of Mathematics and Statistics, The University of Melbourne, Parkville, Victoria, Australia
e-mail: ryan.murphy@unimelb.edu.au

Adelle C.F. Coster
School of Mathematics and Statistics, University of New South Wales, Sydney, NSW 2052, Australia

Joseph P. Ndenda, Mary R. Myerscough
School of Mathematics and Statistics, University of Sydney, Carslaw Building, Eastern Avenue, Camperdown, Sydney, NSW 2006, Australia

Helen M. Byrne
Wolfson Centre for Mathematical Biology, Mathematical Institute, Andrew Wiles Building, University of Oxford, Woodstock Road, Oxford, Oxfordshire OX2 6GG, UK; and
Ludwig Institute for Cancer Research, University of Oxford, Old Road Campus Research Build, Roosevelt Dr, Headington, Oxford, Oxfordshire, OX3 7DQ, UK

D. R. Wood et al. (eds.), *2024 MATRIX Annals, Part II*, MATRIX Book Series 8,
https://doi.org/10.1007/978-3-032-16206-9_12

1 Introduction

Macrophages are a crucial cell type in the innate immune system. They are highly plastic cells that ingest pathogens and cellular debris, produce immune signals and promote healing. They exhibit a variety of phenotypes, some inflammatory and others anti-inflammatory. Macrophages are not only implicated in atherosclerosis [2] but also other inflammatory diseases such as chronic obstructive pulmonary disease [3], Alzheimer's disease [4], kidney disease [5] and cancer [6, 7].

Macrophages drive the growth of atherosclerotic plaque through inflammatory signalling, phagocytosis (ingestion) of modified low density lipoproteins (mod LDL) and efferocytosis (ingestion and removal) of apoptotic cells [2, 8]. They accumulate internalised lipid (mostly cholesterol) via phagocytosis of modLDL and efferocytosis of dead (apoptotic) cells, including their lipid membranes. Lipid-laden macrophages are usually described as foam cells since the accumulated internalised lipid gives them foamy appearance under the microscope. If a foam cell becomes apoptotic, then when it is ingested by another macrophage, its internalised lipid is added to the internalised lipid of the ingesting cell. In this way, macrophages can become very heavily lipid-laden. In a nascent atherosclerotic plaque, intercellular lipid accumulation via macrophage efferocytosis can drive the formation of necrotic regions in the plaque and cause the plaque to become increasingly pathological.

Macrophage efferocytosis in atherosclerotic plaques cannot be directly observed *in vivo*. *In vitro* experiments however shed useful light on how internalised lipid accumulates in plaque macrophages. Ford *et al* [1] performed experiments using latex microbeads to monitor the accumulation of apoptotic cell contents via efferocytosis.

Populations of macrophages were plated onto microscope slides after treatment with interferon γ (IFNγ) and lipopolysaccharide (LPS) to activate an inflammatory, efferocytic phenotype. Before the IFNγ and LPS treatment, macrophages were incubated with latex beads at a ratio of approximately one bead per macrophage. Hence the expected number of beads per macrophage at the start of the experiment was one. As the experiment progressed, some macrophages became apoptotic and were ingested or engulfed by living macrophages, along with any beads that were contained in the apoptotic macrophage. Some bead-carrying macrophages underwent proliferation. When this occurred beads were divided between the daughter cells, although not necessarily equally.

Panel A of Fig. 1 shows micrographs of bead-containing macrophages, at the start of the experiment, after 24 [hr] and after 40 [hr]. Initially, macrophages contained between zero and three beads each. By 24 [hr] most macrophages have acquired more beads and there are fewer macrophages in the field of view. After 48 [hr] macrophages are heavily laden with beads, but some beads are not contained in cells which suggests that efferocytosis has started to become defective [8].

Panel B, Fig. 1, shows two cells at $t = 0$. The right hand cell undergoes apoptosis ($t = 8$ [hr]) and is engulfed by the cell on the left ($t = 16$ [hr]). Panel C shows an example of cell division. The cell in the centre at $t = 0$ divides to give two daughter cells ($t = 8$ [hr]). These daughter cells, in turn, each divide, the right one at $t = 12$ [hr] and the left one at $t = 24$ [hr].

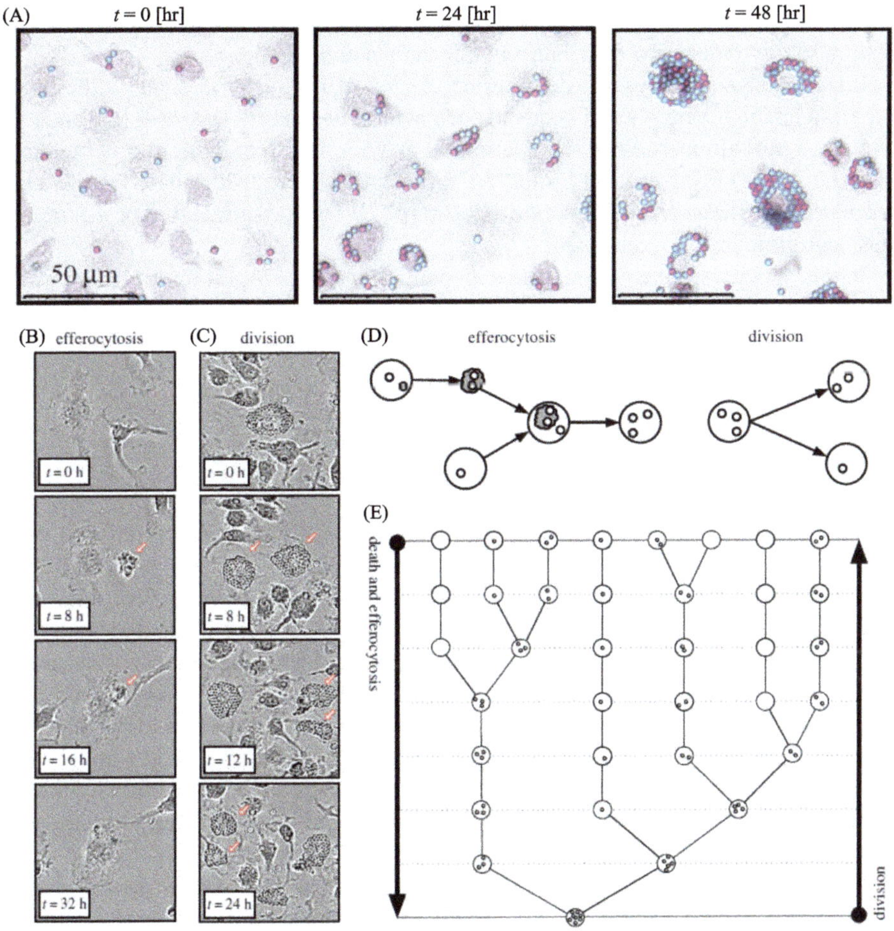

Fig. 1: An *in-vitro* experiment to quantify bead accumulation inside a cannibalistic macrophage population. (A) Experimental images showing bead accumulation in macrophages. (B) Experimental images showing redistribution of beads during macrophage apoptosis (during times 0 and 8 [hr]) and efferocytosis (during 16 and 32 h). (C) Experimental images showing redistribution of beads during macrophage division of one cell (during 0 and 8 [hr]) and of both daughter cells (during 8 and 12 [hr], and 12 and 24 [hr]). (D-E) Schematics showing the role of efferocytosis, cell death, and cell division on bead accumulation. Panel (A) reproduced from Figure 3A in [1] (CC BY 4.0). Panels (B)-(E) reproduced from Figure 1 from in [1] (CC BY 4.0).

Panel D, Fig. 1 shows schematics of bead accumulation via efferocytosis and bead distribution via proliferation for a pair of cells, while Panel E shows these processes on a population level. Panel E can either be read with time running from top to bottom to describe accumulation via efferocytosis, or it can be read from bottom to top to describe distribution of beads via cell division.

Image analysis was used to determine the number of macrophages in the bead experiment and the number of beads in each macrophage at $t = 0, t = 24$ [hr] and $t = 48$ [hr]. A plot of macrophage populations is given in Fig. 4(a) in the supplementary material of [1]. Ford *et al* [1] fitted an exponential curve to the data by eye and estimated that the average death rate at $1/60$ [h^{-1}]. They assumed that proliferation was negligible in this experiment.

In this paper we revisit the simple exponential model, assess it for model identifiability and then re-examine its fit to synthetic and experimental data using objective statistical parameter estimation and prediction techniques.

2 Data

We re-digitise previously published data from Figure 3 in [1]. The published data were generated from *in vitro* experiments designed to quantify the accumulation of latex beads, each of diameter 3 µm, within primary murine bone marrow-derived macrophage cells. In the experiment, 10^5 beads were incubated with 10^5 macrophage cells. The cells were then stimulated into an inflammatory state using interferon γ (IFNγ) and lipopolysaccharide (LPS). The number of beads per cell was counted for every cell using an image recognition algorithm applied to whole-slide photomicrographs. The total number of cells and the proportion of cells with n beads per cell were computed at times 0, 24, and 48 [hr], where time 0 [hr] corresponds to immediately prior to stimulation with IFNγ and LPS. While the published data represent averages from four experiments performed with macrophages derived from different mice on different days, we do not have access to these data at present. From the re-digitised data describing the total number of cells, we obtain three measurements at each time point which correspond to the lower, central, and upper points from Figure 3B in [1]. For the re-digitised data for the number of beads per cell, for each timepoint we obtain at most one measurement for each possible value of number of beads per cell from Figure 3D in [1].

3 Methods

3.1 General mathematical model

Following [1], we consider an ordinary differential equation-based mathematical model that captures the key processes in the experiment: apoptosis (a form of cell death), efferocytosis (where one cell consumes another dead cell), and division (where a parent cell divides into two daughter cells). The total number of beads is conserved throughout these processes. The redistribution of beads within the living and apoptotic macrophage cells may then be viewed as a coagulation–fragmentation process [9, 10]. Following [1], the number density of living macrophage cells containing n beads at time t [hr] is denoted $\phi_n(t)$ [-], and the number density of apoptotic macrophage cells at time t [hr] containing n beads is denoted $\phi_n^\dagger(t)$ [-]. The governing equations for the densities of the living and apoptotic cells are,

$$\frac{\mathrm{d}\phi_n(t)}{\mathrm{d}t} = \underbrace{\eta \sum_{i=0}^{n} \phi_i^\dagger(t)\phi_{n-i}(t)}_{\text{efferocytosis source}} - \underbrace{\eta \phi_n(t) \sum_{i=0}^{\infty} \phi_i^\dagger(t)}_{\text{efferocytosis sink}} - \underbrace{\beta \phi_n(t)}_{\text{apoptosis sink}} + \underbrace{2\alpha \sum_{i=n}^{\infty} \binom{i}{n} \frac{\phi_i(t)}{2^i}}_{\text{division source}} - \underbrace{\alpha \phi_n(t)}_{\text{division sink}}, \tag{1}$$

$$\frac{\mathrm{d}\phi_n^\dagger(t)}{\mathrm{d}t} = \underbrace{-\eta \phi_n^\dagger(t) \sum_{i=0}^{\infty} \phi_i(t)}_{\text{efferocytosis sink}} + \underbrace{\beta \phi_n(t)}_{\text{apoptosis source}}, \tag{2}$$

where α [hr^{-1}] is the cell division rate, β [hr^{-1}] is the apoptosis rate, and η [hr^{-1}] is the efferocytosis rate.

The total number of living macrophage cells, $N(t)$, and the total number of apoptotic macrophage cells, $N^\dagger(t)$, are given by,

$$N(t) = \sum_{i=0}^{\infty} \phi_i(t) \quad \text{and} \quad N^\dagger(t) = \sum_{j=0}^{\infty} \phi_j^\dagger(t). \tag{3}$$

3.2 *Mathematical model with no cell division and instantaneous efferocytosis*

Following [1], we assume that the impact of cell division is negligible in this experiment, and therefore we fix $\alpha = 0$ in Eqs. (1)-(2). We also assume that apoptotic cells are instantaneously consumed, corresponding to $\eta \to \infty$ in Eqs. (1)-(2) (see Supplementary Material of [1] for further details). Under these assumptions, the total number of live cells at time t is

$$N(t) = N(0)\exp(-\beta t), \tag{4}$$

where $N(0)$ [-] is the initial number of living cells that we treat as a model parameter and β [hr^{-1}] is the rate of apoptosis. These two parameters are globally structurally identifiable meaning that each parameter can be uniquely identified given continuous noise-free observations of $N(t)$ [11]. This can be shown by taking the logarithm of Eq. (4) to obtain a line. From this line one can determine $N(0)$ and β from the vertical intercept and gradient, respectively. In what follows, we will explore if these parameters can be estimated given finite noisy data.

3.3 Statistical parameter estimation, identifiability analysis, and prediction

We use statistical modelling to connect the mathematical model in Eq. (4) to the experimental data. The experimental data that we focus on comprise three measurements of the total number of living cells at three times $t_i = 0, 24, 48$ [hr]. We assume that at each time t_i each experimental observation j, denoted $N^{\mathrm{o}}_{i,j}$, is distributed about the mathematical model solution $N(t_i; N(0), \beta)$ that depends on the values of $N(0)$ and β. Specifically, we assume that

$$N^{\mathrm{o}}_{i,j} \sim \mathcal{N}\left(N(t_i; N(0), \beta), \sigma\right), \tag{5}$$

where $\mathcal{N}(\mu, \sigma)$ represents a Gaussian probability distribution with mean μ and standard deviation σ. This assumption of additive Gaussian noise, also often referred to as normally distributed noise, is widely used in the literature and appears appropriate for this preliminary analysis [12]. We denote the entire dataset with $N^{\mathrm{o}}_{1:9} = \{N^{\mathrm{o}}_{i,j} \mid i \in \{1,2,3\}, j \in \{1,2,3\}\}$. In this report, we estimate the mathematical model parameters, $N(0)$ and β, and the statistical model parameter, σ, from the data simultaneously.

To estimate the mathematical and statistical model parameters, $\theta = (\beta, N(0), \sigma)$, we employ an established profile likelihood-based workflow for statistical parameter estimation, practical identifiability analysis, and prediction [12]. Throughout we work with the log-likelihood function,

$$\ell(\theta \mid N^{\mathrm{o}}_{1:9}) = \sum_{i=1}^{3} \sum_{j=1}^{3} \log\left[\phi\left(N^{\mathrm{o}}_{i,j}; N\left(t_i; N(0), \beta\right), \sigma\right)\right], \tag{6}$$

where $\phi(x; \mu, \sigma)$ is the Gaussian probability density function with mean μ and standard deviation σ.

The maximum likelihood estimate (MLE) is then

$$\hat{\theta} = \underset{\theta}{\operatorname{argmax}}\, \ell(\theta \mid N^{\mathrm{o}}_{1:9}). \tag{7}$$

Under the assumption of additive Gaussian noise, the mathematical model solution evaluated at the MLE is the best match to the data, in the sense of the highest likelihood, and is identical to the mathematical model solution obtained via the routinely employed method of least squares.

There may be a range of values of θ that result in similar agreement to the data as the MLE. To determine this range of values, we use an established profile likelihood-based approach detailed in [12]. In particular, we compute profile likelihoods working with the normalised log-likelihood,

$$\hat{\ell}(\theta \mid N^{o}_{1:9}) = \ell(\theta \mid N^{o}_{1:9}) - \ell(\hat{\theta} \mid N^{o}_{1:9}), \tag{8}$$

such that $\hat{\ell}\left(\theta \mid N^{o}_{1:9}\right) \leq 0$ and $\hat{\ell}\left(\hat{\theta} \mid N^{o}_{1:9}\right) = 0$. From these profiles, we compute an approximate 95% confidence interval for each parameter in θ in turn. If an approximate confidence interval is narrow relative to pre-specified parameter bounds, we refer to the parameter as being *practically identifiable*. Otherwise, we refer to the parameter as being *practically non-identifiable*. Informed by experimental observations, and preliminary explorations, we set the following parameter bounds $0.00 \leq \beta \leq 0.05$ [hr^{-1}], $8 \times 10^4 \leq N(0) \leq 12 \times 10^4$ [-], and $100 \leq \sigma \leq 2 \times 10^4$ [-]. As the mathematical model parameters in this study are structurally identifiable, if we observe that a parameter is practically non-identifiable this suggests that additional observations are required to recover practical identifiability of the model parameters. We also propagate forward the uncertainty in the approximate confidence intervals for the model parameters to generate profile-wise confidence sets for deterministic model solutions and profile-wise confidence sets for noisy data realisations (method detailed in [12]).

4 Results

We quantify bead accumulation inside a cannibalistic macrophage population using *in vitro* experimental data, mathematical modelling, and statistical modelling. As described in the mathematical and statistical methods (Section 3), a key parameter that characterises bead accumulation dynamics is the rate of apoptosis, β. Two additional parameters capture the initial number of living cells, $N(0)$, and statistical noise, σ. Before analysing the experimental data we perform a synthetic data study to verify that the method accurately recovers known parameters. We then analyse the experimental data and compare our new objective parameter estimates to previously reported values that were obtained via visual inspection.

4.1 Synthetic data

To verify that our methods recover known parameters, we perform a synthetic data study. We simulate the mathematical model in Eq. (4) using the values reported in [1], specifically $N(0) = 8 \times 10^4$ [-] and $\beta = 1/60 = 0.0167$ [hr^{-1}]. We generate three independent observations at each of three time points, $t = 0, 24, 48$ [hr], and set $\sigma = 10^4$ [-]. By definition, these synthetic data display an exponential decrease in total cell number with time (Fig. 2A). We then compute the MLE, $\hat{\theta} = (\hat{\beta}, \hat{N(0)}, \hat{\sigma}) = (0.0139, 8.02 \times 10^4, 8.03 \times 10^3)$, and find that it is close to the known values used to generate the synthetic data. Evaluating the mathematical model in Eq. (4) at the MLE we observe good agreement to the synthetic data (Fig. 2A). We then generate profile likelihoods for each model parameter (Fig. 2B-D). As each profile likelihood is unimodal and narrow relative to the pre-specified bounds, we conclude that each parameter is practically identifiable. The approximate 95% confidence intervals also capture the known parameter values used to generate the data. Profile-wise confidence sets for deterministic model solutions demonstrate how uncertainty in parameter estimates influences predictions (Fig. 3A-D) and profile-wise confidence sets for noisy data realisations show good agreement with the synthetic data (Fig. 3E-H).

The study in [1] also reports another key quantity: the distribution of the number of beads per cell. We now show how we can use our fitted value for β to predict the distribution of the number of beads per cell. We assume that the initial distribution of beads per cell, $\phi_n(t=0)$ for $n = 1, 2, \ldots, n_{\max}$, is known from the experimental data. We also assume that the maximum number of beads that can be contained in any single living or apoptotic cell is $n_{\max} = 100$. The structured mathematical model in Eqs. (1)-(2) then reduces to $2n_{\max} + 2$ ordinary differential equations, each involving finite, rather than infinite, sums. A simulation of the structured model with the same known value of $\beta = 1/60$ [hr], and $\alpha = 0$, $\eta = 10^4$ is shown in Fig. 4.

We next explore how the uncertainty in the approximate confidence interval for β, obtained from the analysis of synthetic data for the total number of living cells, influences predictions of the distribution of beads per cell. These predictions, in the form of confidence sets for the deterministic model solution of the structured model, are in good agreement with the simulated data (Fig. 4). We do not generate confidence sets for noisy data realisations of the structured model as it is not immediately clear how to attribute the population-level statistical noise, σ, to each compartment, $\phi_n(t)$ for $n = 0, 1, \ldots, n_{\max}$, of the structured mathematical model.

4.2 Experimental data

We now analyse the experimental data using the same workflow as in the synthetic data study. The MLE is found to be $\hat{\theta} = (\hat{\beta}, \hat{N(0)}, \hat{\sigma}) = (0.0231, 9.2 \times 10^4, 6.0 \times 10^3)$. Simulating the mathematical model with the MLE we observe good agreement to the experimental data describing the total number of cells (Fig. 5A). The profile likelihoods suggest that all parameters are practically identifiable (Fig. 5B-D).

Previous work analysing the raw experimental data estimated $\beta = 0.0167$ [hr^{-1}] via visual inspection [1]. This estimate is not contained within the approximate 95% confidence interval for β that we obtain by analysing the re-digitised data. The previously reported value for $N(0)$ from [1] is also not contained within the corresponding approximate 95% confidence interval. Simulating the mathematical model at the values considered in [1] we observe poor agreement with the experimental data (Fig 5A). Our improved results are due to the objective statistical framework that we employ for parameter estimation. However, we caution that we have analysed the

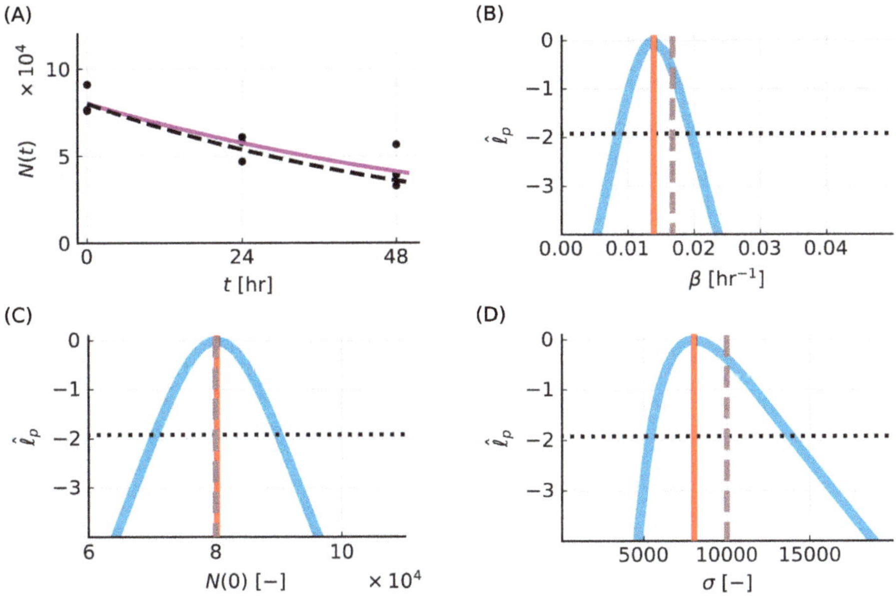

Fig. 2: Best-fit and profile likelihoods for synthetic data study (A) Synthetic data (circles) generated by simulating the mathematical model in Eq. (4) with additive Gaussian noise with known parameters, $\theta = (\beta, N(0), \sigma) = (0.0167, 8 \times 10^4, 10^4)$, Simulation of the mathematical model in Eq. (4) evaluated at the known parameters (black dashed line) and at the MLE (magenta solid line), $\hat{\theta} = (\hat{\beta}, \hat{N(0)}, \hat{\sigma}) = (0.0139, 8.02 \times 10^4, 8.03 \times 10^3)$. (B-D) Profile likelihoods (blue) for (B) β, (C) $N(0)$, (D) σ with MLE (vertical red dashed), known parameters (vertical brown dashed) an approximate 95% confidence interval threshold (horizontal black-dashed). Approximate 95% confidence intervals: $0.0101 \leq \beta \leq 0.0190$, $9.0 \times 10^4 \leq N(0) \leq 1.1 \times 10^4$, and $5.4 \times 10^3 \leq \sigma \leq 13.9 \times 10^3$.

re-digitised data rather than the raw data. Analysing the raw data may yield results that are closer to the previously reported values.

Profile-wise confidence sets for deterministic model solutions reveal how uncertainty in parameter estimates influences predictions (Fig. 6A-D) and profile-wise confidence sets for noisy data realisations show good agreement with the experimental data (Fig. 6E-H). These results are consistent with the synthetic data study.

We next propagate the uncertainty in β to generate predictions of the distribution of beads per cell. These predictions, in the form of confidence sets for the deterministic model solution of the structured model, demonstrate good agreement with the experimental data for most cells (Fig. 6). We observe discrepancies between the simulation of the structured mathematical model and the experimental data for a small proportion of cells that contain many beads. This discrepancy is not unexpected as we compare experimental data to a confidence set for the model solution. It would

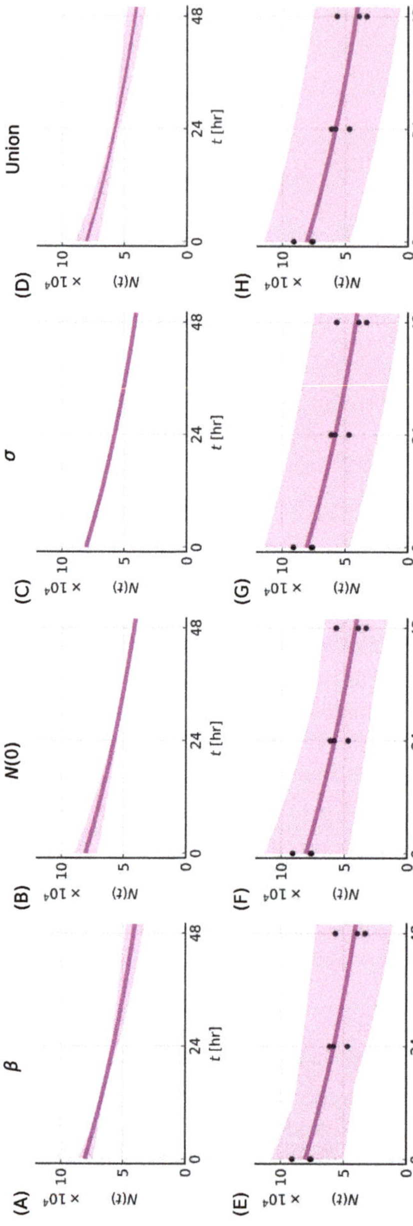

Fig. 3: Predictions for $N(t)$ in the synthetic data study. (A-C) Profile-wise confidence sets for deterministic model solutions for (A) β, (B) $N(0)$, and (C) σ. (D) Union of profile-wise confidence sets in (A-C). (E-G) Profile-wise confidence sets for noisy data realisations (E) β, (F) $N(0)$, and (G) σ. (H) Union of confidence sets for noisy data realisations from (E-G).

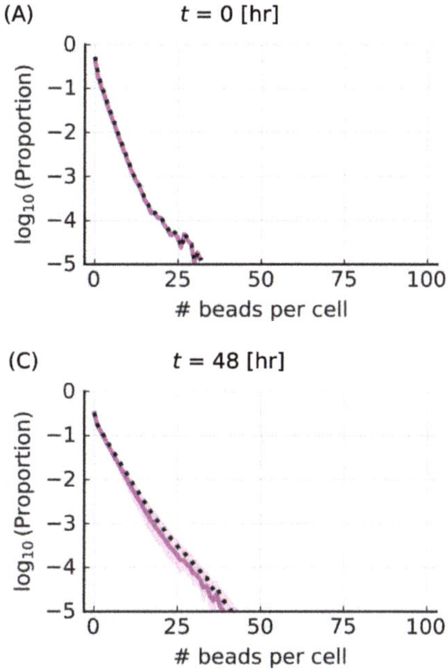

Fig. 4: Predictions for distribution of the number of beads per cell for synthetic data study. Simulation of the structured mathematical model from Eqs. (1)-(2) evaluated at the known value of $\beta = 0.0167$ [hr^{-1}] (black dotted line) and at the MLE $\beta = 0.0139$ (solid magenta). Initial condition at $t = 0$ [hr] set equal to experimental data. Approximate 95% confidence set for the distribution of the number of beads per cell (magenta shaded).

be more sensible to compare the experimental data to a confidence set for noisy data realisations of the structured mathematical model. However, it is unclear which statistical error model should be applied to the structured mathematical model and how this connects to the estimated population-level statistical noise, σ.

5 Discussion

In this study, we quantify bead accumulation inside a cannibalistic population of macrophages. We used an existing ordinary differential equation-based mathematical model designed to quantify the accumulation of latex beads within macrophages [1], and made some simplifying assumptions about cell division and efferocytosis. We then applied a profile likelihood-based parameter estimation, practical identifiability analysis, and prediction workflow using synthetic data (Section 4.1) and then using re-digitised published data from Figure 3 in [1] (Section 4.2). The results for the synthetic data demonstrate good agreement with the simulated data (Figs. 2–4),

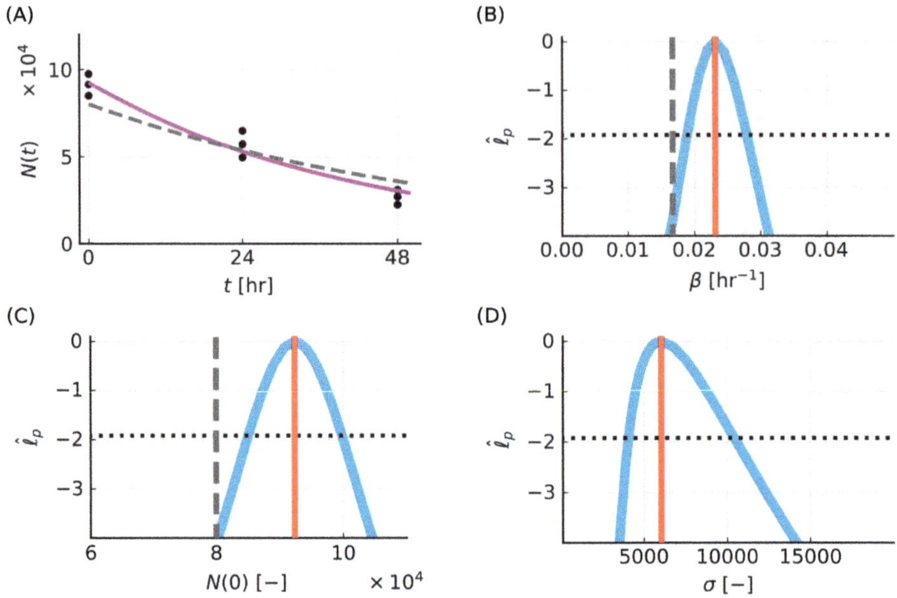

Fig. 5: Best-fit and profile likelihoods for experimental data (A) Experimental data (circles), simulation of the mathematical model in Eq. (4) evaluated at values considered in [1], $(\beta, N(0)) = (0.0167, 8 \times 10^4)$ (grey dashed), and at the MLE (magenta solid line), $\hat{\theta} = (\hat{\beta}, \hat{N(0)}, \hat{\sigma}) = (0.0231, 9.2 \times 10^4, 6.0 \times 10^3)$. (B-D) Profile likelihoods (blue) for (B) β, (C) $N(0)$, (D) σ with MLE (vertical red dashed), values considered in [1] (vertical grey dashed), and an approximate 95% confidence interval threshold (horizontal black-dashed). Approximate 95% confidence intervals: $0.0189 \leq \beta \leq 0.0279$, $8.5 \times 10^4 \leq N(0) \leq 10.0 \times 10^4$, and $4.1 \times 10^3 \leq \sigma \leq 10.4 \times 10^3$.

giving us confidence to apply the workflow to the re-digitised data (Figs. 5–7). Results for the digitised data show that the reported value of β from [1] is outside of the approximate confidence interval obtained in this analysis.

There are many future directions for the work presented here. Firstly, the error model we have adopted in this work for the population size of living cells at time t is a normal distribution, which has infinite support; this is not biologically realistic. The assumption could be relaxed to use a distribution with positive support, for example a log-normal distribution. Secondly, following [1], we assume that the role of cell division is negligible by setting $\alpha = 0$ in Eqs. (1)-(2). The work presented here could be extended to consider non-zero cell proliferation; this would complicate the underlying dynamical system but the statistical parameter estimation, identifiability analysis and prediction process would proceed as presented here. At the scale of the total cell number, additional types of data would be required to decouple the proliferation rate, α, and the rate of apoptosis, β. In the same vein, we could consider the

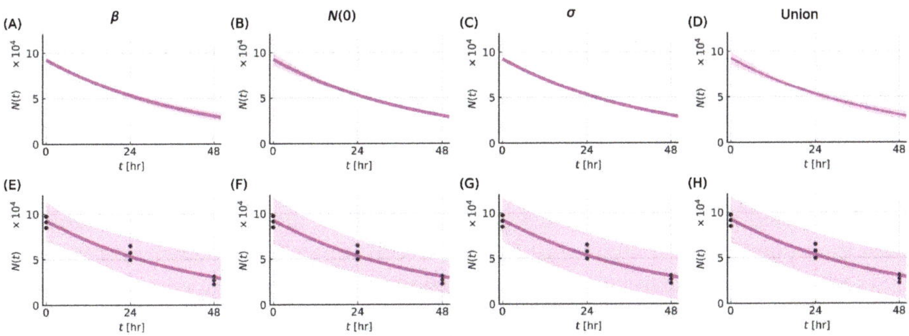

Fig. 6: Predictions for $N(t)$ when analysing experimental data. (A-C) Profile-wise confidence sets for deterministic model solutions for (A) β, (B) $N(0)$, and (C) σ. (D) Union of profile-wise confidence sets in (A-C). (E-G) Profile-wise confidence sets for noisy data realisations (E) β, (F) $N(0)$, and (G) σ. (H) Union of confidence sets for noisy data realisations from (E-G).

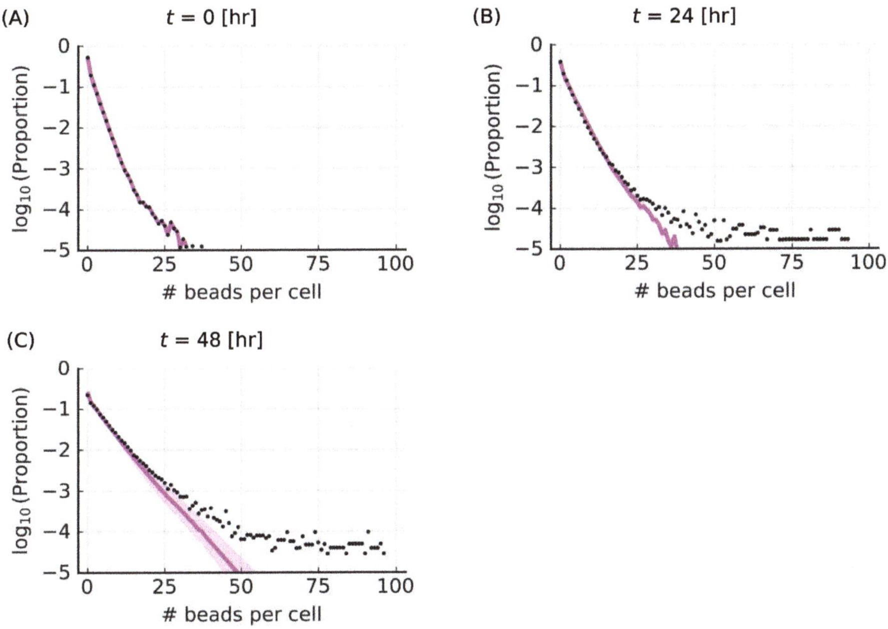

Fig. 7: Predictions for distribution of the number of beads per cell for experimental data. (A-C) Experimental data (black circles) and simulation of the structured mathematical model from Eqs. (1)-(2) evaluated at the MLE $\beta = 0.0231$ (solid magenta). Initial condition at $t = 0$ [hr] set equal to experimental data. Approximate 95% confidence set for the distribution of the number of beads per cell (magenta shaded).

model under various, finite, values for the rate of efferocytosis, η. The original paper by Ford *et al.* [1] presented additional data on the lipid content of the macrophages that we have not used but that could be incorporated into the workflow presented here. In the analysis of experimental data we work with re-digitised data from [1]. This re-digitised data is a representation of the raw data. It would be interesting to revisit this analysis using the complete raw data. A final avenue for future work is to design a single inference framework that uses both the total number of cells at each time point and the number of beads per cell; this would require data integration techniques to construct a joint observational model and hence joint likelihood function for both sets of data.

While there are several limitations to the approach developed here, this report presents a first step to formal calibration and identifiability analysis of an existing model and dataset in the literature of bead accumulation inside macrophages. Our analysis is a useful basis for future studies exploring lipid content in macrophages.

Data and code availability

Re-digitised data and key Julia code that we use to perform this analysis are available on a GitHub repository (https://github.com/ryanmurphy42/Murphy2024AtheroAnnals).

Acknowledgements We thank the mathematical research institute MATRIX in Australia, where this research was performed as part of the MATRIX Research Program: Mathematical Models for Lipids and Cells in Atherosclerotic Plaques.

References

[1] Ford HZ, Zeboudj L, Purvis GS, ten Bokum A, Zarebski AE, Bull JA, Byrne HM, Myerscough MR, Greaves DR. (2019). Efferocytosis perpetuates substance accumulation inside macrophage populations. *Proceedings of the Royal Society B*. 286(1904), 20190730. (doi:10.1098/rspb.2019.0730).

[2] Lusis AJ. (2000) Atherosclerosis. *Nature*. 407, 233–241. (doi:10.1038/35025203).

[3] Vlahos R, Bozinovski S. (2014). Role of alveolar macrophages in chronic obstructive pulmonary disease. *Frontiers in Immunology*. 5, 435. (doi:10.3389/fimmu.2014.00435).

[4] Mammana S, Fagone P, Cavalli E, Basile MS, Petralia MC, Nicoletti F, Bramanti P, Mazzon E. (2018). The role of macrophages in neuroinflammatory and neurodegenerative pathways of Alzheimer's disease, amyotrophic lateral sclerosis, and multiple sclerosis: pathogenetic cellular effectors and potential therapeutic targets. *International Journal of Molecular Sciences*. 19(3), 831. (doi:10.3390/ijms19030831).

[5] Wen Y, Crowley SD. (2020). The varying roles of macrophages in kidney injury and repair. *Current Opinion in Nephrology and Hypertension*. 29(3), 286-292. (doi:10.1097/MNH.0000000000000595).

[6] Zhou L, Zhao T, Zhang R, Chen C, Li J. (2024). New insights into the role of macrophages in cancer immunotherapy. *Frontiers in Immunology*. 15, 1381225. (doi:10.3389/fimmu.2024.1381225).

[7] Zhang J, Hu C, Zhang R, Xu J, Zhang Y, Yuan L, Zhang S, Pan S, Cao M, Qin J, Cheng X, Xu Z. (2023). The role of macrophages in gastric cancer. *Frontiers in Immunology*. 14, 1282176. (doi:10.3389/fimmu.2023.1282176).
[8] Tabas I. (2010). Macrophage death and defective inflammation resolution in atherosclerosis. *Nature Reviews Immunology*. 10(1), 36-46. (doi:10.1038/nri2675).
[9] Aldous DJ. (1999). Deterministic and stochastic models for coalescence (aggregation and coagulation): a review of the mean-field theory for probabilists. *Bernoulli*. 1, 3–48. (doi:10.2307/3318611).
[10] Bertoin J. (2006). Random fragmentation and coagulation processes. Cambridge, UK: Cambridge University Press.
[11] Rey Barreiro X, Villaverde AF. (2023). Benchmarking tools for a priori identifiability analysis. *Bioinformatics*. 39(2), btad065. (doi:10.1093/bioinformatics/btad065).
[12] Murphy RJ, Maclaren OJ, Simpson MJ. (2024). Implementing measurement error models with mechanistic mathematical models in a likelihood-based framework for estimation, identifiability analysis and prediction in the life sciences. *Journal of the Royal Society Interface*. 21(210), 20230402. (doi:10.1098/rsif.2023.0402).

Atherosclerosis: reflections on mathematical modelling, data and other matters *

Keith L. Chambers, Adelle C.F. Coster, Jennifer A. Flegg, Ryan J. Murphy, Mary R. Myerscough, Joseph P. Ndenda, Jichao Pan, Michael G. Watson

Abstract Atherosclerosis is a chronic inflammatory disease that can lead to heart attacks and strokes. Atherosclerotic plaques form in the walls of large arteries. Initially their growth is driven by modified low density lipoproteins, immune cells and inflammatory cytokines. As the plaque matures, smooth muscle cells and necrotic material become important constituents, and as it grows, the plaque starts to distort the artery wall. Modelling has started to address many of these phenomena, but is hampered by difficulties collecting, interpreting and fitting data. In this article we reflect on the enterprise of modelling atherosclerotic plaques, progress so far, what now needs to be done and its relation to cognate areas of mathematical modelling, particularly cancer and wound healing.

Keith L. Chambers
Ludwig Institute, Old Road Campus Research Building, Headington, Oxford OX3 7DQ, UK
e-mail: keith.chambers@ludwig.ox.ac.uk

Adelle C.F. Coster, Michael G. Watson
School of Mathematics and Statistics, UNSW, Sydney, NSW 2052, Australia
e-mail: a.coster,michael.Watson1@unsw.edu.au

Jennifer A. Flegg
School of Mathematics and Statistics, University of Melbourne, Victoria 3010, Australia
e-mail: Jennifer.flegg@unimelb.edu.au

Ryan J. Murphy
UniSA STEM, The University of South Australia, Mawson Lakes, SA 5095, Australia
e-mail: ryan.murphy@unisa.edu.au

Mary R. Myerscough, Joseph P. Ndenda
School of Mathematics and Statistics, University of Sydney, NSW 2006, Australia
e-mail: mary.myerscough@sydney.edu.au e-mail: joseph.ndenda@nm-aist.ac.tz

Jichao Pan
School of Biological Sciences and Medical Engineering, Southeast University, Nanjing 210096, China
e-mail: jc_pan@seu.edu.cn

* Authors listed in alphabetical order.

D. R. Wood et al. (eds.), *2024 MATRIX Annals, Part II*, MATRIX Book Series 8,
https://doi.org/10.1007/978-3-032-16206-9_13

1 What is Atherosclerosis? Why should we care?

Atherosclerosis is a chronic disease of the large arteries. In humans it can lead to heart attacks, strokes and peripheral vascular disease.

An atherosclerotic lesion in the artery wall is initiated when low density lipoprotein (LDL) particles that permeate the artery wall become modified and are retained in the intima, the innermost layer of the artery wall. These retained modified LDL particles (modLDL) generate an immune response which includes stimulating the production of inflammatory cytokines (cell signaling substances). In response to these inflammatory cytokines, monocytes (a type of white blood cell and immune cell) are recruited from the blood stream to the intima where they differentiate into macrophages (another type of immune cell). Macrophages ingest the modLDL particles. Since modLDL carries cholesterol and other fatty molecules, the macrophages acquire internalised stores of cholesterol from the modLDL that they ingest. Cholesterol-filled macrophages are often referred to as foam cells, since this internalised fatty material or lipids gives the cell an appearance of being filled with foam [8].

Lipid from these foam cells can be exported to high density lipoprotein particles (HDL) or the cell can leave the plaque, taking its internalised lipid with it. If neither of these things happens, when a foam cell dies, it leaves behind its load of lipid. This is either ingested by another macrophage or becomes free lipid and part of the fatty necrotic (dead) core that a mature plaque develops.

As a plaque matures, it may acquire a fibrous cap made up of collagen and smooth muscle cells from the artery wall. If this cap is strong and stable, then the material in the plaque, including the necrotic core, is safely sequestered from the bloodstream [1]. If the cap is weak then it is susceptible to rupture. When a plaque ruptures, the material from the necrotic core is released into the blood stream where it promotes the formation of blood clots. These blood clots may block the artery, causing a heart attack, if it is a cardiac artery, or a stroke, if the artery is in the brain.

This is a very simple description of the processes behind the formation of an atherosclerotic plaque. There are many finer details, some of which are critically important in determining the fate of a plaque. Some of these are captured diagrammatically in Figure 1.

Atherosclerosis is the leading cause of death worldwide, exceeding the number of deaths caused by cancer. However, whereas there are many mathematical modelling studies exploring the mechanisms underlying cancer development and progression, there are comparatively few mathematical modelling studies of the biological mechanisms underlying atherosclerotic plaque development and progression.

2 Modelling Atherosclerosis

As atherosclerotic plaque development and progression occur over a timescale of decades with cellular level processes occurring on the time scale of seconds, min-

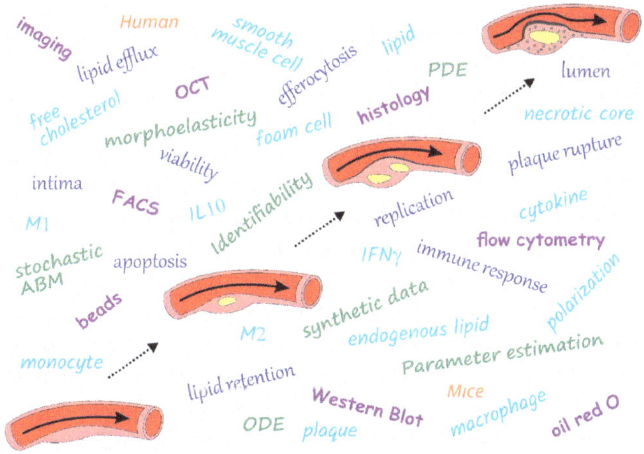

Fig. 1 A schematic diagram of the development of an atherosclerotic plaque. The plaque develops from the lower left hand corner to the upper right hand corner. The surrounding word cloud includes factors that might be incorporated into models at different stages of plaque development or techniques that might be used in modelling.

utes, and hours, most mathematical modelling studies typically focus on one stage of development or progression, typically characterised by early, mid, and late stages. Mathematical modelling studies typically take the form of a system of ordinary differential equations [10] or partial differential equations [4, 7] with, as yet, few agent-based and stochastic-type models (but see [12]). Lipids appear to be fundamental to all mathematical models, with only one model discussed during the workshop explicitly suggesting that lack of oxygen is a key driver of necrosis rather than lipids alone.

It is likely that models for immunological events in plaques will require a suite of different modelling techniques, including multi-scale and discrete models to accurately capture the complex biology of plaque formation that occurs at many scales.

Growing plaques distort the wall of the blood vessel in the latter stages of the disease. The deformation of the artery wall is an important area for classical continuum mechanics modelling, which represents the geometry and morphology of the growing plaque. Much of the current work in this domain is aimed at understanding the changing layered structure of the artery and of the late-stage developing plaque. Linking immunological events with changing plaque morphology via mathematical models is an area of active research [11].

3 Links to experimental and clinical science

Strong collaborations between experimental groups and mathematical modellers are yet to develop in modelling immune events and lipid trafficking in atherosclerosis.

There have certainly been strong collaborations in modelling blood flow and shear stress in atherosclerosis [3], and in modelling cancer and other diseases [14].

There are probably several reasons for this. Modelling the dynamics of lipids and cells in atherosclerosis is a comparatively new endeavour and the field is correspondingly small. Plaque formation, is hidden by its very nature and it is expensive and difficult to do experiments *in vivo*. And finally, it is not clear what exactly modellers can offer experimental scientists and clinicians at this stage of model development.

Most *in vivo* experiments on atherosclerosis are done in genetically modified mice. Generally speaking, mice need to be sacrificed to observe plaques and so only snapshots of plaque progression are possible. For this reason, modelling can provide a useful window into the plaque growth processes and intermediate plaque stages that connect the observed snap shots. However, modellers also need to be aware of the differences in physiology and rate of plaque development between mice (the experimental model) and humans where models, and research generally, will have potential clinical application.

However, as with other biological modelling, a lot of the usefulness of modelling to experimentalists lies in the process of modelling development. "You make me think about my work differently" or a similar statement was made by both experimentalists at the Creswick workshop.

4 Data for models, both real and synthetic

Atherosclerosis is, essentially, a disease of the whole body, so that plaques cannot be isolated and grown *in vitro* for example. Hence it is not surprising that the sort of data required by modellers is far more sparse, than in other cellular systems and cancer. The most scientifically useful models need data for validation and verification. Fitting data to an established model or creating models to fit the experimental data are common in mathematical modelling of biological processes. Extracting meaningful information from data is not merely a technical challenge but also a fundamental scientific question in model development. Modelling of atherosclerosis must address this need for fitting to data and testing against data if it is to progress.

A particular challenge in fitting atherosclerosis models to data is the difficulty in obtaining spatiotemporal *in vivo* data, with snapshot data also challenging to obtain. *In vitro* experiments have been developed to explore specific aspects of atherosclerosis, but they are not able to capture all the features observed *in vivo*. This lack of spatiotemporal data presents challenges for validating and parameterising mechanistic mathematical models, with many parameter values taken from disparate parts of the literature. Consequently, few mathematical models of atherosclerosis have been systematically connected to experimental data, as has been done in the recent cancer mathematical modelling literature. This is a great opportunity for future research.

Synthetic data, generated by simulation models, for example, may be valuable in the absence of experimental data. Such synthetic data can be used in data fitting, as long as we select and verify the data carefully [2]. Unlike real-world data, which may be limited by factors such as sample size, noise, or missing information, synthetic data can be tailored to specific research questions, providing a more comprehensive understanding of biological phenomena. By combining the flexibility of synthetic data with the ability of stochastic models to capture uncertainty, researchers can gain a deeper understanding of complex biological processes and make more accurate predictions.

Model identifiability may be an important tool to ensure model reliability in systems where data is sparse. Structural identifiability addresses whether unique parameter values can be determined with perfect, noise-free data, and practical identifiability examines the ability to estimate parameters accurately given real-world, noisy, and limited data [9]. Understanding these concepts is particularly important in atherosclerosis modelling, where the complexity of biological processes, such as lipid uptake, inflammation, and cellular interactions within arterial plaques, requires careful model construction and validation to predict specific outcomes effectively.

5 Links to other areas of mathematical modelling

The similarities between wound healing and cancer have been made clear, leading to cancers being referred to as "wounds that do not heal" [6]. The modelling of wound healing and cancer has since been referred to as "two sides of the same coin" [5]. The biological dynamics of atherosclerosis is, in many ways, similar to both cancer and wound healing. The field of mathematical modelling of atherosclerosis can draw upon the fields of mathematical modelling in cancer and wound healing, and vice versa.

In terms of the overlaps in modelling approaches, multiphase models of cancer can (and are) being applied to atherosclerosis [13]. Both cancer growth and wound healing processes rely on angiogenesis (the formation of new blood vessels from existing vasculature). Angiogenesis has been modelled extensively in both biological contexts – this also occurs in late-stage atherosclerotic plaques and is starting to gain traction in associated modelling. A link between chronic wound healing and atherosclerosis that is an open problem for modellers, with clinical application, is the formation and treatment of arterial leg ulcers due to underlying arterial blockages due to atherosclerosis.

6 Looking into the future

Recent research has seen the development of sophisticated immunological models of plaque formation. However, these models typically do not consider the growth and morphological change of the plaque, and, moreover, they lack suitable quantitative validation against experimental data. Both of these issues must be addressed, either separately or together, if plaque formation models are to become truly reliable and predictive. We foresee that these techniques will be adopted into our research as we look to build and validate increasingly realistic plaque models that can actively contribute to scientific efforts to reduce the burden of cardiovascular disease.

References

1. Bennett, M.R., Sinha, S. Owens, G.K. Vascular smooth muscle cells in atherosclerosis. Circulation Research (2016) 118, 692-702.
2. Bull, J.A., Byrne, H.M. Quantification of spatial and phenotypic heterogeneity in an agent-based model of tumour-macrophage interactions. PLOS Computational Biology (2023) 19, e1010994
3. Callaghan, F.M. and Grieve, S.M. Spatial resolution and velocity field improvement of 4D-flow. MRI Magnetic Resonance in Medicine (2017) 78, 1959–1968.
4. Chalmers, A.D., Bursill, C.A., Myerscough, M.R. Nonlinear dynamics of early atherosclerotic plaque formation may determine the efficacy of High Density Lipoproteins (HDL) in plaque regression. PLOS ONE (2017) 12, e0187674
5. Chaplain, M.A.J., Byrne, H.M. Mathematical modeling of wound healing and tumor growth: Two sides of the same coin. Wounds—A compendium of Clinical Research and Practice. (1996) 8, 42–48.
6. Dvorak, H.F., Flier, J., Frank, H., Tumors—Wounds that do not heal—Similarities between tumor stroma generation and wound-healing. New England Journal of Medicine (1986) 315, 1650–1659.
7. Ford, H.Z., Byrne H.M., Myerscough M.R. A lipid-structured model for macrophage populations in atherosclerotic plaques. Journal of Theoretical Biology (2019) 479, 48–63.
8. Libby P. Inflammation in atherosclerosis. Nature (2002) 420, 868-874.
9. Murphy, R.J., Maclaren, O.J., Simpson, M.J. Implementing measurement error models with mechanistic mathematical models in a likelihood-based framework for estimation, identifiability analysis and prediction in the life sciences. Journal of the Royal Society Interface (2024) 2, 20230402
10. Lui, G., Myerscough, M.R. Modelling preferential phagocytosis in atherosclerosis: delineating timescales in plaque development. Bulletin of Mathematical Biology (2021) 83, 96.
11. Mirzaei, N.M., Weintraub, W.S., Fok, P.W. An integrated approach to simulating the vulnerable atherosclerotic plaque. American Jounral of Physiology–Heart and Circulatory Physiology (2020) 319:H835–H846.
12. Pan, J.C., Cai, Y., He, H.L., Gu, N., Li, Z.Y. A multiscale modeling study of nanoparticle-based targeting therapy against atherosclerosis. Journal of Biomechanics (2024) 166, 112067.
13. Watson, M.G., Byrne, H.M., Macaskill, C., Myerscough M.R. A two-phase model of early fibrous cap formation in atherosclerosis. Journal of Theoretical Biology (2018) 456, 123-136.
14. West, J., Adler, F., Gallaher, J., Strobl, M., Brady-Nicholls, R., Brown, J. Roberson-Tessi, M., Kim, E., Noble, R., Viossat, Y., Basanta, D., Anderson, A.R.A. A survey of open questions in adaptive therapy: Bridging mathematics and clinical translation. eLife (2023) 12, e84263

Macrophage efferocytosis and apoptosis

Adelle C.F. Coster and Mary R. Myerscough

Abstract Efferocytosis, that is, the removal of dead cells by macrophages, is a key process in the growth, regression and resolution of atherosclerotic plaque. Macrophages acquire internalised lipid when they ingest dying cells. Using data from an *in vitro* experiment measuring macrophage population and average ingested lipid per cell, we explore the processes of efferocytosis and apoptosis in this population. We find that the death rate of macrophages increases with time, which, we suggest is effectively due to their age—at the start of the experiment all cells are the same age. We also find that the rate of efferocytosis is not depndent on the average macrophage lipid load. This agrees with the hypothesis that defective efferocytosis may be a driver of atherosclerosis.

1 Introduction

Macrophages play an important role in atherosclerosis. These immune cells clear apoptotic material such as cellular debris and free lipid from inside apoptotic cells by a processes termed efferocytosis. How this process, and the macrophage life cycle itself, interacts with the development of atherosclerotic plaques is still not completely understood. Here we analyse some experimental data [2] from observations of *in vitro* efferocytosis in murine bone marrow-derived macrophages with a view to determining some parismonious mathematical descriptions for efferocyctosis and

Adelle Coster
School of Mathematics and Statistiscs, University of New South Wales, Sydney NSW 2052, Australia
e-mail: a.coster@unsw.edu.au

Mary Myerscough
School of Mathematics and Statistics, University of Sydney, NSW 2006, Australia
e-mail: mary.myerscough@sydney.edu.au

D. R. Wood et al. (eds.), *2024 MATRIX Annals, Part II*, MATRIX Book Series 8,
https://doi.org/10.1007/978-3-032-16206-9_14

apoptotosis in these cells, by identifying the possible factors implicated in these processes.

2 Experimental Background

Experiments were undertaken in [2] in which primary murine bone marrow-derived macrophages (BMDMs) were differentiated from mouse femur bone marrow hematopoietic stem cells. The BMDMs were cultured in standard growth media and the cells were plated onto slides at approximately 1×10^5 cells per well (wells of two areas in size, 1.8 cm^2 and 2 cm^2 were used), forming a two dimensional layer of cells adhered to the plate surfaces. Cells were incubated for 0, 12, 24, 36, 48, and 60 hours and then fixed with the application of formalin, the cells stained with oil red O (which labels lipid droplets in the cells) and Mayer's hematoxylin (which labels the cells). Microscopy images were obtained for the fixed and stained macrophages. The authors used a machine learning algorithm to determine the number of cells, and the amount of internalised lipid droplet per cell (as an area per cell). In the first of the experiments considered here the cells were incubated in a medium of lipopolysaccharide (LPS) and and a stimulant, interferon γ (IFNγ), which causes the cells to polarize to the classically activated, M1, state. Cells in the M1 state are associated with acute and chronic inflammation. A second experiment, in which the cells were stimulated with LPS and staurosporin (STPN) was also performed. The application of STPN promotes apoptosis in the cells.

Data from [2] was re-digitized to extract the number of macrophages, and the average lipid per cell as a function of time for the two protocols, see Figure 1.

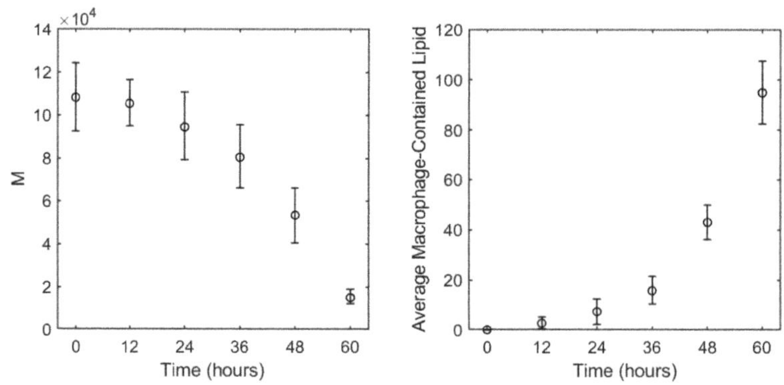

Fig. 1 The macrophage population, M, and the average macrophage-contained lipid load as a function of time, redigitized from [2].

3 Mathematical Modelling

The available data is very sparse, and hence we take a mean field, reductionist modellng approach. The processes that we consider in the model are: replication of the cells, apoptosis (death) of the cells, and efferocytosis, in which apoptotic lipid is ingested by macrophages. This increases the quantity of internalised lipid that they carry.

We note that in the experiments were such that there was no cell immigration or emigration and no other materials in the experimental system apart from the macrophages. Hence the only source of lipid was from the macrophages themselves (largely from their cell membranes). Additionally, all the macrophages were also derived simultaneously from the same source at the start of each experiment and so the cells were all the same age. Given the medium in which the cells were cultured it is expected that the replication rate of the cells was very low.

The state variables we consider for our model of lipid uptake in macrophages are: the live-cell macrophage population, M, the population of apoptotic macrophages (prior to ingestion by efferocytosis), P, the total apoptotic lipid in the system, A_P, and the total macrophage-contained lipid, A_M, which includes any endogenous lipid in the cell membranes. All of these are functions of time, t, taken to be zero at the start of the experiment. Note, as the cells are all the same age, t is also a proxy for the age of the macrophages.

The system of equations for the live and apoptotic cells in the system is as follows:

$$\frac{dM}{dt} = \rho M - fM, \quad \text{and} \tag{1}$$

$$\frac{dP}{dt} = \qquad fM - hMP. \tag{2}$$

The corresponding total lipid levels associated with the live and apoptotic cells are given by:

$$\frac{dA_M}{dt} = \rho M a_0 - fA_M + hMA_P, \quad \text{and} \tag{3}$$

$$\frac{dA_P}{dt} = \qquad + fA_M - hMA_P. \tag{4}$$

Here exponential or Malthusian growth is assumed for cells, with a linear replication rate, ρ, which is assumed to be near zero. Replication increases the macrophage population, M, as well as increasing the total macrophage-contained lipid, A_M, where a_0 is the amount of lipid created to make the cell membranes for the new daughter cell.

The initial conditions are

$$
\begin{aligned}
M(0) &= M_0, \\
P(0) &= 0, \\
A_M(0) &= a_0 M_0, \quad \text{and} \\
A_P(0) &= 0.
\end{aligned}
$$

The rate of apoptosis or death of the cells is proportional to the macrophage population, M. The function for the rate of apoptosis, $f = f(M,P,A_M,A_P,t)$, may depend on the state variables or time. When the cells die, they move from the live-cell macrophage population, M, to the apoptotic population, P. The internalised lipid in the cell, atogether with lipid in its cell membrane, is moved from the total macrophage-contained lipid pool, A_M, to the total apoptotic lipid pool, A_P.

Efferocytosis, or the ingestion of apoptotic material by macrophages removes apoptotic cells. It is assumed that this removal is at a rate proportional to the product MP, the probability of a macrophage finding an apoptotic cell. The rate at which this probability is modified, and full ingestion takes place is given by the function $h = h(M,P,A_M,A_P,t)$, which may depend upon the state variables or time. When apoptotic cells are digested they are removed from the system. The apoptotic lipid associated with the cell is removed from the apoptotic lipid pool, and added to the total macrophage-contained lipid.

Thus the average internalised lipid load per (live) macrophage, $\bar{L}$ is given by

$$
\bar{L} = \frac{A_M - a_0 M}{M}. \tag{5}
$$

3.1 Models for the macrophage and lipid dynamics

Naturally, models that include more degrees of freedom tend to be better able to replicate observations. The functional form of the processes and the resulting dynamics of the state variables have to have appropriate evolutions to match the characteristics of the data however. Here several variants of the rate functions for apoptosis and efferocytosis are explored. We used least-square fitting and constrained the parameters of the models using the data for the macrophage population and the average lipid per live macrophage as functions of time, weighting these two data sets equally. The inferred parameter values for the models are listed in Table 1. The goodness of fit for each of the models is shown in Table 2 – the adjusted R^2 allows for model comparison adjusting for the number of degrees of freedom. Given that the data is extremely sparse, the goodness of fit was poor for all model variants, however the purpose here was not to infer the parameter values with any specificity but rather understand whether the model could reproduce the observed behaviour.

3.1.1 Linear apoptosis and efferocytosis

The data for the macrophage population shows a decrease in this population as a function of time, as might be expected for this experimental system. If we assume the apoptosis rate is linear and constant, $f = \beta$, then the live-cell population is given by an equation for exponential decay, $M = M_0 \exp((\rho - \beta)t)$, where M_0 is the initial macrophage population. If we additionally assume a linear efferocytosis rate function, $h = \eta$ in Equations 1-4, and use least-squares optimisation to fit the parameters to this linear model, the outputs are as shown in Figure 2. It can be seen that the curvature of the plot of the model population as a function of time and the data as a function of time, is at odds and thus a constant apoptosis rate appears to be insufficient to describe the dynamics. The average lipid load is also not well described by this linear model with the model prediction being higher than the data for mid-range values (12 – 36 hours) and very much lower than the last data point (60 hours).

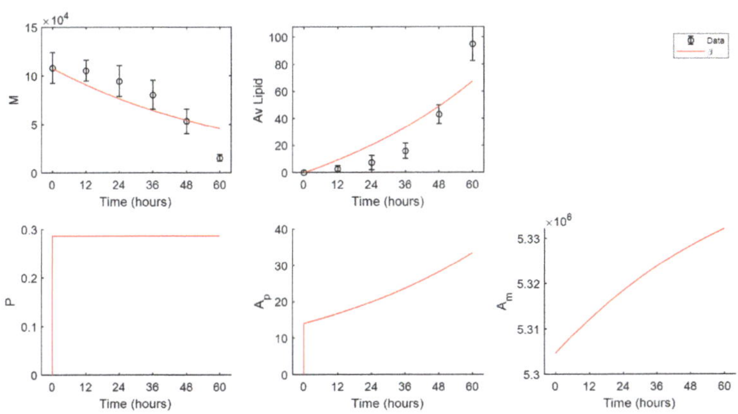

Fig. 2 The macrophage population, M, and the average lipid load per macrophage as a function of time, redigitised from [2]. The solid line indicates the least-squares of the model (Equations 1-4) with with constant apoptotosis, $f = \beta$, and efferocytosis rate functions, $h = \eta$. The fit was equally constrained by both data sets. Note the curvature of the macrophage population data and the curve do not match.

3.1.2 Age-driven apoptosis and linear efferocytosis

From the macrophage population data, shown in Figure 1, it appears that the apoptosis rate increases with time. This could be due to the age of the cells (recall that the age of the cells is synchronised in this system) or due to the increasing lipid load that the cells carry as time increases. For this model variant with apoptosis increasing with time (or equivalently age) we heuristically set

$$f = \beta \exp(kt), \tag{6}$$

where β and k are constants, and again assume a linear efferocytosis rate function, $h = \eta$. Optimising the parameters of the system (Equations 1-4) constrained both by the macrophage population and the average lipid load data, the trend for the population decay is captured, (Figure 3). However, the late time dynamics of the average lipid load is underestimated. Therefore, the population dynamics are well-modelled if we assume a death rate that increases with time, but the average lipid load per cell as predicted by the model does not ageee well with the data at 48 and 60 hours.

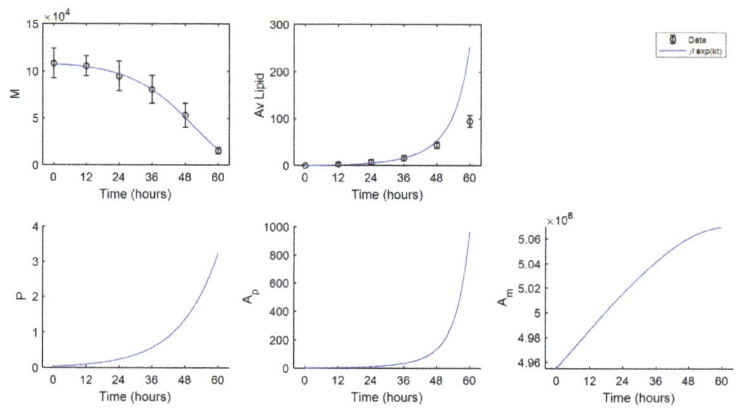

Fig. 3 The macrophage population, M, and the average lipid load per macrophage as a function of time, redigitized from [2]. The solid line indicates the least-squares fit of the model (Equations 1-4) with age-dependent apoptotosis and efferocytosis rate function, Equation 6 and constant efferocytosis rate function, $h = \eta$. The fit was equally constrained by both data sets. Note the model trend matches macrophage population data, but overestimates the average lipid load at long times.

As the time (or age) dependent apoptosis rate, Equation 6, captures the macrophage population dynamics it is used all subsequent variants of the model, below.

3.1.3 Age-driven apoptosis and lipid-dependent efferocytosis

The average lipid load at longer times is not well represented by a constant efferocytosis rate, (Figure 3). This could be due to the age of the cells (recall that time can be a proxy for age in this system) or due to the increasing lipid load that the cells carry as time increases. Here we consider the efferocytosis rate to be a decreasing function of lipid load,

$$h = \eta \left(1 - \frac{(A_M)^n}{M^n C^n + (A_M)^n} \right), \tag{7}$$

where η, C, and n are constants. This gives an efferocytosis rate which is maximal at low levels of internalised lipid (which will occur at early times) and then decreases with a change to a minimal rate. The timing and rapidity of transition of efferocytosis as a function of A_M is controlled by the parameters C and n.

The optimised model outputs, constrained both by the macrophage population data and the average lipid per macrophage data are shown in Figure 4 for $n = 2 \dots 10$. The assumption of an age-dependent apoptosis rate function, Equation 6, ensures the macrophage population dynamics are captured. It can be seen, however, that irrespective of the value of n (or the other parameter values), the late time dynamics in the data for the average lipid load cannot be replicated.

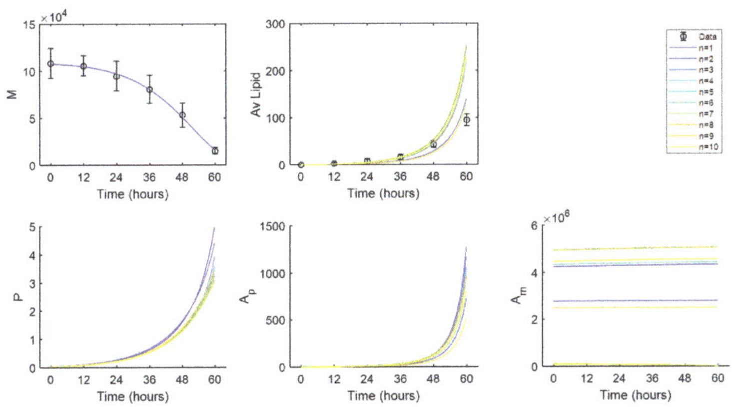

Fig. 4 The macrophage population, M, and the average lipid load per macrophage as a function of time, redigitized from [2]. The solid lines indicates the least-squares of the model (Equations 1-4) with age-dependent apoptotosis, Equation 6, for $n = 1 \dots 10$ and lipid-dependent efferocytosis rate function, Equation 7. The fit was equally constrained by both data sets. Note the model trend matches macrophage population data, although it does not match the average lipid levels at long times.

3.1.4 Age-driven apoptosis and efferocytosis

As a lipid-dependent efferocytosis rate appears to be unable to represent the observations, particularly at later times, we test here whether the average lipid content per cell can be explained by efferocytosis becoming less effective as time (and the age of the macrophages) increases. To do this, we make the efferocytosis rate time dependent:

$$h = \eta \exp(-gt). \tag{8}$$

The macrophage dynamics continue to be well described by an age-dependent apoptosis rate, Equation 6, and it can be seen that this, in combination with an age-dependent efferocytosis rate function, is also able to capture the dynamics of the average lipid load, Figure 5.

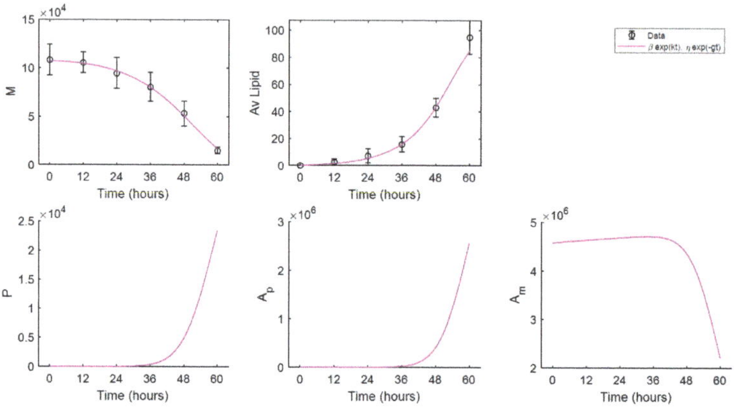

Fig. 5 The macrophage population, M, and the average lipid load per macrophage as a function of time, redigitized from [2]. The solid line indicates the least-squares fit of the model (Equations 1-4) with age-dependent apoptotosis and efferocytosis rate functions, Equations 6 and 8. The fit was equally constrained by both data sets. Note the model trends match both the macrophage population data and average lipid timecourse.

4 Discussion and Conclusion

In this study, we have identified that apoptosis for a particular population of macrophages cultured *in vitro* depends on age. We also found that death rate depended on age. Since all cells were the same age at the start of the experiment, we can use time t to represent cell age from the start of the experiment. We also found that internalised lipid load in macrophages appears to be best captured by a decreasing efferocytosis rate with time/age. This is in agreement with the theory that atherosclerosis progresses, in part, due to a rise in defective efferocytosis [1, 3].

If macrophages' efferocytic abilities decline with age, then more apoptotic material will be left behind in the tissue. Inside a plaque this will, in due course, undergo secondary necrosis and contribute to a core of free lipid and cellular debris in the plaque. In an *in vitro* experiment, we would expect to see lipid droplets become evident outside of cells. Adding further observations at late time, both between 48 and 60 hours and after 60 hours may give more information about what is occurring and, with suitable modifications to the machine learning protocol, free lipid may be able to observed and quantified.

f	β	Eqn 6	Eqn 6	Eqn 6
h	η	η	Eqn 7, $n = 2$	Eqn 8
C			639.9 (-9.031e+06, 9.033e+06)	
M_0	1.0960E+05 (9.196e+04, 1.272e+05)	1.0770E+05 (9.838e+04, 1.17e+05)	1.0770E+05 (9.78e+04, 1.177e+05)	1.0790E+05
a_0	45.9 (-2.456e+06, 2.456e+06)	46 (-1694, 1786)	32.88 (-1805, 1871)	45.21
β	0.01545 (-827, 827)	0.001731 (-0.004167, 0.00763)	0.001913 (-0.03905, 0.04288)	0.001575
η	8.6400 (-331.9, 349.2)	0.0457 (0.009066, 0.08225)	0.0214 (-1.552, 1.594)	0.2358 (-0.3742, 0.8458)
k	-	0.07309 (0.004731, 0.1415)	0.07093 (-0.3321, 0.474)	0.07486 (0.01586, 0.1339)
ρ	0.0007022 (-827, 827)	0.0004957 (-0.01336, 0.01436)	0.0007714 (-0.07182, 0.07337)	2.426E-05 (-0.0135, 0.01355)
g				0.2058 (-0.3795, 0.7912)

Table 1 The parameter values for each least-squares fit for the different models, indicated by the apoptosis rate function, f, efferocytosis function, h, and order, n. The 95% confidence interval is indicated in brackets. Both datasets, the macrophage population and the average lipid load as functions of time, were equally weighted to constrained the parameter estimates. Note that the values are not well constrained.

It would be interesting to test the hypothesis that the effectiveness of efferocytosis declines with the age of the macrophage. One approach may be to perform similar experiments to [2] using macrophages that have differentiated from monocytes of different ages to explore whether cell age, rather than time as a macrophage, influences efferocytic ability.

Future work would link these models to the other investigations of efferocytosis using beads instead of quantifying internalised lipid [2]. Internalised beads structure the macrophage population in a similar way to accumulated internalised lipids but could be used to quantify the number of efferocytic interactions explictly. Such an investigation would have the capacity to explore the processes/parameters which were common across the different experiments and those that were protocol specific. The lipid accumulation experiments in the presence of STPN could also be used for more specificity of the efferocytic parameters, given that STPN is an apoptosis enhancing agent.

References

1. Bäck, M., Yurdagul, A., Tabas, I., Öörni, K., Kovanen, P.: Inflammation and its resolution in atherosclerosis: mediators and therapeutic oopportunities. Nature Reviews: Cardiology **16** (2019)
2. Ford, H.Z., Zeboudj, L., Purvis, G.S., Ten Bokum, A., Zarebski, A.E., Bull, J.A., Byrne, H.M., Myerscough, M.R., Greaves, D.R.: Efferocytosis perpetuates substance accumulation inside

f	h	n	R^2	Adjusted R^2
β	η		0.9056	0.8935
Eqn 6	η	-	0.9721	0.9675
Eqn 6	Eqn 7	2	0.9721	0.9663
Eqn 6	Eqn 7	3	0.9720	0.9662
Eqn 6	Eqn 7	4	0.9720	0.9662
Eqn 6	Eqn 7	5	0.9720	0.9662
Eqn 6	Eqn 7	6	0.9720	0.9662
Eqn 6	Eqn 7	7	0.9720	0.9662
Eqn 6	Eqn 7	8	0.9720	0.9662
Eqn 6	Eqn 7	9	0.9720	0.9662
Eqn 6	Eqn 7	10	0.9720	0.9662
Eqn 6	Eqn 8		0.9837	0.9803

Table 2 The least-squares goodness-of-fit R^2 and Adjusted R^2, adjusting for the number of model parameters, for each of the different models, indicated by the apoptosis rate function, f, and efferocytosis function, h. Both datasets, the macrophage population and the average lipid load as functions of time, were equally weighted to constrained the parameter estimates.

macrophage populations. Proceedings of the Royal Society B: Biological Sciences **286**(1904) (2019). DOI 10.1098/rspb.2019.0730

3. Tabas, I.: Macrophage death and defective inflammation resolution in atherosclerosis. Nature Reviews: Immunology **10** (2010)

Investigating the impact of macrophage lineage on atherosclerotic plaque development using an agent-based model

Adrianne L. Jenner, Angela M. Reynolds, Keith L. Chambers, Andrew J. Murphy, Helen M. Byrne, Mary R. Myerscough

Abstract Atherosclerosis is a chronic inflammatory disease of the arteries that can lead to heart attack and stroke. Atherosclerotic plaques build up over time due to an inflammatory response to the presence of lipids in the blood. A key immune cell in plaque development is monocytes, which differentiate into macrophages as they enter the vessel wall. In this work, we develop an agent-based model of early plaque development to investigate the impact of monocyte lineage on plaque development. Previous research has shown a correlation between monocyte lineage and a differentiated macrophage's ability to efflux lipid and emigrate from the plaque. We investigate this here and show under what conditions plaques arise in the vessel wall. The model presented here can serve as a platform for future hypothesis generation on macrophage involvement in plaque development to be tested.

Adrianne L. Jenner
School of Mathematical Sciences, Queensland University of Technology, Brisbane, Australia,
e-mail: adrianne.jenner@qut.edu.au

Angela Reynolds
Virginia Commonwealth University, Richmond, USA
e-mail: areynolds2@vcu.edu

Keith L. Chambers
Mathematical Institute, The University of Oxford, Oxford, UK

Andrew J. Murphy
Baker Heart and Diabetes Institute, Melbourne, Australia

Helen Byrne
Mathematical Institute and Ludwig Institute for Cancer Research, The University of Oxford, Oxford, UK

Mary R. Myerscough
School of Mathematics and Statistics, The University of Sydney, Sydney, Australia

D. R. Wood et al. (eds.), *2024 MATRIX Annals, Part II*, MATRIX Book Series 8,
https://doi.org/10.1007/978-3-032-16206-9_15

1 Introduction

Atherosclerosis is an inflammatory disease of the arteries that is a major cause of cardiovascular disease and stroke [2]. Atherosclerotic lesions or plaques are characterised by the accumulation of lipids, immune cells, smooth muscle cells and necrotic cell debris in blood vessel walls [2]. These plaques can eventually rupture, releasing thrombogenic material which causes heart attacks or strokes. Cardiovascular disease and stroke caused by atherosclerosis is the leading cause of death in both developed and developing countries [9].

Early atherosclerosis is driven by the emigration of monocytes into the intima in response to immune signals due to the presence of modified low-density lipoproteins (modLDL) and the subsequent differentiation of these monocytes into macrophages. [9]. Macrophages take-up lipid, which reduces their motility and turns them into foam cells [19]. Excessive lipid uptake also causes macrophages to undergo apoptosis [30]. If lipid-laden apoptotic bodies are not efficiently cleared, their accumulation drives the formation of a large necrotic core. Plaques with large necrotic cores can eventually rupture, causing blockages downstream that lead to heart attack or stroke.

Monocytes differentiate into macrophages at the site of inflammation in response to the cytokine macrophage colony stimulating factor (M-CSF) [21]. Monocyte lineage branches are a key feature of the inflammatory process, particularly in atherosclerosis [8]. In humans, the monocyte pool comprises three subsets: classical, intermediate and non-classical [20, 21]. Generally, the time spent in the circulation differentiates between lineages. Classical monocytes circulate for a day, whereas intermediate and non-classical monocytes circulate for on average 4 and 7 days respectively [20]. Recent evidence suggests that macrophages derived from specific monocyte lineages have different behaviours in atherosclerotic plaques [23, 22]. It is hypothesized that macrophages derived from different monocyte subsets behave differentially in atherosclerotic plaques. Evidence suggests that murine Ly6-Chi monocytes, the equivalent to human classical monocytes are required to achieve atherosclerotic plaque regression [22]. We suggest this could be due to superior cholesterol efflux lipid abilities and hence emigrate from the plaque, and macrophages derived from alternate (intermediate and non-classical) monocytes are unable to perform these keys tasks efficiently.

Given the challenge of *in vivo* data collection from plaques, mathematical modelling has the potential to be a useful tool in developing a deeper understanding of atherosclerosis plaque development and rupture. Partial differential equations (PDEs) have been used to describe the spatial distribution of immune cells within a plaque [10, 14, 25]. PDEs have also been used to track the internalised lipid concentration in cells in the plaque [24] as well as their phenotype [4]. Ordinary differential equations (ODEs) have also been used to model atherosclerosis [18], particularly immune cell involvement. One of the few models that incorporates macrophage lineages, uses a system of ODEs to capture how macrophage phenotype derives from monocyte phenotype [1]. Agent-based models (ABMs) are another useful tool for capturing biological phenomena. ABMs allow for the direct encoding of biological rules into agents representing cells [16, 28]. ABMs have previously been employed to model atherosclerotic plaque development [3, 5, 6, 7, 26, 27], although this is still a relatively new area of modelling.

To date, the effects of monocyte lineage, in terms of classical and alternate monocytes, on atherosclerotic plaque development has not been incorporated into any models. In this paper, we present an ABM of early-stage atherosclerotic plaque development using the open-source software PhysiCell [13]. We focus this model on the role of macrophages in plaque development and examine the impact of monocyte lineage. This model is presented in a way that future researchers could download and simulate hypotheses in the platform. All aspects of the ABM are detailed below, with some example simulations of the model in the results section. The code for this model, can be accessed through Github `https://github.com/AdrianneJennerQUT/macrophage-athero-model`.

2 Methods

We use the open-source cell-based simulator software PhysiCell [13] to develop a 2-dimensional agent-based model (ABM) of atherosclerotic plaque development on a fixed rectangular domain. We simplify the biology of atherosclerosis and consider mainly the early stages of the plaque development. We assume macrophages are the main driver of early plaque development and the key processes are immigration, emigration, lipid uptake, lipid efflux, proliferation, apoptosis, and efferocytosis/phagocytosis. Descriptions of the domain, of the PDE that models the lipid distribution, and of the agents and their characteristics are given below with a summary of parameters in Table 1 and rules in Fig. 1. For more details on the PhysiCell software, see [12, 13] or previous applications of PhysiCell in cancer modelling [15] or COVID-19 modelling [11].

2.1 Domain and simulation time

We consider a 2-dimensional Cartesian coordinate system with domain length L μm and height H μm. We assume our domain models the lumen and has a rectangular shape, see Fig. 1, with $H < L$. We chose a domain length of $L = 1000$ μm and height of $H = 300$ μm to simulate the size of the vessel wall. We consider the horizontal axis to be parallel to the artery wall, i.e. the top of the domain representing the endothelium, and the vertical axis to be perpendicular to the artery wall.

Cells are initialised in the domain and interact with a discretised version of a PDE that describes the lipid concentration (see Section 2.2). To achieve this in PhysiCell, the domain is discretised into grids of size $20\mu m \times 20\mu m$. This discretisation is used to define how cells interact with the lipid concentration described in Section 2.2, see Ghaffarizadeh *et al.* [12] for more details.

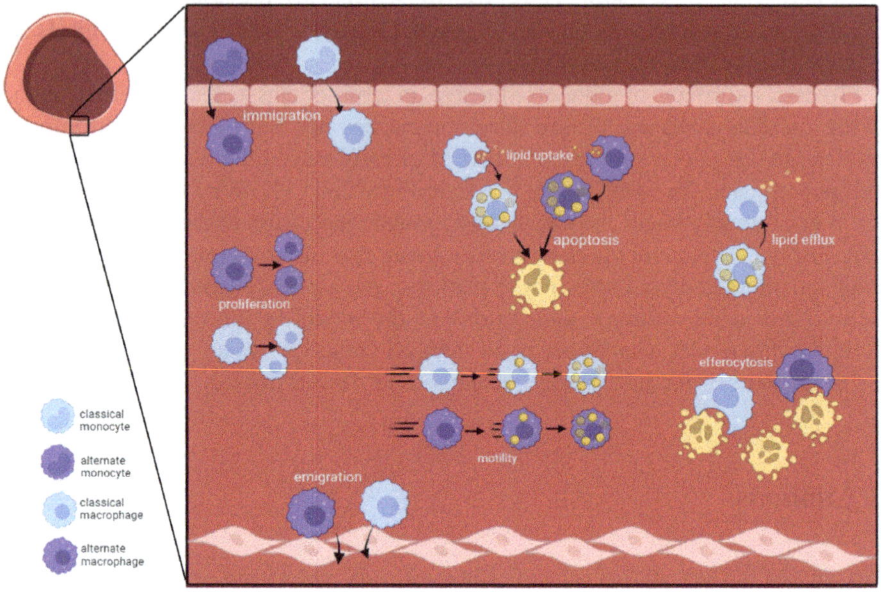

Fig. 1 Schematic summarising the major components of the model. Early-stage atherosclerotic plaque development is modelled on a fixed domain considered to represent a small segment of the vessel wall. Macrophages from both lineages enter the plaque. Once in the plaque they are able to undergo proliferation and uptake lipid. Only classical macrophages are able to efflux lipid. Macrophages will move more slowly as their amount of internalised lipid increases until they eventually undergo apoptosis from taking up too much lipid. Macrophages can efferocytose or phagocytose apoptotic macrophages. If macrophages reach the bottom of the domain, they are also able to emigrate. Created using biorender.

PhysiCell has three inbuilt time-steps: a diffusion time-step $\Delta t_{diffusion}$, a mechanics time-step $\Delta t_{mechanics}$ and a phenotype time-step $\Delta t_{phenotype}$. We use the inbuilt time-stepping choices in PhysiCell [13] and set a total time for the simulation as T. We assume the time-unit of our simulation is minutes. To evaluate computational cost, we tracked the simulation time for the parameters in Table 1 for multiple final times T, see Fig. 2, and found a linear relationship between the computational time and the number of days being simulated.

2.2 PDE for lipid concentration

PhysiCell uses BioFVM [12] to simulate the chemical microenvironment. BioFVM solves the PDE in Eq. 1. We leverage this PDE to model lipid movement in the domain. This model is briefly described below, but refer to [12, 13] for more details.

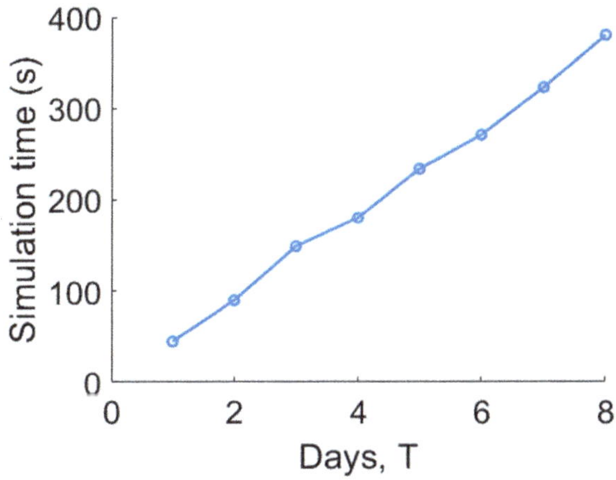

Fig. 2 The computational time in seconds required to simulate a single instance of the model for the parameters in Table 1 for varying final time T.

Consider the density of a substrate or chemical species $\rho(x,y,t)$ as a function of time t and space (x,y) in the domain $H \times L$, is given by:

$$\frac{\partial \rho}{\partial t} = D\nabla^2 \rho - \lambda \rho + \sum_{\text{cells k}} \delta(\mathbf{x}-\mathbf{x}_k)W_k[S_k(\rho_k^* - \rho) - U_k \rho]. \tag{1}$$

Here $\delta(x)$ is the Dirac delta function, $\mathbf{x}_k$ is the kth cell's position, W_k is the kth cell's volume, S_k is the kth cell's source rates, U_k is the kth cell's uptake rate and ρ_k^* is the saturation density of the kth cell. The constants D, and λ are the diffusion coefficients and decay rates of the substrate respectively.

We assume $\rho(x,y,t)$ represents lipid in our domain. Lipid accumulation first occurs deep in the vessel wall [17]. As such, we assume a gradient of lipid that increases as we move deeper into the vessel wall. To achieve this, the lipid gradient is maintained by introducing "ghost cells" that secrete lipid at a rate dependent on their Euclidean distance from the lumen wall (i.e. the top of the domain):

$$S_k = \frac{g_{base}}{H} y_k,$$

where S_k is the secretion rate of the kth cell, g_{base} is the base secretion rate of lipid by "ghost cells", and y_k is the y-component of the Cartesian coordinate for the position of the kth cell, i.e. $\mathbf{x}_k = [x_k, y_k]^T$. We similarly modified the secretion target ρ_k^* of the kth cell so that this was also dependent on the distance from the vessel wall, i.e.

$$\rho_k^* = \frac{\rho_{base}^*}{H} y_k,$$

where ρ^*_{base} is the base or maximum saturation density for "ghost cells". We allow lipid to diffuse slowly in the domain, $D = 10^{-3}\ \mu m^2/min$, but assume no lipid decay $\lambda = 0$.

2.3 Macrophage lineage class: classical and alternate

Macrophages are represented by cells (circle agents) in the model. We model two classes of lineages: classical and alternate (see Fig. 1. These two lineages refer to the monocyte lineages, where classical monocytes are monocytes that have spent little time in the blood and alternate monocytes have been in circulation longer [20, 21]. We assume that when a monocyte differentiates it passes its lineage characteristics onto the macrophage that it becomes. As such, we assume there are two types of macrophages in the model which we will call classical and alternate to reflect their monocyte origin. These macrophages only differ by their lipid efflux property. All other properties (emigration, movement, uptake, phagocytosis, proliferation etc) are the same. We assume that classical macrophages efflux lipid, i.e. transfer lipid to high density lipoprotein (HDL) particles, at a rate $S_k = s_{base}$ and alternate macrophages do not efflux lipid, i.e. $S_k = 0$.

2.4 Macrophage immigration

We assume macrophages arrive from the blood at a rate m_a cells/min. Estimating $m_a = 0.003$, means every 333 mins on average one cell arrives (roughly every 5.55 hours). To implement this stochastically, we assume, in a single time step $\Delta t_{phenotype}$, that a macrophage has a probability of $m_a \times \Delta t_{phenotype}$ of arriving. If a macrophage arrives, we randomly allocate it a position in the top section of the domain between y_{min} and the top of the domain. We also randomly allocate the arriving macrophage as either coming from the classical lineage class with probability p_{lin} or from the alternate lineage class with probability $1 - p_{lin}$. We assumed that at homeostasis 92% of macrophages arriving into the lumen come from the classical monocyte lineage and so $p_{lin} = 0.92$.

2.5 Macrophage emigration

We know that macrophage emigration from the plaque is a critical part of the plaque life-cycle. We do not introduce a bias into cell movement to model this. Instead, we assume that if a macrophage is in the bottom section of the domain (the lowest 20μm) then it leaves the domain. When a macrophage emigrates, it is removed from the simulation and cannot return.

2.6 Macrophage lipid uptake

We assume all macrophages uptake lipid from the extracellular microenvironment at a rate u_{lipid}. This parameter feeds directly into Eq. 1. We use the inbuilt PhysiCell tracking to keep account of the intracellular amount of lipid and also uptake from the lipid substrate ρ. Classical and alternate macrophages uptake the lipid at the same rate. Macrophages stop up-taking lipid once they undergo apoptosis, i.e. die, see Section 2.10.

2.7 Macrophage lipid efflux

We assume macrophages efflux lipid at a rate that is dependent on their lineage class. Macrophages efflux lipid at a rate s_{base} if they are classical macrophages. We assume alternate macrophages do not efflux lipid. We assume for simplicity that the lipid effluxed by macrophages onto HDL particles is rapidly cleared from the lesion by HDL diffusion, and, consequently, does not contribute to the extracellular lipid density, ρ. As such, we model lipid removal from classical macrophages as a continuous decrease in intracellular lipid content at rate $\Delta t \times s_{base} \times L_k$, where L_k is the kth cell's internal amount of lipid.

2.8 Macrophage movement

We assume macrophage movement is random (i.e. Brownian motion without chemotaxis). We represent the speed of macrophage movement, v, as a linear function of the internal lipid L_k, i.e.

$$v(L) = \begin{cases} v_{max}\left(1 - \frac{L_k}{L_T}\right), & L_k < L_T \\ 0, & L_k \geq L_T \end{cases}$$

where v_{max} is the maximum macrophage speed, L_k is the kth cell's internal lipid, and L_T is the internal lipid threshold. Once the internalised lipid has exceeded this threshold, the macrophage stops moving. All other motility rules are set by PhysiCell [13].

2.9 Macrophage proliferation

Macrophages undergo proliferation at a rate b. To model this, we use PhysiCell's inbuilt proliferation model called "live". This converts proliferation to a probabilistic rate. The population-scale model is a simple exponential growth model. Once a cell divides it initially starts small and then grows to its homeostatic volume, again using the inbuilt PhysiCell rules. If a cell divides, its internal lipid concentration is divided evenly between its daughter cells.

2.10 Macrophage apoptosis

Macrophages may undergo apoptosis, i.e. programmed cell death, either as a result of excessive lipid accumulation or naturally. Firstly, we assume there is a natural apoptosis rate of a_{base}, and use the inbuilt "apoptosis" model in PhysiCell to capture this. Similar to proliferation, this natural apoptosis is modelled probabilistically. We impose a long duration of disintegration t_{dis} minutes and after this period the apoptotic cell will disappear from the simulation. We introduced this into the model based on the assumption that an apoptotic cell will not reside in the tissue forever. The second way that macrophages can undergo apoptosis is through exceeding an internalised lipid threshold of L_a. Once a macrophage has exceeded this internalised lipid threshold then it undergoes apoptosis, following the "apoptosis" model inbuilt in PhysiCell. This is modelled deterministically, i.e. once a macrophage's internal lipid exceeds this threshold it must undergo apoptosis.

2.11 Macrophage phagocytosis/efferocytosis

At any point in time, if a macrophage that has not undergone apoptosis encounters an apoptotic macrophage, it has a probability p_e of undergoing efferocytosis, whereby the healthy macrophage phagocytoses (ingests) the apoptotic macrophage. To model this, we first check the distance between a live macrophage and nearby apoptotic macrophages is within a Euclidean distance of 1.05 times the radius of the two macrophages. The macrophage will then either internalise the apoptotic macrophage or nibble at the apoptotic macrophage dependent on the amount of internalised lipid in that macrophage.

Let L_{apop} be the internal lipid in an apoptotic macrophage and L_{live} be the internal lipid in a live macrophage attempting to eat that apoptotic macrophage. If $L_{apop} + L_{live} < L_a$, then the live macrophage is able to ingest all of the apoptotic macrophage and we use the "ingest cell" built into PhysiCell. If $L_{apop} + L_{live} = L_a$, the live macrophage will still ingest the apoptotic macrophage but will undergo apoptosis immediately afterwards. If $L_{apop} + L_{live} > L_a$ then the live macrophage will nibble at the apoptotic body until it has ingested an amount of internal lipid so that it's internal lipid equals the L_a threshold, i.e. it will ingest a fraction of the apoptotic macrophage:

$$f = \frac{L_a - L_{live}}{L_{apop}}.$$

The nibbling macrophage will then undergo apoptosis as its internalised lipid will now be $L_{live} = L_a$, and the apoptotic macrophage that was being nibbled will have

$$L_{apop} = (1-f)L_{apop},$$

internalised lipid remaining.

2.12 Parameterisation and evaluation of stochasticity

In Table 1, we summarise all parameters in the model and their estimated value. Some of these values were estimated based on literature. We sought to evaluate the impact of the stochasticity in the ABM on the model output and this is summarised in Fig. 3A-B, where we average the number of dead cells and classical macrophages at day 6 for the number of simulations N and the parameters in Table 1. We find that as the number of simulations increases to 100, we slowly approach a stable average. The mean and standard deviation for the dead cells and classical macrophages over time for 100 simulations are given as well in Fig. 3C-D.

Table 1 Table of parameter values in ABM model of atherosclerotic plaque development. Where references have been found in the literature, these have been provided in the final column. The units of lipid have been left undefined as the lipid parameters have not been tied directly to literature.

Param	Units	Description	Value	Ref
L	μm	Domain length	1000	
H	μm	Domain height	300	[17]
T	min	Total simulation time	8,640	
D	μm^2/min	Lipid diffusion	10^{-3}	
λ	1/min	Lipid decay	0	
m_a	cells/min	Macrophage arrival rate	3×10^{-3}	
y_{min}	μ	Arrival distance from top	80	
p_{lin}	prob.	Prob. of being from classical lineage	0.92	[20]
u_{lipid}	1/min	Rate of lipid uptake	0.008	
v_{max}	1/min	Max speed	2	
L_T	lipid	Lipid speed threshold	450	
L_a	lipid	Lipid apop. threshold	500	
b	1/min	Proliferation rate	0.00034	
a_{base}	1/min	Natural apoptosis rate	0.00002	[29, 31]
t_{dis}	min	Apoptosis duration	5516	
p_e	prob.	Prob. of phagocytosis	1	
s_{base}	1/min	Lipid efflux rate	0.1	
g_{base}	1/min	Ghost cells secretion rate	0.01	
ρ^*_{base}	lipid units	Max. lipid saturation density	1	

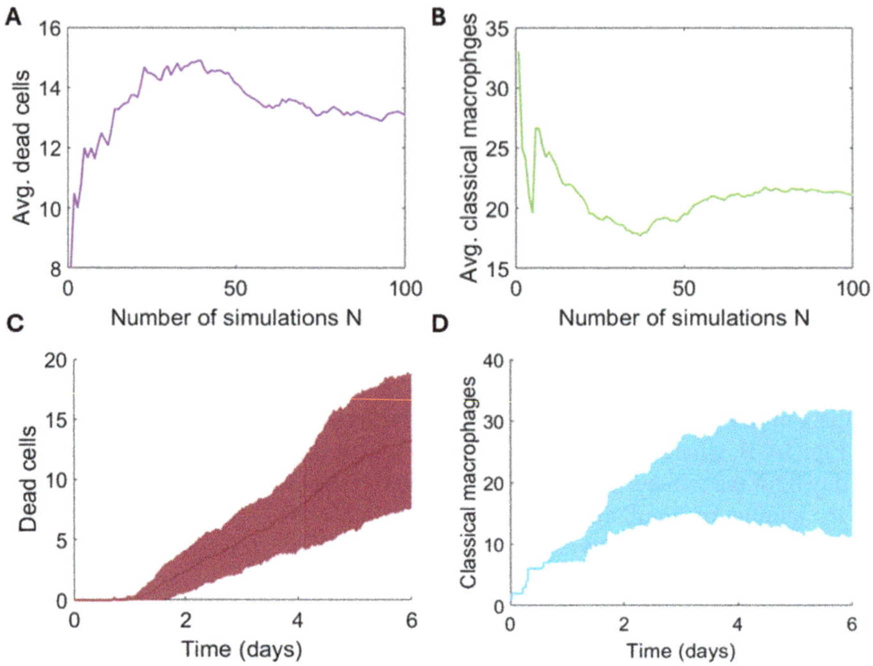

Fig. 3 Evaluating the stochasticity in the ABM for parameters in Table 1. For a number of simulations N the (A) average dead cells and (B) average classical lineage macrophages was calculated at time T. The average and standard deviation of $N = 100$ simulations of the model over 6 days is plotted for (C) dead cells and (D) classical macrophages. The average is represented as a solid line and the standard deviation is the shaded region.

3 Results

3.1 *Summary of the outputs of the ABM for early plaque development*

There are numerous outputs of the ABM that are useful for evaluating plaque development. A single simulation of the model for the parameters in Table 1 is plotted in Fig. 4. The spatial distribution of macrophages has been plotted at days 1, 3 and 6 (Fig. 4A). A clear increase in the number of apoptotic cells is seen at later time points. The density of lipid at day 3 is plotted in Fig. 4B, where the gradient of lipid increases deeper into the vessel wall and away from the endothelium (i.e. top of the domain). Sinks are also evident in the domain and this is caused by macrophages that have taken up lipid. The total number of macrophages from each lineage, as well as dead macrophages and average internalised lipid has been plotted in Fig. 4C. A video simulation using the parameters in Table 1 is also available in the supplementary material.

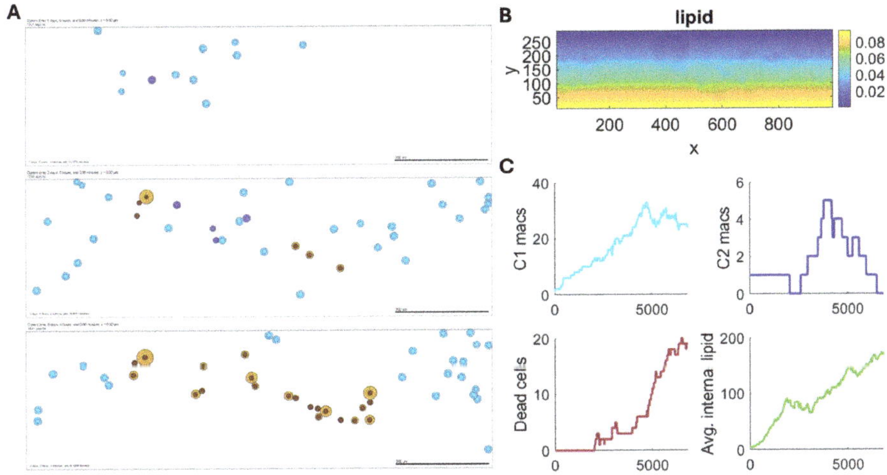

Fig. 4 Plot of the PhysiCell outputs from a single simulation using the base parameters in Table 1. (A) Still-shots of the cell spatial arrangement at 1 day (top), 3 days (middle) and 6 days (bottom). Classical macrophages are plotted in light blue, alternate macrophages are plotted in purple and apoptotic cells are brown. The size of the cytoplasm and nucleus for the apoptotic cells scales with the amount of the internalised lipid. (B) Spatial density of lipid at 3 days given by Eq. 1, corresponding to the simulation domain in (A - middle). (C) Plots over time of the number of macrophages from the classical lineage (C1 macs) (top left), macrophages from the alternate lineage (C2 macs) (top right), dead cells (bottom left) and average internal lipid (i.e. total internal lipid/number of cells) (bottom right).

Over the 6 days, we see a small number of apoptotic macrophages accumulate in the vessel wall. This is due to the large quantity of lipid in the domain, which promotes large macrophage lipid loads and, consequently, lipid-induced apoptosis. There is a noticable number of classical macrophages that remain alive in the vessel wall, and this is due to the high probability of macrophages arriving from this lineage class and their inherent ability to efflux lipid that they uptake.

3.2 Impact of macrophage immigration and lineage

We sought to investigate the impact of varying the macrophage immigration rate m_a and the lineage probability p_{lin} on the presence of a plaque. We first simulated two changes to the base parameterisation of the model: $p_{lin} = 0.5$ (Fig. 5A) and $p_{lin} = 0.5$ and $m_a = 0.03$ (Fig. 5B). In other words, the probability of an immigrating macrophage being classical or alternate was equal (Fig. 5A) and we increased the arrival rate of macrophages (Fig. 5B).

Decreasing the proportion of classical macrophages that arrive (i.e. $p_{lin} = 0.92$ to $p_{lin} = 0.5$, Fig. 5A) does not appear to have a major impact on the number of dead cells (comparing Fig. 4 and 5A). However, increasing the arrival rate of cells, $m_a = 0.03$, and decreasing the proportion of classical macrophages immigrating, $p_{lin} = 0.5$, does seem to result in more apoptotic cells over time (Fig. 5B). This suggests that large macrophage influx rates promote apoptotic cell accumulation.

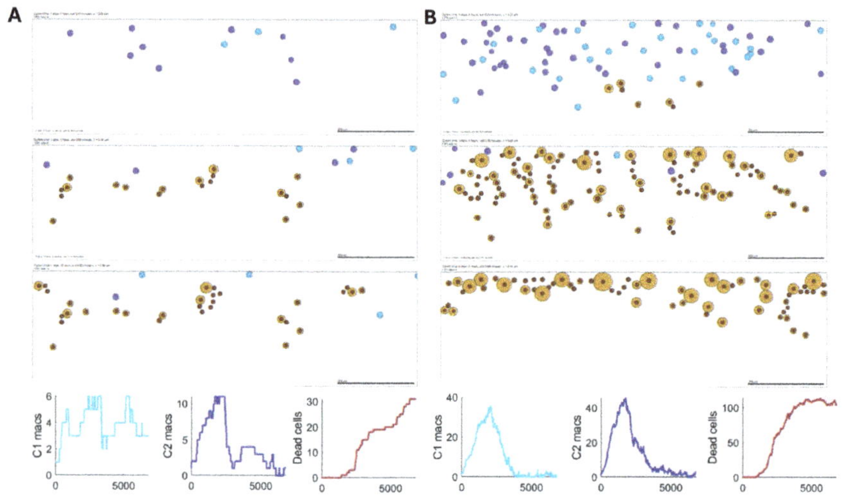

Fig. 5 Plot of the PhysiCell outputs from two simulations using the base parameters in Table 1 with (A) $p_{lin} = 0.5$ and (B) $p_{lin} = 0.5$ and $m_a = 0.03$. Still-shots of the cell spatial arrangement at 1 day (top), 3 days (middle) and 6 days (bottom). Cells from the classical lineage are plotted in light blue, cells from the alternate lineage class are plotted in purple and apoptotic cells are plotted in brown. (bottom row). The size of the nucleus and cytoplasm is scaled to represent the amount of internalised lipid. In the bottom row are plots over time of the number of macrophages from the classical lineage (C1 macs), alternate lineage class (C2 macs) and dead cells.

For the higher arrival rate of macrophages (Fig. 5B) initially apoptotic cells accumulate further into the lumen, then over time they seem to increase throughout the vessel wall and eventually are only present at the vessel wall edge (top section of the domain). This is likely due to two factors. The first is that arriving macrophages are dying nearer to the vessel wall as they phagocytose cells with internalised lipid that exceeds the internalised lipid threshold. The second reason is that cells further into the vessel wall are likely exceeding their duration of disintegration t_{dis}. This parameter determins the number of minutes after a cell has died that its apoptotic body remains. If we increased this parameter to be larger, we would expect to see a large presence of apoptotic cells further into the vessel wall.

To investigate further the impact of macrophage arrival rates and their lineage, we conducted a two-parameter sensitivity analysis, whereby we varied m_a and p_{lin} and recorded the number of dead cells and macrophages at day 6. We also introduced a binary metric *plaque* and assumed a plaque is present if the number of dead cells is greater than the number of live cells, see Fig. 6. Having a higher probability of macrophages from the classical lineage immigrating into the vessel wall, allows for the apoptotic bodies to be cleared quickly and a plaque to be avoided, i.e. $p_{lin} \geq 0.7$. However, when the arrival rate is rapid, $m_a = 0.03$, plaques can still emerge with a 90% classical lineage. This is likely due to the large influx of macrophages resulting in a high turnover of apoptotic bodies. Interestingly, the number of macrophages at day 4.5 and day 6 is largest for an intermediate arrival rate, i.e. $m_a = 0.0165$ compared to $m_a = 0.003$ or $m_a = 0.03$. We have only averaged 10 simulations, so this may change as we increase the number of simulations to 100 (see Fig 3) although it could suggest that there may be an optimal arrival rate for macrophages for which they are able to better manage the accumulation of apoptotic cells.

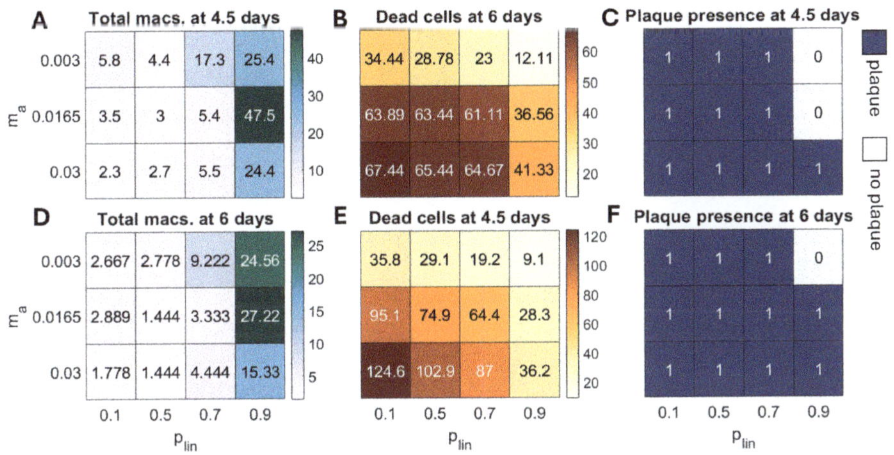

Fig. 6 Heatmaps capturing the sensitivity of the total number of macrophages, dead cells and presence of a plaque for combinations of m_a and p_{lin} at day (A-C) $T = 4.5$ and day (D-F) $T = 6$. The total number of (A,D) classical and alternate macrophages and (B,E) dead cells is plotted for pairs of (m_a, p_{lin}). (C,F) The presence of a plaque is given by a binary output (1 or 0) where a plaque is present if the total macrophages are less than the dead cells. Note, these heatmaps are the average of 10 simulations and noting the stochasticity in Fig. 3 the numeric values are likely to change as the number of simulations N increases.

We then investigated the stochasticity in the results of Fig. 6 for two parameterisations: $p_{lin} = 0.7$ and $p_{lin} = 0.92$, where $m_a = 0.003$, see Fig. 7. Decreasing the dominance of classical macrophages in the arriving macrophage pool increases the number of alternate macrophages slightly. This slight increase is enough to cause a significant decrease in the classical macrophage count, resulting in an increase in apoptotic cells. This is likely due to the fact that there are fewer classical macrophages arriving and hence fewer to proliferate and then clear apoptotic cells. This is further supported by investigating the ratio of dead cells to macrophages, where clearly the macrophages are able to stay on top of the apoptotic cell build up when $p_{lin} = 0.92$, i.e. ratio < 1, but are unable to cope when $p_{lin} = 0.7$, i.e. ratio > 1. This does not hold true for all simulations of $p_{lin} = 0.7$ as in some cases we can see that the ratio of dead cells to live macrophages is less than 1, suggesting the macrophages were able to control the build up of apoptotic cells in the plaque.

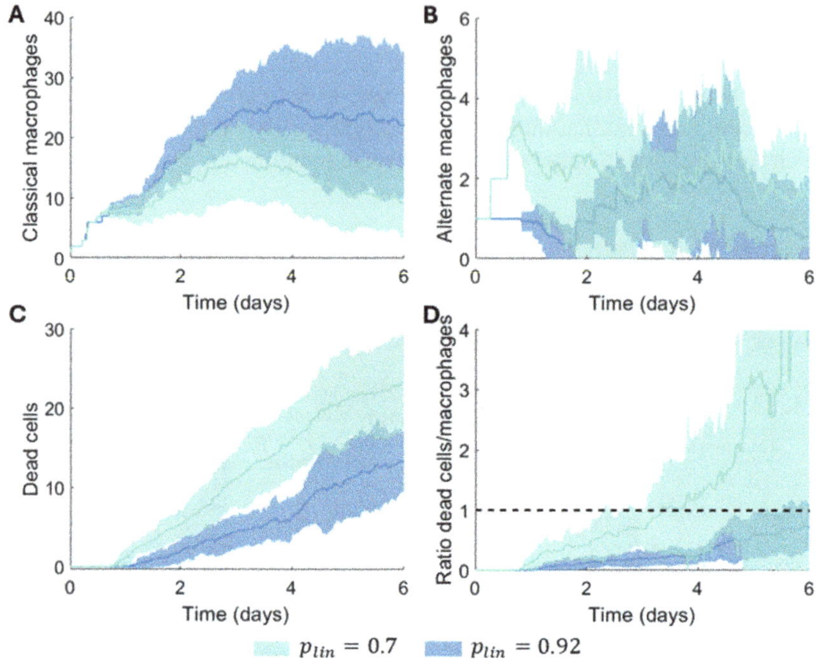

Fig. 7 Further investigating the stochasticity in the model simulations for $p_{lin} = 0.7$ and $p_{lin} = 0.92$. The mean and standard deviation of 20 model simulations over 6 days is plotted for two different p_{lin} values, with all other values fixed as in Table 1. Total numbers of (A) classical macrophages, (B) alternative macrophages and (C) dead cells are given. In (D) the ratio of dead cells to live macrophages is plotted with a dashed line at 1, representing the threshold for *plaque* used in Fig. 6. The mean is represented by a solid line and the standard deviation is represented by the shaded region.

3.3 Impact of internal lipid threshold on plaque development

We then investigated the impact of the internal lipid threshold L_a on plaque formation. Fixing the parameters to the base values in Table 1 and increasing the arrival rate of macrophages to $m_a = 0.03$ (see Fig. 6), we simulated varying values of the internal lipid threshold L_a, see Fig. 8. We found that for all values chosen (L_a from 100 to 2000) we did not avoid the creation of a *plaque* (assuming a plaque occurs when the number of dead cells is more than the live cells at 6 days). Interestingly, increasing the internal lipid threshold only delayed the peak in macrophage concentration, with the number of dead cells being similar across simulations for $L_a = 500$ to 2000.

This suggests that a macrophage's ability to process lipid and avoid apoptosis does not avoid the eventual plaque creation because the macrophages will inevitably take up too much lipid and die. It also suggests that the size of the plaque, i.e. the number of dead cells, isn't overly sensitive to the choice of the internal lipid threshold, unless this value is very low, e.g. $L_a = 100$.

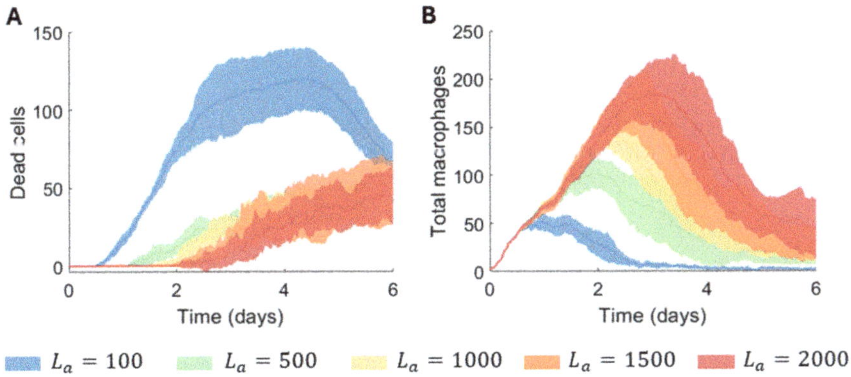

Fig. 8 Mean and standard deviation of 10 trajectories of the model for varying internal lipid threshold L_a. The dead cells (A) and total macrophages (B) is plotted for each L_a noted and the base parameters in Table 1 with $m_a = 0.03$. The mean is represented by the solid line and the standard deviation is represented by the shaded region.

4 Conclusion

Atherosclerosis is a common condition that can cause life-threatening problems. The establishment of atherosclerotic plaques or lesions is driven by inflammation of innate immune cells, primarily macrophages. Macrophages immigrate into the vessel wall, take up lipids and may contribute to necrotic plaque formation via apoptosis. Plaques with necrotic cores may rupture, causing heart attacks and strokes. Understanding the early stages of plaque onset is crucial to developing effective intervention methods. In this work, we develop a simple agent-based model (ABM) in PhysiCell that captures the early establishment of apoptotic macrophages in the vessel wall.

Using our model, we explored the relationship between macrophage lineage and plaque formation. We found that increases in the arrival of classical macrophages decreased the likelihood of plaque development. Furthermore, a slow arrival rate of macrophages helps to prevent large apoptotic clusters that are difficult to clear. While intuition may suggest that increasing the influx of classical macrophages helps to clear the plaque, this in fact causes an increase in uncleared apoptotic clusters. We also found that increasing the tolerance of macrophages to internalised lipid does not stop the development of a plaque, but instead delays its progression.

The ABM presented here is open-source and provides a platform for future researchers to test hypotheses on macrophage dynamics in early plaque formation. The model is computationally efficient and the inherent stochasticity, both spatially and at the population level, provides a nice representation of what we might expect in human plaques. This model could easily be extended to account for other aspects of atherosclerosis, such as smooth muscle cells, chemotaxis and other inflammatory cells.

Acknowledgements We acknowledge the MATRIX workshop funding support.

References

1. Bezyaev, V., Sadekov, N., Volpert, V.: A model of chronic inflammation in atherosclerosis. In: ITM Web of Conferences, vol. 31, p. 04002. EDP Sciences (2020)
2. Björkegren, J.L., Lusis, A.J.: Atherosclerosis: recent developments. Cell **185**(10), 1630–1645 (2022)
3. Blagojevic, A., Sustersic, T., Filipovic, N.: Simulation of atherosclerotic plaque development using agent-based modelling (2022)
4. Chambers, K.L., Myerscough, M.R., Watson, M.G., Byrne, H.M.: Blood lipoproteins shape the phenotype and lipid content of early atherosclerotic lesion macrophages: a dual-structured mathematical model. Bulletin of Mathematical Biology **86**(9), 112 (2024)
5. Corti, A., Chiastra, C., Colombo, M., Garbey, M., Migliavacca, F., Casarin, S.: A fully coupled computational fluid dynamics–agent-based model of atherosclerotic plaque development: multiscale modeling framework and parameter sensitivity analysis. Computers in biology and medicine **118**, 103,623 (2020)
6. Corti, A., Colombo, M., Migliavacca, F., Rodriguez Matas, J.F., Casarin, S., Chiastra, C.: Multiscale computational modeling of vascular adaptation: a systems biology approach using agent-based models. Frontiers in Bioengineering and Biotechnology **9**, 744,560 (2021)
7. Corti, A., Colombo, M., Rozowsky, J.M., Casarin, S., He, Y., Carbonaro, D., Migliavacca, F., Rodriguez Matas, J.F., Berceli, S.A., Chiastra, C.: A predictive multiscale model of in-stent restenosis in femoral arteries: linking haemodynamics and gene expression with an agent-based model of cellular dynamics. Journal of the Royal Society Interface **19**(188), 20210,871 (2022)
8. Davies, L.C., Rosas, M., Jenkins, S.J., Liao, C.T., Scurr, M.J., Brombacher, F., Fraser, D.J., Allen, J.E., Jones, S.A., Taylor, P.R.: Distinct bone marrow-derived and tissue-resident macrophage lineages proliferate at key stages during inflammation. Nature communications **4**(1), 1886 (2013)
9. Fan, J., Watanabe, T.: Atherosclerosis: Known and unknown. Pathology international **72**(3), 151–160 (2022)
10. Friedman, A., Hao, W.: A mathematical model of atherosclerosis with reverse cholesterol transport and associated risk factors. Bulletin of mathematical biology **77**, 758–781 (2015)
11. Getz, M., Wang, Y., An, G., Asthana, M., Becker, A., Cockrell, C., Collier, N., Craig, M., Davis, C.L., Faeder, J.R., et al.: Iterative community-driven development of a sars-cov-2 tissue simulator. BioRxiv pp. 2020–04 (2020)
12. Ghaffarizadeh, A., Friedman, S.H., Macklin, P.: Biofvm: an efficient, parallelized diffusive transport solver for 3-d biological simulations. Bioinformatics **32**(8), 1256–1258 (2016)

13. Ghaffarizadeh, A., Heiland, R., Friedman, S.H., Mumenthaler, S.M., Macklin, P.: Physicell: An open source physics-based cell simulator for 3-d multicellular systems. PLoS computational biology **14**(2), e1005,991 (2018)
14. Hao, W., Friedman, A.: The ldl-hdl profile determines the risk of atherosclerosis: a mathematical model. PloS one **9**(3), e90,497 (2014)
15. Jenner, A.L., Smalley, M., Goldman, D., Goins, W.F., Cobbs, C.S., Puchalski, R.B., Chiocca, E.A., Lawler, S., Macklin, P., Goldman, A., et al.: Agent-based computational modeling of glioblastoma predicts that stromal density is central to oncolytic virus efficacy. Iscience **25**(6) (2022)
16. Metzcar, J., Wang, Y., Heiland, R., Macklin, P.: A review of cell-based computational modeling in cancer biology. JCO clinical cancer informatics **2**, 1–13 (2019)
17. Nakashima, Y., Fujii, H., Sumiyoshi, S., Wight, T.N., Sueishi, K.: Early human atherosclerosis: accumulation of lipid and proteoglycans in intimal thickenings followed by macrophage infiltration. Arteriosclerosis, thrombosis, and vascular biology **27**(5), 1159–1165 (2007)
18. Ougrinovskaia, A., Thompson, R.S., Myerscough, M.R.: An ode model of early stages of atherosclerosis: mechanisms of the inflammatory response. Bulletin of mathematical biology **72**, 1534–1561 (2010)
19. Pataki, M., Lusztig, G., Robenek, H.: Endocytosis of oxidized ldl and reversibility of migration inhibition in macrophage-derived foam cells in vitro. a mechanism for atherosclerosis regression? Arteriosclerosis and Thrombosis: A Journal of Vascular Biology **12**(8), 936–944 (1992). DOI 10.1161/01.atv.12.8.936
20. Patel, A.A., Zhang, Y., Fullerton, J.N., Boelen, L., Rongvaux, A., Maini, A.A., Bigley, V., Flavell, R.A., Gilroy, D.W., Asquith, B., et al.: The fate and lifespan of human monocyte subsets in steady state and systemic inflammation. Journal of Experimental Medicine **214**(7), 1913–1923 (2017)
21. Poznyak, A.V., Nikiforov, N.G., Starodubova, A.V., Popkova, T.V., Orekhov, A.N.: Macrophages and foam cells: Brief overview of their role, linkage, and targeting potential in atherosclerosis. Biomedicines **9**(9), 1221 (2021)
22. Rahman, K., Vengrenyuk, Y., Ramsey, S.A., Vila, N.R., Girgis, N.M., Liu, J., Gusarova, V., Gromada, J., Weinstock, A., Moore, K.J., et al.: Inflammatory ly6c hi monocytes and their conversion to m2 macrophages drive atherosclerosis regression. The Journal of clinical investigation **127**(8), 2904–2915 (2017)
23. Tacke, F., Alvarez, D., Kaplan, T.J., Jakubzick, C., Spanbroek, R., Llodra, J., Garin, A., Liu, J., Mack, M., Van Rooijen, N., et al.: Monocyte subsets differentially employ ccr2, ccr5, and cx3cr1 to accumulate within atherosclerotic plaques. The Journal of clinical investigation **117**(1), 185–194 (2007)
24. Thon, M.P., Ford, H.Z., Gee, M.W., Myerscough, M.R.: A quantitative model of early atherosclerotic plaques parameterized using in vitro experiments. Bulletin of mathematical biology **80**, 175–214 (2018)
25. Thon, M.P., Myerscough, M.R., Gee, M.W.: A spatially resolved and quantitative model of early atherosclerosis. Bulletin of mathematical biology **81**(10), 4022–4068 (2019)
26. Tomasevic, S., Filipović, N.: Use case: Agent-based modeling for atherosclerotic plaque progression in carotid arteries. In: In Silico Clinical Trials for Cardiovascular Disease: A Finite Element and Machine Learning Approach, pp. 245–270. Springer (2024)
27. Tsompou, P.I., Potsika, V.T., Petrovic, N., Pezoulas, V.C., Siogkas, P.K., Tsakanikas, V.D., Pleouras, D.S., Papafaklis, M., Nikopoulos, S., Sakellarios, A.I., et al.: Computational modeling of atherosclerotic plaque progression through an efficient 3d agent-based modeling approach. In: 2022 IEEE-EMBS International Conference on Biomedical and Health Informatics (BHI), pp. 01–04. IEEE (2022)
28. West, J., Robertson-Tessi, M., Anderson, A.R.: Agent-based methods facilitate integrative science in cancer. Trends in cell biology **33**(4), 300–311 (2023)
29. Williams, J.W., Zaitsev, K., Kim, K.W., Ivanov, S., Saunders, B.T., Schrank, P.R., Kim, K., Elvington, A., Kim, S.H., Tucker, C.G., et al.: Limited proliferation capacity of aortic intima resident macrophages requires monocyte recruitment for atherosclerotic plaque progression. Nature immunology **21**(10), 1194–1204 (2020)

30. Yao, P.M., Tabas, I.: Free cholesterol loading of macrophages is associated with widespread mitochondrial dysfunction and activation of the mitochondrial apoptosis pathway. Journal of Biological Chemistry **276**(45), 42,468–42,476 (2001). DOI 10.1074/jbc.M101419200
31. Yona, S., Kim, K.W., Wolf, Y., Mildner, A., Varol, D., Breker, M., Strauss-Ayali, D., Viukov, S., Guilliams, M., Misharin, A., et al.: Fate mapping reveals origins and dynamics of monocytes and tissue macrophages under homeostasis. Immunity **38**(1), 79–91 (2013). DOI 10.1016/j.immuni.2012.12.001

Chapter 4

MATRIX Program: The Geometry of Moduli Spaces in String Theory

Quantum aspects of heterotic G_2 systems

Xenia de la Ossa, Magdalena Larfors, Matthew Magill and Eirik E. Svanes

Abstract Compactifications of the heterotic string, to first order in the α' expansion, on manifolds with integrable G_2 structure give rise to three-dimensional $\mathcal{N} = 1$ supergravity theories that admit Minkowski and AdS ground states. As shown in [49], such vacua correspond to critical loci of a real superpotential W. We perform a perturbative study around a supersymmetric vacuum of the theory, which confirms that the first order variation of the superpotential, δW, reproduces the BPS conditions for the system, and furthermore shows that $\delta^2 W = 0$ gives the equations for infinitesimal moduli. This allows us to identify a nilpotent differential, and a symplectic pairing, which we use to construct a bicomplex, or a double complex, for the heterotic G_2 system. Using this complex, we determine infinitesimal moduli and their obstructions in terms of related cohomology groups. Finally, by interpreting $\delta^2 W$ as an action, we compute the one-loop partition function of the heterotic G_2 system and show it can be decomposed into a product of one-loop partition functions of Abelian and non-Abelian instanton gauge theories.

Xenia de la Ossa
Mathematical Institute, University of Oxford
e-mail: delaossa@maths.ox.ac.uk

Magdalena Larfors
Department of Physics and Astronomy, Uppsala University
Department of Mathematics, Uppsala University
e-mail: magdalena.larfors@physics.uu.se

Matthew Magill
Department of Mathematics, Uppsala University
e-mail: matthew.magill@math.uu.se

Eirik E. Svanes
Department of Mathematics and Physics, University of Stavanger
e-mail: eirik.svanes@uis.no

D. R. Wood et al. (eds.), *2024 MATRIX Annals, Part II*, MATRIX Book Series 8,
https://doi.org/10.1007/978-3-032-16206-9_16

1 Introduction

There has been a lot of recent interest in both the physics and mathematics literature in studying coupled systems of gauge theory and gravity (geometry). The heterotic string provides a fertile ground for this endeavour. Heterotic strings come with a gauge bundle, whose connection should, for supersymmetric solutions, satisfy an instanton condition that depends on the geometry. Moreover, the coupling of the gauge sector with the geometry through the heterotic Bianchi identity forces the geometry to be generically torsional - a further layer of complication which has made understanding mathematical properties of these solutions, such as the moduli problem, more challenging.

In this paper we study the moduli of heterotic string compactifications on manifolds endowed with a G_2 structure, leading to a system of equations called the heterotic G_2 system [53]. This system corresponds to minimally supersymmetric, three-dimensional solutions of the heterotic string theory to first order in the α' expansion, as explored in e.g. [34, 22, 28, 35, 45, 31, 32, 41, 33, 50, 30]. An interesting aspect of these systems is that they admit an $\mathcal{N} = 1$ three dimensional effective field theory on both anti de Sitter (AdS_3) and Minkowski ($\mathbb{M}_3$) spacetimes. The curvature radius of the AdS_3 space is related to a (scalar) torsion class which vanishes only in the case of $\mathbb{M}_3$ [49]. In contrast, $\mathcal{N} = 1$ compactifications of the heterotic strings to four dimensions, which are known as Hull-Strominger systems [36, 60], do not admit AdS_4 solutions.

In the last decade, progress has been made towards a better understanding of heterotic systems. For Hull–Strominger systems, the infinitesimal moduli were parameterised as the first cohomology of a double extension sequence [44, 1, 54]. The moduli problem has been shown to be elliptic, and connected to moduli spaces of solutions of suitable Killing spinor equations on a (holomorphic) Courant algebroid [25]. Finite deformations have been explored [3], alongside the moduli space geometry, moment maps, stability, flow equations, and further connections with generalised geometry, see e.g. [11, 26, 5, 27, 43, 24, 39] and references therein.

For heterotic G_2 systems, the infinitesimal moduli were shown to be parametrised by the first cohomology of an operator $\check{\mathcal{D}}$, which operates on a bundle $\mathcal{Q}$ which couples geometric and gauge degrees of freedom [51, 53, 52]. For $\check{\mathcal{D}}$ to be nilpotent, and thus define a cohomology, it is important that its curvature satisfies the G_2 instanton condition. When this is the case, the related $\check{\mathcal{D}}$ complex is elliptic [51, 53]. An interesting consequence of this is that the $\mathcal{Q}$ bundle of the heterotic G_2 system satisfies the same instanton condition as does the heterotic gauge bundle, and, by dimensional oxidation, a similar result can be shown for the Hull–Strominger system [52]. This is an example of what has come to be known as "instantons for instantons" and has been explored in depth in [59]. Further investigations of heterotic G_2 geometry and its moduli, and their connections to generalised geometry has since been explored, both from a mathematical and physical standpoint, see [13, 20, 14, 42]. Explicit local and global solutions to the system have also been constructed, e.g. [17, 7, 38, 40, 46, 48, 19, 23]. In spite of this progress, the mathematical structures governing these G_2 geometries are more mysterious than their six-dimensional $SU(3)$ counterparts, and tools for understanding them are sorely needed.

As history has shown, physics can provide useful insights into the understanding geometric structures. Examples of this include mirror symmetry [12], and the use of topological field theory to study geometric invariants, such as Chern–Simons theory and the Jones polynomial of knots [62], or holomorphic Chern–Simons theory and Donaldson–Thomas invariants [61, 16]. In the case of heterotic solutions, physics provides us with a useful functional, the superpotential, whose critical points are exactly the solutions of the heterotic BPS equations [49, 47]. In the six-dimensional case, the superpotential was essential in the derivation of the Maurer–Cartan equation of the L_∞ algebra governing finite deformations [3]. It is natural to ask if similar progress can be made in the G_2 case. However, the six-dimensional $SU(3)$ case has an advantage over the G_2 solutions: it preserves more supersymmetry, and as a result, the superpotential is required to be a holomorphic functional of the fields. This provides a useful parametrisation of the moduli which can be exploited in understanding higher order deformations. For the G_2 case, an analogous parametrisation is not yet known. Yet, we may explore the deformations to quadratic order, and we will do this in the present paper.

Following this perturbative strategy, we will, in this paper, study classical and quantum aspects of the three-dimensional $\mathcal{N} = 1$ supergravity with superpotential W. We will work perturbatively around a supersymmetric vacuum of the theory. To first order, the variation of the superpotential reproduces the BPS conditions for the system, and, as we will see, the second order variations give the equations for moduli. Moreover, the analysis allows us to construct a bicomplex, or a double complex, which also encodes the gauge symmetry of the theory. Using this double complex, we can determine obstructions to infinitesimal deformations, and make remarks about finite deformations.

Finally, by interpreting the second order variation of the superpotential as an action, we can explore quantum aspects of the system. This direct computation of the quadratic action has two advantages, compared to the recent generalised geometry analysis of Refs. [39]. Firstly, it becomes clear that the heterotic G_2 moduli problem is governed by a double complex, as was suggested in [39]. Secondly, we will see that upon a field redefinition, the total complex can be diagonalised into three separate complexes. This makes it easier to compute and express the one-loop partition function in terms of determinants of Laplacians on more familiar bundles, as has been done in [4] in the case of type II compactifications on $SU(3)$ structure manifolds. We will compute the absolute value of the one-loop partition function here, and leave the study of its phase and η-invariant, and the extent to which it defines topological invariants, to future work.

The paper is organised as follows. In section 2 we introduce the heterotic G_2 superpotential, compute its first and second order variations and study its relevance for the moduli problem. We then explore the deformation theory to quadratic order, and define a double complex, or bicomplex, that encodes the action and symmetries of this theory, in section 3. With this in hand, we discuss infinitesimal moduli and their obstructions. In section 4 we compute the one-loop partition function of the quadratic theory. Section 5 contains our conclusions and a discussion of open problems. Two proofs of propositions are provided in Appendix 6 and Appendix 7.

2 Variations of the superpotential and equations for moduli

Our aim in this paper is to study the three-dimensional $\mathcal{N}=1$ supergravity with superpotential W, which is associated to the heterotic G_2 system. We will work perturbatively around a supersymmetric vacuum of the theory. To first order, the variation of the superpotential reproduces the BPS conditions for the system, and the second order variations give the equations for moduli, which we compute in the latter half of this section. To start, we recapitulate how the superpotential is defined.

2.1 The superpotential and the heterotic G_2 system

Let Y be a seven dimensional manifold with a G_2 structure determined by a well defined nowhere vanishing three form, φ. In [49], the authors obtained the necessary and sufficient constraints for $\mathcal{N}=1$ preserving heterotic string compactifications on Y as the critical points of an effective three dimensional superpotential W. This superpotential was obtained directly from dimensional reduction of the ten dimensional heterotic superstring action, keeping terms up to first order in α'. In this paper we use this superpotential to obtain an action whose equations of motion are equivalent to those describing the infinitesimal deformations of the G_2 heterotic system. In this subsection we summarise briefly the results of [49] as they are the starting point of the research presented in this paper.

Consider, then, the superpotential [49]

$$W=\frac{1}{2}\int_Y e^{-2\phi}\left((H+h\varphi)\wedge\psi+\frac{1}{2}\mathrm{d}\varphi\wedge\varphi\right), \tag{1}$$

where H is the flux on the internal 7-manifold Y, h is a constant which determines the AdS_3 curvature scale (which is also the three dimensional flux), ϕ is the dilaton, and $\psi=*_\varphi\varphi$. In this work we consider the space $\mathcal{M}$ of all quadruples

$$[(Y,\varphi),(V,A),(TY,\Theta),H], \tag{2}$$

where (V,A) is a vector bundle over Y with connection A, (TY,Θ) is the tangent bundle of Y with connection Θ and H is the three form flux on Y given by

$$H=\mathrm{d}B+\frac{\alpha'}{4}(CS(A)-CS(\Theta)), \tag{3}$$

where $CS(A)$ and $CS(\Theta)$ represent the Chern-Simons forms for the connections A and Θ respectively.

The G_2 structure on Y determined by φ has intrinsic torsion classes τ_i determined uniquely by φ. Indeed, they are given by the unique decomposition of $\mathrm{d}\varphi$ and $\mathrm{d}\psi$ into G_2 irreducible representations

$$\mathrm{d}\varphi = -i_T(\varphi) = \tau_0\,\psi + 3\tau_1 \wedge \varphi + *\tau_3\,, \tag{4}$$

$$\mathrm{d}\psi = -i_T(\psi) = 4\,\tau_1 \wedge \psi + *\tau_2\,, \tag{5}$$

where one can write the intrinsic torsion T as a two form valued in TY as

$$T^a = \widehat{T}^a + \frac{1}{6}\,\tau_2^{ab}\,\varphi_b\,, \tag{6}$$

and $\widehat{T}$ represents the totally antisymmetric part of the torsion

$$\widehat{T} = -\frac{1}{6}\,\tau_0\,\varphi + \tau_1 \lrcorner \psi + \tau_3\,. \tag{7}$$

A compactification of the heterotic string preserves at least one supersymmetry if the fields in the quadruple (2) satisfy the conditions derived from

$$W = 0\,, \qquad \delta W = 0\,.$$

As it was shown in [49], these conditions give precisely the heterotic G_2 system of differential equations for the quadruple (2).

To compute δW we describe the infinitesimal deformations of the G_2 structure in terms of a vector valued one form M as follows:

$$\delta\varphi = i_M(\varphi)\,, \quad \delta\psi = i_M(\psi)\,, \quad M = \frac{1}{7}\,\mathrm{tr}M\,\mathbf{1} + m + \mathring{\Delta} \in \Omega^1(Y, TY)\,, \tag{8}$$

where $m \in \Omega^2_7(X)$ and $\mathring{\Delta}$ is traceless symmetric. The variations of the three form flux H are

$$\delta H = \mathrm{d}\mathscr{B} + \frac{\alpha'}{2}\Big(\mathrm{tr}(\alpha \wedge F) - \mathrm{tr}(\kappa \wedge R)\Big)\,, \tag{9}$$

where $\alpha \in \Omega^1(Y, \mathrm{End}(V))$ and $\kappa \in \Omega^1(Y, \mathrm{End}(TY))$ and $\mathrm{d}\mathscr{B}$ corresponds to the (α' corrected) variations of $\mathrm{d}B$. The α'-corrected variations $\mathscr{B}$ of the B-field are such that $\mathrm{d}\mathscr{B}$ is a well defined three form on Y.

The first order variation of the superpotential is given by

$$\delta W = \frac{1}{2}\int_Y e^{-2\phi}\Big(-2\,\delta\phi\,\mathfrak{w} + \delta\mathfrak{w}\Big)\,,$$

where

$$\mathfrak{w} = (H + h\,\varphi) \wedge \psi + \frac{1}{2}\,\mathrm{d}\varphi \wedge \varphi\,.$$

In [49] it was found that

$$\begin{aligned} \delta\mathfrak{w} = {} & i_{\mathring{M}}(\psi)\wedge(H-\widehat{T}) - (\mathscr{B}-2m)\wedge\big(\mathrm{d}\psi - 2\,\mathrm{d}\phi\wedge\psi\big) \\ & + \frac{\alpha'}{2}\Big(\mathrm{tr}(\alpha\wedge F) - \mathrm{tr}(\kappa\wedge R)\Big)\wedge\psi + \left(\delta h + \frac{3}{7}\,\mathrm{tr}M\left(h+\frac{1}{3}\,\tau_0\right)\right)\varphi\wedge\psi\,. \end{aligned} \tag{10}$$

where $\mathring{M}$ is the traceless part of M. Thus, with reference to equation (8), $\mathring{M} = \mathring{\Delta} + m$.

Noting that each term in the expression (10) must vanish separately, we obtain the equations of motion, that is, the heterotic G_2 system:

$$\begin{aligned} & \mathrm{d}(e^{-2\phi}\,\psi) = 0 \qquad \Longleftrightarrow \qquad \tau_2 = 0\,, \qquad 2\,\tau_1 = \mathrm{d}\phi\,, \\ & H = \widehat{T}\,, \\ & F\wedge\psi = 0\,, \qquad R(\Theta)\wedge\psi = 0\,, \\ & h = -\frac{1}{3}\,\tau_0\,, \qquad \delta h = 0\,. \end{aligned} \tag{11}$$

The vanishing of τ_2 is the necessary and sufficient condition for there to exist a totally antisymmetric torsion (which is then unique) [21]. The second equation identifies the torsion with the three form flux H. The third line means that the connections A and Θ are instantons. Finally, the three dimensional flux h is identified with the torsion class, τ_0, and moreover their values are fixed.

2.2 Second order variations of W

We have that

$$\delta_2\delta_1 W = \frac{1}{2}\int_Y e^{-2\phi}\big(2\,(2\,\delta_1\phi\;\delta_2\phi - \delta_2\delta_1\phi)\,\mathfrak{w} - 2\,(\delta_2\phi\;\delta_1\mathfrak{w} + \delta_1\phi\;\delta_2\mathfrak{w}) + \delta_2\delta_1\mathfrak{w}\big)\,.$$

We will show in this section that

$$\begin{aligned} \frac{1}{2}\,(\delta_2\delta_1 W + \delta_1\delta_2 W)\big|_0 &= \frac{1}{4}\int_Y e^{-2\phi}\,(\delta_2\delta_1\mathfrak{w} + \delta_1\delta_2\mathfrak{w})\,|_0 \\ &= \langle \mathscr{Z}_1, \mathbb{D}\,\mathscr{Z}_2\rangle_{\mathscr{E}}\,, \end{aligned} \tag{12}$$

where the symbol $|_0$ means that after varying we enforce the equations of motion, and

$$\mathscr{Z}\in\mathscr{E}^0 = \Omega^0(Y)\oplus\Omega^1(Y,\mathscr{Q})\oplus\Omega^2_7(Y)\,.$$

As we will show in section 3, this defines a shifted symplectic pairing $\langle\cdot,\cdot\rangle_{\mathscr{E}}$, and a nilpotent differential operator $\mathbb{D}$, which is self-adjoint with respect to $\langle\cdot,\cdot\rangle_{\mathscr{E}}$. The differential operator is defined such that the equations for infinitesimal moduli of the heterotic system correspond to

$$\check{\mathbb{D}}\mathscr{Z} = 0 ,$$

where $\check{\mathbb{D}}$ denotes projection to G_2 irreducible representations that will be defined below. These definitions will allow us, in the ensuing sections, to define a bicomplex that captures the gauge symmetries of the 3-dimensional theory, and quantize the theory. This furthermore requires the present definition of $\mathscr{Z}$, which is an extension of the infinitesimal deformations defined in [53]. In that reference, the infinitesimal moduli of the heterotic G_2 system were identified with $H^1(Y,\mathscr{Q})$. As will become evident in section 3, this does not suffice to provide a complete description of the gauge degrees of freedom, which we need in order to quantize the theory.

For the second order variation of W, we start from equation (10) and compute $\delta_2\delta_1 \mathfrak{w}\,|_0$. For ease of exposition, we relegate most of this analysis to Appendix 6. We also refer the reader to section 2.2.1 of [53] for definitions of the insertion operator i_α, and further explanation of our notation. The result of the computation is summarised in the following proposition

Proposition 1. *Let*

$$\beta = \frac{1}{2}(\pi_7(\mathscr{B}) - 2m) , \tag{13}$$

$$y = M + \frac{1}{2}\pi_{\mathbf{14}}(\mathscr{B}) , \tag{14}$$

and

$$\hat{\delta}\phi = \delta\phi - \frac{2}{7}\mathrm{tr}M . \tag{15}$$

Then

$$\begin{aligned}\frac{1}{2}\,\delta_2\delta_1\mathfrak{w}\,|_0 = \Big\{ \; & i_{y_1}\Big[\mathrm{d}_\zeta y_2 - \frac{\alpha'}{4}\Big(\mathrm{tr}(\alpha_2\wedge F) - \mathrm{tr}(\kappa_2\wedge R)\Big)\Big] - i_{\mathring{M}_1}(\mathrm{d}\beta_2) \\ & + \frac{\alpha'}{4}\Big[\mathrm{tr}\big(\alpha_1\wedge(\mathrm{d}_A\alpha_2 - i_{M_2}(F)\big) - \mathrm{tr}\big(\kappa_1\wedge(\mathrm{d}_\Theta\kappa_2 - i_{M_2}(R)\big)\big)\Big] \\ & - \beta_1\wedge\Big(\frac{1}{3}\psi\lrcorner\Big(i_{\mathrm{d}_\zeta\mathring{M}_2}(\psi)\Big) + i_{\mathring{M}_2}(4\,\tau_1) - 2\,\hat{\delta}_2(\mathrm{d}\phi)\Big)\Big\}\wedge\psi ,\end{aligned} \tag{16}$$

where ζ is the Hull connection given by

$$\mathrm{d}_\zeta V^a = \mathrm{d}V^a + \Gamma^a{}_{bc}\mathrm{d}x^b\wedge V^c , \quad \forall V\in\Omega^0(Y,TY) ,$$

and $\Gamma^a{}_{bc}$ are related to the symbols of a G_2 compatible connection by

$$\nabla_b V^a = \partial_b V^a + \Gamma^a{}_{cb}V^c , \quad \forall V\in\Omega^0(Y,TY) .$$

Proof. See Appendix 6.

We have only defined the Hull connection acting on vector valued 0-forms. Since this is a metric connection, it is straight forward to generalize the definition to other tensor valued p-forms [51, 53].

Looking back at equation (12), we now need to compute the *symmetric* second order variation of the superpotential. The result of symmetrising the second order variation in proposition 1 is summarised in the following proposition.

Proposition 2.

$$
\begin{aligned}
&\frac{1}{2}(\delta_2\delta_1 W + \delta_1\delta_2 W)\,|_0 \\
&= \int_Y \Big\{ \; i_{y_1}\Big[\mathrm{d}_\zeta y_2 - \frac{\alpha'}{4}\Big(\mathrm{tr}(\alpha_2\wedge F) - \mathrm{tr}(\kappa_2\wedge R)\Big)\Big] - i_{y_1}(\hat{\mathrm{d}}\beta_2) \\
&\quad + \frac{\alpha'}{4}\Big[\mathrm{tr}\big(\alpha_1\wedge(\mathrm{d}_A\alpha_2 - i_{M_2}(F))\big) - \mathrm{tr}\big(\kappa_1\wedge(\mathrm{d}_\Theta\kappa_2 - i_{M_2}(R))\big)\Big] \\
&\quad - \beta_1\wedge\Big[\frac{1}{3}\psi\lrcorner\Big(i_{\mathrm{d}_\zeta \mathring{M}_2}(\psi)\Big) + i_{\mathring{M}_2}(4\,\tau_1) - \mathrm{d}(\hat{\delta}_2\phi)\Big] - \hat{\delta}_1\phi\,\mathrm{d}\beta_2\Big\}\wedge\Psi\,,
\end{aligned}
\tag{17}
$$

where we recall that $\Psi = e^{-2\phi}\,\psi$ *is a closed 4-form, and*

$$
\hat{\mathrm{d}}\beta = \mathrm{d}\beta - \pi_{\mathbf{1}}(\mathrm{d}\beta) = \mathrm{d}\beta - \check{\mathrm{d}}\beta\,.
$$

Proof. Most terms in (16) are symmetric except those which contain β. Indeed, up to total derivatives and supersymmetry conditions:

$$
\begin{aligned}
i_{y_2}(\mathrm{d}_\zeta y_1)\wedge\Psi &= i_{y_1}(\mathrm{d}_\zeta y_2)\wedge\Psi\,,\\
\mathrm{tr}(\alpha_2\wedge\mathrm{d}_A\alpha_1)\wedge\Psi &= \mathrm{tr}(\alpha_1\wedge\mathrm{d}_A\alpha_2)\wedge\Psi\,,\\
-\mathrm{tr}(\alpha_2\wedge i_{y_1}(F))\wedge\Psi &= -i_{y_1}(\mathrm{tr}(\alpha_2\wedge F))\wedge\Psi\,,\\
i_{\mathring{M}_2}\big(\mathrm{d}\beta_1\big)\wedge\Psi &= \beta_1\wedge\Big(\frac{1}{3}\psi\lrcorner\Big(i_{\mathrm{d}_\zeta\mathring{M}_2}(\psi)\Big) + i_{\mathring{M}_2}(4\,\tau_1)\Big)\wedge\Psi\,,\\
\beta_2\wedge\mathrm{d}(\hat{\delta}\phi_1)\wedge\Psi &= -\,\hat{\delta}\phi_1\,\mathrm{d}\beta_2\wedge\Psi\,.
\end{aligned}
$$

Here we only remark on the fourth equation, as it is straightforward to prove the others. To prove this equation have

$$
i_{\mathring{M}_2}(\mathrm{d}\beta_1)\wedge\Psi = i_{\mathring{M}_2}(\mathrm{d}\beta_1\wedge\Psi) - \mathrm{d}\beta_1\wedge i_{\mathring{M}_2}(\Psi)\,.
$$

The first term vanishes as $\mathring{M}$ is traceless. Then

$$
\begin{aligned}
i_{\mathring{M}_2}(\mathrm{d}\beta_1)\wedge\Psi &= -\mathrm{d}\big(\beta_1\wedge i_{\mathring{M}_2}(\Psi)\big) + \beta_1\wedge\mathrm{d}i_{\mathring{M}_2}(\Psi)\\
&= \beta_1\wedge\big(i_{\mathrm{d}_\zeta\mathring{M}_2}(\Psi) + i_{\mathring{M}_2}(4\tau_1)\wedge\Psi\big)\,,
\end{aligned}
$$

from which the result follows.

Finally, to see how the the second term on the right hand side of equation (17) comes about we have

$$i_{\mathring{M}_1}(\mathrm{d}\beta_2)\wedge\Psi = \left(i_{M_1}(\hat{\mathrm{d}}\beta_2)+\frac{1}{2}i_{\pi_{\mathbf{14}}(\mathscr{B}_1)}(\hat{\mathrm{d}}\beta_2)\right)\wedge\Psi = i_{y_1}(\hat{\mathrm{d}}\beta_2)\wedge\Psi\,,$$

where

$$\hat{\mathrm{d}}\beta = \mathrm{d}\beta - \pi_{\mathbf{1}}(\mathrm{d}\beta) = \mathrm{d}\beta - \check{\mathrm{d}}\beta\,.$$

and we have used

$$\hat{\mathrm{d}}\beta\wedge\psi = 0\,.$$

□

We are now ready to read off from equation (17) the set of differential equations that the infinitesimal deformations must satisfy:

$$\mathrm{d}(\hat{\delta}\phi)-\frac{1}{3}\psi\lrcorner\left(i_{\mathrm{d}_\zeta\mathring{M}}(\psi)\right)-i_{\mathring{M}}(4\tau_1)=0\,. \tag{18}$$

$$\pi_{\mathbf{7}}\left[\mathrm{d}_\zeta y_a+\frac{\alpha'}{4}\Big(\mathrm{tr}(\alpha\wedge F_a)-\mathrm{tr}(\kappa\wedge R_a)\Big)-(\hat{\mathrm{d}}\beta)_a\right]=0\,, \tag{19}$$

$$\pi_{\mathbf{7}}\big(\mathrm{d}_A\alpha - i_M(F)\big)=0\,, \tag{20}$$

$$\pi_{\mathbf{7}}\big(\mathrm{d}_\Theta\kappa - i_M(R)\big)=0\,, \tag{21}$$

$$\pi_{\mathbf{1}}(\mathrm{d}\beta)=0\,. \tag{22}$$

These equations include and extend the equations for infinitesimal moduli presented by some of the present authors in [53]; equation (18) is equation (3.10) in [53], and the equations (19)-(21) reduce, upon setting β to zero, to equations (5.10)-(5.13) in that reference. To comment on these differences, we note that while the constraint $\beta=0$ was an allowed and simplifying assumption in the study of infinitesimal moduli of the heterotic G_2 system done in Ref. [53], it is by no means necessary. As we will see below, for the analysis in this paper, β and $\delta\hat{\phi}$ will play a prominent role.

2.3 The pairing $\langle\cdot,\cdot\rangle_{\mathscr{E}}$ and the operator $\mathbb{D}$

In this subsection, we will repackage the infinitesimal deformations in a manner that allows us to write equation (17) in the form of equation (12), i.e. to identify the operator $\mathbb{D}$ and the pairing on $\mathscr{E}$.

Let

$$\mathscr{X} = \begin{pmatrix} \hat{\delta}\phi \\ Z \\ \beta \end{pmatrix} \in \mathscr{E}^0 , \qquad Z = \begin{pmatrix} y \\ \alpha \\ \kappa \end{pmatrix} \in \Omega^1(Y, \mathscr{Q}) , \tag{23}$$

where $\beta \in \Omega^2_7(Y)$, $\hat{\delta}\phi \in \Omega^0(Y)$, and $\mathscr{Q}$ is the bundle defined by

$$\mathscr{Q} = TY \oplus \mathrm{End}(V) \oplus \mathrm{End}(TY) . \tag{24}$$

Let $\mathscr{D}$ be the differential operator acting on $\Omega^p(Y, \mathscr{Q})$ defined by [53]

$$\mathscr{D} = \begin{pmatrix} \mathrm{d}_\zeta & \mathscr{F}^\dagger & \mathscr{R}^\dagger \\ \mathscr{F} & \mathrm{d}_A & 0 \\ \mathscr{R} & 0 & \mathrm{d}_\Theta \end{pmatrix} , \tag{25}$$

where

$$\begin{aligned} \mathscr{F} \ : \quad \Omega^p(Y, TY) &\longrightarrow \Omega^{p+1}(Y, \mathrm{End}(V)) \\ y^a &\longmapsto \mathscr{F}(M) = (-1)^p \, i_y(F) , \end{aligned} \tag{26}$$

$$\begin{aligned} \mathscr{F}^\dagger \ : \ \Omega^p(Y, \mathrm{End}(V)) &\longrightarrow \Omega^{p+1}(Y, TY) \\ \alpha &\longmapsto (\mathscr{F}^\dagger \alpha)^a = -\mathrm{tr}(\alpha \wedge F)^a , \end{aligned} \tag{27}$$

with analogous definitions for $\mathscr{R}$ and $\mathscr{R}^\dagger$. Note that the same labelling $\mathscr{F}$ and $\mathscr{F}^\dagger$ may be used, without ambiguity, for the action of these maps on different p-forms.

We also define the following new operations on $\mathscr{E}^0$:

$$\begin{aligned} \tau \ : \quad \Omega^1(Y, TY) &\longrightarrow \Omega^1(Y) \\ y &\longmapsto \tau(\mathring{M}) = -i_{\mathring{M}}(4\tau_1) - \frac{1}{3} \psi \lrcorner i_{\mathrm{d}_\zeta \mathring{M}}(\psi) , \end{aligned} \tag{28}$$

$$\begin{aligned} \tau^\dagger \ : \ \Omega^2_7(Y) &\longrightarrow \Omega^2(Y, TY) \\ \beta &\longmapsto (\tau^\dagger \beta)_a = -(\hat{\mathrm{d}}\beta)_a . \end{aligned} \tag{29}$$

With these definitions at hand, equation (17) can be written as

$$\begin{aligned} \frac{1}{2}(\delta_2\delta_1 W + \delta_1\delta_2 W)\,|_0 = \langle Z_1, \mathscr{D} Z_2 \rangle_{\mathscr{Q}} + \int_Y \Big\{ &- \hat{\delta}_1\phi \, \mathrm{d}\beta_2 \\ &+ \Big(i_{y_1}\big(\tau^\dagger(\beta_2)\big) + \beta_1 \wedge \big[\tau(\mathring{M}_2) + \mathrm{d}\hat{\delta}_2\phi\big] \Big) \Big\} \wedge \Psi , \end{aligned} \tag{30}$$

where we recall that $\Psi = e^{-2\phi}\psi$, and $\langle \cdot, \cdot \rangle_{\mathscr{Q}}$ is the pairing on $\mathscr{Q}$ (defined as in [53])

$$\langle\cdot,\cdot\rangle_{\mathscr{D}} = \int g_{\mathscr{D}}(\cdot,\cdot)\,, \quad g_{\mathscr{D}}(\cdot,\cdot) = \begin{pmatrix} i.(\cdot) & 0 & 0 \\ 0 & \frac{\alpha'}{4}\operatorname{tr}_V(\cdot\wedge\cdot) & 0 \\ 0 & 0 & -\frac{\alpha'}{4}\operatorname{tr}_{TY}(\cdot\wedge\cdot) \end{pmatrix} \wedge\Psi\,. \tag{31}$$

It is easy to check that $\mathscr{F}^\dagger$ is indeed the adjoint of $\mathscr{F}$ with respect to $\langle\cdot,\cdot\rangle_{\mathscr{D}}$. This follows from the the symmetry of (17) discussed in the proof of proposition 2. Similar considerations apply for the operators $\mathscr{R}$ and $\mathscr{R}^\dagger$. Furthermore, these considerations imply that $\mathscr{D}$ is a self adjoint operator with respect to $\langle\cdot,\cdot\rangle_{\mathscr{D}}$, that is,

$$\langle Z_1, \mathscr{D}Z_2\rangle_{\mathscr{D}} = \langle Z_2, \mathscr{D}Z_1\rangle_{\mathscr{D}}\,.$$

As mentioned before, the proof is in fact equivalent to the process of finding the symmetric second order variation of the superpotential.

Finally, we can define a new pairing $\langle\cdot,\cdot\rangle_{\mathscr{E}}$, and rewrite equation (30) as

$$\frac{1}{2}\left(\delta_2\delta_1 W + \delta_1\delta_2 W\right)|_0 = \langle \mathscr{Z}_1, \mathbb{D}\mathscr{Z}_2\rangle_{\mathscr{E}}\,, \tag{32}$$

where $\mathscr{E}^0 = \Omega^0(Y)\oplus\Omega^1(Y,\mathscr{Q})\oplus\Omega^2_7(Y)$, and $\mathbb{D}$ is the differential operator which acts on $\mathscr{E}^0$ given by

$$\mathbb{D} = \begin{pmatrix} \mathrm{d} & \tau & 0 & 0 & 0 \\ 0 & \mathrm{d}_\zeta & \mathscr{F}^\dagger & \mathscr{R}^\dagger & \tau^\dagger \\ 0 & \mathscr{F} & \mathrm{d}_A & 0 & 0 \\ 0 & \mathscr{R} & 0 & \mathrm{d}_\Theta & 0 \\ 0 & 0 & 0 & 0 & -\mathrm{d} \end{pmatrix} = \begin{pmatrix} \mathrm{d} & \tau & 0 \\ \mathbf{0} & \mathscr{D} & \tau^\dagger \\ 0 & \mathbf{0} & -\mathrm{d} \end{pmatrix}\,. \tag{33}$$

From the computations in section 2.2, in particular from the symmetrisation of the second order variation of the superpotential, it is clear that $\mathbb{D}$ is self adjoint with respect to $\langle\cdot,\cdot\rangle_{\mathscr{E}}$ and that $\tau^\dagger$ is precisely the adjoint of τ with respect to this pairing.

Clearly, the equations for moduli, (18)-(22), can be recast in terms of $\mathbb{D}\mathscr{Z}$, provided that the required projections to G_2 irreducible representations are enforced. Defining $\check{\mathbb{D}}$ to be the projected differential operator, (18)-(22) are equivalent to

$$\check{\mathbb{D}}\mathscr{Z} = \begin{pmatrix} \mathrm{d} & \check{\tau} & 0 \\ \mathbf{0} & \check{\mathscr{D}} & \check{\tau}^\dagger \\ 0 & \mathbf{0} & -\check{\mathrm{d}} \end{pmatrix} \mathscr{Z} = 0\,,$$

where the check denotes projection to the **1** and **7** G_2 representations, as appropriate, and

$$\check{\tau}(y) = \tau(\mathring{M})\,, \qquad \check{\tau}^\dagger(\beta) = -\pi_7(\hat{\mathrm{d}}\beta)_a\,.$$

In chapter 3.1, we will give a definition of $\check{\mathbb{D}}$ acting on other tensor-valued p-forms.

2.4 Trivial deformations and a first look at infinitesimal moduli

For completeness, in this subsection we comment briefly on the trivial deformations of the system. A full discussion is deferred to the next chapter.

The infinitesimal variations $\mathscr{X} \in \mathscr{E}^0$ are not yet the infintesimal moduli of the heterotic G_2 system. We need to quotient out by the trivial deformations of the system, corresponding to diffeomorphisms and gauge symmetries. As we will see later, these trivial deformations can be shown to be

$$\begin{aligned} 2(\hat{\delta}\phi)_{triv} &= 2\left(\mathscr{L}_V\varphi - \frac{2}{7}\mathrm{tr}M_{triv}\right) \\ y^a_{triv} &= \mathrm{d}_\zeta V^a + \mathscr{F}^\dagger(\varepsilon)^a + (\hat{\mathrm{d}}\mu)^a\,, \\ \alpha_{triv} &= \mathrm{d}_A\varepsilon + \mathscr{F}(V) \\ \kappa_{triv} &= \mathrm{d}_\Theta\lambda + \mathscr{R}(V) \\ \beta_{triv} &= \mathrm{d}\mu \end{aligned} \tag{34}$$

where $V \in \Omega^0(Y, TY)$, $\mu \in \Omega^1(Y)$, $\varepsilon \in \Omega^0(Y, \mathrm{End}(V))$, $\lambda \in \Omega^0(Y, \mathrm{End}(TY))$. There is also a gauge-of-gauge symmetry since the B field is a 2-gerbe:

$$\mu_{triv} = \mathrm{d}\nu\,,\ \nu \in \Omega^0(Y)\,. \tag{35}$$

The trivial deformations of the heterotic G_2 system corresponding to the $\mathscr{Q}$-valued elements $Z = (y, \alpha, \kappa)^T$ were explained in detail in Ref. [53], where it was shown that there exists a canonical G_2 complex

$$0 \to \Omega^0(\mathscr{Q}) \to \Omega^1(\mathscr{Q}) \to \Omega^2_7(\mathscr{Q}) \to \Omega^3_1(\mathscr{Q}) \to 0 \tag{36}$$

which is elliptic. The nilpotent differential of this complex is the operator $\check{\mathscr{D}}$ defined in the previous subsection. Focusing on the $\mathscr{Q}$-valued elements of the deformations, it can be shown that the infinitesimal moduli are captured by the cohomology group $H^1(Y, \mathscr{Q})$ [53].

In the next section, we will write down an analogous complex that also includes the degrees of freedom related to the dilaton and B-field.

3 Defining the complex

In the last section, we showed that the equations satisfied by the infinitesimal deformations for the heterotic G_2 system correspond to

$$\check{\mathbb{D}}\mathscr{X} = 0\,,$$

a constraint which ensures the vanishing of the second order variation of the heterotic G_2 superpotential. We also made some observations about the trivial deformations of the system.

In the rest of this paper, we will interpret this constraint as the equations of motion of an action, which we will write as

$$S = \langle \mathscr{X}, \check{\mathbb{D}} \mathscr{X} \rangle_{\mathscr{E}} \,. \tag{37}$$

This will allow us to formalize and extend the observations made for infinitesimal deformations. To do so, we start in this section, by constructing a complex

$$0 \to \mathscr{E}^{-2} \to \mathscr{E}^{-1} \to \mathscr{E}^{0} \to \mathscr{E}^{1} \to \mathscr{E}^{2} \to \mathscr{E}^{3} \to 0 \tag{38}$$

where the arrows correspond to action by the operator $\check{\mathbb{D}}$, which extends the operator defined above to act on elements in $\mathscr{E}^d$, vector spaces of homological degree d. In more detail,

$$\mathscr{E}^* = \check{\Omega}^*[2] \oplus \check{\Omega}^*(\mathscr{Q})[1] \oplus \check{\Omega}^{4+*} \,.$$

where the numbers within the brackets denote the shift of the homological degree, and the checks denote the appropriate projections to G_2 representations,[1] e.g.

$$\mathscr{E}^0 = (\check{\Omega}^*[2] \oplus \check{\Omega}^*(\mathscr{Q})[1] \oplus \check{\Omega}^{4+*})^0 = \check{\Omega}^2 \oplus \check{\Omega}^1(\mathscr{Q}) \oplus \check{\Omega}^4 = \Omega^2_7 \oplus \Omega^1(\mathscr{Q}) \oplus \Omega^4_1 \,.$$

This complex encodes the equations of moduli and symmetries of the heterotic G_2 system. The aim of this section is to work out the properties of this complex in detail. To that end, let us first write out the complex in the following form:

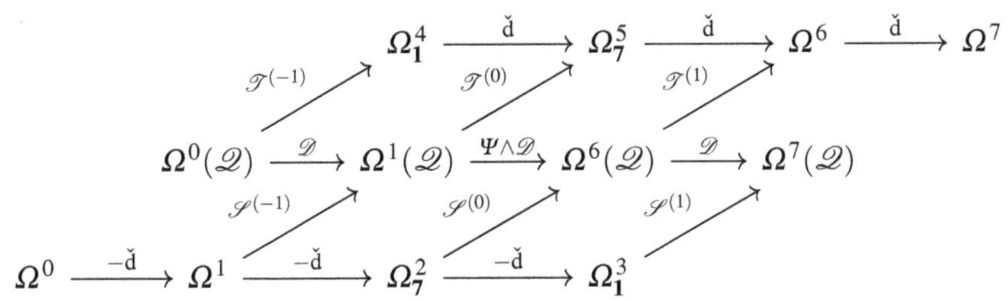

(39)

The presentation of the complex is chosen in order to simplify our analysis below, which possibly makes the connection to the heterotic G_2 system, and the $\mathscr{Q}$ complex (36) a bit opaque.[2] Each row of this total complex is an elliptic complex, that are all modelled on the canonical G_2 complex introduced in [57, 58, 18] and the $\mathscr{Q}$ complex of [53]. This implies, in particular, that the complex (39) has nilpotent horizontal differentials. In more detail, we have that

---

[1] Although it will not be of relevance for the present analysis, we note that (38) is in fact the total complex of a bicomplex, see section 3.3.

[2] An alternative, isomorphic, presentation would be to work with $\check{\mathscr{D}}$ differentials in the middle row, as in (36).

- The nilpotent differential $\check{\mathrm{d}}$ is the canonical G_2 differential, obtained by projecting the image of the de Rham differential to the relevant G_2 irreducible representation;
- $\Psi = e^{-2\phi}\psi$ is the closed 4-form of the integrable G_2 structure, and $\mathscr{D}$ is the differential defined in equation (25);
- The bottom complex is, up to a sign, the canonical G_2 complex of [57, 58, 18];
- The top complex is isomorphic to the canonical G_2 complex (wedge with Ψ);
- The middle complex is isomorphic (again by wedge with Ψ) to the $\mathscr{Q}$-valued complex defined in (36), with nilpotent differential $\check{\mathscr{D}}$.
- The maps $\mathscr{S}^{(p)}$ are the components of the de Rham differential that are projected out in the canonical G_2 differential. $\mathscr{T}^{(-p)}$ are their adjoints, with respect to the shifted symplectic pairing of the complex. These maps are determined by the symmetries and equations of the heterotic G_2 system, as detailed below.

The main benefit with this presentation of the total complex, is that it has a very simple shifted symplectic pairing (just wedge and integrate):

$$\begin{aligned}\omega(\xi,\eta) &= \langle \xi,\eta\rangle_{\mathscr{E}} = \int \xi\wedge\eta \ , \quad \xi,\eta\in\check{\Omega}^{4+*}\oplus\check{\Omega}^*[2] \\ \omega(X,Y) &= \langle X,Y\rangle_{\mathscr{E}} = \int g_{\mathscr{Q}}(X\wedge Y) \ , \ X,Y\in\check{\Omega}^*(\mathscr{Q})[1]\,.\end{aligned} \tag{40}$$

where $g_{\mathscr{Q}} = g\oplus \mathrm{tr}$ is a natural metric on the $\mathscr{Q}$-bundle, defined in (31) using the G_2 metric, g, and taking the trace on vector bundle indices. We note that $\mathscr{D}$ is metric with respect to $g_{\mathscr{Q}}$. Moreover, with respect to this pairing, the $\mathscr{Q}$-row is orthogonal to the top and bottom rows of the complex.

The shifted symplectic form induces the signs displayed in the complex. Since ω is closed under the differential $\check{\mathbb{D}}$ of the complex (here $|\alpha|$ denotes homological degree)

$$0 = \check{\mathbb{D}}\omega(\alpha,\beta) = \omega(\check{\mathbb{D}}\alpha,\beta) + (-1)^{|\alpha|}\omega(\alpha,\check{\mathbb{D}}\beta) \tag{41}$$

we find that there is a relative sign between the differentials in the top and bottom row of the complex, and relative signs between the vertical maps.

As a side remark, by defining the Lie algebroid, $\mathfrak{g}$:

$$\mathfrak{g} := \Omega^0(\mathscr{Q})[1] \xrightarrow{\mathscr{D}} \Omega^1(\mathscr{Q})\,, \tag{42}$$

we can regard $\check{\Omega}\oplus\check{\Omega}[2] \cong T^*[-1]\check{\Omega}$ and $\check{\Omega}(\mathscr{Q})[1] \cong T^*[-1](\mathfrak{g})$. From this perspective, the symplectic form defined above is the canonical symplectic form on a cotangent bundle.

3.1 Defining the differential operator $\check{\mathbb{D}}$ of the complex

We will now return to the precise definition of the differential

$$\check{\mathbb{D}}^i : \mathscr{E}^i \to \mathscr{E}^{i+1}$$

of the total complex. To define the correct differential $\check{\mathbb{D}}$ we are guided by the following principles:

1. It must be nilpotent;
2. It must be self-adjoint with respect to ω;
3. At homological degree zero, we should recover the action (37) from $\omega\left(\mathscr{X}, \check{\mathbb{D}}\mathscr{X}\right)$;
4. The full complex must be elliptic.

In addition, we can interpret the $\mathscr{E}^{(-1)} \to \mathscr{E}^{(0)}$ as a gauge transformation, so we are here guided by the symmetries of the theory. In particular, the trivial deformations (34) should be recovered at homological degree -1.

Our main task now is to determine the differential $\check{\mathbb{D}}^{(i)}$, with the properties just listed. We start at homological degree zero, which should match the results of the last section (condition 3 on the above list). This is in part immediate: we recognize the infinitesimal deformations $\mathscr{X}$ of the heterotic G_2 system as elements in

$$\mathscr{E}^0 = \Omega^4_{\mathbf{1}} \oplus \Omega^1(\mathscr{Q}) \oplus \Omega^2_{\mathbf{7}} .$$

Also, the horizontal parts of $\check{\mathbb{D}}^{(0)}$ match, up to wedging with Ψ, the diagonal terms of self-dual differential operator defined in (33). In particular, the signs of the terms in the action, as presented in (30) are consistent with the signs in the top and bottom row of the complex. Below, we will show that the maps $\mathscr{T}^{(0)}, \mathscr{S}^{(0)}$ match the off-diagonal terms of (33) in a similar fashion. This ensures that the homological degree zero part of the complex encodes the same equations of motion as the action (37).

At homological degree -1, we are guided by the symmetries of the theory. These are captured by $\mathscr{E}^i$ with $i < 0$ in the total complex. On formal grounds, for nilpotent $\check{\mathbb{D}}$, the action (37) should have a gauge symmetry corresponding to $\check{\mathbb{D}}$-exact variations, and this should match the trivial deformations of the heterotic G_2 system, (34). We thus expect:

$$\mathscr{X}^{(0)}_{triv} = \check{\mathbb{D}}^{(-1)} \mathscr{X}^{(-1)}, \text{ where } \mathscr{X}^{(-1)} \in \mathscr{E}^{-1} = \Omega^0(Y, \mathscr{Q}) \oplus \Omega^1(Y) . \tag{43}$$

where the first of the two summands contribute to the trivial deformations of the $\mathscr{Q}$-bundle, and the second summand contain trivial, exact $\beta = \mathrm{d}\mu$, parametrised by 1-forms μ. These 1-forms have gauge-for-gauge symmetries, (35) which we expect correspond to

$$\mathscr{X}^{(-1)}_{triv} = \check{\mathbb{D}}^{(-2)} \mathscr{X}^{(-2)}, \text{ where } \mathscr{X}^{(-2)} \in \mathscr{E}^{-2} = \Omega^0(Y) . \tag{44}$$

These spaces match the first columns of the total complex (or the lower left corner of the bicomplex), but we have yet to construct the differential operators $\check{\mathbb{D}}^{(-1)}, \check{\mathbb{D}}^{(-2)}$.

Since the horizontal derivatives have already been determined, we can immediately deduce that $\check{\mathbb{D}}^{(-2)} = -\check{\mathrm{d}}$. For $\check{\mathbb{D}}^{(-1)}$, we must also determine the maps $\mathscr{S}^{(-1)}, \mathscr{T}^{(-1)}$. Once these maps are determined, we can use self-duality and nilpotency to construct $\mathscr{S}^{(1)}, \mathscr{T}^{(1)}$; this is the topic of the ensuing subsections. For ease of presentation we will first deduce the three $\mathscr{S}$ maps, and then the $\mathscr{T}$, keeping an eye on the symmetries and equations of the heterotic theory along the way, so that these relevant constraints are reproduced by our formalism.

3.1.1 The $\mathscr{S}^{(p)}$ maps

The map $\mathscr{S}^{(-1)}: \Omega^1 \to \Omega^1(\mathscr{Q})$ encodes a gauge symmetry of the variations β.

In our complex, the summand $\check{\Omega}[2]$, in degree zero, corresponds to a linear combination of the **7**-part of the B-field and the G_2 structure variations, while the **14** part of the B-field is embedded inside $\Omega^1(\mathscr{Q})$. The degree -1 entry of $\check{\Omega}[2]$ is the space of gauge transformations of the full B-field variation and, consequentially, there is a component of our differential $\mathscr{S}^{(-1)}: \Omega^1 \to \Omega^1(TY) \subset \Omega^1(\mathscr{Q})$ which, on an arbitrary one form μ, acts by:

$$(\mathscr{S}^{(-1)}\mu)^a = -(\hat{\mathrm{d}}\mu)^a\,, \tag{45}$$

where we have defined $\hat{\mathrm{d}}$ by the general formula $\mathrm{d} = \check{\mathrm{d}} + \hat{\mathrm{d}}$ and the sign is motivated by the sign in the bottom row of the complex. In particular, in this case $\hat{\mathrm{d}} = \pi_{\mathbf{14}}\mathrm{d}$.

Turning now to $\mathscr{S}^{(0)}: \Omega^2_{\mathbf{7}} \to \Omega^6(\mathscr{Q})$, we can now be guided by the explicit computation of the second-order action. On an arbitrary section $\beta \in \Omega^2_{\mathbf{7}}$ we have

$$(\mathscr{S}^{(0)}\beta)^a = -(\hat{\mathrm{d}}\beta)^a \wedge \Psi\,, \tag{46}$$

where $\hat{\mathrm{d}} = \pi_{\mathbf{7}\oplus\mathbf{27}}\mathrm{d}$. This matches the map $\tau^\dagger$ wedged with Ψ, cf. equation (29), as required to reproduce the action (30).

Finally we have $\mathscr{S}^{(1)}: \Omega^3_{\mathbf{1}} \to \Omega^7(\mathscr{Q})$. Guided by comparison with the maps in lower degree, we posit that

$$(\mathscr{S}^{(1)}\chi)^a = -(\hat{\mathrm{d}}\chi)^a \wedge \Psi = -(\mathrm{d}\chi)^a \wedge \Psi \tag{47}$$

where, noting that $\check{\mathrm{d}}\chi = 0$, for $\chi \in \Omega^3_{\mathbf{1}}$ we have $\hat{\mathrm{d}} = \mathrm{d}$. The correctness of this assertion will be verified when we compute its adjoint and see that it is, indeed, encoding a symmetry. We turn to these computations now.

3.1.2 The $\mathscr{T}^{(p)}$ maps

Given our symplectic form, the maps $\mathscr{T}^{(p)}$ are completely determined by the maps, $\mathscr{S}$, which we defined above. The most important, for us, is that the adjoints $\mathscr{T}^{(-1)}$ and $\mathscr{T}^{(0)}$ are physically sensible.

We will begin with the map $\mathscr{T}^{(-1)}$ which we demand is adjoint to $\mathscr{S}^{(1)}$. First note that the gauge sector of $\mathscr{Q}$ is irrelevant to this part of the story, and so will be suppressed. Then, computing explicitly, we have for arbitrary $\sigma \in \Omega^0(TY)$, $\chi \in \Omega^3_{\mathbf{1}}$:

$$\begin{aligned}
\omega(\mathscr{T}^{(-1)}\sigma,\chi) &:= +\omega(\sigma,\mathscr{S}^{(1)}\chi) && \text{since } |\sigma| = -1 \text{ and } \check{\mathbb{D}}\omega(\alpha,\beta)=0 \\
&= -\int \sigma^a (\mathrm{d}\chi)_a \wedge \Psi && \text{sign in } \mathscr{S}^{(1)} \text{ and } \sigma \in \Omega^0(TY) \\
&= +\int (\sigma^a \Psi)_a \wedge \mathrm{d}\chi && \text{since } \mathrm{d}\chi \wedge \Psi = 0 \\
&= +\int \mathrm{d}((\sigma^a \Psi)_a) \wedge \chi \\
&= +\int \pi_1 \mathrm{d}(\sigma^a \Psi_a) \wedge \chi\,.
\end{aligned}$$

Since this holds for all χ, we conclude that $\mathscr{T}^{(-1)}\sigma = +\pi_1(\mathrm{d}\sigma^a\Psi_a) \in \Omega^4_{\mathbf{1}}$.

Since Ψ is d-closed, we can re-express this map in terms of the Lie derivative, using Cartan's magic formula,

$$\mathscr{T}^{(-1)}\sigma = +\pi_1 \mathscr{L}_\sigma \Psi\,.$$

The physics interpretation of the space $\Omega^4_{\mathbf{1}}$ is that it is the space of dilaton variations. It is manifestly convenient to use the dilaton to redefine the G_2 four form, ψ to the closed four form Ψ, and thus dilaton variations lead to variations in this form - it is thus more precisely variations of Ψ which are parametrised by $\Omega^4_{\mathbf{1}}$. The group of diffeomorphisms has a natural action on the space of functions (which is where the dilaton lives) and this naturally induces a map out of the Lie algebra of diffeomorphisms to the space of dilaton variations, here encoded by $\Omega^4_{\mathbf{1}}$. This is *precisely* the map $\mathscr{T}^{(-1)}$ that we have identified, so we are satisfied.

Next we turn to $\mathscr{T}^{(0)}$, the adjoint to $\mathscr{S}^{(0)}$. Similar computations to the above yield:

$$\begin{aligned}
\omega(\mathscr{T}^{(0)}M,\beta) &:= -\omega(M,\mathscr{S}^{(0)}\beta) && \text{since } |M| = 0 \text{ and } \check{\mathbb{D}}\omega(\alpha,\beta)=0 \\
&= \int M^a \wedge (\hat{\mathrm{d}}\beta)_a \wedge \Psi && \text{recall sign in } \mathscr{S}^{(0)} \\
&= \int (M^a \wedge \Psi)_a \wedge \hat{\mathrm{d}}\beta \\
&= \int \pi_R (M^a \wedge \Psi)_a \wedge \mathrm{d}\beta \\
&= -\int \mathrm{d}\pi_R (M^a \wedge \Psi)_a \wedge \beta\,,
\end{aligned}$$

where we defined $\pi_R : \mathbf{1} \oplus \mathbf{7} \oplus \mathbf{27} \to \mathbf{7} \oplus \mathbf{27}$. This shows that at homological degree 0, $\omega(\mathscr{X},\check{\mathbb{D}}^{(0)}\mathscr{X})$ is equivalent to the action (37). Furthermore, through a straightforward, but somewhat tedious, computation one may show that $\mathscr{T}^{(0)}$ agrees with the map τ in (28); we omit the details here.

Finally, we must compute $\mathscr{T}^{(1)}$:

$$\begin{aligned}\omega(\mathscr{T}^{(1)}\eta,\mu) &= \omega(\eta,\mathscr{S}^{(-1)}\mu)\\ &= -\int \eta^a \wedge (\hat{\mathrm{d}}\mu)_a \qquad \text{recall sign in } \mathscr{S}^{(-1)}\\ &= +\int \mathrm{d}\pi_{\mathbf{14}}(\eta^a_{\ a}) \wedge \mu\,.\end{aligned}$$

This map does not have a physical motivation, but this is not an issue, as it is deduced from $\mathscr{S}^{(-1)}$, which has been shown encodes a gauge symmetry of β.

3.2 Nilpotency and ellipticity

Having defined the components of a self-adjoint differential, we must now ensure that it is, in fact, nilpotent. We can write our differential as a matrix:

$$\check{\mathbb{D}} = \begin{pmatrix} \check{\mathrm{d}} & \mathscr{T} & 0 \\ 0 & \check{\mathscr{D}} & \mathscr{S} \\ 0 & 0 & -\check{\mathrm{d}} \end{pmatrix} \tag{48}$$

where we redefine our notation so that, in order to match the middle complex of (39), we have $\check{\mathscr{D}} = \{\mathscr{D}, \Psi\wedge\mathscr{D}, \mathscr{D}\}$ at homological degree $\{-1,0,1\}$. This squares to:

$$\check{\mathbb{D}}^2 = \begin{pmatrix} \check{\mathrm{d}}^2 & \check{\mathrm{d}}\mathscr{T} + \mathscr{T}\check{\mathscr{D}} & \mathscr{T}\mathscr{S} \\ 0 & \check{\mathscr{D}}^2 & \check{\mathscr{D}}\mathscr{S} - \mathscr{S}\check{\mathrm{d}} \\ 0 & 0 & \check{\mathrm{d}}^2 \end{pmatrix}. \tag{49}$$

We know well that the contributions on the diagonal vanish, which means that we need to verify the remaining entries are identically zero. That means verifying that the compositions $\mathscr{T}^{(p+1)} \circ \mathscr{S}^{(p)}$ vanish, and that each of the squares in total complex (39) anticommute.

3.2.1 The $\mathscr{T}\circ\mathscr{S}$ compositions

Nilpotency requires that the compositions $\mathscr{T}^{(1)}\circ\mathscr{S}^{(0)}$ and $\mathscr{T}^{(0)}\circ\mathscr{S}^{(-1)}$ both vanish. In fact, it suffices to check only one of these compositions, because they are adjoint to each other. The argument we give will make this adjointness manifest.

Indeed, non-degeneracy of the symplectic form implies that $\mathscr{T}^{(1)}\mathscr{S}^{(0)}\beta = 0$ if and only if $\omega(\mu, \mathscr{T}^{(1)}\mathscr{S}^{(0)}\beta) = 0$ for all $\mu \in \Omega^1$. By definition, this is true if and only if $\omega(\mathscr{S}^{(-1)}\mu, \mathscr{S}^{(0)}\beta) = 0$. It can be noted that starting with $\mathscr{T}^{(0)}\mathscr{S}^{(-1)}$ and taking the adjoint of $\mathscr{T}$ would give precisely the same pairing here. Expanding this out, we have:

$$\omega(\mathscr{S}^{(-1)}\mu, \mathscr{S}^{(0)}\beta) = \int (\hat{d}\mu)^a \wedge (\hat{d}\beta)_a \wedge \Psi\,,$$

which is a G_2 equivariant map $\mathbf{14} \otimes (\mathbf{7} \oplus \mathbf{27}) \to \mathbf{1}$ and must, therefore, be vanishing.

3.2.2 Anticommuting squares

The final check of nilpotency is that each of the squares in diagram (39) anticommute. While the computations are straightforward, and similar to the ones presented above (i.e. we will make use of G_2 representation theory as above), we include them for completeness.

Gauge contributions to the squares

The maps $\mathscr{S}$ and $\mathscr{T}$ act trivially on the gauge sector, so it may appear that gauge fields only appear in the $\check{\mathscr{D}}^2$ computation. This is not true, however, because $\check{\mathscr{D}}$ intermixes the gauge and vector field sectors. Nevertheless, the only possible problems are in checking anticommutativity of the squares; composing the operators we see that one leg trivially vanishes, while the vanishing of the other must be verified.

The trivial directions arise because the $\mathscr{T}$ and $\mathscr{S}$ maps are identically zero on gauge fields. The nontriviality arises in two ways: the $\check{\mathscr{D}}$ operator maps gauge sector to the vector field sector (relevant for the $\mathscr{T}$ squares); and, the $\check{\mathscr{D}}$ operator maps vector fields to the gauge sector (relevant for the $\mathscr{S}$ squares).

We will start with a section $\alpha \in \Omega^0(\mathfrak{g}^{ad}) \subset \Omega^0(\mathscr{Q})$ and must verify that $\mathscr{T}^{(0)}\check{\mathscr{D}}\alpha = \mathscr{T}^{(0)}\mathscr{D}\alpha = 0$. Recall from section 2.3 that the vector field component of $\mathscr{D}\alpha$ is given, up to overall factors, by:

$$(\mathscr{D}\alpha)^a \propto \mathrm{tr}(\alpha F^a), \tag{50}$$

where F is the gauge curvature, which is a G_2 instanton by the equations of motion.

Then:

$$\mathscr{T}^{(0)}\mathscr{D}\alpha \propto \pi_7 d\pi_R(\mathrm{tr}(\alpha \wedge F^a) \wedge \psi'_a) = 0.$$

The result follows trivially from the observation that, for a one form valued in the tangent bundle, say N, we have $i_N\psi'$ is independent of the $\mathbf{14}$, while F^a is a one form valued in the tangent bundle comprised only of the $\mathbf{14}$. In other words, this map vanishes because there are no nontrivial G_2 maps $\mathbf{14} \otimes \mathbf{1} \to \mathbf{1} \oplus \mathbf{7} \oplus \mathbf{27}$.

Consider now $\kappa \in \Omega^1(\mathfrak{g}^{ad})$ and its contribution to $\check{\mathbb{D}}^2$:

$$\begin{aligned}\mathscr{T}^{(1)}(\mathscr{D} \wedge \Psi)\kappa &\propto d\pi_{\mathbf{14}}(\mathrm{tr}(\kappa \wedge F^a) \wedge \Psi)_a \\ &= d\pi_{\mathbf{14}}(\mathrm{tr}(\kappa_a \wedge F^a) \wedge \Psi) = 0,\end{aligned}$$

where we used that $F^a \wedge \Psi_a = 0$ to get to the second line, and then used that the resulting expression is clearly in the $\mathbf{7}$, so killed by the projection.

We now turn to the possible contributions in the $\mathscr{S}$-squares, where the fact that $\mathscr{D}$ maps a vector field to a gauge field is the new concern. In particular:

$$\begin{aligned}\check{\mathscr{D}}\mathscr{S}^{(-1)}\mu &\propto (\hat{d}\mu)^a \wedge F_a \wedge \Psi \\ &= (\hat{d}\mu)^a{}_a \wedge F \wedge \Psi - ((\hat{d}\mu)^a \wedge F \wedge \Psi)_a - (\hat{d}\mu)^a \wedge F \wedge \Psi_a \\ &= 0\,.\end{aligned}$$

where we used that contracting an index obeys the Leibniz rule, and that each term vanishes separately due to both F and $\hat{d}\mu$ being in the **14**.

The final gauge theoretic concern is:

$$\check{\mathscr{D}}\mathscr{S}^{(0)}\beta = \mathscr{D}\mathscr{S}^{(0)}\beta = (\hat{d}\beta)^a \wedge F_a \wedge \Psi \ = 0\,,$$

where we have made the simple observation that there are no non-zero G_2 maps $(\mathbf{7} \oplus \mathbf{27}) \otimes \mathbf{14} \to \mathbf{1}$.

Possible vector field contributions

The remaining obstructions to nilpotency are vector field contributions, i.e. $\check{\mathscr{D}}\mathscr{S} - \mathscr{S}\check{\mathrm{d}}$ and $\check{\mathrm{d}}\mathscr{T} + \mathscr{T}\check{\mathscr{D}}$. Adjointness implies that we need only check the $\mathscr{S}$, or respectively $\mathscr{T}$, type squares, but we shall show both as a non-trivial check of our prior computations. This means that there are four possible contributions to check.

We will start in the top-left corner of Diagram 39:

$$\begin{aligned}\check{\mathrm{d}}\mathscr{T}^{(-1)}\sigma + \mathscr{T}^{(0)}\mathscr{D}\sigma &= \check{\mathrm{d}}(\pi_1 \mathrm{d}(\sigma^a \Psi_a)) - \check{\mathrm{d}}(\pi_R(\mathrm{d}_\zeta \sigma^a \wedge \Psi)_a) \\ &= \check{\mathrm{d}}\left(\pi_1\left(\mathrm{d}_\zeta \sigma^a \Psi_a + \sigma^a \mathrm{d}_\zeta \Psi_a\right) - \pi_R((\mathrm{d}_\zeta \sigma^a)_a \Psi - \mathrm{d}_\zeta \sigma^a \wedge \Psi_a)\right) \\ &= \check{\mathrm{d}}\left((\mathrm{d}_\zeta \sigma)^a \Psi_a + \pi_1 \sigma^a \mathrm{d}_\zeta \Psi_a\right) \\ &= \check{\mathrm{d}}\left((\mathrm{d}_\zeta \sigma)^a \Psi_a + \sigma^a \mathrm{d}_\zeta \Psi_a\right) \\ &= \check{\mathrm{d}}\mathrm{d}(\sigma^a \Psi_a) = 0\,,\end{aligned}$$

where in the 2nd to last equality we use that $\pi_1 \mathrm{d}_\zeta \Psi_a = \mathrm{d}_\zeta \Psi_a$. The vanishing of this term is the first check that the signs in the complex are correctly defined, given our starting point (minus signs in top row and definition of $\mathscr{S}^{(0)}$).

Moving to the next square at the top, the two terms are $\check{\mathrm{d}}\mathscr{T}^{(0)}$ and $\mathscr{T}^{(1)}\check{\mathscr{D}}$. Computing the first:

$$\begin{aligned}\check{\mathrm{d}}\mathscr{T}^{(0)}M &= -\mathrm{d}\pi_7 \mathrm{d}\pi_R(M^a \wedge \Psi)_a = \mathrm{d}\pi_7 \mathrm{d}\pi_R(M^a \wedge \Psi_a) \\ &= \mathrm{d}\pi_7 \mathrm{d}(M^a \wedge \Psi_a - \pi_1(M^a \wedge \Psi_a)) = \mathrm{d}\pi_7 \mathrm{d}(M^a \wedge \Psi_a)\,, \text{ since} \\ \mathrm{d}\pi_1(M^a \wedge \Psi_a) &\propto \mathrm{d}(\mathrm{tr}M \wedge \Psi) \propto \mathrm{dtr}M \wedge \Psi\,.\end{aligned}$$

and the second:

$$
\begin{aligned}
\mathscr{T}^{(1)}\check{\mathscr{D}}M = \mathscr{T}^{(1)}\mathscr{D}M\wedge\Psi &= +\mathrm{d}\pi_{\mathbf{14}}\big((\mathrm{d}_\zeta M^a\wedge\Psi)_a\big)\\
&= \mathrm{d}\pi_{\mathbf{14}}\left((\mathrm{d}_\zeta M)^a{}_a\wedge\Psi+\mathrm{d}_\zeta M^a\wedge\Psi_a\right)\\
&= \mathrm{d}\pi_{\mathbf{14}}(\mathrm{d}_\zeta M^a\wedge\Psi_a) \qquad\text{since first term in } 7;\\
\mathrm{d}(M^a\wedge\Psi_a) &= \mathrm{d}_\zeta M^a\wedge\Psi_a - M^a\wedge\mathrm{d}_\zeta\Psi_a\\
&= \mathrm{d}_\zeta M^a\wedge\Psi_a - M^a\wedge\nabla_a\Psi\\
&\equiv \mathrm{d}_\zeta M^a\wedge\Psi_a \quad \mathrm{mod}\ \ker\pi_{\mathbf{14}} \qquad (51)\\
\implies \mathscr{T}^{(1)}\mathscr{D}M\wedge\Psi &= +\mathrm{d}\pi_{\mathbf{14}}\mathrm{d}(M^a\wedge\Psi_a)
\end{aligned}
$$

Consequently,

$$
\check{\mathrm{d}}\mathscr{T}^{(0)} + \mathscr{T}^{(1)}\check{\mathscr{D}} = +\mathrm{d}^2(M^a\wedge\Psi_a) = 0
$$

The $\mathscr{S}$ squares also work out as they should. I.e.

$$
(\check{\mathscr{D}}S^{(-1)}\mu)_a = -\mathrm{d}_\zeta((\hat{\mathrm{d}}\mu))_a\wedge\Psi = -[(\nabla_a\hat{\mathrm{d}}\mu\Psi) - (\mathrm{d}\hat{\mathrm{d}}\mu)_a\wedge\Psi] = +(\mathrm{d}\hat{\mathrm{d}}\mu)_a\wedge\Psi;
$$

and

$$
S^{(0)}\check{\mathrm{d}}\mu = -(\hat{\mathrm{d}}\check{\mathrm{d}}\mu)_a\wedge\Psi = -(\mathrm{d}\check{\mathrm{d}}\mu)_a\wedge\Psi\,.
$$

Thus, clearly $\check{\mathscr{D}}S^{(-1)}\mu - S^{(0)}\check{\mathrm{d}}\mu = 0$. In the above we have used Lemma 2 of [53]

$$
\mathrm{d}_\zeta z_a = -\mathrm{d}z_a + \nabla_a z\,,\ z\in\Omega^1(T^*Y)\,.
$$

And finally, we have the bottom-right square contributions, which come from $S^{(1)}\circ\check{d}$ and $\check{\mathscr{D}}\circ S^{(0)}$. The first of these can be computed to be (using the lemma again)

$$
\begin{aligned}
\mathscr{D}S^{(0)}\beta &= -\mathrm{d}_\zeta((\hat{\mathrm{d}}\beta)_a\wedge\Psi)\\
&= -\mathrm{d}_\zeta(\hat{\mathrm{d}}\beta)_a\wedge\Psi\\
&= -\nabla_a\hat{\mathrm{d}}\beta\wedge\Psi + (\mathrm{d}(\hat{\mathrm{d}}\beta))_a\wedge\Psi\\
&= (\mathrm{d}\hat{\mathrm{d}}\beta)_a\wedge\Psi;
\end{aligned}
$$

and the second:

$$
S^{(1)}\check{\mathrm{d}}\beta = -\mathrm{d}(\check{\mathrm{d}}\beta)_a\wedge\Psi\,.
$$

Again, adding the terms with a relative minus sign gives zero.

We conclude that the sequence written down in (39), with $\mathscr{S}$ defined as in (45), (46),(47) and $\mathscr{T}$ defined by duality, has a nilpotent operator $\check{\mathbb{D}}$.[3]

We now have all the information to prove the following proposition

[3] We also remark that the nilpotency condition for $\check{\mathbb{D}}$, corresponds to a G_2 instanton condition for $\mathbb{D}$ defined in (33), though strictly speaking $\mathbb{D}$ is not a connection. Thus the "instanton for instanton" property that has been observed for complexes related to heterotic systems [52, 59] prevails also for the $\mathscr{E}$ complex.

Proposition 3. *The sequence* (39) *is an elliptic complex with shifted symplectic form* ω.

Proof. See Appendix 7.

3.3 The bicomplex

On formal grounds, the above computations show that we can associate a bicomplex to the theory described by the action (37). This was suggested, but not explicitly demonstrated in [39]. The bicomplex is constructed from the following components

- bigraded vector spaces $\mathscr{B}^{i,j}$
- horizontal differentials d_h
- vertical differentials d_v
- the differentials anti-commute $\mathrm{d}_h\mathrm{d}_v + \mathrm{d}_v\mathrm{d}_h = 0$
- a shifted symplectic pairing $\omega(\cdot,\cdot)$.

We write the natural bicomplex of the heterotic G_2 system as

$$
\begin{array}{ccccccccc}
 & -1 & & 0 & & 1 & & 2 \\
1 & \Omega^4_{\mathbf{1}} & \longrightarrow & \Omega^5_{\mathbf{7}} & \longrightarrow & \Omega^6 & \longrightarrow & \Omega^7 \\
 & \uparrow & & \uparrow & & \uparrow & & \uparrow \\
0 & \Omega^0(\mathscr{Q}) & \longrightarrow & \Omega^1(\mathscr{Q}) & \longrightarrow & \Omega^6(\mathscr{Q}) & \longrightarrow & \Omega^7(\mathscr{Q}) \\
 & \uparrow & & \uparrow & & \uparrow & & \uparrow \\
-1 & \Omega^0 & \longrightarrow & \Omega^1 & \longrightarrow & \Omega^2_{\mathbf{7}} & \longrightarrow & \Omega^3_{\mathbf{1}}
\end{array}
$$

where we have defined

$$
\begin{aligned}
\mathscr{B}^{*,1} &:= \check{\Omega}^{4+*}[1]\,; \\
\mathscr{B}^{*,0} &:= \check{\Omega}^{*}(\mathscr{Q})[1]\,; \\
\mathscr{B}^{*,-1} &:= \check{\Omega}^{*}[1]\,,
\end{aligned}
$$

where the number in the bracket denotes the shift and the check denotes taking the relevant G_2 irreducible representation, for example $\mathscr{B}^{0,1} = \Omega^5_{\mathbf{7}}$. The horizontal arrows are given by the relevant differentials associated to each sub-complex, while the vertical arrows are given by the $\mathscr{S}$ and $\mathscr{T}$ maps, except the vertical arrows $\Omega^0 \to \Omega^0(\mathscr{Q})$ and $\Omega^7(\mathscr{Q}) \to \Omega^7$ which are the zero maps.

Associated to the bicomplex is a total complex given by

$$
0 \to \mathscr{E}^{-2} \to \mathscr{E}^{-1} \to \mathscr{E}^{0} \to \mathscr{E}^{1} \to \mathscr{E}^{2} \to \mathscr{E}^{3} \to 0 \tag{52}
$$

where

$$\mathscr{E}^i := \mathrm{Tot}(\mathscr{B})^* = \bigoplus_{j+k=i} \mathscr{B}^{j,k},$$

Of course, this total complex is nothing but the complex (38) defined in section 3.

3.4 Infinitesimal moduli and obstruction theory

As mentioned at the end of section 2, with the constraint $\beta = 0 = \hat{\delta}\phi$, the infinitesimal moduli of the heterotic G_2 system are captured by $H^1(Y, \mathscr{Q})$ [53]. The action (37), and associated gauge symmetries given by $\check{\mathbb{D}}$-exact forms, make it clear that the full infinitesimal moduli are captured by

$$H^0_{\check{\mathbb{D}}}(Y, \mathscr{E}) = \frac{\{\mathscr{Z}^0 \in \mathscr{E}^0 \mid \check{\mathbb{D}}\mathscr{Z}^0 = 0\}}{\{\mathscr{Z}^0 \in \mathscr{E}^0 \mid \mathscr{Z}^0 = \check{\mathbb{D}}\mathscr{Z}^{-1}\}}$$

where the superscript denotes homological degree (note that the fields in $\mathscr{E}^0$, say, have different form degree).

To understand the higher order deformations requires analysing finite order deformations $\mathscr{Z}$ of the superpotential of a background solution. The corresponding equation of motion is then a Maurer-Cartan equation of the form

$$\check{\mathbb{D}}\mathscr{Z} + \mathscr{O}(\mathscr{Z}^2) = 0. \tag{53}$$

Solutions to this equation parameterises finite deformations of the system, i.e. which of the infinitesimal moduli can be integrated to a finite deformation. Figuring out the explicit form of the higher order terms $\mathscr{O}(\mathscr{Z}^2)$ requires an investigation of the full deformation theory of heterotic G_2 systems and its associated L_∞-algebra. This is beyond the scope of this paper. We can however make some remarks.

Note first that a perturbative expansion of the Maurer-Cartan equation (53) together with general deformation theory arguments leads to an obstruction, or Kuranishi, map

$$\mathscr{O} : H^0_{\check{\mathbb{D}}}(Y, \mathscr{E}) \to H^1_{\check{\mathbb{D}}}(Y, \mathscr{E}),$$

where the un-obstructed deformations lie in the kernel of this map. The details of this map depends of course on the deformation problem at hand. For "generic" Maurer-Cartan equations, the expectation is that this map will be injective for $h^0_{\check{\mathbb{D}}}(Y, \mathscr{E}) \leq h^1_{\check{\mathbb{D}}}(Y, \mathscr{E})$ and surjective for $h^0_{\check{\mathbb{D}}}(Y, \mathscr{E}) \geq h^1_{\check{\mathbb{D}}}(Y, \mathscr{E})$. In particular, for $h^0_{\check{\mathbb{D}}}(Y, \mathscr{E}) = h^1_{\check{\mathbb{D}}}(Y, \mathscr{E})$ the map is "generically" a bijection, just like a generic quadratic matrix is invertible. The expected, or "virtual", dimension of the moduli space in this case is zero, consisting of (possibly infinitely many) points. If this holds, and the moduli space can be compactified in some appropriate sense, there is then a natural enumerate invariant for topological spaces admitting heterotic G_2 solutions. Namely, how many such solutions exist for a given topology?

We thus want to establish the following:

Proposition 4. *For the heterotic G_2 system, $h^0_{\check{\mathbb{D}}}(Y,\mathscr{E}) = h^1_{\check{\mathbb{D}}}(Y,\mathscr{E})$.*

Proof. That this is true follows from the existence of the non-degenerate pairing ω, and Hodge theory. To invoke Hodge theory, we need to introduce a positive definite pairing $\tilde{g}$ on the complex. Using the same notation as in (40) for the elements in $\mathscr{E}$, we pick the following pairing

$$\tilde{g}(\mathscr{Z}_1,\mathscr{Z}_2) = \int \tilde{g}_{\mathscr{Q}}(X_1 \wedge *X_2) + \xi_1 \wedge *\xi_2 + \eta_1 \wedge *\eta_2 = \int \tilde{g}_{AB}\,\mathscr{Z}_1^A \wedge *\mathscr{Z}_2^B\,, \quad (54)$$

where $\tilde{g}_{\mathscr{Q}}$ is the pairing on $\mathscr{Q}$ defined in (31), but with the sign of the trace flipped on the $\mathrm{End}(TY)$ part, so that it is positive definite. The reason for this choice will become clear below.

Let us first compute how the adjoint of $\check{\mathbb{D}}$ acts with respect to this inner product. This is a bit subtle, as we need to invoke ω to perform the integration by parts. Let ω_{AB} denote the point-wise components of the pairing ω, with inverse ω^{AB}. We compute

$$\begin{aligned}\tilde{g}(\check{\mathbb{D}}\mathscr{Z}_1,\mathscr{Z}_2) &= \int \check{\mathbb{D}}\mathscr{Z}_1^A \wedge *\mathscr{Z}_2^B\,\tilde{g}_{AB}\\ &= \int \check{\mathbb{D}}\mathscr{Z}_1^A \wedge *\mathscr{Z}_2^B\,\tilde{g}_{BC}\,\omega^{CD}\,\omega_{DA}\\ &= (-1)^{|\mathscr{Z}_1|}\int \mathscr{Z}_1^A \wedge \check{\mathbb{D}}\left(*\mathscr{Z}_2^B\,\tilde{g}_{BC}\,\omega^{CD}\right)\omega_{DA}\,. \end{aligned} \quad (55)$$

It is then clear that for $\mathscr{Z}$ in the kernel of the adjoint requires

$$\check{\mathbb{D}} * \left(\omega^{-1}\circ\tilde{g}\cdot\mathscr{Z}\right) = 0\,. \quad (56)$$

Let us define the map

$$\Theta\,:\;\; \mathscr{E}^0 \to \mathscr{E}^1\,, \quad \Theta(\mathscr{Z}) = *\left(\omega^{-1}\circ\tilde{g}\cdot\mathscr{Z}\right)\,, \quad (57)$$

which is an isomorphism. If we now take $\mathscr{Z}$ to be harmonic, then $\Theta(\mathscr{Z})$ is $\check{\mathbb{D}}$-closed. To check coclosure, we must compute the adjoint action on $\Theta(\mathscr{Z})$. We get

$$\check{\mathbb{D}} * \left(\omega^{-1}\circ\tilde{g}\cdot\Theta(\mathscr{Z})\right) = \check{\mathbb{D}}\left(\omega^{-1}\circ\tilde{g}\circ\omega^{-1}\circ\tilde{g}\cdot\mathscr{Z})\right)\,. \quad (58)$$

For our particular choice of $\tilde{g}$, it is straight forward to check that $\omega^{-1}\circ\tilde{g}\circ\omega^{-1}\circ\tilde{g}$ is the identity. Hence

$$\check{\mathbb{D}} * \left(\omega^{-1}\circ\tilde{g}\cdot\Theta(\mathscr{Z})\right) = \check{\mathbb{D}}\mathscr{Z} = 0\,, \quad (59)$$

so $\Theta(\mathscr{Z})$ is both closed and co-closed.

It follows that the map Θ maps harmonic forms to harmonic forms. This establishes the above isomorphism between $H^0_{\check{\mathbb{D}}}(Y,\mathscr{E})$ and $H^1_{\check{\mathbb{D}}}(Y,\mathscr{E})$.

As a corollary, by a similar argument, one can also establish the more general Serre duality-like result

$$h^p_{\check{\mathbb{D}}}(Y,\mathscr{E}) = h^{1-p}_{\check{\mathbb{D}}}(Y,\mathscr{E})\,. \tag{60}$$

The corresponding Euler characteristic, or "index"

$$\chi(\check{\mathbb{D}},\mathscr{E}) = \sum_{p=-2}^{3} h^p_{\check{\mathbb{D}}}(Y,\mathscr{E})\,, \tag{61}$$

therefore vanishes. This shows that the moduli problem of the heterotic G_2 system is an "index zero moduli problem". This is encouraging, as such moduli problems have a tendency to lead to mathematically well-behaved topological invariant theories and enumerative geometry, such as, for example, in the corresponding moduli problems of ordinary Chern-Simons theory [62], holomorphic Chern-Simons theory [16, 61], and more speculatively G_2 instantons [15].

3.5 Field redefinitions and diagonalization of $\check{\mathbb{D}}$

While the $\mathscr{E}^*$ complex, and the associated bicomplex, provide a good description of the deformation theory, the fact that $\check{\mathbb{D}}$ has off-diagonal components will complicate our analysis in section 4. Thus, it is convenient to look for a field-redefinition which decouples the rows of the $\mathscr{E}^*$ complex. In this section we show how to accomplish this. To do so, we first enlarge the $\mathscr{E}^*$ complex to a complex which does not include projections in the top and bottom rows. We call this complex $\mathscr{F}^*$. It takes the form

$$\begin{array}{ccccccccccccc}
 & & \cdots & \xrightarrow{\mathrm{d}} & \Omega^3 & \xrightarrow{\mathrm{d}} & \Omega^4 & \xrightarrow{\mathrm{d}} & \Omega^5 & \xrightarrow{\mathrm{d}} & \Omega^6 & \xrightarrow{\mathrm{d}} & \Omega^7 \\
 & & & & & \overset{\mathscr{T}^{(-1)}}{\nearrow} & & \overset{\mathscr{T}^{(0)}}{\nearrow} & & \overset{\mathscr{T}^{(1)}}{\nearrow} & & & \\
 & & & & \Omega^0(\mathscr{Q}) & \xrightarrow{\mathscr{D}} & \Omega^1(\mathscr{Q}) & \xrightarrow{\Psi\wedge\mathscr{D}} & \Omega^6(\mathscr{Q}) & \xrightarrow{\mathscr{D}} & \Omega^7(\mathscr{Q}) & & \\
 & & & \overset{\mathscr{S}^{(-1)}}{\nearrow} & & \overset{\mathscr{S}^{(0)}}{\nearrow} & & \overset{\mathscr{S}^{(1)}}{\nearrow} & & & & & \\
\Omega^0 & \xrightarrow{-\mathrm{d}} & \Omega^1 & \xrightarrow{-\mathrm{d}} & \Omega^2 & \xrightarrow{-\mathrm{d}} & \Omega^3 & \xrightarrow{-\mathrm{d}} & \Omega^4 & \xrightarrow{-\mathrm{d}} & \cdots & &
\end{array} \tag{62}$$

For this to still give a bicomplex, we need to be careful with how we define the $\mathscr{S}$ and $\mathscr{T}$ maps. The $\mathscr{T}$ maps are now defined with additional projectors onto the correct irreducible representations in the top complex, while the $\mathscr{S}$ maps only act on the appropriate irreducible representations on the bottom complex. Specifically,

$$\mathscr{T}_{\mathrm{new}} = \pi \circ \mathscr{T}_{\mathrm{old}} \tag{63}$$

$$\mathscr{S}_{\mathrm{new}} = \mathscr{S}_{\mathrm{old}} \circ \pi\,, \tag{64}$$

where π denotes the projection onto the appropriate representation in the old complex $\mathscr{E}^*$.

Using the new complex $\mathscr{F}^*$, we would now like to find a field redefinition which diagonalizes the complex. We will focus on the degree zero fields, as these are the classical fields in the action (37). The operator of the complex $\mathscr{F}^*$ now reads

$$\check{\mathbb{D}} = \begin{pmatrix} \mathrm{d} & \mathscr{T} & 0 \\ 0 & \check{\mathscr{D}} & \mathscr{S} \\ 0 & 0 & -\mathrm{d} \end{pmatrix} = \begin{pmatrix} \mathbf{d} & \mathscr{T} & 0 \\ 0 & \mathbf{d} & \mathscr{S} \\ 0 & 0 & \mathbf{d} \end{pmatrix}, \tag{65}$$

where we have denoted the differential on each row as $\mathbf{d}$ for a smoother computation (we can reinsert the proper differentials at any stage in the computation below). The pairing on $\mathscr{E}$ extends naturally to $\mathscr{F}$, so we can write the action as

$$S = \langle \mathscr{X}, \check{\mathbb{D}} \mathscr{X} \rangle_{\mathscr{F}}, \tag{66}$$

for $\mathscr{X}$ in the image of the natural inclusion from $\mathscr{E}$.

We would like to find maps G and G^{-1} so that the operator diagonalizes as

$$G^{-1} \circ \check{\mathbb{D}} \circ G = \begin{pmatrix} \mathbf{d} & 0 & 0 \\ 0 & \mathbf{d} & 0 \\ 0 & 0 & \mathbf{d} \end{pmatrix}. \tag{67}$$

The map G then corresponds to a field redefinition of $\mathscr{X}$, which will follow from what we show below. Let us postulate the following form of the map G

$$G = \begin{pmatrix} 1 & B & C \\ 0 & 1 & A \\ 0 & 0 & 1 \end{pmatrix}. \tag{68}$$

Then the inverse can easily be checked to be

$$G^{-1} = \begin{pmatrix} 1 & -B & BA - C \\ 0 & 1 & -A \\ 0 & 0 & 1 \end{pmatrix}. \tag{69}$$

Let us then compute

$$G^{-1} \circ \check{\mathbb{D}} \circ G = \begin{pmatrix} \mathbf{d} & \mathscr{T} + [\mathbf{d}, B] & [\mathbf{d}, C] + \mathscr{T} \circ A + B \circ ([A, \mathbf{d}] - \mathscr{S}) \\ 0 & \mathbf{d} & \mathscr{S} + [\mathbf{d}, A] \\ 0 & 0 & \mathbf{d} \end{pmatrix}. \tag{70}$$

We will show below that the map A can be chosen so that

$$[\mathbf{d}, A] + \mathscr{S} = 0. \tag{71}$$

Furthermore, we can also choose the map B so that

$$[\mathbf{d}, B] + \mathscr{T} = 0. \tag{72}$$

With these choices for A and B, the expression then simplifies to

$$G^{-1} \circ \check{\mathbb{D}} \circ G = \begin{pmatrix} \mathbf{d} & 0 & [\mathbf{d}, C] - [\mathbf{d}, B] \circ A \\ 0 & \mathbf{d} & 0 \\ 0 & 0 & \mathbf{d} \end{pmatrix} = \begin{pmatrix} \mathbf{d} & 0 & [\mathbf{d}, C - BA] - B \circ \mathscr{S} \\ 0 & \mathbf{d} & 0 \\ 0 & 0 & \mathbf{d} \end{pmatrix}, \quad (73)$$

where we have used (71).

Let us now see that we can trivialise the $\mathscr{T}^{(0)}$ map as in (72). Recall that for $Z = (M, \alpha) \in \Omega^1(\mathscr{Q})$ we have

$$\begin{aligned} \mathscr{T}^{(0)} Z &= -\pi_7 \circ \mathrm{d}\left(\pi_{7+27} M^a \wedge \Psi_a\right) \\ &= -\mathrm{d}\left(\pi_{7+27}(M^a \wedge \Psi_a)\right) + \pi_{14} \mathrm{d}\left(\pi_{7+27} M^a \wedge \Psi_a\right) \\ &= -\mathrm{d}\left(\pi_{7+27}(M^a \wedge \Psi_a)\right) + \pi_{14} \mathrm{d}\left(M^a \wedge \Psi_a\right) \\ &= -\mathrm{d}\left(\pi_{7+27}(M^a \wedge \Psi_a)\right) + \pi_{14}\left((\mathrm{d}_\zeta M^a) \wedge \Psi_a\right) \\ &= -\mathrm{d}\left(\pi_{7+27}(M^a \wedge \Psi_a)\right) + \pi_{14}\left((\mathrm{d}_\zeta M^a \wedge \Psi)_a\right) \\ &= -\mathrm{d}\left(\pi_{7+27}(M^a \wedge \Psi_a)\right) + \pi_{14}\left(((\mathscr{D} Z)^a \wedge \Psi)_a\right) \\ &= -\mathrm{d}\left(\pi_{7+27}(Z^a \wedge \Psi_a)\right) + \pi_{14}\left(((\check{\mathscr{D}} Z)^a)_a\right), \end{aligned} \quad (74)$$

where in the third equality we have used that the singlet drops out from the second term, the fourth equality follows the computation in (51), the **7**-part again drops out in the fifth equality, while the gauge contribution drops out in the sixth equality. This then defines the maps $B^{(0)}$ and $B^{(1)}$ so that we may write

$$\mathscr{T}^{(0)} = -[\mathbf{d}, B]^{(0)} = -\mathbf{d} \circ B^{(0)} + B^{(1)} \circ \mathbf{d}. \quad (75)$$

Note that the projectors in $B^{(0)}$ and $B^{(1)}$ go outside of the original complex $\mathscr{E}^*$. This is why we needed to expand our complex to $\mathscr{F}^*$ in order to diagonalise the differential. The $\mathscr{S}^{(0)}$ can similarly be trivialised as in (71). Indeed, this follows readily as $\mathscr{S}^{(0)}$ is again the adjoint of $\mathscr{T}^{(0)}$ with respect to the pairing on $\mathscr{F}^*$.

Consider next the upper right-hand corner of the last matrix in (73). We have for $\beta \in \Omega^2$

$$B^{(1)} \circ \mathscr{S}^{(0)} \beta = B^{(1)}\left(-(\hat{\mathrm{d}} \pi_7(\beta))\right) = 0. \quad (76)$$

This follows as $\hat{\mathrm{d}}$ projects onto the $\mathbf{7} + \mathbf{27}$ irreducible representation when acting on 2-forms, while $B^{(1)}$ projects onto the **14** irreducible representation. It follows that

$$G^{-1} \circ \check{\mathbb{D}} \circ G = \begin{pmatrix} \mathbf{d} & 0 & [\mathbf{d}, C - B \circ A] \\ 0 & \mathbf{d} & 0 \\ 0 & 0 & \mathbf{d} \end{pmatrix}. \quad (77)$$

We may then simply pick $C = B \circ A$ to get a diagonalised differential.

We have succeeded in diagonalising the differential $\check{\mathbb{D}}$, at least in degree zero, but at a cost. We needed to expand our complex to $\mathscr{F}^*$. Furthermore, the field redefinitions we do, given by G and G^{-1}, add parts of irreducible representations to our fields which are not in the original complex $\mathscr{E}^*$. Hence, we are not quite done yet; we need to ensure that the $\mathscr{F}^*$ complex does not introduce unphysical degrees of freedom. We will resolve this question in section 4.2, by showing that the classical theory described in terms of the diagonal operator is equivalent to the original theory, and that the result also holds for the one-loop partition function.

4 1-loop partition function

We have seen that the second order variations heterotic G_2 superpotential can be interpreted as the action (37), which we repeat here for convenience,

$$S = \langle \mathscr{Z}, \check{\mathbb{D}}\, \mathscr{Z} \rangle_{\mathscr{E}} \,. \tag{78}$$

We will now use this quadratic action to compute the absolute value of the one-loop partition function of the heterotic G_2 system. For a quadratic action, this can be done using formal path integral manipulations, resulting in a rigorous end result. Our analysis will be modelled on the corresponding computation for three-dimensional Chern-Simons theory [62], which we repeat for the reader's convenience. The reader is referred to [56] for a more extended review.

4.1 Warm up: Abelian Chern-Simons theory

To set the stage, we start by recalling Abelian Chern-Simons theory on a three-manifold $\mathscr{M}_3$ [62]

$$S[\alpha] = \int_{\mathscr{M}_3} \alpha \wedge \mathrm{d}\alpha \,. \tag{79}$$

The absolute value of the partition function is formally defined as the Euclidean path integral

$$|Z| = \frac{1}{\mathrm{Vol}(G)} \int \mathscr{D}\alpha\, e^{-S[\alpha]} \,, \tag{80}$$

where $\mathrm{Vol}(G)$ denotes the infinite volume of gauge transformations to be determined.

Picking a metric, we can filter the path-integral under the Hodge-decomposition as

$$\begin{aligned} |Z| &= \frac{1}{\mathrm{Vol}(G)} \int_{\mathrm{d}\Omega^0} \mathscr{D}\alpha \int_{\mathrm{d}^\dagger \Omega^2} \mathscr{D}\alpha\, e^{-S[\alpha]} = \frac{\mathrm{Vol}(\mathrm{d}\Omega^0)}{\mathrm{Vol}(G)} \int_{\mathrm{d}^\dagger \Omega^2} \mathscr{D}\alpha\, e^{-S[\alpha]} \\ &= \frac{\mathrm{Vol}(\mathrm{d}\Omega^0)}{\mathrm{Vol}(G)} \frac{1}{\sqrt{\det(\mathrm{d}|_{\mathrm{d}^\dagger \Omega^2})}} \,, \end{aligned} \tag{81}$$

where we use that $S[\alpha]$ is independent of exact modes, and we ignore potential finite-dimensional contributions from harmonic on-shell modes. The determinant is defined as

$$\det\left(\mathrm{d}|_{\mathrm{d}^\dagger\Omega^2}\right) := \det\left(\mathrm{d}^\dagger\mathrm{d}|_{\mathrm{d}^\dagger\Omega^2}\right)^{\frac{1}{2}} . \tag{82}$$

Note then that, due to standard Hodge theory, the determinant of the Laplacian on one-forms, Δ^1, decomposes as

$$\det\left(\Delta^1\right) = \det\left(\mathrm{d}\mathrm{d}^\dagger|_{\mathrm{d}\Omega^0}\right)\det\left(\mathrm{d}^\dagger\mathrm{d}|_{\mathrm{d}^\dagger\Omega^2}\right) . \tag{83}$$

The partition function can therefore be written as

$$|Z| = \frac{\mathrm{Vol}(\mathrm{d}\Omega^0)}{\mathrm{Vol}(G)}\frac{\det\left(\mathrm{d}\mathrm{d}^\dagger|_{\mathrm{d}\Omega^0}\right)^{\frac{1}{4}}}{\det\left(\Delta^1\right)^{\frac{1}{4}}} . \tag{84}$$

Let us then consider $\det\left(\mathrm{d}\mathrm{d}^\dagger|_{\mathrm{d}\Omega^0}\right)$. If $\alpha = \mathrm{d}\gamma$ is an eigenvector of $\mathrm{d}\mathrm{d}^\dagger|_{\mathrm{d}\Omega^0}$, then

$$\mathrm{d}\mathrm{d}^\dagger\mathrm{d}\gamma = \lambda\mathrm{d}\gamma . \tag{85}$$

We can assume that $\gamma \in \mathrm{Im}(\mathrm{d}^\dagger)$, and d is invertible on this image, so this equation is equivalent to

$$\mathrm{d}^\dagger\mathrm{d}\gamma = \Delta^0\gamma = \lambda\gamma . \tag{86}$$

It follows that $\det\left(\mathrm{d}\mathrm{d}^\dagger|_{\mathrm{d}\Omega^0}\right) = \det(\Delta^0)$.

Next let's consider $\mathrm{Vol}(\mathrm{d}\Omega^0)$. Elementary linear algebra states that

$$\mathrm{Vol}(\mathrm{d}\Omega^0) = \frac{\det\left(\mathrm{d}|_{\mathrm{d}^\dagger\Omega^1}\right)}{\ker(\mathrm{d}:\Omega^0\to\Omega^1)}\mathrm{Vol}(\mathrm{d}\Omega^0) = \det\left(\Delta^0\right)^{\frac{1}{2}}\mathrm{Vol}(\Omega^0) , \tag{87}$$

where we have ignored the finite-dimensional kernel of constant functions. Collecting everything, one finds

$$|Z| = \frac{\mathrm{Vol}(\Omega^0)}{\mathrm{Vol}(G)}\frac{\det\left(\Delta^0\right)^{\frac{3}{4}}}{\det\left(\Delta^1\right)^{\frac{1}{4}}} . \tag{88}$$

By choosing $\mathrm{Vol}(G) = \mathrm{Vol}(\Omega^0)$, as the volume of gauge transformations, we get

$$|Z| = \frac{\det\left(\Delta^0\right)^{\frac{3}{4}}}{\det\left(\Delta^1\right)^{\frac{1}{4}}} . \tag{89}$$

Upon regularisation of the determinants of the Laplacians, this is precisely the Ray-Singer torsion of the manifold $\mathscr{M}_3$.

4.2 1-loop partition function of the heterotic G_2 system

The computation of the absolute value of the partition function of the theory (78) proceeds in exactly the same fashion as for three-dimensional Chern-Simons theory. The only difference being that the theory (78) is one-reducible, i.e. it has gauge of gauge symmetries, in addition to gauge symmetries. Keeping this in mind, and the fact that our fields take values in level zero $\mathscr{E}^0$ of the complex (38), the formal computation of the (absolute value of the) partition function gives

$$|Z| = \frac{\det\left(\Delta_{\check{\mathbb{D}}}^{(-1)}\right)^{\frac{3}{4}}}{\det\left(\Delta_{\check{\mathbb{D}}}^{(-2)}\right)^{\frac{5}{4}}\det\left(\Delta_{\check{\mathbb{D}}}^{(0)}\right)^{\frac{1}{4}}}. \tag{90}$$

This is the same expression as the partition function computed in [39] using generalised geometry techniques, except that the differential $\check{\mathbb{D}}$ is slightly different from the one appearing in [39]. We expect the results to agree upon an appropriate field redefinition of the quadratic action.

Here the Laplacians are defined as

$$\Delta_{\check{\mathbb{D}}} = \check{\mathbb{D}}\check{\mathbb{D}}^\dagger + \check{\mathbb{D}}^\dagger\check{\mathbb{D}}, \tag{91}$$

where the adjoint $\check{\mathbb{D}}^\dagger$ is defined using a *positive definite* inner product on $\mathscr{E}^*$. For example, we can use the inner product given by (54)

$$(\mathscr{X}_1, \mathscr{X}_2) = \int \left(\tilde{g}_{\mathscr{Q}}(X_1 \wedge *X_2) + \xi_1 \wedge *\xi_2 + \eta_1 \wedge *\eta_2\right), \tag{92}$$

where $\mathscr{X}_i = (\xi_i, X_i, \eta_i) \in \mathscr{B}^{*,1} \oplus \mathscr{B}^{*,0} \oplus \mathscr{B}^{*,-1}$. Note that this is not the pairing ω, which is not positive definite, and therefore does not give rise to an appropriate Hodge decomposition.

We would like to be a bit more explicit, and write this partition function in terms of determinants of more familiar operators. To do so, it is convenient to use the field-redefinition introduced in section 3.5, which decouples the rows of the $\mathscr{E}^*$ complex. To proceed, recall that the original action (78) is invariant under shifts by $\check{\mathbb{D}}$-exact forms of the fields. This is in fact the gauge symmetry of the heterotic theory. In particular, fields in the lowest row of the $\mathscr{E}^*$ complex are shifted by $\check{\mathrm{d}}$-exact forms. Let $\beta \in \Omega^2_7$ correspond to this part of the of the field $\mathscr{X}$. If we ignore a finite-dimensional part given by harmonic forms, we can Hodge decompose β as

$$\beta = \check{\mathrm{d}}^{\tilde{\dagger}}(f\Phi) + \check{\mathrm{d}}\gamma, \tag{93}$$

where $\gamma_1 \in \Omega^1$, and $\Phi = \tilde{*}\Psi$ is the Hodge-dual of Ψ with respect to the metric given by Ψ. The adjoint is also defined with respect to this metric. Without loss of generality, we can pick a gauge where β does not have any $\check{\mathrm{d}}$-exact part (this does not change the action). Since the field redefinitions corresponding to G and G^{-1} do not touch the bottom row of $\mathscr{E}^*$, it leaves β unchanged. It follows that the gauge choice for β is preserved: the $\check{\mathrm{d}}\gamma$-term cannot appear in the action after the field redefinition.

Let $\alpha \in \Omega^4_{\mathbf{1}}$ be the top-row part of $\mathscr{Z}$, and $\tilde{\alpha} \in \Omega^4$ be the part of the field after the field redefinition. This $\tilde{\alpha}$ need not be a singlet. The part of the action involving β after a field redefinition is then

$$S[\beta,\tilde{\alpha}] = \int \beta \wedge \mathrm{d}\tilde{\alpha}\,. \tag{94}$$

Since the $\check{\mathrm{d}}$-exact part of β does not appear, this term is the same as

$$\begin{aligned} S[\beta,\tilde{\alpha}] &= \int \check{\mathrm{d}}^{\tilde{\dagger}}(f\Phi)\wedge \mathrm{d}\tilde{\alpha} = (\mathrm{d}^{\tilde{\dagger}}(f\Phi),\tilde{*}\mathrm{d}\tilde{\alpha}) = (\tilde{*}\mathrm{d}(f\Psi),\tilde{*}\mathrm{d}\tilde{\alpha}) \\ &= (\mathrm{d}(f\Psi),\mathrm{d}\tilde{\alpha}) = (f\Psi,\mathrm{d}^{\tilde{\dagger}}\mathrm{d}\tilde{\alpha}) = (f,\Psi \lrcorner \mathrm{d}^{\tilde{\dagger}}\mathrm{d}\tilde{\alpha})\,, \end{aligned} \tag{95}$$

where (,) is the positive definite pairing induced by the metric $\tilde{g}$ of Ψ, and we have not been careful with signs. We can assume without loss of generality that f does not have a constant part (as it would drop out of (93)). If we Hodge decompose $\Psi \lrcorner \mathrm{d}^{\tilde{\dagger}}\mathrm{d}\tilde{\alpha}$, we get

$$\Psi \lrcorner \mathrm{d}^{\tilde{\dagger}}\mathrm{d}\tilde{\alpha} = \mathrm{d}^{\tilde{\dagger}}\mathrm{d}g + \text{constant}\,. \tag{96}$$

Note that while g is a singlet, it can get contributions from the $(\mathbf{7}+\mathbf{27})$-representations of the middle row of the complex, via the field redefinition.

We thus get

$$\begin{aligned} S[\beta,\tilde{\alpha}] &= (f,\mathrm{d}^{\tilde{\dagger}}\mathrm{d}g) = (\mathrm{d}f,\mathrm{d}g) = (\Psi\wedge \mathrm{d}f,\Psi\wedge \mathrm{d}g) \\ &= (\tilde{*}\mathrm{d}(f\Psi),\tilde{*}\mathrm{d}(g\Psi)) = \int \beta\wedge \mathrm{d}(g\Psi)\,. \end{aligned} \tag{97}$$

This shows that the $\mathbf{7}$-valued two-form β can only couple to a singlet, also after the field redefinition! As the singlet has no gauge transformations, except for a finite-dimensional space of translations by constants which we ignore, we will take this as our redefined singlet in the top row of the original complex $\mathscr{E}^*$, now using the diagonal differential on $\mathscr{E}^*$. By slight abuse of notation we shall also refer to this singlet as $\alpha \in \Omega^4_{\mathbf{1}}$.

The above computations shows that the original classical action (78) can be written as

$$S = \langle Z, \check{\mathscr{D}} Z\rangle_{\mathscr{Q}} + \int \beta\wedge \check{\mathrm{d}}\alpha\,, \tag{98}$$

for $\alpha \in \Omega^4_{\mathbf{1}}$. The gauge transformations in the current form of the action the β-fields are precisely as before, shifts by $\check{\mathrm{d}}$-exact forms. The gauge symmetries of (98) are given by $\mathbf{d}$-exact forms, where $\mathbf{d}$ refers to the diagonal differential introduced in section 3.5.

We had to pick a particular $\check{\mathbb{D}}$-gauge to write the action in the form of (98). However, the field redefinitions we have done only include translations corresponding to the upper-triangular matrices G and G^{-1}, which therefore have a trivial Jacobian in the change of the path integral measure. A general $\check{\mathbb{D}}$-gauge transformation will add terms to the action (98), but these terms can, as the above computations show, be absorbed again by field redefinitions, which again, do not change the path integral measure. The theories are therefore equivalent as quantum theories.

The action (98) is simpler to quantise in terms of determinants of more familiar Laplacians. Furthermore, it is easier to study its BV quantisation, η-invariants, etc. We will leave this for future work, but compute the (absolute value of the) partition function here. A straight-forward computation then gives

$$|Z| = \frac{\left(\det \Delta_{\check{\mathscr{D}}}^{(-1)}\right)^{\frac{3}{4}}}{\det\left(\Delta_{\check{\mathscr{D}}}^{(0)}\right)^{\frac{1}{4}}} \times \frac{(\det \Delta_{\mathbf{7}})^{\frac{1}{2}}}{(\det \Delta_{\mathbf{1}})^{\frac{3}{2}}}\,, \tag{99}$$

where the superscript denotes the Laplacian at the corresponding entry in the $\mathscr{E}^*$ complex, and the subscripts denote the irreducible representation. Here we have used that the Laplacians acting on one-forms and $\mathbf{7}$-valued two-forms are equivalent, and similarly with functions and singlet valued four-forms [2].

It is now straight-forward to compare the factors in (99) with the terms in the action (98). The first factor in (99) derives from the first term in (98), and is the one-loop partition function of an instanton gauge theory [9], with instanton connection $\mathscr{D}$ on the bundle $\mathscr{Q}$. The second factor derives from the second term in (98). It is the inverse of the square of the result of an Abelian instanton gauge theory. The square comes from the fact that we have two fields, β and α, while the inverse is due to the fields being shifted by one form degree. The result may also be compared with previous quantization of type II strings on G_2 manifolds [10, 8, 2].

5 Conclusion and outlook

In this paper, we have used the superpotential we defined in [49], for three-dimensional, $\mathscr{N}=1$ compactifications of heterotic supergravity on 7-manifolds Y with G_2 structure, to develop a deformation theory of such heterotic G_2 systems. We have shown that the vanishing of the second order variation of the superpotential encodes the equations for infinitesimal moduli of these string configurations, as the constraint

$$\check{\mathbb{D}}\mathscr{Z} = 0\,.$$

Here $\mathscr{Z} \in \mathscr{E}^0 = \Omega^4_{\mathbf{1}} \oplus \Omega^1(\mathscr{Q}) \oplus \Omega^2_{\mathbf{7}}$ is constructed from the deformations of the supergravity fields (dilaton, metric, gauge field and B-field). We have shown, by explicitly constructing the relevant nilpotent differential $\check{\mathbb{D}}$, and the required symplectic pairing, that this constraint can be encoded in an elliptic double complex (or the associated total complex). This is in accordance with more formal reasoning, based on generalised geometry, for heterotic G_2 compactifications [39]. We have also noted that the nilpotency of $\check{\mathbb{D}}$ can be interpreted as a G_2 instanton condition, as has been observed previously for complexes related to heterotic systems [52, 59].

Our bicomplex encodes the equations for infinitesimal moduli of heterotic G_2 systems at zero homological degree. At lower homological degree one finds the gauge, and gauge for gauge, symmetries of the theory. It thus follows that the infinitesimal moduli of heterotic G_2 systems are captured by $H^0_{\check{\mathbb{D}}}(Y,\mathscr{E})$ where the superscript encodes homological degree. Moreover, the obstruction space is given by $H^1_{\check{\mathbb{D}}}(Y,\mathscr{E})$. We have shown that the dimensions of these two cohomology groups are the same, showing that the expected, or "virtual", dimension of the infinitesimal moduli space is zero. This generalizes to a Serre-like duality and shows that the complex has vanishing Euler characteristic $\chi(\check{\mathbb{D}},\mathscr{E})$. This indicates that there might be natural enumerative invariants for heterotic G_2 systems. Computing $H^p_{\check{\mathbb{D}}}(Y,\mathscr{E})$ in explicit solutions, such as [46, 40], is an interesting avenue for future work.

As we discussed already in [49], one motivation for our study of heterotic superpotentials is that these functionals allow us to explore finite deformations of heterotic systems. This has been established for the Hull-Strominger system in [3]: by choosing a parametrisation for finite deformations suggested by holomorphy, it was shown that the superpotential expansion truncates at cubic order, and that the deformation algebra is an L_3 algebra. In [49] we conjectured that there should be a similar convenient parametrisation for the finite deformations of heterotic G_2 systems, which should lead to a superpotential expansion that truncates at fourth order. While we have not been able to confirm this suspicion in this paper, we have taken an important step towards its resolution by identifying a field redefinition that diagonalises the differential appearing in the second order variation of the superpotential.

By interpreting the second order variation of the superpotential as a classical quadratic action in $\mathscr{Z}$, we have initiated a study of quantum aspects of the heterotic G_2 system. Using the formal analogy of the action with three-dimensional Chern-Simons theory, we could express the absolute value of the one-loop partition function in terms of determinants of Laplacians $\Delta^p_{\check{\mathbb{D}}}$. This result has the same functional form as obtained in [39] using generalised geometry techniques. With our explicit double complex in hand, we can go further: we have shown that we may perform a field redefinition that diagonalizes $\check{\mathbb{D}}$. Using this, the one-loop partition function can be rewritten in terms of more familiar Laplacians. This is along the lines of what is obtained from quantization of type II strings on G_2 spaces [10, 8, 2].

In this paper, we have not performed the full BV quantisation of the heterotic G_2 systems. This is a natural next step to take. We expect that, by using the double complex in its diagonalised format, it will be relatively straight forward to compute the phase of the one-loop partition function (i.e. the η invariant). We may then compare with the results obtained from topological G_2 strings [9, 8]. Since, in contrast to these topological studies, our bicomplex encodes both gauge and geometric degrees of freedom, we may be able to provide new insights to possible geometric anomalies of the heterotic G_2 system. The intricate coupling between these degrees of freedom may also help expose to what extent topological invariants and enumerative geometry may be defined for instantons in the G_2 context [15, 37].

Acknowledgements We would like to thank Mario Garcia-Fernandez, Spiro Karigiannis, Julian Kupka, Sebastien Picard, Jock McOrist, Martin Sticka, Charles Strickland-Constable, David Tennyson, Fridrich Valach, for enlightening discussions. ML and MM were supported in part by Vetenskapsrådet, grant no. 2020-03230 and grant no. 2023-00508, respectively. ML thanks the Erwin Schrödinger Institute in Vienna, and the organisers of the program *The Landscape vs the Swampland*, for hospitality during some parts of this work. XD and ES would like to thank the organisers of the scientific programmes *The Geometry of Moduli Spaces in String Theory* and *New Deformations of Quantum Field and Gravity Theories* respectively, hosted by the MATRIX Institute, for the hospitality during later stages of this work. XD thanks The Sydney Mathematical Institute for support and hospitality while this work was being completed. For the purpose of open access, the authors have applied a CC-BY public copyright license to any Author Accepted Manuscript (AAM) version arising from this submission.

6 Appendix: Proof of proposition 1

In this appendix we give a detailed proof of proposition 1. We begin by varying (10) which gives

$$
\begin{aligned}
\delta_2\delta_1\mathfrak{w}|_0 = & \, i_{\mathring{M}_1}(\psi)\wedge\delta_2\left(H-\widehat{T}\right) \\
& + \frac{\alpha'}{2}\Big(\mathrm{tr}(\alpha_1\wedge\delta_2(F\wedge\psi)) - \mathrm{tr}(\kappa_1\wedge\delta_2(R\wedge\psi))\Big) \\
& - (\mathscr{B}_1 - 2m_1)\wedge\delta_2(\mathrm{d}\psi - 2\mathrm{d}\phi\wedge\psi) \\
& + \delta_2\left(\delta_1 h + \frac{3}{7}\,\mathrm{tr}M_1\left(h+\frac{1}{3}\,\tau_0\right)\right)\varphi\wedge\psi\,.
\end{aligned}
\tag{100}
$$

We now analyse each term.

<u>First term of equation (100)</u>

$$
\begin{aligned}
& i_{\mathring{M}_1}(\psi)\wedge\delta_2\left(H-\widehat{T}\right) \\
& \quad = i_{\mathring{M}_1}(\psi)\wedge\left(\mathrm{d}\mathscr{B}_2 + \frac{\alpha'}{2}\Big(\mathrm{tr}(\alpha_2\wedge F) - \mathrm{tr}(\kappa_2\wedge R)\Big) - \delta_2\widehat{T}\right)\,.
\end{aligned}
$$

To continue we use the following identity:

$$
i_M(\psi)\wedge\lambda = \mathrm{tr}M\,\lambda\wedge\psi - i_M(\lambda)\wedge\psi\,,\qquad \forall\lambda\in\Omega^3(Y) \tag{101}
$$

Then, noting that $\mathring{M}$ is traceless, we find (on shell)

$$
\begin{aligned}
& i_{\mathring{M}_1}(\psi)\wedge\delta_2\left(H-\widehat{T}\right) \\
&\qquad = -i_{\mathring{M}_1}\left[\mathrm{d}\mathscr{B}_2+\frac{\alpha'}{2}\Big(\mathrm{tr}(\alpha_2\wedge F)-\mathrm{tr}(\kappa_2\wedge R)\Big)-\delta_2\widehat{T}\right]\wedge\psi\,.
\end{aligned}
\tag{102}
$$

We need to express $\delta\widehat{T}$ in terms of M by varying the structure equations (4) and (5) being careful to only set $\tau_2=0$ on shell while retainnig $\delta\tau_2$. This computation we done in section 3 of [53] where the authors found

$$
\delta\widehat{T}_a\lrcorner\varphi=\big(2\,(\mathrm{d}m)_a+2\,\mathrm{d}_\zeta M_a\big)\lrcorner\varphi+\delta\tau_{2\,ab}\,\mathrm{d}x^b\,,\qquad \pi_7(\delta\tau_2)=0\,. \tag{103}
$$

Since only $\pi_7\big((\delta T)_a\big)$ contributes to (102) we have

$$
\begin{aligned}
i_{\mathring{M}_1}\left(\mathrm{d}\mathscr{B}_2-\delta_2\widehat{T}\right)\wedge\psi &= i_{\mathring{M}_1}\Big(\mathrm{d}(\pi_7(\mathscr{B}_2)-2m_2)+\mathrm{d}_\zeta\big(2M_2+\pi_{\mathbf{14}}(\mathscr{B}_2)\big)\Big)\wedge\psi \\
&\quad +\mathring{M}_1^a\wedge*\big((\delta_2\tau_2)_{ab}\mathrm{d}x^b\big)\,,
\end{aligned}
\tag{104}
$$

where we have used the identity

$$
\big(\mathrm{d}_\zeta z_a+(\mathrm{d}z)_a\big)\wedge\psi=0\,,\qquad \forall z\in\Omega^2_{\mathbf{14}}(Y)\,.
$$

As $\pi_7(\delta\tau_2)=0$, the last term in (104) vanishes

$$
\mathring{M}_1^a\wedge*\big((\delta_2\tau_2)_{ab}\mathrm{d}x^b\big)=2*(m_1\lrcorner\delta_2\tau_2)=0\,.
$$

Then (102) becomes

$$
\begin{aligned}
& i_{\mathring{M}_1}(\psi)\wedge\delta_2\left(H-\widehat{T}\right) \\
&\qquad = 2\,i_{\mathring{M}_1}\left(\mathrm{d}_\zeta y_2-\frac{\alpha'}{4}\Big(\mathrm{tr}(\alpha_2\wedge F)-\mathrm{tr}(\kappa_2\wedge R)\Big)-\mathrm{d}\beta_2\right)\wedge\psi\,,
\end{aligned}
\tag{105}
$$

where β and y are given in (13) and (14).

Second term of equation (100): Recall that [6]

$$
\delta F=\mathrm{d}_A\alpha-i_M(F)\,,
$$

with a similar equation for δR. Then, on shell we obtain

$$
\begin{aligned}
& \delta_2\left(\frac{\alpha'}{2}\Big(\mathrm{tr}(\alpha_1\wedge F)-\mathrm{tr}(\kappa_1\wedge R)\Big)\wedge\psi\right) \\
&\qquad = \frac{\alpha'}{2}\Big(\mathrm{tr}\big(\alpha_1\wedge(\mathrm{d}_A\alpha_2-i_{M_2}(F))\big)-\mathrm{tr}\big(\kappa_1\wedge(\mathrm{d}_\Theta\kappa_2-i_{M_2}(R))\big)\Big)\wedge\psi
\end{aligned}
$$

Third term of equation (100) Recall [53]

$$\delta \mathrm{d}\psi = i_{\mathrm{d}_\zeta(M)}(\psi) + i_M(\mathrm{d}\psi) = i_{\mathrm{d}_\zeta(M)}(\psi) + i_M(4\tau_1)\wedge\psi + 4\tau_1\wedge i_M(\psi)\,.$$

Then

$$(\mathscr{B}_1 - 2m_1)\wedge\delta_2(\mathrm{d}\psi - 2\,\mathrm{d}\phi\wedge\psi)$$
$$= (\mathscr{B}_1 - 2m_1)\wedge\left(i_{\mathrm{d}_\zeta M_2}(\psi) + \left(i_{M_2}(4\,\tau_1) - 2\,\mathrm{d}\delta_2\phi\right)\wedge\psi + (4\,\tau_1 - 2\,\mathrm{d}\phi)\wedge i_{M_2}(\psi)\right)\,.$$

The last term vanishes on shell. Then, separating $\mathscr{B}$ into its **7** and **14** parts we have

$$\begin{aligned}(\mathscr{B}_1 - 2m_1)&\wedge\delta_2(\mathrm{d}\psi - 2\,\mathrm{d}\phi\wedge\psi)\\ &=\pi_{\mathbf{14}}(\mathscr{B}_1)\wedge i_{\mathrm{d}_\zeta M_2}(\psi)\\ &\quad + 2\,\beta_1\wedge\left(\frac{1}{3}\,\psi\lrcorner\left(i_{\mathrm{d}_\zeta M_2}(\psi)\right) + i_{M_2}(4\,\tau_1) - 2\,\delta_2(\mathrm{d}\phi)\right)\wedge\psi\,.\end{aligned} \tag{106}$$

The first term of (**??**) can written as

$$\begin{aligned}\pi_{\mathbf{14}}(\mathscr{B}_1)\wedge i_{\mathrm{d}_\zeta M_2}(\psi) &= -\pi_{\mathbf{14}}(\mathscr{B}_1)^a\wedge i_{\mathrm{d}_\zeta M_{2a}}\wedge\psi\\ &= -\left(i_{\pi_{\mathbf{14}}(\mathscr{B}_1)}(\mathrm{d}_\zeta y_2) - \frac{1}{2}\,i_{\pi_{\mathbf{14}}(\mathscr{B}_1)}\left(\mathrm{d}_\zeta\pi_{\mathbf{14}}(\mathscr{B}_2)\right)\right)\wedge\psi\\ &= -\left(i_{\pi_{\mathbf{14}}(\mathscr{B}_1)}(\mathrm{d}_\zeta y_2) - \frac{1}{2}\,i_{\pi_{\mathbf{14}}(\mathscr{B}_1)}\left(\mathrm{d}\pi_{\mathbf{14}}(\mathscr{B}_2)\right)\right)\wedge\psi\,,\end{aligned}$$

where the last term vanishes. Then the third term of (100) is given by

$$-(\mathscr{B}_1 - 2m_1)\wedge\delta_2(\mathrm{d}\psi - 2\,\mathrm{d}\phi\wedge\psi)$$
$$= \left(i_{\pi_{\mathbf{14}}(\mathscr{B}_1)}(\mathrm{d}_\zeta y_2) - 2\,\beta_1\wedge\left(\frac{1}{3}\,\psi\lrcorner\left(i_{\mathrm{d}_\zeta M_2}(\psi)\right) + i_{M_2}(4\,\tau_1) - 2\,\delta_2(\mathrm{d}\phi)\right)\right)\wedge\psi\,.$$

Fourth term of equation (100): On shell, this term gives

$$\delta_2\left(\delta_1 h + \frac{3}{7}\,\mathrm{tr}M_1\left(h + \frac{1}{3}\,\tau_0\right)\right)\varphi\wedge\psi = \frac{1}{7}\,\mathrm{tr}M_1\,\delta_2\tau_0\,\varphi\wedge\psi \tag{107}$$

To compute $\delta\tau_0$ in terms of M, recall

$$\tau_0 = \frac{1}{7}\,\mathrm{d}\varphi\lrcorner\psi\,.$$

Taking the Hodge-dual

$$\tau_0\,\varphi\wedge\psi = \mathrm{d}\varphi\wedge\varphi\,,$$

and varying this equation we find

$$(\delta\tau_0+\tau_0\,\mathrm{tr}M)\varphi\wedge\psi=\mathrm{d}i_M(\varphi)\wedge\varphi+\mathrm{d}\varphi\wedge i_M(\varphi)=i_{\mathrm{d}_\zeta M}(\varphi)\wedge\varphi+i_M(\mathrm{d}\varphi\wedge\varphi)$$
$$=i_{\mathrm{d}_\zeta M}(\varphi)\wedge\varphi+\tau_0\,\mathrm{tr}M\varphi\wedge\psi\,.$$

Then

$$\delta\tau_0=\frac{1}{7}\,\psi\lrcorner i_{\mathrm{d}_\zeta M}(\varphi)\,.$$

It is not hard to prove that

$$\delta\tau_0=\frac{1}{7}\,\psi\lrcorner i_{\mathrm{d}_\zeta M}(\varphi)=\frac{2}{7}\big(-2\,\mathrm{d}m+i_M(\widehat{T})\big)\lrcorner\varphi\,, \tag{108}$$

which implies

$$\delta\tau_0\,\varphi\wedge\psi=2\big(-2\,\mathrm{d}m+i_M(\widehat{T})\big)\wedge\psi\,.$$

Then, the fourth term of (100) becomes

$$\delta_2\left(\delta_1 h+\frac{3}{7}\,\mathrm{tr}M_1\left(h+\frac{1}{3}\,\tau_0\right)\right)\varphi\wedge\psi=\frac{2}{7}\,\mathrm{tr}M_1\big(-2\,\mathrm{d}m_2+i_{M_2}(T)\big)\wedge\psi\,. \tag{109}$$

Putting together the results we have so far, we find

$$\begin{aligned}\frac{1}{2}\,\delta_2\delta_1\mathfrak{w}\,|_0=\Big\{&i_{\mathring{M}_1}\Big[\,\mathrm{d}_\zeta y_2-\frac{\alpha'}{4}\Big(\mathrm{tr}(\alpha_2\wedge F)-\mathrm{tr}(\kappa_2\wedge R)\Big)\Big]-i_{\mathring{M}_1}(\mathrm{d}\beta_2)\\&+\frac{1}{7}\,\mathrm{tr}M_1\,(-2\,\mathrm{d}m_2+i_{M_2}(T))+\frac{1}{2}\,i_{\pi_{\mathbf{14}}(\mathscr{B}_1)}(\mathrm{d}_\zeta y_2)\\&+\frac{\alpha'}{4}\Big[\mathrm{tr}\big(\alpha_1\wedge(\mathrm{d}_A\alpha_2-i_{M_2}(F))\big)-\mathrm{tr}\big(\kappa_1\wedge(\mathrm{d}_\Theta\kappa_2-i_{M_2}(R))\big)\Big]\\&-\beta_1\wedge\Big[\frac{1}{3}\,\psi\lrcorner\,\Big(i_{\mathrm{d}_\zeta M_2}(\psi)\Big)+i_{M_2}(4\,\tau_1)-2\,\delta_2(\mathrm{d}\phi)\Big]\Big\}\wedge\psi\,.\end{aligned} \tag{110}$$

We need to somewhat reorganize the terms to obtain the equation (16) claimed in the proposition. Note that only the third term in (109) contains $\mathrm{tr}M_1$. This term precisely combines with the first term giving

$$\Big\{i_{\mathring{M}_1}\Big[\,\mathrm{d}_\zeta y_2-\frac{\alpha'}{4}\Big(\mathrm{tr}(\alpha_2\wedge F)-\mathrm{tr}(\kappa_2\wedge R)\Big)\Big]+\frac{1}{7}\,\mathrm{tr}M_1\,(-2\mathrm{d}m_2+i_{M_2}(T))\Big\}\wedge\psi$$

$$=i_{M_1}\Big[\mathrm{d}_\zeta y_2-\frac{\alpha'}{4}\Big(\mathrm{tr}(\alpha_2\wedge F)-\mathrm{tr}(\kappa_2\wedge R)\Big)\Big]\wedge\psi$$

The computation to see this is straightforward.

$$i_{\frac{1}{7}\,\mathrm{tr}M_1\mathbf{1}}(\mathrm{d}_\zeta y_2)\wedge\psi=\frac{1}{7}\,\mathrm{tr}M_1\,\mathrm{d}x^a\wedge\mathrm{d}_\zeta y_{2a}\wedge\psi=\frac{1}{7}\,\mathrm{tr}M_1\,\mathrm{d}x^a\wedge\mathrm{d}_\zeta M_{2a}\wedge\psi\,,$$

where $\pi_{\mathbf{14}}(\mathscr{B})$ drops out because of the product with ψ:

$$dx^a \wedge d_\zeta\big(\pi_{\mathbf{14}}(\mathscr{B}))_a\big) \wedge \psi = -dx^a \wedge d\big(\pi_{\mathbf{14}}(\mathscr{B})\big)_a \wedge \psi = -d\big(\pi_{\mathbf{14}}(\mathscr{B})\big) \wedge \psi = 0\,,$$

as $\pi_{\mathbf{1}} d(\pi_{\mathbf{14}}(\mathscr{B}) = 0$ when $\tau_2 = 0$. Then

$$\begin{aligned} i_{\frac{1}{7}\mathrm{tr}M_1 \mathbf{1}}(d_\zeta y_2) \wedge \psi &= \frac{2}{7}\mathrm{tr}M_1 \left(\partial_b m_{2\,ac} + \frac{1}{2}\hat{T}^e{}_{ab} M_{2\,ec}\right) dx^{abc} \wedge \psi \\ &= \frac{1}{7}\mathrm{tr}M_1 \left(-2\,dm_2 + i_{M_2}(\hat{T})\right) \wedge \psi\,. \end{aligned}$$

Note also that the third term of equation (109) can also be combined with the first. Then, we have

$$\begin{aligned} \frac{1}{2}\delta_2\delta_1 \mathfrak{w}\,|_0 = \Big\{ & i_{y_1}\Big[d_\zeta y_2 - \frac{\alpha'}{4}\Big(\mathrm{tr}(\alpha_2 \wedge F) - \mathrm{tr}(\kappa_2 \wedge R)\Big)\Big] - i_{\mathring{M}_1}(d\beta_2) \\ & + \frac{\alpha'}{4}\Big[\mathrm{tr}\big(\alpha_1 \wedge (d_A \alpha_2 - i_{M_2}(F))\big) - \mathrm{tr}\big(\kappa_1 \wedge (d_\Theta \kappa_2 - i_{M_2}(R))\big)\Big] \\ & - \beta_1 \wedge \Big[\frac{1}{3}\psi \lrcorner \Big(i_{d_\zeta M_2}(\psi)\Big) + i_{M_2}(4\,\tau_1) - 2\,\delta_2(d\phi)\Big]\Big\} \wedge \psi\,. \end{aligned} \tag{111}$$

Finally, we observe that the last term can be written as

$$\begin{aligned} & \beta_1 \wedge \Big[\frac{1}{3}\psi \lrcorner \Big(i_{d_\zeta M_2}(\psi)\Big) + i_{M_2}(4\,\tau_1) - 2\,d\delta_2\phi\Big] \wedge \psi \\ & \qquad = \beta_1 \wedge \Big[\frac{1}{3}\psi \lrcorner \Big(i_{d_\zeta \mathring{M}_2}(\psi)\Big) + i_{\mathring{M}_2}(4\,\tau_1) - 2\,d\hat{\delta}_2\phi\Big] \wedge \psi\,, \end{aligned} \tag{112}$$

where we have defined

$$\hat{\delta}\phi = \delta\phi - \frac{2}{7}\mathrm{tr}M\,.$$

In passing from the first line to the second in this equation we have used

$$\begin{aligned} i_{d_\zeta(\mathrm{tr}M\,\mathbf{1})}(\psi) &= \big(d(\mathrm{tr}M) \wedge dx^a - \mathrm{tr}M\,T^a\big) \wedge \psi_a = 4\,d(\mathrm{tr}M) \wedge \psi - i_T(\psi) \\ &= 4\,\big(d(\mathrm{tr}M) - \mathrm{tr}M\,\tau_1\big) \wedge \psi\,. \end{aligned}$$

This proves the proposition. □

7 Appendix: Ellipticity

In this appendix, we prove that the $(\mathscr{E}, \check{\mathbb{D}})$ complex is elliptic. Good introductions to the notions of ellipticity of operators and complexes, can be found in the book of

Gilkey [29], and the lecture notes of Pati [55]. A very brief summary of most of the notions we need can also be found in appendix B of [51].

Recall first that a complex (E,D) is elliptic if the induced complex $(E,\sigma_L(D))$ of leading (or principal) symbols $\sigma_L(D)$ is exact. Reyes-Carrion [58, 57] proved this is the case for the canonical G_2 complex with differential $\check{\mathrm{d}}$, and in [51, 53] this was used to prove ellipticity of the $\check{\mathscr{D}}$-complex.

Second, we will make use of

Lemma 1. *Let $E := \bigoplus L_i$, $F := \bigoplus M_j$ two vector bundles, given by a direct sum with the same index set. Let $D : E \to F$ a differential operator such that, with respect to the above decomposition: D is upper triangular; all entries, $D_{ji} : L_i \to M_j$, have the same degree; and, the diagonal entries D_{ii} are elliptic. Then, the total operator, D, is elliptic.*

Proof. Given our assumptions, the symbol of D is a matrix $\sigma_\xi(D)$ with entries $\sigma_\xi(D)_{ij} = \sigma_\xi(D_{ij})$, so the result follows by linear algebra. □

Clearly, $\check{\mathbb{D}}$ is upper triangular, and by the results quoted before the Lemma, the diagonal entries of $\check{\mathbb{D}}$ are elliptic. Moreover, the off-diagonal elements are given by the $\mathscr{S}$ and $\mathscr{T}$ maps, which are all degree-1 operators. Thus, the conditions for Lemma 1 are fulfilled, and we conclude the $(\mathscr{E},\check{\mathbb{D}})$ complex is elliptic.

Lemma 1 on the level of the Laplacians gives the same result.

References

1. Anderson, L.B., Gray, J., Sharpe, E.: Algebroids, Heterotic Moduli Spaces and the Strominger System. JHEP **1407**, 037 (2014). DOI 10.1007/JHEP07(2014)037
2. Ashmore, A., Coimbra, A., Strickland-Constable, C., Svanes, E.E., Tennyson, D.: Topological G_2 and Spin(7) strings at 1-loop from double complexes. JHEP **02**, 089 (2022). DOI 10.1007/JHEP02(2022)089
3. Ashmore, A., De La Ossa, X., Minasian, R., Strickland-Constable, C., Svanes, E.E.: Finite deformations from a heterotic superpotential: holomorphic Chern-Simons and an L_∞ algebra. JHEP **10**, 179 (2018). DOI 10.1007/JHEP10(2018)179
4. Ashmore, A., Ibarra, J.J.M., McNutt, D.D., Strickland-Constable, C., Svanes, E.E., Tennyson, D., Winje, S.: A heterotic Kodaira-Spencer theory at one-loop. JHEP **10**, 130 (2023). DOI 10.1007/JHEP10(2023)130
5. Ashmore, A., Strickland-Constable, C., Tennyson, D., Waldram, D.: Heterotic backgrounds via generalised geometry: moment maps and moduli. JHEP **11**, 071 (2020). DOI 10.1007/JHEP11(2020)071
6. Atiyah, M.F.: Complex analytic connections in fibre bundles. Transactions of the American Mathematical Society **85**(1), 181–207 (1957). URL http://www.jstor.org/stable/1992969
7. del Barco, V., Moroianu, A., Raffero, A.: Purely coclosed $\mathbf{G_2}$-structures on 2-step nilpotent Lie groups (2020)
8. de Boer, J., de Medeiros, P., El-Showk, S., Sinkovics, A.: G(2) Hitchin functionals at one loop. Class. Quant. Grav. **25**, 075,006 (2008). DOI 10.1088/0264-9381/25/7/075006
9. de Boer, J., de Medeiros, P., El-Showk, S., Sinkovics, A.: Open G(2) strings. JHEP **02**, 012 (2008). DOI 10.1088/1126-6708/2008/02/012
10. de Boer, J., Naqvi, A., Shomer, A.: The Topological G(2) string. Adv. Theor. Math. Phys. **12**(2), 243–318 (2008)

11. Candelas, P., De La Ossa, X., McOrist, J., Sisca, R.: The Universal Geometry of Heterotic Vacua. JHEP **02**, 038 (2019). DOI 10.1007/JHEP02(2019)038
12. Candelas, P., De La Ossa, X.C., Green, P.S., Parkes, L.: A Pair of Calabi-Yau manifolds as an exactly soluble superconformal theory. Nucl. Phys. B **359**, 21–74 (1991). DOI 10.1016/0550-3213(91)90292-6
13. Clarke, A., Garcia-Fernandez, M., Tipler, C.: Moduli of G_2 structures and the Strominger system in dimension 7 (2016)
14. Clarke, A., Garcia-Fernandez, M., Tipler, C.: T-dual solutions and infinitesimal moduli of the G_2-Strominger system. Adv. Theor. Math. Phys. **26**(6), 1669–1704 (2022). DOI 10.4310/ATMP.2022.v26.n6.a3
15. Donaldson, S., Segal, E.: Gauge theory in higher dimensions, ii. arXiv preprint arXiv:0902.3239 (2009)
16. Donaldson, S., Thomas, R.: Gauge theory in higher dimensions. The geometric universe (Oxford, 1996) pp. 31–47 (1998)
17. Fernandez, M., Ivanov, S., Ugarte, L., Villacampa, R.: Compact supersymmetric solutions of the heterotic equations of motion in dimensions 7 and 8. Adv. Theor. Math. Phys. **15**(2), 245–284 (2011). DOI 10.4310/ATMP.2011.v15.n2.a1
18. Fernández, M., Ugarte, L.: Dolbeault cohomology for g2-manifolds. Geometriae Dedicata **70**(1), 57–86 (1998)
19. Fino, A., Martín-Merchán, L., Raffero, A.: The twisted G_2 equation for strong G_2-structures with torsion. arXiv e-prints arXiv:2306.07128 (2023).
20. Fiset, M.A., Quigley, C., Svanes, E.E.: Marginal deformations of heterotic G_2 sigma models. JHEP **02**, 052 (2018). DOI 10.1007/JHEP02(2018)052
21. Friedrich, T., Ivanov, S.: Parallel spinors and connections with skew-symmetric torsion in string theory. ArXiv Mathematics e-prints (2001)
22. Friedrich, T., Ivanov, S.: Killing spinor equations in dimension 7 and geometry of integrable G(2) manifolds. J. Geom. Phys. **48**, 1 (2003). DOI 10.1016/S0393-0440(03)00005-6
23. Galdeano, M., Stecker, L.: The heterotic G_2 system with reducible characteristic holonomy. arXiv e-prints (2024)
24. Garcia-Fernandez, M., Molina, R.G., Streets, J.: Pluriclosed flow and the Hull-Strominger system (2024)
25. Garcia-Fernandez, M., Rubio, R., Tipler, C.: Infinitesimal moduli for the Strominger system and generalized Killing spinors (2015)
26. Garcia-Fernandez, M., Rubio, R., Tipler, C.: Holomorphic string algebroids. Trans. Am. Math. Soc. **373**(10), 7347–7382 (2020). DOI 10.1090/tran/8149
27. Garcia-Fernandez, M., Rubio, R., Tipler, C.: Gauge theory for string algebroids. J. Diff. Geom. **128**(1), 77–152 (2024). DOI 10.4310/jdg/1721075260
28. Gauntlett, J.P., Martelli, D., Pakis, S., Waldram, D.: G structures and wrapped NS5-branes. Commun.Math.Phys. **247**, 421–445 (2004). DOI 10.1007/s00220-004-1066-y
29. Gilkey, P.: Invariance theory, the heat equation, and the Atiyah-Singer index theorem. Paper version originally published by Publish or Perish Inc., USA, 1984, Electronic reprint c© 1996 Peter B. Gilkey. (1996)
30. Gran, U., Gutowski, J.B., Papadopoulos, G.: On supersymmetric Anti-de-Sitter, de-Sitter and Minkowski flux backgrounds (2016)
31. Gran, U., Lohrmann, P., Papadopoulos, G.: The Spinorial geometry of supersymmetric heterotic string backgrounds. JHEP **0602**, 063 (2006). DOI 10.1088/1126-6708/2006/02/063
32. Gran, U., Papadopoulos, G., Roest, D.: Supersymmetric heterotic string backgrounds. Phys.Lett. **B656**, 119–126 (2007). DOI 10.1016/j.physletb.2007.09.024
33. Gray, J., Larfors, M., Lüst, D.: Heterotic domain wall solutions and SU(3) structure manifolds. JHEP **1208**, 099 (2012). DOI 10.1007/JHEP08(2012)099
34. Gunaydin, M., Nicolai, H.: Seven-dimensional octonionic Yang-Mills instanton and its extension to an heterotic string soliton. Phys. Lett. **B351**, 169–172 (1995). DOI 10.1016/0370-2693(95)00375-U. [Addendum: Phys. Lett.B376,329(1996)]
35. Gurrieri, S., Lukas, A., Micu, A.: Heterotic on half-flat. Phys.Rev. **D70**, 126,009 (2004). DOI 10.1103/PhysRevD.70.126009

36. Hull, C.: Compactifications of the Heterotic Superstring. Phys.Lett. **B178**, 357 (1986). DOI 10.1016/0370-2693(86)91393-6
37. Joyce, D.: Conjectures on counting associative 3-folds in G_2-manifolds (2016)
38. Klaput, M., Lukas, A., Matti, C.: Bundles over Nearly-Kahler Homogeneous Spaces in Heterotic String Theory. JHEP **1109**, 100 (2011). DOI 10.1007/JHEP09(2011)100
39. Kupka, J., Strickland-Constable, C., Svanes, E.E., Tennyson, D., Valach, F.: BPS complexes and Chern–Simons theories from G-structures in gauge theory and gravity (2024)
40. Lotay, J.D., Earp, H.N.S.: The heterotic G_2 system on contact Calabi Yau 7-manifolds. Trans. Am. Math. Soc. Ser. B **10**(26), 907–943 (2023). DOI 10.1090/btran/129
41. Lukas, A., Matti, C.: G-structures and Domain Walls in Heterotic Theories. JHEP **1101**, 151 (2011). DOI 10.1007/JHEP01(2011)151
42. McOrist, J., Sticka, M., Svanes, E.E.: The physical moduli of heterotic G_2 string compactifications (2024)
43. McOrist, J., Svanes, E.E.: Heterotic quantum cohomology. JHEP **11**, 096 (2022). DOI 10.1007/JHEP11(2022)096
44. Melnikov, I.V., Sharpe, E.: On marginal deformations of (0,2) non-linear sigma models. Phys.Lett. **B705**, 529–534 (2011). DOI 10.1016/j.physletb.2011.10.055
45. Micu, A.: Heterotic compactifications and nearly-Kahler manifolds. Phys.Rev. **D70**, 126,002 (2004). DOI 10.1103/PhysRevD.70.126002
46. de la Ossa, X., Galdeano, M.: Families of solutions of the heterotic G_2 system (2021)
47. de la Ossa, X., Hardy, E., Svanes, E.E.: The Heterotic Superpotential and Moduli. JHEP **01**, 049 (2016). DOI 10.1007/JHEP01(2016)049
48. de la Ossa, X., Larfors, M., Magill, M.: Almost contact structures on manifolds with a G_2 structure. Adv. Theor. Math. Phys. **26**(1), 143–216 (2022). DOI 10.4310/ATMP.2022.v26.n1.a3
49. de la Ossa, X., Larfors, M., Magill, M., Svanes, E.E.: Superpotential of three dimensional $\mathcal{N}$ = 1 heterotic supergravity. JHEP **01**, 195 (2020). DOI 10.1007/JHEP01(2020)195
50. de la Ossa, X., Larfors, M., Svanes, E.E.: Exploring $SU(3)$ structure moduli spaces with integrable G_2 structures. Adv. Theor. Math. Phys. **19**, 837–903 (2015). DOI 10.4310/ATMP.2015.v19.n4.a5
51. de la Ossa, X., Larfors, M., Svanes, E.E.: Infinitesimal moduli of G2 holonomy manifolds with instanton bundles. JHEP **11**, 016 (2016). DOI 10.1007/JHEP11(2016)016
52. de la Ossa, X., Larfors, M., Svanes, E.E.: Restrictions of Heterotic G_2 Structures and Instanton Connections. In: Proceedings, Nigel Hitchin's 70th Birthday Conference : Geometry and Physics : A Festschrift in honour of Nigel Hitchin : 2 volumes: Aarhus, Denmark, Oxford, UK, Madrid, Spain, September 5-16, 2016 (2017)
53. de la Ossa, X., Larfors, M., Svanes, E.E.: The Infinitesimal Moduli Space of Heterotic G_2 Systems. Commun. Math. Phys. **360**(2), 727–775 (2018). DOI 10.1007/s00220-017-3013-8
54. de la Ossa, X., Svanes, E.E.: Holomorphic Bundles and the Moduli Space of N=1 Supersymmetric Heterotic Compactifications. JHEP **10**, 123 (2014). DOI 10.1007/JHEP10(2014)123
55. Pati, V.: Elliptic complexes and index theory. Https://www.isibang.ac.in/ statmath/resource/elliptic.pdf
56. Pestun, V., Witten, E.: The Hitchin functionals and the topological B-model at one loop. Lett. Math. Phys. **74**, 21–51 (2005). DOI 10.1007/s11005-005-0007-9
57. Reyes Carrion, R.: Some special geometries defined by Lie groups. Ph.D. thesis, University of Oxford (1993)
58. Reyes Carrion, R.: A generalization of the notion of instanton. Differ. Geom. Appl. **8**, 1–20 (1998). DOI 10.1016/S0926-2245(97)00013-2
59. Silva, A.A.d., Garcia-Fernandez, M., Lotay, J.D., Earp, H.N.S.: Coupled G_2-instantons (2024)
60. Strominger, A.: Superstrings with Torsion. Nucl.Phys. **B274**, 253 (1986). DOI 10.1016/0550-3213(86)90286-5
61. Thomas, R.: Gauge theory on calabi-yau manifolds. Ph.D. thesis, University of Oxford (1997)
62. Witten, E.: Quantum Field Theory and the Jones Polynomial. Commun. Math. Phys. **121**, 351–399 (1989). DOI 10.1007/BF01217730

Calabi-Yau threefolds across quadratic singularities

Sébastien Picard

Abstract We examine aspects of non-Kähler geometry in the context of Calabi-Yau degenerations and resolutions. We begin by reviewing some basics of Calabi-Yau geometry in Section 1, describe topological features of the conifold transition in Section 2, and survey recent developments on the geometrization of conifold transitions in Section 3.

1 Calabi-Yau Threefolds

1.1 Definitions

The topic of this survey is the Calabi-Yau threefold. A central theme is how natural questions about these spaces can lead to non-Kähler outcomes. Arriving at these non-Kähler structures sets a departure point for various new possible directions of research. We begin in Section 1 with a review of the key properties of Calabi-Yau threefolds.

Definition 1. We take the definition of a Calabi-Yau threefold X to be a compact complex manifold with $\dim_{\mathbb{C}} X = 3$ which is projective, satisfies $H^1(X,\mathbb{C}) = 0$, and admits a holomorphic volume form Ω.

Recall that a holomorphic volume form on X is given by a $(3,0)$-form Ω such that

$$\Omega \overset{\text{loc}}{=} f(z)\, dz^1 \wedge dz^2 \wedge dz^3$$

where $f(z)$ is a local nowhere vanishing holomorphic function.

Sébastien Picard
UBC, 1984 Mathematics Rd, Vancouver, Canada, V6T 1Z2
e-mail: spicard@math.ubc.ca

D. R. Wood et al. (eds.), *2024 MATRIX Annals, Part II*, MATRIX Book Series 8,
https://doi.org/10.1007/978-3-032-16206-9_17

Remark 1. The definition of a Calabi-Yau threefold is not standardized in the literature, and some setups do not require the vanishing of $H^1(X,\mathbb{C})$ or generalize the existence of a holomorphic volume form Ω to $c_1(X)=0$. A consequence of the Calabi-Yau [13, 113] theorem is that a Kähler manifold X with $c_1(X)=0$ must have K_X holomorphically torsion, but not necessarily trivial. Projectivity in Definition 1 is redundant because $H^1(X,\mathbb{C})=0$. This then implies $h^{0,2}=0$, and projectivity then follows from the Kodaira embedding theorem.

Example 1. A basic example of a Calabi-Yau threefold is the following quintic threefold:

$$X=\left\{\sum_{i=0}^{4} Z_i^5=0\right\}\subset \mathbb{P}^4.$$

This defines a smooth complex manifold by the implicit function theorem, since a hypersurface $\{P=0\}\subseteq \mathbb{P}^n$ with P a homogeneous polynomial is smooth if there is no non-zero point where simultaneously $P=0$ and $DP=0$. The holomorphic volume form in this case is

$$\Omega \overset{\text{loc}}{=} \frac{dw_1\wedge dw_2\wedge dw_3}{w_4^4}, \qquad w_i=\frac{Z_i}{Z_0}$$

over the open set $U=\{Z_0\neq 0,\ w_4\neq 0\}$. This formula can be verified to glue on overlaps of similar local open sets to define a global section $\Omega\in H^0(X,\Omega^3)$. That X satisfies $H^1(X,\mathbb{C})=0$ follows from the Lefschetz hyperplane theorem.

Example 2. Any homogeneous degree 5 polynomial P will also define a Calabi-Yau threefold $\{P=0\}\subseteq\mathbb{P}^4$ provided the zero locus is smooth.

Example 3. For more examples of Calabi-Yau threefolds beyond quintics $\{P=0\}\subseteq \mathbb{P}^4$, we refer to for example Hübsch's book [60].

The Hodge diamond of a Calabi-Yau threefold has two parameters:

$$\begin{array}{ccccccc}
 & & & 1 & & & \\
 & & 0 & & 0 & & \\
 & 0 & & h^{1,1} & & 0 & \\
1 & & h^{2,1} & & h^{2,1} & & 1 \\
 & 0 & & h^{1,1} & & 0 & \\
 & & 0 & & 0 & & \\
 & & & 1 & & &
\end{array}$$

Mirror symmetry [31, 55, 17, 73, 69, 101] predicts that Calabi-Yau threefolds come in pairs $(X,\check{X})$ exchanging the two parameters $(h^{1,1},h^{2,1})$.

$$h^{1,1}(X)=h^{2,1}(\check{X}),\quad h^{1,1}(\check{X})=h^{2,1}(X).$$

A first indication of the role of non-Kähler manifolds comes from the inherent asymmetry of Calabi-Yau threefolds where $h^{2,1}$ may vanish while $h^{1,1}$ cannot. Indeed, a Kähler metric $\omega>0$ creates a non-zero Kähler class $[\omega]\in H^{1,1}(X)$. Kontsevich

[70] has suggested that the mirror theory of curve enumeration on a threefold X with $h^{2,1}(X)=0$ should involve Hodge structures on a non-Kähler complex threefold $\check{X}$ with $h^{1,1}(\check{X})=0$.

We now review the significance of the two parameters $(h^{1,1},h^{2,1})$.

1.2 Discussion of $h^{1,1}$

Calabi-Yau threefolds are studied in differential geometry because they support solutions to the Ricci-flat equation. Recall that on a Riemannian manifold (M,g), the Riemann curvature tensor is a second order invariant of g_{ij} given by

$$R_{pq}{}^{k}{}_{j} = \partial_p \Gamma_q{}^{k}{}_{j} + \Gamma_p{}^{k}{}_{r}\Gamma_q{}^{r}{}_{j} - (p \leftrightarrow q),$$

with

$$\Gamma_i{}^{k}{}_{j} = \frac{g^{kp}}{2}(-\partial_p g_{ij} + \partial_i g_{pj} + \partial_j g_{ip}).$$

The Ricci tensor is given by

$$R_{ij} = -R_{ik}{}^{k}{}_{j},$$

and a fundamental equation in geometry and physics is the Ricci-flat equation

$$R_{ij} = 0.$$

On a Calabi-Yau threefold X, one can look for solutions of the following form. Choose a background reference Kähler metric $g_{\mu\bar{\nu}}$ on X such as the pullback of the Fubini-Study metric in the embedding $\iota : X \to \mathbb{P}^N$, and set the ansatz

$$\tilde{g}_{\mu\bar{\nu}} = g_{\mu\bar{\nu}} + \partial_\mu \partial_{\bar{\nu}} u > 0 \tag{1}$$

for an unknown potential function $u \in C^\infty(X)$. Let Greek indices denote holomorphic coordinates $\{z^\alpha\}$ on the complex manifold X. Kähler [64] computed the Ricci tensor for ansatz (1) and derived the equation

$$R_{\alpha\bar{\beta}} = \partial_\alpha \partial_{\bar{\beta}} \log |\Omega|^2_{\tilde{g}}, \quad R_{\alpha\beta} = 0$$

where

$$|\Omega|^2_{\tilde{g}} \overset{\text{loc}}{=} \frac{f(z)\overline{f(z)}}{\det \tilde{g}_{\mu\bar{\nu}}}.$$

We can then look for solutions of the Ricci-flat equation by setting $|\Omega|_{\tilde{g}} = 1$, so that the equation to solve becomes

$$\det(g_{\alpha\bar{\beta}} + u_{\alpha\bar{\beta}}) = e^f \det g_{\alpha\bar{\beta}} \tag{2}$$

where $e^f = |\Omega|_g^2$ is given. This geometric complex Monge-Ampère equation is an adaptation of the fundamental PDE

$$\det D^2 u = \psi > 0,$$

on a domain in $\mathbb{R}^n$. The complex Monge-Ampère equation (2) was solved by Yau.

Theorem 1 (Yau's theorem [113]). *There exists a unique solution* $u \in C^\infty(X)$ *to* (2) *with* $\sup_X u = 0$.

Consequently, there exists Ricci-flat metrics on a Calabi-Yau threefold X. The ansatz (1) is interpreted as finding a unique Kähler Ricci-flat metric in a given Kähler class $[\omega] \in H^{1,1}(X,\mathbb{R})$, where a Kähler metric g defines a form $\omega \in \Omega^{1,1}(X,\mathbb{R})$ via $\omega = i g_{\mu\bar{\nu}}\, dz^\mu \wedge d\bar{z}^\nu$ and so by the $\partial\bar{\partial}$-lemma the ansatz (1) is equivalent to setting $[\tilde{\omega}] = [\omega]$.

The construction comes in families parametrized by the choice of background Kähler metric $g_{\mu\bar{\nu}}$, but as noted above really the parameter is the Kähler class $[\omega] \in H^{1,1}(X,\mathbb{R})$. Thus $h^{1,1}$ counts the dimension of the space of Kähler Ricci-flat metrics on X with fixed complex structure.

1.3 Discussion of $h^{2,1}$

Having discussed the parameter $h^{1,1}$, we now give the interpretation of the parameter $h^{2,1}$.

Recall that a family of complex manifolds over a base $B \subseteq \mathbb{C}^k$ with $0 \in B$ is defined as a proper holomorphic submersion

$$\pi : \mathscr{X} \to B$$

where $\mathscr{X}$ is a complex manifold. The fibers $X_b = \pi^{-1}(b)$ for $b \in B$ are then all compact complex manifolds varying holomorphically in the parameter b. Ehresmann's theorem (see e.g. [68] for an exposition) states that after possibly replacing B with a neighborhood of 0, there exists a diffeomorphism

$$\Psi : X_0 \times B \xrightarrow{\cong} \mathscr{X}$$

such that $\pi \circ \Psi = \mathrm{pr}_2$, with pr_2 the projection to the second factor. Therefore all manifolds X_b are diffeomorphic. Each X_b comes with a complex structure tensor J_b, and via the diffeomorphism we obtain a family of complex structures also denoted J_b over the fixed differentiable manifold X_0.

Remark 2. Recall that given a complex manifold X with holomorphic coordinates $\{z^\alpha\}$, the corresponding complex structure tensor $J \in \Gamma(\mathrm{End}\, T_{\mathbb{C}}X)$ is defined by

$$J^\alpha{}_\beta = i\delta^\alpha{}_\beta \quad J^{\bar{\alpha}}{}_{\bar{\beta}} = -i\delta^\alpha{}_\beta, \quad J^\alpha{}_{\bar{\beta}} = J^{\bar{\alpha}}{}_\beta = 0,$$

in a holomorphic coordinate system. Here we use index notation for components of a tensor $A \in \Gamma(\mathrm{End}\, T_{\mathbb{C}}X)$ so that e.g.

$$A\left(\frac{\partial}{\partial z^{\beta}}\right) = A^{\alpha}{}_{\beta}\frac{\partial}{\partial z^{\alpha}} + A^{\bar{\alpha}}{}_{\beta}\frac{\partial}{\partial \bar{z}^{\bar{\alpha}}}.$$

In a smooth real coordinate system, a complex structure tensor $J \in \Gamma(\mathrm{End}\, TX)$ is characterized by the property

$$J^2 = -id, \quad N(J) = 0 \tag{3}$$

where $N \in \Omega^2(TM)$ is the Nijenhuis tensor given by

$$N^k{}_{ij} = \frac{1}{4}\Big(J^k{}_p \partial_j J^p{}_i + \partial_p J^k{}_j J^p{}_i - (i \leftrightarrow j)\Big).$$

The Newlander-Nirenberg theorem (see e.g. [29]) states that the existence of a $J \in \Gamma(\mathrm{End}\, TX)$ satisfying (3) on a smooth manifold X is equivalent to the existence of holomorphic coordinate charts. Such a J splits $T_{\mathbb{C}}X = T^{1,0}X \oplus T^{0,1}X$ into the $+i$ and $-i$ eigenvalues of J, and the interpretation of N is that $N = 0$ if and only if $[U,V] \in T^{1,0}X$ for all $U,V \in \Gamma(T^{1,0}X)$.

Given a family of complex structures J_t varying smoothly with a parameter $t \in (-\varepsilon, \varepsilon)$ on a fixed smooth manifold X, differentiation along the path defines the fluctuation tensor

$$\eta = \frac{d}{dt}\Big|_{t=0} J_t.$$

One can verify the following calculations :

1. Differentiating $J_t^2 = -id$ shows that
$$\eta \in \Omega^{0,1}(T^{1,0}X),$$
so that for example $\eta^{\bar{\alpha}}{}_{\bar{\beta}} = 0$ in holomorphic coordinates at $t = 0$.
2. Differentiating $N(J_t) = 0$ leads to the constraint
$$\bar{\partial}\eta = 0,$$
so that $[\eta]$ defines a class in $H^{0,1}(X, T^{1,0}X)$.
3. If the family J_t comes from $J_t = (\Theta_t)_*^{-1} J_0 (\Theta_t)_*$ where $\Theta_t : X \to X$ is a smoothly varying family of diffeomorphisms, then
$$\eta = \bar{\partial}V, \quad V \in \Gamma(T^{1,0}X)$$
so that $[\eta] = [0]$ in $H^{0,1}(X, T^{1,0}X)$.

In summary, a family of complex structures produces a cohomology class in $H^{0,1}(X, T^{1,0}X)$, and deformations just coming from families of diffeomorphisms are not counted.

Remark 3. On a Calabi-Yau threefold, $K_X \cong \mathcal{O}_X$ and this parameter space has dimension

$$h^{0,1}(T^{1,0}X) = h^{0,1}(T^{1,0}X \otimes K_X) = h^{2,1}.$$

For further analysis of the parameter space of complex structures on a Calabi-Yau threefold, see Candelas-de la Ossa [16].

The natural question that arises is the inverse problem: does a given class $[\eta] \in H^{0,1}(X, T^{1,0}X)$ come from a family $(X, J(t))$ of complex structures with $[\frac{d}{dt}|_{t=0}J] = [\eta]$? It does, and this is known as the BTT theorem (see e.g. [87] for a recent exposition).

Theorem 2 (Bogomolov-Tian-Todorov Theorem [102, 104]). *Let X be a compact Kähler manifold admitting a holomorphic volume form. There is a family of complex manifolds*

$$\pi : \mathscr{X} \to B, \quad \pi^{-1}(0) = X$$

where B is a neighborhood of the origin in $H^{0,1}(X, T^{1,0}X)$. Hence given any $[\eta] \in H^{0,1}(X, T^{1,0}X)$, there exists a family J_t of complex structures varying smoothly with $t \in (-\varepsilon, \varepsilon)$ such that $[\frac{d}{dt}|_{t=0}J] = [\eta]$.

We note that there is an analog of this theorem in the non-Kähler setting [87, 112] provided X satisfies the $\partial\bar{\partial}$-lemma in addition to admitting a holomorphic volume form. Later in our discussion, we will see that the $\partial\bar{\partial}$-lemma is preserved throughout the connected web of threefolds [74, 39, 71] even across Kähler to non-Kähler deformations.

1.4 Rational curves

A key topic in the study of Calabi-Yau threefolds is its set of rational curves. These special submanifolds can be studied from various different points of view. For example, the extraordinary work of Candelas-de la Ossa-Green-Parkes [17] shocked the algebraic geometry community by predicting a formula for the number of rational curves of degree d by methods of string theory.

Let X be a compact complex threefold with holomorphic volume form. Let $C \subseteq X$ be a smooth complex submanifold of complex dimension 1 with $C \cong \mathbb{P}^1$. The exact sequence defining the normal bundle is

$$0 \to TC \to TX|_C \to \mathscr{N}_C \to 0.$$

The first Chern class of this sequence satisfies

$$c_1(TX|_C) = c_1(TC) + c_1(\mathscr{N}_C)$$

and so $c_1(\mathscr{N}_C) = -2$. Grothendieck's classification of holomorphic bundles over $\mathbb{P}^1$ implies that the normal bundle of C must be isomorphic to

$$\mathscr{O}(a) \oplus \mathscr{O}(-a-2) \to \mathbb{P}^1.$$

The case of $\mathscr{O}(-1)^{\oplus 2}$ is distinguished by its symmetry and that it is the only case in which the curve is rigid, as the other bundles admit infinitesimal deformations in $H^0(\mathscr{N}_C)$.

Such $(-1,-1)$-curves are expected to exist in abundance on Calabi-Yau threefolds [37, 70]. Work of Clemens [24] and Katz [66] guarantees the existence of $(-1,-1)$-curves on the generic quintic threefold.

Example 4. Here is an explicit example of a $(-1,-1)$-curve following [66]. Consider the quintic threefold given by

$$X = \{f_2Z_2 + f_3Z_3 + f_4Z_4 = 0\} \subseteq \mathbb{P}^4,$$

and the embedded holomorphic curve

$$C = \{Z_2 = Z_3 = Z_4 = 0\} \subseteq X,$$

where we take the specific f_i to be

$$X = \left\{(Z_0^4 + Z_2^4)Z_2 + (Z_0^2Z_1^2 + Z_3^4)Z_3 + (Z_1^4 + Z_4^4)Z_4 = 0\right\} \subseteq \mathbb{P}^4,$$

so that X is smooth. We will compute the normal bundle of $C \subset X$. Recall the definition of the normal bundle: let $p \in C$, and suppose $p \in U \subset X$ is a coordinate chart such that

$$C \cap U = \{y = 0\}$$

where (x_1, y_1, y_2) are holomorphic coordinates on U. Suppose also $p \in \tilde{U}$ with another such submanifold coordinate system $(\tilde{x}_1, \tilde{y}_1, \tilde{y}_2)$. Then on overlaps $U \cap \tilde{U} \cap C$, the transition function

$$\left.\frac{\partial \tilde{y}}{\partial y}\right|_C$$

defines the data of the normal bundle $\mathscr{N}_C$.

- Let U be an open set in $\{Z_0 \neq 0\} \subseteq X$ near C. Here we use local coordinates on X given by (w_1, w_3, w_4) where $w_i = Z_i/Z_0$. The equation of X is

$$(1 + w_2^4)w_2 + (w_1^2 + w_3^4)w_3 + (w_1^4 + w_4^4)w_4 = 0, \tag{4}$$

  and the curve appears as

$$C \cap U = \begin{cases} w_3 = 0 \\ w_4 = 0. \end{cases}$$

  By the inverse function theorem, $w_2(w_1, w_3, w_4)$ is a holomorphic function of the remaining coordinates.

- Let $\tilde{U}$ be an open set in $\{Z_1 \neq 0\} \subseteq X$ near C. Here we use local coordinates on X given by $(\tilde{w}_0, \tilde{w}_2, \tilde{w}_3)$ where $\tilde{w}_i = Z_i/Z_1$. The curve appears as

$$C \cap \tilde{U} = \begin{cases} \tilde{w}_2 = 0 \\ \tilde{w}_3 = 0. \end{cases}$$

- We have covered the curve by open charts: $C \subseteq U \cup \tilde{U}$. We conform to earlier conventions by setting $(x, y_1, y_2) = (w_1, w_3, w_4)$ and $(\tilde{x}, \tilde{y}_1, \tilde{y}_2) = (\tilde{w}_0, \tilde{w}_2, \tilde{w}_3)$. Then we recognize x and $\tilde{x}$ as the coordinates along C and since $\tilde{x} = x^{-1}$ we see $C = \mathbb{P}^1$. The transition function of the normal bundle may be computed by implicit differentiation of (4) which gives

$$\left.\frac{\partial \tilde{y}}{\partial y}\right|_C = \begin{bmatrix} -x & -x^3 \\ x^{-1} & 0 \end{bmatrix}.$$

This transition function is a disguise of $\mathcal{O}(-1) \oplus \mathcal{O}(-1)$. Recall that two sets of transition functions $\{g_{\alpha\beta}, U_\alpha \cap U_\beta\}$ and $\{\hat{g}_{\alpha\beta}, U_\alpha \cap U_\beta\}$ define isomorphic bundles if there exists $\lambda_\alpha : U_\alpha \to GL(k, \mathbb{R})$ with

$$\hat{g}_{\alpha\beta} = \lambda_\alpha g_{\alpha\beta} \lambda_\beta^{-1}.$$

In this example, one can find 2×2 matrices λ_1, λ_2 such that

$$\begin{bmatrix} x & 0 \\ 0 & x \end{bmatrix} = \lambda_1 \begin{bmatrix} -x & -x^3 \\ x^{-1} & 0 \end{bmatrix} \lambda_2^{-1}$$

with the matrix on the left the familiar transition function defining $\mathcal{O}(-1) \oplus \mathcal{O}(-1) \to \mathbb{P}^1$.

We will later use a local model for the holomorphic curve $C \subset X$, but we remark that the holomorphic tubular neighborhood is generally false: if $S \subset X$ is a compact holomorphic submanifold, there is not necessarily a neighborhood $U \subset X$ of S which is biholomorphic to a neighborhood of the zero section in the total space of the normal bundle $\mathcal{N}_S$. When this is true, S is sometimes said to satisfy the formal neighborhood principle. There is much literature on this subject starting with foundational work of Grauert [51] in the case of codimension one submanifolds. In general codimension, there is work of [1] where the condition for a formal neighborhood involves vanishing of $H^1(S, T_S \otimes \mathrm{Sym}^k(\mathcal{N}_S^*))$ and $H^1(S, \mathcal{N}_S \otimes \mathrm{Sym}^{k+1}(\mathcal{N}_S^*))$.

Returning to $(-1,-1)$ curves C on a Calabi-Yau threefold X, it is well-known (e.g. [70]) that there exists a neighborhood $U \subset X$ which is biholomorphic to a neighborhood of the zero section in $\mathcal{O}(-1)^{\oplus 2} \to \mathbb{P}^1$.

2 Crossing singularities

This section describes a mechanism known as a conifold transition, and how its study naturally leads to non-Kähler geometry.

2.1 Rolling in the landscape

Consider the set of all possible Calabi-Yau threefolds. It was noticed by string theorists Green-Hübsch and Candelas-Green-Hübsch [20, 53, 54] that there is a way to travel in this landscape by a process known as a conifold transition. This led to the idea of viewing Calabi-Yau threefolds not as isolated objects, but as part of a unified moduli space. We will define a conifold transition momentarily, but let us first state Reid's conjecture.

Conjecture 1. [90] All Calabi-Yau threefolds can be connected by a sequence of conifold transitions.

Early work on the algebro-geometric foundations of conifold transitions goes back to Clemens [25] and Friedman [36]. A conifold transition, which we will denote by

$$\hat{X} \to X_0 \rightsquigarrow X_t,$$

is a two step process. First, N holomorphic $(-1,-1)$-curves on $\hat{X}$ are contracted to points producing a complex analytic space X_0 with N ordinary double point singularities. Second, the complex structure of X_0 is deformed in a family such that X_t are smooth complex manifolds for $t \neq 0$.

Remark 4. One can also study an extension of Reid's conjecture where $\hat{X} \to X_0 \rightsquigarrow X_t$ more generally denotes a birational contraction followed by smoothing. In this generality, X_0 admits more complicated singularities than ordinary double points. We focus our attention on conifold transition in this survey, and refer to Gross [57] for an alternate version of Reid's conjecture on connecting Calabi-Yau threefolds by general geometric transitions.

2.1.1 Topological description

Broadly speaking, a conifold transition is a type of topological surgery of 6-manifolds. We start with the topological implications and return later to the complex analytic definition. At the level of topology, N sets of the form $(D^4 \times S^2)_i$ are removed from one manifold and replaced by $(S^3 \times D^3)_i$ in the other by gluing along the common boundary $S^3 \times S^2$. Call the first manifold $\hat{X}$ and the second X_t. The first Betti number b_1 does not change.

Let $[C_i] \in H_2(D^4 \times S^2)_i$ and $[L_i] \in H_3(S^3 \times D^3)_i$ be generators of homology. By the excision principle for Euler characteristic, [25, 93, 72] the topological change is captured by the formula

$$N = \rho + \mu \tag{5}$$

where:

$$\begin{aligned} N &= \text{number of } (D^4 \times S^2)_i \text{ removed from } \hat{X}, \\ \rho &= \text{number of independent 2-cycles } [C_i] \in H_2(\hat{X}, \mathbb{C}), \\ \mu &= \text{number of independent 3-cycles } [L_i] \in H_3(X_t, \mathbb{C}). \end{aligned}$$

We also have

$$\begin{aligned} b_2 &\mapsto b_2 - \rho \\ b_3 &\mapsto b_3 + 2\mu. \end{aligned}$$

We see that as $\hat{X}$ deforms to X_t across a conifold transition $\hat{X} \to X_0 \rightsquigarrow X_t$, its Betti numbers jump as $b_2 \downarrow$ and $b_3 \uparrow$.

2.1.2 The definition of a conifold transition

We now give the definition of a conifold transition following Friedman [37]. Let $\hat{X}$ be a Calabi-Yau threefold. The deformation process of a conifold transition, denoted $\hat{X} \to X_0 \rightsquigarrow X_t$, is defined as follows:

1. Find disjoint $(-1,-1)$-curves $C_i \subseteq \hat{X}$.
2. Contract the curves C_i to points p_i to form a complex analytic space X_0 with ordinary double point singularities at p_i.
3. Realize X_0 as the central fiber of a smoothing $\pi : \mathscr{X} \to \Delta$.

By a smoothing of X_0, we mean a proper flat map

$$\pi : \mathscr{X} \to \Delta, \quad \pi^{-1}(0) = X_0$$

with $\mathscr{X}$ a smooth complex fourfold, $\Delta \subseteq \mathbb{C}$ the unit disk, and $\pi^{-1}(t) = X_t$ smooth complex manifolds for $t \neq 0$. Furthermore, we require that a neighborhood $\mathscr{U} \subseteq \mathscr{X}$ of each point p_i is locally analytically isomorphic to the local model

$$\pi : \mathscr{V} \to \mathbb{C}, \quad \pi(z,t) = t,$$

where

$$\mathscr{V} = \{z_1^2 + z_2^2 + z_3^2 + z_4^2 = t\} \subseteq \mathbb{C}^4 \times \mathbb{C}.$$

Recall that the notion of a flat map $\varphi : X \to Y$ generalizes the notion of a holomorphic submersion by allowing singular fibers. A holomorphic map $\varphi : X \to Y$ between connected complex manifolds is flat if and only if it is an open map. If φ is flat in p and the fiber $X_{\varphi(p)}$ is a manifold at p, then φ is a submersion at p. We refer to Fischer's book [35] for these statements and more.

Remark 5. Realizing step (1) is already part of Reid's conjecture. It is an open problem whether every Calabi-Yau threefold X admits $(-1,-1)$-curves, and Reid-Friedman [90, 37] speculate on whether there is a collection of such curves generating $H_2(X,\mathbb{Z})$.

We will explain Step (2) in §2.2.1, but we note for now that it follows automatically from Step (1). Step (3) requires a global condition on the configuration of the curves C_i. This global condition is Friedman's condition:

$$\sum_{i=1}^{N} \lambda_i [C_i] = 0 \quad \text{in} \quad H^4(\hat{X},\mathbb{C}) \tag{6}$$

with each $\lambda_i \neq 0$. That a linear relation such as this must hold can be deduced from the topological change formula (5) which implies $N > \rho$ since $\mu \geq 1$, and with additional work one can show that all λ_i are non-zero [37, 70, 91]. For extensions of Friedman's condition to nodal Calabi-Yau n-folds in higher dimensions n, see Rollenske-Thomas [91].

The Friedman-Kawamata-Ran-Tian theorem asserts that if Friedman's condition (6) is satisfied, then step (3) can be realized. Namely, X_0 admits a smoothing. The full statement of the theorem takes the form of a singular version of the BTT theorem for ordinary double point singularities. We state the version here taken from Tian ([103] p. 476).

Theorem 3 (Friedman-Kawamata-Ran-Tian Theorem [36, 67, 89, 103]). *Let $\hat{X}$ be a Calabi-Yau threefold. Let C_i be a collection of N disjoint $(-1,-1)$ curves satisfying Friedman's condition (6). Let $\mu : \hat{X} \to X_0$ be the holomorphic contraction of the C_i with $\mu(C_i) = p_i$ resulting in a complex analytic space X_0 with ordinary double point singularities p_i.*

Then there exists a proper flat map $\pi : \mathscr{X} \to B$ where $B \subseteq \mathbf{T}$ is a open neighborhood of the origin in $\mathbf{T} = H^1(X_{0,reg}, T^{1,0})$. Here $\mathscr{X}$ is a smooth complex manifold and $\pi^{-1}(0) = X_0$.

The vector space of infinitesimal deformations $\mathbf{T}$ can be understood by the exact sequence

$$0 \to H^1(\hat{X}, T^{1,0}) \to \mathbf{T} \to \bigoplus_{i=1}^{N} \mathbb{C}[p_i] \to H^4(\hat{X},\mathbb{C})$$

where the last map is $[p_i] \mapsto [C_i]$. Deformations in direction $H^1(\hat{X},T^{1,0})$ preserve all singular points. Deformations mapped to a vector

$$\sum_i \lambda_i [p_i] \in \bigoplus_{i=1}^{N} \mathbb{C}[p_i] = \mathbb{C}^N$$

with all $\lambda_i \neq 0$ deform X_0 to a family of smooth complex manifolds.

In summary, one needs to check Friedman's condition (6) on a configuration of curves on an initial threefold, and then general theory produces a conifold transition to a new threefold.

Corollary 1. *[37, 103] Let $\hat{X}$ be a Calabi-Yau threefold, and let $\{C_i\} \subseteq \hat{X}$ be a collection of disjoint $(-1,-1)$-curves satisfying Friedman's condition* (6). *Then there exists a conifold transition $\hat{X} \to X_0 \rightsquigarrow X_t$.*

We note that realizing the local model $\mathcal{V} \to \Delta$ on the global flat family $\mathcal{X} \to \Delta$ is work of [65].

2.2 Local model

Consider the singular point $0 \in V_0$ where

$$V_0 = \left\{ z_1^2 + z_2^2 + z_3^2 + z_4^2 = 0 \right\} \subseteq \mathbb{C}^4.$$

This singularity is sometimes called a conifold singularity, nodal singularity, quadratic singularity, or ordinary double point. It is distinguished by the holomorphic Morse lemma [7]: a holomorphic function

$$f : \mathbb{C}^{n+1} \to \mathbb{C}, \quad \text{with } f(0) = 0,\ Df(0) = 0$$

and non-degenerate holomorphic Hessian matrix $f_{\alpha\beta}(0)$ admits local holomorphic coordinates w near the origin such that

$$f(w) \overset{\text{loc}}{=} \sum_{i=1}^{n+1} w_i^2.$$

The singularity $0 \in V_0$ can be desingularized in two distinct ways:

1. **By small resolution:** This is a resolution of singularities

   $$\mu : \hat{V} \to V_0$$

   such that $\hat{V}$ is a smooth complex manifold with $\mu : \hat{V} \backslash E \to V_0 \backslash \{0\}$ a biholomorphism away from the exceptional set

   $$\mu^{-1}(0) = \mathbb{P}^1.$$

   The space $\hat{V}$ is biholomorphic to the total space of $\mathcal{O}(-1)^{\oplus 2} \to \mathbb{P}^1$. We will give further details on this small resolution in §2.2.1 below.
2. **By smoothing:** The complex analytic space V_0 can be deformed into a smooth complex manifold by adding a parameter t:

   $$V_t = \left\{ z_1^2 + z_2^2 + z_3^2 + z_4^2 = t \right\} \subseteq \mathbb{C}^4.$$

The spaces V_t for $t \neq 0$ are smooth complex manifolds diffeomorphic to T^*S^3. We can insert V_0 as the central fiber of a family

$$\mathscr{V} \xrightarrow{\pi} \mathbb{C}, \quad \pi(z,t) = t$$

where

$$\mathscr{V} = \left\{ (z,t) \in \mathbb{C}^4 \times \mathbb{C} : z \in V_t \right\}.$$

We note that $\|z\|^2 \geq |t|$ on V_t, and the set $L_t = \{\|z\|^2 = |t|\} \subseteq V_t$ is called the vanishing cycle. It is diffeomorphic to a 3-sphere, so that $L_t \cong S^3$ and these collapse to a point as $t \to 0$.

The local model of the conifold transition is then

$$\hat{V} \xrightarrow{\mu} V_0 \rightsquigarrow V_t. \tag{7}$$

It should be checked that this description is compatible with the local topological surgery of replacing a neighborhood of the form $D^4 \times S^2$ with one of the form $S^3 \times D^3$. This is done in [25] Lemma 1.11, where in particular the following diffeomorphisms are shown:

$$\hat{V} \cong \mathbb{R}^4 \times S^2, \quad V_0 \backslash \{0\} \cong (0,\infty) \times (S^3 \times S^2), \quad V_t \cong S^3 \times \mathbb{R}^3.$$

See also [93] or [72] for alternate expositions of these diffeomorphisms. There is a classic diagram depicting the local model of the conifold transition which one can find in Figure 1 of Candelas-Green-Hübsch [20] (see also Figure D.1. of [60]).

2.2.1 More on small resolutions

We now provide more details on the small resolution $\hat{V} \xrightarrow{\mu} V_0$. We define

$$\hat{V} = \mathscr{O}(-1)^{\oplus 2} \to \mathbb{P}^1.$$

The space $\hat{V} = U \cup \tilde{U}$ is covered by two coordinate charts with coordinates (u,v,λ) satisfying the coordinate transformation

$$\tilde{\lambda} = \lambda^{-1}, \quad \tilde{u} = \lambda u, \quad \tilde{v} = \lambda v. \tag{8}$$

The coordinate λ is in the $\mathbb{P}^1$-direction while the u,v are along the fibers of the vector bundle. Let $E \cong \mathbb{P}^1$ be the zero section $\{u = v = 0\}$. There is a biholomorphism

$$\mu : \hat{V} \backslash E \to V_0 \backslash \{0\},$$

where

$$V_0 = \left\{ z_1^2 + z_2^2 + z_3^2 + z_4^2 = 0 \right\} \subseteq \mathbb{C}^4.$$

This can be constructed as follows. First, by unitary change of coordinates we identify V_0 with

$$\hat{V}_0 = \{w_1 w_2 - w_3 w_4 = 0\} \subseteq \mathbb{C}^4.$$

Next, we desingularize this space by blowing-up $\mathbb{C}^4$ along $\{w_2 = w_4 = 0\}$ and taking the proper transform of $\hat{V}_0$. Recall that this means

$$\mathrm{Bl}\,\mathbb{C}^4_{\{w_2=w_4=0\}} = \Big\{ ((w_1,w_2,w_3,w_4),[u,v]) \in \mathbb{C}^4 \times \mathbb{P}^1 : (w_2,w_4) \in [u,v] \Big\}.$$

The proper transform of $\hat{V}_0$ is biholomorphic to $\hat{V}$. Indeed, over the chart $U = \{u = 1\}$ there are coordinates (v, w_2, w_3), while over the chart $V = \{v = 1\}$ there are coordinates (u, w_1, w_4), and the change of coordinate relation is

$$u = v^{-1}, \quad w_1 = v w_3, \quad w_4 = v w_2,$$

which can be identified with (8). The blow-up map of $\hat{V}_0$ induced by $\mu(w,[u]) = w$, is denoted

$$\mu : \hat{V} \to \hat{V}_0,$$

with exceptional set $\mu^{-1}(0) = E \cong \mathbb{P}^1$ so that $\mu : \hat{V} \backslash E \to \hat{V}_0 \backslash \{0\}$ is a biholomorphism.

Remark 6. The resolution of singularities

$$\mu : \hat{V} \to V_0, \quad \mu^{-1}(0) = \mathbb{P}^1$$

is called a small resolution. It is not unique [7] as we could have alternatively blown-up $\hat{V}_0 \subseteq \mathbb{C}^4$ along $\{w_2 = w_3 = 0\}$.

Remark 7. If X_0 is a complex analytic space with ordinary double point singularities, then a neighborhood of each singularity may be identified with a neighborhood of $0 \in V_0$ and this local procedure defines a small resolution $\mu : \hat{X} \to X_0$. This construction also defines the contraction of a $(-1,-1)$ curve C on a Calabi-Yau threefold $\hat{X}$ to a singular point of the local form $0 \in V_0$. Indeed, let $C \subseteq \hat{X}$ be a $(-1,-1)$-curve. Then there exists a neighborhood $U \subseteq \hat{X}$ of C biholomorphic to a neighborhood of the zero section in $\hat{V}$, and the local construction defines a contraction $\mu : \hat{X} \to X_0$.

2.3 Examples

Complete intersection Calabi-Yau threefolds can all be connected by conifold transitions; see [53, 54] for work in the string theory literature and [111] in the mathematics literature. From another perspective, rather than connecting known examples, conifold transitions can also be used to construct new examples of projective Calabi-Yau threefolds; see e.g. [8, 14]. We list here some simple explicit examples of conifold transitions.

Example 5. This example can be found in Candelas-Green-Hübsch [20]. Consider the singular quintic

$$X_0 = \left\{ Z_3 G(Z_0, \dots, Z_4) + Z_4 H(Z_0, \dots, Z_4) = 0 \right\} \subseteq \mathbb{P}^4$$

with $G = Z_3^4 + Z_2^4 - Z_0^4$ and $H = -Z_4^4 - Z_1^4 - Z_0^4$. There are 16 singular points. Near each singularity there exists holomorphic local coordinates such that the local model of the singularity is

$$\{uv - xy = 0\} \subseteq \mathbb{C}^4,$$

and this is biholomorphic to $V_0 = \{\sum_i z_i^2 = 0\} \subseteq \mathbb{C}^4$. We can desingularize X_0 in two different ways:

- By small resolution $\hat{X} \to X_0$. Define $\hat{X} \subseteq \mathbb{P}^4 \times \mathbb{P}^1$ by

  $$UZ_4 - VZ_3 = 0, \quad UG(Z) + VH(Z) = 0,$$

  with $[U, V] \in \mathbb{P}^1$ and $[Z_0, \dots, Z_4] \in \mathbb{P}^4$.
- By smoothing

  $$X_t = \{Z_3 G + Z_4 H = t Z_0 Z_1 Z_2 Z_3 Z_4\} \subseteq \mathbb{P}^4.$$

  The X_t are smooth quintics for $t \neq 0$.

This provides an explicit example of a conifold transition

$$\hat{X} \to X_0 \rightsquigarrow X_t.$$

One can verify that the topological change formula (5) becomes $16 = 1 + 15$ by direct calculation of the Betti numbers of $\hat{X}$ and X_t, which results in $b_2 \mapsto b_2 - 1$ and $b_3 \mapsto b_3 + (2)(15)$. See [93] for a detailed analysis of this example.

Example 6. The following example was found by Friedman [36]. Let $\hat{X} \subseteq \mathbb{P}^4$ be a smooth quintic threefold. Then $b_2(\hat{X}) = 1$, so a pair of disjoint $(-1,-1)$-curves C_1, C_2 satisfy Friedman's condition (6). Choose C_1, C_2 to generate $H_2(\hat{X}, \mathbb{Z})$; see [24, 37] for such curves. General theory (Theorem 3) gives the existence of a conifold transition

$$\hat{X} \to X_0 \rightsquigarrow X_t.$$

We notice that something has gone wrong. The topological change formula (5) implies $b_2(X_t) = 0$ and so the complex analytic manifolds X_t cannot admit a Kähler structure. In fact, X_t is a simply connected 6-manifold with $H_2(X_t, \mathbb{Z}) = 0$ and $H_3(X_t, \mathbb{Z})$ torsion-free. By a theorem of Wall [110], the X_t are diffeomorphic to connected sums of $S^3 \times S^3$.

Example 7. We note that Lu-Tian [77, 78] has constructed examples of conifold transitions from Kähler Calabi-Yau threefolds to non-Kähler complex manifolds diffeomorphic to $\#_k(S^3 \times S^3)$ for $k \geq 2$. We refer to their work for these explicit examples.

Remark 8. The above examples show that a conifold transition may deform a Kähler Calabi-Yau threefold into a non-Kähler complex manifold. We will see in the sequel that these complex analytic threefolds inherit some properties of Kähler Calabi-Yau manifolds. It is argued by Reid [90], Friedman [37], Kontsevich [70], Yau [114, 41], Tian [103], etc that these analytic threefolds ought to be included into the mathematical theory developed for the study of projective Calabi-Yau threefolds. Friedman [36] and Reid [90] propose a handle-body decomposition for Calabi-Yau threefolds by contraction and smoothing of rational curves spanning $H_2(X,\mathbb{Z})$ so that the resulting threefold is non-Kähler and diffeomorphic to a connected sum $\#_k(S^3\times S^3)$. This program would lead to a sort of uniformization theorem for complex threefolds.

2.3.1 Reversing the arrow

The directionality of a conifold transition is not standardized in the literature. In all cases, the main feature is the desingularization of ordinary double point singularities by two mechanisms: either small resolution or smoothing of complex structure. Following the conventions established here, we refer to a reverse conifold transition as a deformation where the first step is degeneration of complex structure followed by small resolution:

$$X_t \rightsquigarrow X_0 \to \hat{X}.$$

This direction of travel may also potentially connect a projective Calabi-Yau threefold to a non-Kähler threefold. Indeed, it is well-known that Moishezon manifolds [83] are not necessarily Kähler.

Example 8. Here is a classic example of a quintic with a single ordinary double point. We take the explicit coefficients from §1.8 in [20], and this particular example can also be found in Hübsch's book ([60] Ch. D, section D.3.3). Consider the family

$$X_t = \left\{ Z_5^3 \left(\sum_{i=1}^4 Z_i^2 \right) + \sum_{i=1}^4 a_i Z_i^5 + t Z_5^5 \right\} \subseteq \mathbb{P}^4$$

where a_i are non-zero generic constants, and $t\in\mathbb{C}$ is a parameter. As

$$X_t \overset{t\to 0}{\rightsquigarrow} X_0$$

the family of smooth quintics degenerates to a singular space X_0 with a single ordinary double point at $P=[0,0,0,0,1]$. Let

$$\mu : \hat{X} \to X_0$$

be a small resolution with $\mu^{-1}(P) = C \cong \mathbb{P}^1$. The formula $N=\rho+\mu$ becomes $1=0+1$, and we see that

$$[C] = 0 \in H_2(\hat{X},\mathbb{R}).$$

The existence of a Kähler metric ω on $\hat{X}$ would lead to a contradiction:

$$0 < \int_C \omega = \int_{\partial\Omega} \omega = 0.$$

We see that the process of degeneration and resolution intertwines Kähler and non-Kähler manifolds.

2.4 Summary

A conifold transition is a mechanism to connect two distinct Calabi-Yau threefolds. We have noted that a conifold transition starting from a Kähler manifold may lead to a non-Kähler manifold. We may explore the ways to transfer mathematical structures from one manifold to the other. The following section, Section §3, will build on this theme from the point of view of differential geometry. One can hope that these extra structures can be used to constrain the space of non-Kähler manifolds connected to projective Calabi-Yau and potentially bound these threefolds in a suitable sense. For studies on how algebro-geometric or symplectic structures deform through a conifold transition, we refer to e.g. Li-Ruan [75], Lee-Lin-Wang [72], Lin-Wang [76], and references therein.

3 Geometric structures across singularities

3.1 Geometry of local conifold transitions

A theme in differential geometry is to understand metric constraints preserved through surgery. We will study conifold transitions in this context. Our starting point is the geometrization of the local model of a conifold transition, and afterwards we will move on to global compact geometries. On the non-compact local model, Ricci-flat metrics were constructed by Candelas-de la Ossa [15] and we now review their construction. For more details in the presentation style similar to the one taken here, we refer to e.g. [23, 27, 41].

3.1.1 Small resolution

On the total space

$$\hat{V} = \mathcal{O}(-1)^{\oplus 2} \to \mathbb{P}^1,$$

the Candelas-de la Ossa ansatz [15] with parameter $a > 0$ is

$$\omega_{co,a} = i\partial\bar{\partial} f_a(r^3) + 4a^2 \omega_{\mathrm{FS}},$$

where r is a power of the distance function to the zero section E given by

$$r^3(u,v) = |u|^2_{\omega_{\rm FS}} + |v|^2_{\omega_{\rm FS}},$$

with u, v fiber coordinates as before. The power of r will be motivated later in (9) below. This ansatz is substituted into the Ricci-flat equation where the equation becomes the following ODE

$$x(f_a'(x))^3 + 6a^2(f_a'(x))^2 = 1, \quad f_a(x) = f_a(r^3).$$

The parameter a measures the volume of $\{r=0\} = E \cong \mathbb{P}^1$, so that

$$\mathrm{Vol}(\mathbb{P}^1, \omega_{co,a}) \to 0$$

as $a \to 0$. A holomorphic volume form $\hat{\Omega}$ on $\hat{V}$ is given by extending $\mu^*\Omega_0$ across E by Hartog's theorem, where $\mu : \hat{V} \to V_0$ is a small resolution and Ω_0 is given below in (11).

3.1.2 The singular space

Setting $a = 0$, we find the following explicit solution on $\hat{V}\backslash E$:

$$\omega_{co,0} = \frac{3}{2} i\partial\bar{\partial} r^2, \quad f_0(x) = \frac{3}{2}x^{2/3}.$$

We noted earlier that $\hat{V}\backslash E$ is biholomorphic to the complement of the origin in

$$V_0 = \left\{ \sum_{i=1}^4 z_i^2 = 0 \right\} \subseteq \mathbb{C}^4,$$

and the correspondence identifies the radius as $r^3 = \|z\|^2$ on $V_0 \subseteq \mathbb{C}^4$. After rescaling the metric to neglect the factor of 3, the limit of $\omega_{co,a}$ as $a \to 0$ can be identified with the Ricci-flat geometry

$$(V_0, \omega_{co,0}), \quad \omega_{co,0} = \frac{i}{2}\partial\bar{\partial} r^2.$$

We note that $V_0 \subseteq \mathbb{C}^4$ comes with a scaling action $z \mapsto \lambda z$ and is diffeomorphic to a cone

$$V_0 \backslash \{0\} \cong (0,\infty) \times \Sigma, \quad \Sigma = V_0 \cap \{r=1\}$$

where the link is $\Sigma = S^2 \times S^3$ [15]. Indeed, if we write $z = x + iy$ then the equation for V_0 becomes

$$\|x\|^2 = \frac{\|z\|^2}{2}, \quad \|y\|^2 = \frac{\|z\|^2}{2}, \quad \langle x, y \rangle = 0.$$

We interpret the link $\Sigma = \{\|z\| = 1\}$ as a sphere bundle $S(TS^3)$ associated to $TS^3 \to S^3$, and as TS^3 is a trivial bundle we have $\Sigma = S^2 \times S^3$.

The metric $\omega_{co,0} = \frac{i}{2}\partial\bar{\partial}r^2$ is a conical metric (see e.g. [98]) as it can be written in polar coordinates

$$g_{co,0} = dr \otimes dr + r^2 g_\Sigma, \tag{9}$$

where g_Σ is a metric on the link Σ. Thus the Candelas-de la Ossa metric $\omega_{co,0}$ is an example of a Calabi-Yau cone metric.

Remark 9. The power of $r = \|z\|^{2/3}$ for the Calabi-Yau cone metric can be anticipated by scaling: the Kähler-Ricci flat equation is

$$\omega_{co,0}^3 = i\Omega_0 \wedge \bar{\Omega}_0 \tag{10}$$

where

$$\Omega_0 \overset{\text{loc}}{=} \frac{dz_1 \wedge dz_2 \wedge dz_3}{z_4}. \tag{11}$$

The scaling $z \mapsto \lambda z$ leaves (10) invariant.

3.1.3 Smoothing

Next, we equip the smoothings

$$V_t = \left\{ \sum_{i=1} z_i^2 = t \right\} \subseteq \mathbb{C}^4,$$

with Kähler Ricci-flat metrics. The holomorphic volume form Ω_t has the same expression as (11). The Candelas-de la Ossa ansatz [15] for the Calabi-Yau metric is

$$\omega_{co,t} = i\partial\bar{\partial} f_t(r^3), \quad r^3 = \|z\|^2,$$

and the Ricci-flat equation on V_t leads to the following ODE for the potential $f_t(x) = f_t(r^3)$:

$$(f_t'(x))^3 x + (f_t'(x))^2 f_t''(x)(x^2 - |t|^2) = 1/6.$$

Setting $t = 0$ recovers the solution $f_0(x) = c_0 x^{2/3}$. The submanifold

$$L_t = \{\|z\|^2 = |t|\} \cong S^3$$

is the vanishing cycle $L_t \subseteq V_t$, and it is special Lagrangian with respect to $(V_t, \omega_{co,t}, \Omega_t)$. Assuming $t > 0$ for simplicity, the special Lagrangian equations take the form

$$\omega|_L = 0, \quad \operatorname{Im}\Omega|_L = 0.$$

The parameter t measures the volume of the special Lagrangian 3-spheres, so that

$$\operatorname{Vol}(L_t, \omega_{co,t}) \to 0$$

as $t \to 0$.

3.1.4 Summary

The local model of a conifold transition is thus geometrized by Kähler Ricci-flat metrics. We have

$$(\hat{V}, g_{co,a}) \to (V_0, g_{co,0}) \leftarrow (V_t, g_{co,t})$$

where convergence is uniform as $a \to 0$, $t \to 0$ on compact sets away from the singularities. The convergence is also continuous in the Gromov-Hausdorff sense [40]. The process replaces a holomorphic $\mathbb{P}^1$ by a special Lagrangian S^3. For a uniqueness result on Kähler Ricci-flat metrics asymptotic to the cone $(V_0, g_{co,0})$, see [28].

3.2 Global Kähler geometry

Next, we consider the global setting of a conifold transition of compact Calabi-Yau manifolds. Let $\hat{X}$ be a smooth compact projective Calabi-Yau threefold, and let us degenerate a collection of holomorphic curves to deform $\hat{X}$ by conifold transition

$$\hat{X} \to X_0 \rightsquigarrow X_t.$$

As a first step, we insert the additional hypothesis into our setup that the smoothings X_t happen to be projective. Then both $\hat{X}$ and X_t admit Kähler Ricci-flat metrics by Yau's theorem [113]. The question becomes understanding these metrics and their degenerations in families. The study of this problem was instigated by Ruan-Zhang [94] and Rong-Zhang [92]. Let us recall their setup.

- Let
$$\pi : \mathscr{X} \to \Delta, \quad X_t = \pi^{-1}(t)$$
be a projective smoothing of Calabi-Yau threefolds such that X_t is smooth for $t \neq 0$ and the singular set of X_0 consists of finitely many ordinary double points. Let $\mathscr{L} \to \mathscr{X}$ be an ample line bundle. We consider the family of Calabi-Yau metrics (X_t, ω_t) defined by
$$\omega_t \in c_1(\mathscr{L})|_{X_t}, \quad t \neq 0$$
where ω_t is Kähler Ricci-flat.
- Let
$$\mu : \hat{X} \to X_0$$
be a small resolution. Let $\hat{\omega}$ be a reference Kähler on $\hat{X}$. We consider the family of Calabi-Yau metrics $(\hat{X}, \hat{\omega}_a)$ defined by
$$\hat{\omega}_a \in a[\hat{\omega}] + \mu^* c_1(\mathscr{L})|_{X_0}, \quad a \neq 0$$
where $\hat{\omega}_a$ is Kähler Ricci-flat.

The theorem of Ruan-Zhang [94], Rong-Zhang [92], and J. Song [97] is that

$$(\hat{X}, \hat{\omega}_a) \to (X_0, d_0) \leftarrow (X_t, \omega_t)$$

in the Gromov-Hausdorff sense, (X_0, d_0) is a compact metric length space induced by a singular Kähler Ricci-flat metric ω_0, and the metrics converge smoothly to ω_0 on compact sets away from the singularities.

The significance of this result is the manifestation of continuity in a process which is plainly discontinuous topologically as the Betti numbers b_i jump across the conifold transition. Nevertheless, string theory [100, 56] realizes passing through the conifold singularity in the moduli space as a continuous process. The theorem of [94, 92, 97] is a mathematical counterpart, where the continuity is realized by using families of Kähler Ricci-flat metrics and Gromov-Hausdorff convergence.

The singular Calabi-Yau metric (X_0, ω_0) exists by Eyssidieux-Guedj-Zeriahi [34]. Better estimates for the singular metric in this context were derived by Hein-Sun [59], who showed that the limit (X_0, g_0) inducing the metric space (X_0, d_0) is a singular Calabi-Yau metric with conical singularities, meaning that

$$\sup_{\{r<\delta\}} |g_0 - c g_{co,0}|_{g_{co,0}} \leq C r^{\lambda}$$

in a small neighborhood of an ordinary double point singularity identified with $V_0 \subseteq \mathbb{C}^4$ and $(V_0, r, \omega_{co,0})$ are as defined above. This is a kind of local uniqueness, where the global metric ω_0 is well-approximated by the explicit Candelas-de la Ossa local model near the singular points. For more results on convergence of global Calabi-Yau metrics to local models, see [22, 115].

A consequence of [59] is that the vanishing 3-cycles can be given the structure of a special Lagrangian 3-sphere with respect to the global Calabi-Yau structure $(X_t, \omega_t, \Omega_t)$. Therefore global Calabi-Yau conifold transitions exchange holomorphic $\mathbb{P}^1$s with a special Lagrangian S^3s.

3.3 Global non-Kähler geometry

We have seen that despite taking initial data $\hat{X}$ to be a projective Calabi-Yau threefold, when deforming across a quadratic singularity

$$\hat{X} \to X_0 \rightsquigarrow X_t$$

the complex threefold X_t will not necessarily remain Kähler. We conclude that when considering Calabi-Yau threefolds in families, we are forced to enlarge the space of objects considered by including appropriate limiting manifolds. These limiting objects are certain non-Kähler complex manifolds.

We look for geometric structures on X_t to help understand it. The Kähler-Ricci flat (2) equation is no longer applicable. Instead, the results of Fu-Li-Yau [41] and joint work with Collins and Yau [27] find the following structure:

Theorem 4. *[41, 27] Let $\hat{X}$ be a Calabi-Yau threefold, and let*

$$\hat{X} \to X_0 \rightsquigarrow X_t$$

be a conifold transition. For small enough t, the complex manifold X_t admits the structure (Ω_t, g_t, h_t) solving

$$d(|\Omega_t|_{\omega_t} \omega_t^2) = 0 \tag{12}$$
$$F_{h_t} \wedge \omega_t^2 = 0 \tag{13}$$
$$d\Omega_t = 0, \tag{14}$$

where:

- *Ω_t is a holomorphic volume form*
- *(g_t, h_t) is a pair of hermitian metric on $T^{1,0}X$*
- *$\omega_t = g_t(J_t \cdot, \cdot)$*
- *$F_h = \bar{\partial}(h^{-1}\partial h)$ is the Chern curvature.*

Furthermore, near the vanishing cycles $(S^3)_i$, the global metrics (g_t, h_t) satisfy

$$|g_t - a_i g_{co,t}|_{g_{co,t}} \leq C|t|^\lambda,$$
$$|h_t - b_i g_{co,t}|_{g_{co,t}} \leq C|t|^\lambda$$

for constants $C > 1$ and $\lambda > 0$ and scaling constants a_i, b_i. Here $g_{co,t}$ are the Kähler Ricci-flat metrics obtained by Candelas-de la Ossa on the local model V_t.

We now provide some context for these equations.

Remark 10. On a complex manifold of dimension n, a dual notion to the Kähler condition $d\omega = 0$ is the balanced metric condition $d \star \omega = 0$, or

$$d\omega^{n-1} = 0. \tag{15}$$

The geometry of such metrics was studied by Michelsohn [82]. A theorem of Alessandrini-Bassanelli [2] states that the existence of balanced metrics is invariant under modifications. Here X_t is not birational to $\hat{X}$, and Fu-Li-Yau's [41] result is the construction of balanced metrics on X_t. A conformal change of the metric can be used to go between (15) and (12).

Remark 11. The equation

$$F_h \wedge \omega^{n-1} = 0$$

is the Hermitian-Yang-Mills equation on the tangent bundle $T^{1,0}X$. This equation was solved for general stable holomorphic vector bundles $E \to X$ over a Kähler manifold by Donaldson-Uhlenbeck-Yau [33, 109]. On the complex manifolds $\#_k S^3 \times S^3$ from Example 6, [11] proved stability of the tangent bundle by algebraic methods.

In principle, this equation could also be solved through conifold transitions for bundles other than the tangent bundle, though it is not evident how a conifold transition creates a stable holomorphic vector bundle $E_t \to X_t$. There is a proposal for this mechanism in the string theory literature [3]. In the mathematics literature, Chuan [23] solved the Hermitian-Yang-Mills equation with the simplifying assumption that the bundle to carry through the transition is holomorphically trivial in a neighborhood of the contracted curves.

Remark 12. Though $(X_t, \omega_t, \Omega_t)$ is non-Kähler, the vanishing cycles in X_t are also special Lagrangian [26] in the sense that they are calibrated cycles with respect to a conformal change of metric and the closed 3-form calibration

$$\mathrm{Re}\, e^{-i\hat{\theta}} \Omega$$

for a constant angle $\hat{\theta}$. The concept of such special cycles on (not necessarily Kähler) complex manifolds was introduced by Harvey-Lawson [58]. Thus from the perspective of special submanifolds, conifold transitions exchange holomorphic $\mathbb{P}^1$s with special Lagrangian S^3s regardless of whether X_t is Kähler or not.

Remark 13. Strominger [99] derived equations (12), (13), (14) from constraints of supersymmetry [10]. A possible interpretation of the theorem is that supersymmetry is preserved through conifold transitions, even though the Kähler condition is relaxed.

Remark 14. The system (12) (13) (14) is not expected to be the final word on geometrization of conifold transitions. To rigify this structure and for example obtain a finite dimensional moduli space, or alternatively a sort of uniqueness theorem, one needs to impose an additional equation. For example, in heterotic string theory, the heterotic Bianchi identity is coupled to the supersymmetric equations (12) (13) (14), and this gives the finite dimensionality of the moduli space [18, 19, 30, 46, 48, 86].

3.3.1 The geometric system

We may wonder whether X_t admits Ricci-flat metrics, even though it is non-Kähler. From the perspective of string theory, this is not the right equation once the 3-form flux H is non-zero [62]. Candelas-Horowitz-Strominger-Witten [21] found the Kähler Ricci-flat equation by setting $H = 0$, but in general the equations of motion couple the Ricci tensor to a 3-form H and a scalar function Φ (see e.g. [79] for a modern exposition).

In the setting of conifold transitions, computing the Ricci tensor of the metric g satisfying (12) with $\omega = i g_{\mu\bar{\nu}}\, dz^\mu \wedge d\bar{z}^\nu$ leads to (see e.g. [4, 85] for this calculation)

$$R_{mn} + 2\nabla_m \nabla_n \Phi - \frac{1}{4} H_{mpq} H_n{}^{pq} = \frac{1}{2} (dH)^p{}_{pmn}, \tag{16}$$

where $H = i(\partial - \bar{\partial})\omega$ and $\Phi = -\frac{1}{2} \log |\Omega|_\omega$.

Although Theorem 4 constructs solutions of (16) through conifold transitions, the system of equations of heterotic string theory [61, 99] require an additional equation on dH, the heterotic Bianchi identity, to ensure the cancellation of anomalies [52], and this additional equation has not yet been solved through conifold transitions.

S.-T. Yau has conjectured that the complete set of equations appearing in Strominger's paper [99] is solvable through conifold transitions.

Conjecture 2 (Yau's conjecture [114]). Let $\hat{X}$ be a Calabi-Yau threefold, and let

$$\hat{X} \to X_0 \rightsquigarrow X_t$$

be a conifold transition. For small enough t, there exists a triple $(\Omega_t, \omega_t, h_t)$ solving (12), (13), (14) and

$$i\partial\bar{\partial}\omega_t = \alpha_t'(\operatorname{Tr} R_{\omega_t} \wedge R_{\omega_t} - \operatorname{Tr} F_{h_t} \wedge F_{h_t}) \tag{17}$$

where R_ω is the Chern curvature of ω and $\alpha_t' > 0$.

Remark 15. The conjecture is stated here in its simplest form where the unknown to solve for is a pair of metrics (g_t, h_t) on $T^{1,0}X_t$. Conceivably there could be a setup where h_t is a metric on an auxiliary bundle $E_t \to X_t$.

Remark 16. There are other proposed versions of (17). In string theory, the equation is a formal expansion about the parameter α' which is only valid in certain regimes, and at higher order in α' the curvature R_ω should be computed using the Hull connection [61, 79, 81]. There is an alternate version of (17) compatible with the formalism of generalized geometry where the equation is

$$i\partial\bar{\partial}\omega = \langle F_h \wedge F_h \rangle \tag{18}$$

where the right-hand side generalizes the special case of $\langle \cdot, \cdot \rangle = \operatorname{Tr}_{V_0} - \operatorname{Tr}_{V_1}$ with F_h the curvature of $h = h_0 \oplus h_1$ on $V_0 \oplus V_1$ for a pair of holomorphic vector bundles V_0, V_1. For the interpretation of this equation as a natural structure in generalized geometry, we refer to [43, 44, 45, 47].

Assuming this additional equation, either (17) or (18), can be solved, what are the implications? Returning to the theme of understanding Kähler to non-Kähler conifold transitions, a major direction for future work in this area is to understand what we learn about X_t from (17). By associating to X_t the moduli space $\mathcal{M}(X_t)$ of solutions to the equations, we may hope to learn about X_t from $\mathcal{M}(X_t)$. For the implications of these equations in string theory, see e.g. [5, 6, 18, 30, 80] and references therein.

3.3.2 Degenerations of the Hermitian-Yang-Mills equation

We compare this setup to the compact Kähler case described in Section §3.2. The analogous non-Kähler theorem is best understood as a result on degenerations of

Hermitian-Yang-Mills metrics. The background geometry is set by the conformally balanced metrics ω constructed by Fu-Li-Yau [41] along the conifold transition $\hat{X} \to X_0 \rightsquigarrow X_t$. These satisfy

$$\mathrm{Vol}(C_i, \hat{\omega}_a) \overset{a\to 0}{\to} 0, \quad \mathrm{Vol}(L_t, \omega_t) \overset{t\to 0}{\to} 0$$

and the geometry causes submanifolds to degenerate: holomorphic curves $C_i \subseteq \hat{X}$ and special Lagrangian cycles $L_t \subseteq X_t$ are tending to zero volume. The task is to solve the Hermitian-Yang-Mills equation on this degenerating background and analyze its limiting singularities. Combining joint work with T. Collins and S.-T. Yau [27], and B. Friedman and C. Suan [40], we have the following theorem:

Theorem 5. *[27, 40] Let*

$$(\hat{X}, \hat{\omega}_a) \to (X_0, \omega_0) \leftarrow (X_t, \omega_t)$$

be the path of reference Fu-Li-Yau metrics. Then there exists a unique normalized sequence of Hermitian-Yang-Mills metrics h on $T^{1,0}X$ with respect to this degenerating geometry such that

$$(\hat{X}, \hat{h}_a) \to (X_0, h_0) \leftarrow (X_t, h_t)$$

in the Gromov-Hausdorff sense. Furthermore the limit (X_0, ω_0, h_0) is a singular solution to the Hermitian-Yang-Mills equation with conical singularities, in the sense that

$$\sup_{\{r<\delta\}} |h_0 - c g_{co,0}|_{g_{co,0}} \le C r^{\lambda} \tag{19}$$

near the singularities.

This theorem shows that the Candelas-de la Ossa local model described in §3.1 still holds in the global compact case. Namely, the compact metrics h are well-approximated by the Kähler Ricci-flat local model $g_{co,0}$ near the singularities, though globally it receives non-Kähler corrections. This is an analog of the Hein-Sun theorem [59] for the Hermitian-Yang-Mills equation in the non-Kähler case. The two main steps of proof in [27] are:

1. Obtain uniform estimates

$$|\hat{h}_a|_{g_{co,a}} + |\hat{h}_a^{-1}|_{g_{co,a}} \le C$$

near the exceptional curves by adapting the Uhlenbeck-Yau method [109]. This estimate requires the global condition that $T^{1,0}\hat{X}$ is a stable vector bundle over $\hat{X}$. Taking a limit then yields

$$C^{-1} g_{co,0} \le h_0 \le C g_{co,0}.$$

2. Upgrade uniform equivalence to polynomial decay (19). That is, show that $g_{co,0}^{-1} h_0$ decays to $c\mathrm{Id}_{T^{1,0}X}$. The toy model for this sort of phenomenon in PDE

is the following: suppose u is a scalar function on a cone V_0 with $\Delta_{g_{cone}} u = 0$ and $|u| \leq C_1$, and show the decay

$$|u - u(0)| \leq C_2 r^{\lambda}.$$

This is elementary for harmonic functions, and that the analogous sort of statement holds for the nonlinear Hermitian-Yang-Mills equation depends on a certain Poincaré inequality invoking a stability condition on $T^{1,0}V_0$. We refer to [27] for details, and see [63] for another instance of this technique.

3.3.3 Alternate setups

We now note some of the alternate approaches to the geometrization of conifold transitions.

- The equations (12), (13), (14) have not yet been solved in the reverse direction.

  $$X_t \rightsquigarrow X_0 \rightarrow \hat{X}$$

  Here X_t is a degenerating family of projective threefolds and $\hat{X}$ is a small resolution of singularities. The complex manifold $\hat{X}$, though possibly non-Kähler, admits balanced metrics [49, 50].
- There are alternate non-Kähler equations which may be relevant. One of these is the balanced Chern-Ricci flat equations

  $$d\omega^2 = 0, \quad |\Omega|_\omega = const.$$

  The analysis of these equations was developed in [42, 106, 107]. The balanced Chern-Ricci flat equations were solved in [50] across reverse conifold transitions. They are still unsolved in the direction $\hat{X} \rightarrow X_0 \rightsquigarrow X_t$.
- There are other options inspired by Type IIB string theory with flux [108] [105].

  $$d\omega^2 = 0, \quad i\partial\bar{\partial}\omega = \rho_B.$$

  As noted in [43], it is not clear how a conifold transition creates a 4-form ρ_B which is Poincaré dual to a linear combination of holomorphic curves for a forward conifold transition. In the reverse direction, the small resolution creates holomorphic curves satisfying Friedman's relation and the Type IIB equation may be well-suited. See also [45] for more on this idea.
- As X_t cannot be simultaneously complex analytic and symplectic, another alternative is to let go of the complex analytic structure of X_t but preserve the symplectic structure. This point of view was developed by Smith-Thomas-Yau [96]. In other words, although the initial projective threefold solves $d\omega = 0$ and $d\Omega = 0$, across the singularity we may choose to either: preserve $d\omega = 0$ but allow $d\Omega \neq 0$, or preserve $d\Omega = 0$ and allow $d\omega \neq 0$.

In all cases, the fundamental question is how to use the above geometric structures to constrain the possible manifolds appearing on the other side of a degeneration and resolution.

3.4 Departure from Kähler geometry

3.4.1 Complex analytic threefolds

We have seen that conifold transitions can take us out of Kähler geometry. Generally speaking, there are sometimes advantages to working with complex analytic threefolds instead of restricting to projective threefolds or manifolds admitting a Kähler structure. One such example is J. Pardon's resolution of the MNOP conjecture [84]. Pardon's theory of enumeration of holomorphic curves is formulated in the analytic category. His proof of the MNOP conjecture uses the generality of complex analytic families rather than algebraic geometry. The central object is Pardon's Grothendieck group

$$H_c^*(\mathscr{Z}, \mathrm{Cpx}_3)$$

which is the total homology of the double complex

$$C_{-p}(\mathrm{Cpx}_3, C_c^q(\mathscr{Z})) = \bigoplus_{\mathscr{X} \to \Delta^p} C_c^q(\mathscr{Z}(\mathscr{X}/\Delta^p))$$

with direct sum over all complex analytic families $\mathscr{X} \to \Delta^p$ of threefolds over a p-simplex $\Delta^p \subseteq \mathbb{R}^p$. The two differentials are the differential of cohomology with compact support $C_c^*(Z)$ and the other is an alternating sum of restriction to the boundary of the simplex. Here $\mathscr{Z}(\mathscr{X}/\Delta^n)$ is the space of compact 1-cycles lying entirely in the fibers of $\mathscr{X} \to \Delta^n$.

In Pardon's formalism [84], curve enumeration theories such as Gromov-Witten theory a la Behrend-Fatechi [9] are homomorphisms out of the Grothendieck group

$$\mathrm{GW} : H_c^*(\mathscr{Z}, \mathrm{Cpx}_3) \to \mathbb{Q}((u))$$

where powers of u keep track of the genus of the curve. Given a projective threefold X and denoting by $Z(X,\beta)$ the space of curves C with $[C] = \beta$, the constant function $\mathbf{1}_{X,Z(X,\beta)}$ defines an element in $[\alpha] \in H_c^0(\mathscr{Z}, \mathrm{Cpx}_3)$. This produces an enumerative invariant

$$\mathrm{GW}([\alpha]).$$

Deformation invariance comes from connecting a pair of threefolds by a family $\mathscr{X} \xrightarrow{\pi} \Delta^1$ to define the same class in $H_c^0(\mathscr{Z}, \mathrm{Cpx}_3)$. Indeed, equivalence in Pardon's homology is a way to package a vast generalization of deformation invariance.

This formalism for studying curve enumeration invariants uses cohomology with compact support, and enables the exploration of possibly non-compact spaces of curves on general complex analytic threefolds. In Pardon's application to the MNOP conjecture, the framework allows him to extract open neighborhoods of holomorphic curves and separately deform them to break the curve into a union of isolated rigid curves, and then deduce the general conjecture from the case of local curves [12].

3.4.2 The web of threefolds

Consider the set $\mathscr{W}$ of all complex threefolds connected to projective Calabi-Yau threefolds by conifold transitions. It is unknown what is the precise category of complex threefolds constituting this set. We have seen in Example 6 that a conifold transition may connect a Kähler Calabi-Yau threefold to a non-Kähler complex threefold. It is not necessary to collapse b_2 to zero to do this: contract for example only two curves on $\hat{X}$ in Example 5 to obtain a non-Kähler X_t with $b_2(X_t) = 1$.

An open question is whether there is a sense in which $\mathscr{W}$ is bounded. One could hope to constrain this space of complex threefolds by equipping them with special geometric structures. Currently, we know that the threefolds X_t with small t linked to a projective Calabi-Yau threefold $\hat{X}$ by conifold transition $\hat{X} \to X_0 \rightsquigarrow X_t$ satisfy the following properties:

- There exists a holomorphic volume form Ω [37].
- There holds $h^{1,0} = h^{0,1} = 0$ [37].
- The $\partial\bar{\partial}$-lemma is satisfied [74, 39, 71]. Moreover, the Hodge-Riemann bilinear relation holds on $H^{2,1}$.
- The tangent bundle $T^{1,0}$ is stable with respect to a balanced metric ω [41, 27].

We see that the analytic threefolds in the web of conifold transitions $\mathscr{W}$ inherit some of the properties of Kähler geometry, even though they may or may not actually be Kähler.

Remark 17. The general theory of balanced Calabi-Yau $\partial\bar{\partial}$-manifolds and their deformation theory has been developed by Wu [112], Popovici [88], and Lee [71].

A consequence of Yau's theorem [113] is that Kähler Calabi-Yau manifolds have stable tangent bundle. The result of [41, 27] is a generalization of Yau's theorem to the web of threefolds in regions where Kähler Ricci-flat metrics cannot exist. As a corollary, we may rule out complex threefolds with unstable $T^{1,0}X$ from appearing as limits of Kähler to non-Kähler conifold transitions. Recall that given a hermitian metric ω satisfying $d\omega^2 = 0$ on a complex threefold X, stability of $T^{1,0}X$ is the property

$$\int_X c_1(F) \wedge \omega^2 < 0$$

for all torsion-free coherent subsheaves $F \subseteq T^{1,0}X$ of rank 1,2. It is possible in this generality that $[\omega^2] = 0$; this is for example the case in Example 6 and it was shown in [11] that for this particular example $T^{1,0}$ has no holomorphic subbundles.

It is well-known (e.g. [38]) that stability implies

$$H^0(\operatorname{End} T^{1,0}X) = \{\lambda \operatorname{Id} : \lambda \in \mathbb{C}\}. \tag{20}$$

Stability is also relevant when associating a moduli space to a holomorphic bundle, which can illuminate the topology of the underlying manifold [32]. It is hoped that the geometric structures presented in this survey will help improve our understanding of the possible complex analytic threefolds appearing as limits of degenerations and resolutions of Calabi-Yau threefolds.

Acknowledgements I owe much of my understanding of Calabi-Yau threefolds to frequent discussions with J. Bryan, T. Collins, and S.-T. Yau. I thank the MATRIX-Simons Scholarship program for the opportunity to travel and participate in the MATRIX research program on "The geometry of moduli spaces in string theory". Thanks to C. Suan, B. Friedman, R. Friedman, T.-J. Lee, T. Hübsch and C.-L. Wang, for discussions and comments. Thank you to the reviewer for helpful suggestions.

References

1. M. Abate, F. Bracci and F. Tovena: Embeddings of submanifolds and normal bundles. Advances in Mathematics **220** (2009), no. 2, 620-656.
2. L. Alessandrini and G. Bassanelli: Metric properties of manifolds bimeromorphic to compact Kahler spaces. J. Differential Gometry **37** (1993), 95-121.
3. L. Anderson, C. Brodie, and J. Gray: Branes and bundles through conifold transitions and dualities in heterotic string theory. Phys. Rev. D (2023), no. 103.
4. A. Ashmore, R. Minasian, and Y. Proto: Geometric flows and supersymmetry. Commun. Math. Phys. (2024), **405**:16.
5. A. Ashmore, X. de la Ossa, R. Minasian, C. Strickland-Constable and E.E. Svanes: Finite deformations from a heterotic superpotential: holomorphic Chern–Simons and an L infinity algebra. J. High Energ. Phys. **179** (2018).
6. A. Ashmore, J. Ibarra, D. McNutt, C. Strickland-Constable, E.E. Svanes, D. Tennyson, S. Winje: A heterotic Kodaira-Spencer theory at one-loop. Preprint.
7. M.F. Atiyah: On analytic surfaces with double points. Proceedings of the Royal Society of London A **247** (1958), no. 1249, 237-244.
8. V. Batyrev and M. Kreuzer: Constructing new Calabi-Yau 3-folds and their mirrors via conifold transitions. Adv. Theor. Math. Phys. **14** (2010), 879-898.
9. K. Behrend and B. Fantechi: The intrinsic normal cone. Invent. Math. **128** (1997), no. 1, 45–88.
10. E. Bergshoeff and M. de Roo: The quartic effective action of the heterotic string and supersymmetry. Nuclear Physics B **328** (1989), no. 2, 439–468.
11. Y.D. Bozhkov: Specific complex geometry of certain complex surfaces and three-folds. PhD Thesis at the University of Warwick, 1992.
12. J. Bryan and R. Pandharipande: The local Gromov-Witten theory of curves. J. Amer. Math. Soc. **21** (2008), no. 1, 101–136.
13. E. Calabi: On Kahler manifolds with vanishing canonical class. Algebraic geometry and topology. A symposium in honor of S. Lefschetz, pp. 78–89. Princeton University Press, Princeton, N. J., 1957.
14. P. Candelas and R. Davies: New Calabi-Yau manifolds with small Hodge numbers. Fortschritte der Physik **58** (2010), 383-466.
15. P. Candelas and X. de la Ossa: Comments on conifolds. Nuclear Physics B **342** (1990), no. 1, 246-268.

16. P. Candelas and X. de la Ossa: Moduli space of Calabi-Yau manifolds. Nuclear Physics B **355** (1991), 455-481.
17. P. Candelas, X. de la Ossa, P.S. Green, L. Parkes: A pair of Calabi-Yau manifolds as an exactly soluble superconformal theory. Nuclear Physics B **359** (1991), 21-74.
18. P. Candelas, X. de la Ossa, and J. McOrist: A metric for heterotic moduli. Commun. Math. Phys. **356**, no. 2 (2017), 567-612.
19. P. Candelas, X. de la Ossa, J. McOrist, and R. Sisca: The universal geometry of heterotic vacua. J. High Energ. Phys. **38** (2019).
20. P. Candelas, P. Green, and T. Hübsch: Rolling among Calabi-Yau vacua. Nuclear Phys. B **330** (1990), no. 1, 49–102.
21. P. Candelas, G. Horowitz, A. Strominger, and E. Witten: Vacuum configurations for superstrings. Nuclear Phys. B **258** (1985), no. 1, 46–74.
22. S.K. Chiu and G. Szekelyhidi: Higher regularity for singular Kahler–Einstein metrics. Duke Mathematical Journal **172** (2023), no. 18, 3521–3558.
23. M.-T. Chuan: Existence of Hermitian-Yang-Mills metrics under conifold transitions. Comm. Anal. Geom. **20** (2012), no. 4, 677–749.
24. H. Clemens: Homological equivalence modulo algebraic equivalence is not finitely generated. Publ. Math. IHES **58** (1983), 19-38.
25. H. Clemens: Double solids. Advances in Mathematics **47** (1983), 107-230.
26. T. Collins, S. Gukov, S. Picard, and S.-T. Yau: Special Lagrangian cycles and Calabi-Yau transitions. Commun. Math. Phys. **401** (2023), 769–802.
27. T. C. Collins, S. Picard, and S. -T. Yau: Stability of the tangent bundle through conifold transitions. Comm. Pure Appl. Math. **77** (2024), 284-371.
28. R.J. Conlon and H.J. Hein: Classification of asymptotically conical Calabi–Yau manifolds. Duke Math. J. **173** (2024), no. 5, 947-1015.
29. J.P. Demailly: Complex Analytic and Differential Geometry. Book available on the author's website. https://www-fourier.ujf-grenoble.fr/ demailly/manuscripts/agbook.pdf
30. X. de la Ossa, and E. E. Svanes: Holomorphic bundles and the moduli space of N = 1 supersymmetric heterotic compactifications. J. High Energ. Phys. **123** (2014).
31. L. Dixon: Some world-sheet properties of superstring compactifications, or orbifolds and otherwise. ICTP Ser. Theoret. Phys. **4** (1988), 67–126.
32. S.K. Donaldson: An application of gauge theory to four dimensional topology. J. Differential Geom. **18** (1983), 279-315.
33. S.K. Donaldson: Anti self-dual Yang-Mills connections over complex algebraic surfaces and stable vector bundles. Proc. London Math. Soc. **50** (1985), no. 1, 1-26.
34. P. Eyssidieux, V. Guedj, and A. Zeriahi: Singular Kahler-Einstein metrics. Journal of the American Mathematical Society **22** (2009), no. 3, 607–639.
35. G. Fischer: Complex Analytic Geometry. Springer Berlin, Heidelberg 1970.
36. R. Friedman: Simultaneous resolution of threefold double points. Math. Ann. **274** (1986), no. 4, 671–689.
37. R. Friedman: On threefolds with trivial canonical bundle. Proc. Sympos. Pure Math. **53** (1991), 103-134.
38. R. Friedman: Algebraic Surfaces and Holomorphic Vector Bundles. Springer-Verlag 1998.
39. R. Friedman: The $\partial\bar{\partial}$-lemma for general Clemens manifolds. Pure and Applied Mathematics Quarterly **15** (2019), 1001–1028.
40. B. Friedman, S. Picard and C. Suan: Gromov-Hausdorff continuity of non-Kahler Calabi-Yau conifold transitions. Preprint.
41. J. Fu, J. Li, and S.-T. Yau: Balanced metrics on non-Kahler Calabi-Yau threefolds. J. Differential Geom. **90** (2012), no. 2, 81-129.
42. J. Fu, Z. Wang, and D. Wu: Form-type Calabi-Yau equations. Math. Res. Lett. **17** (2010), no. 5, 887–903.
43. M. Garcia-Fernandez: Lectures on the Strominger system. Travaux Mathematiques Vol. **XXIV** (2016) 7-61.
44. M. Garcia-Fernandez and R. Gonzalez Molina: Futaki invariants and Yau's conjecture on the Hull-Strominger system. Preprint.

45. M. Garcia-Fernandez, R. Gonzalez Molina and J. Streets: Pluriclosed flow and the Hull-Strominger system. Preprint.
46. M. Garcia-Fernandez, R. Rubio, C. Tipler: Infinitesimal moduli for the Strominger system and Killing spinors in generalized geometry. Mathematische Annalen **369** (2017) 539-595.
47. M. Garcia-Fernandez, R. Rubio, C. Tipler: Holomorphic string algebroids. Trans. Amer. Math. Soc. **373** (2020) 7347-7382.
48. M. Garcia-Fernandez, R. Rubio, C. Tipler: Gauge theory for string algebroids. J. Differential Geom. **128** (2024), no. 1, 77-152.
49. F. Giusti and C. Spotti: A Kummer construction for Chern-Ricci flat balanced manifolds. Preprint.
50. F. Giusti and C. Spotti: Chern-Ricci flat balanced metrics on small resolutions of Calabi-Yau threefolds. Preprint.
51. H. Grauert: Uber Modifikationen und exzeptionelle analytische Mengen. Math. Ann. **146** (1962), 331–368.
52. M.B. Green and J.H. Schwarz: Anomaly cancellations in supersymmetric $D = 10$ gauge theory and superstring theory. Physics Letters B. **149** (1984), 117–122.
53. P.S. Green and T. Hübsch: Connecting Moduli Spaces of Calabi-Yau Threefolds. Commun. Math. Phys. **119** (1988), 431-441.
54. P.S. Green and T. Hübsch: Possible Phase Transitions among Calabi-Yau Compactifications. Phys. Rev. Lett. **61** (1988), 1163.
55. B.R. Greene and M.R. Plesser: Duality in Calabi-Yau moduli space. Nuclear Physics B **338** (1990), no. 1, 15-37.
56. B.R. Greene, D.R. Morrison, and A. Strominger: Black hole condensation and the unification of string vacua. Nuclear Physics B **451**, no. 1-2 (1995), 109-120.
57. M. Gross: Primitive Calabi–Yau threefolds. J. Differential Geom. **45** (1997), 288–318.
58. R. Harvey and H.B. Lawson: Calibrated geometries. Acta Mathematica **148** (1982), 47-157.
59. H.J. Hein and S. Sun: Calabi-Yau manifolds with isolated conical singularities. Publ. Math. IHES **126** (2017), no. 1, 73–130.
60. T. Hübsch: Calabi-Yau Manifolds: A Bestiary for Physicists. World Scientific (1992).
61. C. Hull: Compactifications of the Heterotic Superstring. Phys.Lett. B **178** (1986) 357.
62. C. M. Hull and P. K. Townsend: The two loop beta function for sigma models with torsion. Phys. Lett. B **191**, 115 (1987).
63. A. Jacob, T. Walpuski: Hermitian-Yang-Mills metrics on reflexive sheaves over asymptotically cylindrical Kahler manifolds. Comm. Partial Differential Equations **43** (2018), no. 11, 1566–1598.
64. E. Kahler: Uber eine bemerkenswerte Hermitesche Metrik. Abh. Math. Semin. Univ. Hambg. **9** (1933), 173–186.
65. A. Kas and M. Schlessinger: On the versal deformation of a complex space with an isolated singularity. Math. Ann. **196** (1972), 23–29.
66. S. Katz: On the finiteness of rational curves on quintic threefolds. Compositio Mathematica **60** (1986), no. 2, 151-162.
67. Y. Kawamata: Unobstructed deformations. A remark on a paper of Z. Ran: "Deformations of manifolds with torsion of negative canonical bundle". J. Algebraic Geom. **1** (1992), no. 2, 183–190.
68. K. Kodaira: Complex Manifolds and Deformation of Complex Structures. Grundlehren der Math. Wiss. **283**, Springer (1986).
69. M. Kontsevich: Homological Algebra of Mirror Symmetry. Proceedings of the International Congress of Mathematicians (1995), 120-139.
70. M. Kontsevich: Mirror symmetry in dimension 3. Asterique **237** (1996), Seminaire Bourbaki no. 801, 275-293.
71. T.-J. Lee: Finite distance problem on the moduli of non-Kahler Calabi-Yau $\partial\bar{\partial}$-threefolds. Preprint.
72. Y.-P. Lee, H.-W. Lin, and C.-L. Wang: Towards A+B theory in conifold transitions for Calabi–Yau threefolds. J. Differential Geom. **110** (2018), no. 3, 495-541.

73. W. Lerche, C. Vafa and N.P. Warner: Chiral rings in N = 2 superconformal theories. Nucl. Phys. B **324** (1989), no. 2, 427–474.
74. C. Li: Polarized Hodge structures for Clemens manifolds. Mathematische Annalen **389** (2024), 525-541.
75. A. Li and Y. Ruan: Symplectic surgery and Gromov–Witten invariants of Calabi–Yau 3-folds. Invent. Math. **145** (2001), no. 1, 151–218.
76. Y. Li and S.-S. Wang: Gromov–Witten/Pandharipande–Thomas correspondence via conifold transitions. Preprint.
77. P. Lu and G. Tian: The complex structure on a connected sum of $S^3 \times S^3$ with trivial canonical bundle. Mathematische Annalen **298** (1994), 761–764.
78. P. Lu and G. Tian: Complex structures on connected sums of $S^3 \times S^3$. Manifolds and geometry, 284-293, Sympos. Math., XXXVI, Cambridge Univ. Press, Cambridge, 1996.
79. D. Martelli and J. Sparks: Non-Kahler heterotic rotations. Advances in Theoretical and Mathematical Physics **15**, no. 1 (2011), 131-174.
80. J. McOrist and E.E. Svanes: Heterotic quantum cohomology. J. High Energ. Phys. **11** (2022).
81. I. Melnikov, R. Minasian, and S. Sethi: Heterotic fluxes and supersymmetry. J. High Energ. Phys. **6** (2014).
82. M.L. Michelsohn: On the existence of special metrics in complex geometry. Acta Math. **149** (1982), 261-295.
83. B. G. Moishezon: On n-dimensional compact complex manifolds having n algebraically independent meromorphic functions. I. Izv. Math. **30**:1 (1966).
84. J. Pardon: Universally counting curves in Calabi–Yau threefolds. Preprint.
85. S. Picard: The Strominger system and flows by the Ricci tensor. Surveys in Differential Geometry **27**, no. 1 (2022), 103-145.
86. S. Picard and P.-L. Wu: Balanced and Aeppli Parameters for the Heterotic Moduli. Preprint.
87. D. Popovici: Non-Kähler Hodge Theory and Deformations of Complex Structures. Book available on the author's website. https://www.math.univ-toulouse.fr/ popovici/hodge-def.pdf
88. D. Popovici: Holomorphic Deformations of Balanced Calabi-Yau $\partial\bar{\partial}$-Manifolds. Annales de l'Institut Fourier **69** (2019), no. 2, 673-728.
89. Z. Ran: Deformations of manifolds with torsion or negative canonical bundle. J. Algebraic Geom. **1** (1992), no. 2, 279–291.
90. M. Reid: The moduli space of 3-folds with K = 0 may nevertheless be irreducible. Math. Ann. **278** (1987), no. 1-4, 329–334.
91. S. Rollenske and R. Thomas: Smoothing nodal Calabi-Yau n-folds. Journal of Topology **2** (2009), no. 2, 405-421.
92. X. Rong and Y. Zhang: Continuity of extremal transitions and flops for Calabi-Yau manifolds. J. Differential Geom. **89** (2011), no. 2, 233–269.
93. M. Rossi: Geometric transitions. J. Geom. Phys. **56** (2006), 1940-1983.
94. W.D. Ruan and Y. Zhang: Convergence of Calabi–Yau manifolds. Advances in Mathematics **228** (2011), no. 3, 1543–1589.
95. Y.-T. Siu: Lectures on Hermitian-Einstein metrics for stable bundles and Kahler-Einstein metrics. Birkhauser Verlag, Basel, 1987.
96. I. Smith, R.P. Thomas, S.-T. Yau: Symplectic Conifold Transitions. J. Differential Geom. **62** (2002), no. 2, 209-242.
97. J. Song: On a Conjecture of Candelas and de la Ossa. Commun. Math. Phys. **334** (2015), no. 2, 697–717.
98. J. Sparks: Sasaki-Einstein manifolds. Surveys in Differential Geometry **16** (2011), 265-324.
99. A. Strominger: Superstrings with torsion. Nuclear Physics B **274** (1986), no. 2, 253–284.
100. A. Strominger: Massless black holes and conifolds in string theory. Nuclear Physics B **451** (1995), no. 1-2, 96–108.
101. A. Strominger, S.-T. Yau, E. Zaslow: Mirror symmetry is T-duality. Nuclear Physics B **479** (1996), 243–259.
102. G. Tian: Smoothness of the universal deformation space of compact Calabi-Yau manifolds and its Petersson-Weil metric. Adv. Ser. Math. Phys. **1** (1987), 629–646.

103. G. Tian: Smooth 3-folds with trivial canonical bundle and ordinary double points. Essays on mirror manifolds, 458–479, Int. Press, Hong Kong, 1992.
104. A Todorov: The Weil-Petersson geometry of the moduli space of SU(n) (Calabi-Yau) manifolds. I. Comm. Math. Phys. **126** (1989), no. 2, 325–346.
105. A. Tomasiello: Reformulating supersymmetry with a generalized Dolbeault operator. JHEP **02** (2008) 010.
106. V. Tosatti: Non-Kähler Calabi-Yau manifolds. Contemp. Math. **644** (2015), 261-277.
107. V. Tosatti and B. Weinkove: The Monge-Ampère equation for (n-1)-plurisubharmonic functions on a compact Kähler manifold. J. Amer. Math. Soc. **30** (2017), no. 2, 311-346.
108. L.-S. Tseng and S.-T. Yau: Non-Kaehler Calabi-Yau manifolds. Proc. Sympos. Pure Math. **85** (2012), 241-254.
109. K. Uhlenbeck and S.T. Yau: On the existence of Hermitian-Yang-Mills connections in stable vector bundles. Comm. Pure Appl. Math. **39** (1986), 257–293.
110. C. T. C. Wall: Classification problems in topology V : On certain 6-manifolds. Invent. Math. **1** (1966), 355-374.
111. S.-S. Wang: On the connectedness of the standard web of Calabi-Yau 3-folds and small transitions. Asian J. Math. **22** (2018), no. 6, 981-1004.
112. C.-C. Wu: On the Geometry of Superstrings with Torsion. Thesis in the Department of Mathematics, Harvard University (2006).
113. S.-T. Yau: On the Ricci curvature of a compact Kähler manifold and the complex Monge-Ampère equation. I. Comm. Pure Appl. Math. **31** (1978), 339-411.
114. S.T. Yau and S. Nadis: The shape of inner space: String theory and the geometry of the universe's hidden dimensions. Basic Books, 2010.
115. J. Zhang: On polynomial convergence to tangent cones for singular Kahler-Einstein metrics. Preprint.

An overview of Bagger-Witten line bundles

Eric Sharpe

Abstract We give a brief overview of recent progress in understanding Bagger-Witten line bundles, which are bundles over moduli spaces of two-dimensional $N = 2$ SCFTs whose existence is a consequence of the global $U(1)_R$ symmetry of the theories. Our overview includes a discussion of applications in supergravities coupled to gauge theories, a proposal for a purely geometric interpretation, and explicit examples over moduli spaces of Calabi-Yau manifolds.

1 Introduction

Briefly, Bagger-Witten line bundles are line bundles over moduli spaces[1] of two-dimensional SCFTs, whose existence is implied by symmetries of the theories, and which are closely interrelated with supersymmetry. They were originally discovered in [22], where they resolved a puzzle in four-dimensional $N = 1$ supergravity, and were later given worldsheet realizations in terms of two-dimensional $N = 2$ SCFTs in e.g. [1, 16].

We can understand Bagger-Witten line bundles as follows. Over a moduli space of conformal field theories, there exist bundles whose structure groups coincide with the symmetries of the theories. As one crosses from one coordinate patch to another over the moduli space, across the overlaps, theories in one patch are related to theories in the other by global symmetries. If one has a family of theories with global symmetry G, one is naturally led to expect a principal G bundle over the moduli space, and/or vector bundles associated to such a principal bundle. In the case of

Eric Sharpe
Department of Physics MC 0435, 850 West Campus Drive, Virginia Tech, Blacksburg, VA 24061, USA
e-mail: ersharpe@vt.edu

[1] All moduli 'spaces' here will be stacks, usually Deligne-Mumford stacks, as we shall see explicitly in examples, but for simplicity and readability we will usually refer to them as 'spaces.'

D. R. Wood et al. (eds.), *2024 MATRIX Annals, Part II*, MATRIX Book Series 8,
https://doi.org/10.1007/978-3-032-16206-9_18

two-dimensional $N = 2$ SCFTs, generically the only global symmetry is the $U(1)_R$ symmetry, hence a moduli space of two-dimensional $N = 2$ SCFTs will naturally carry a line bundle, and as it is an R symmetry, that line bundle should somehow be related to supersymmetry. Bagger-Witten line bundles are of this form.

Although Bagger-Witten line bundles have often been discussed and applied (see e.g. [17]), their properties have historically not been well-understood, and until recently concrete examples were not known.

The purpose of this talk is to quickly review Bagger-Witten line bundles, culminating in some recent developments, a geometric definition, and concrete examples.

2 Appearance in supergravity

In this section we briefly review the original argument of [22], in four-dimensional $N = 1$ supergravity, for the existence of a line bundle over the moduli space[2], and other supergravity implications discussed since.

Let $\mathscr{M}$ denote the moduli space of scalar field vevs. (In a string compactification, $\mathscr{M}$ is a large patch on the moduli space of SCFTs, corresponding to the weakly-coupled large-radius regime.) Let K denote the Kähler potential on M, which defines the kinetic terms of the supergravity action.

In Kähler geometry, across coordinate patches,

$$K \mapsto K + f + \overline{f}, \tag{1}$$

where f is a holomorphic function. This also defines a symmetry of rigidly symmetric four-dimensional $N = 1$ theories. There, the kinetic terms can be written in superspace as [21, equ'n (22.1)]

$$\int d^4\theta\, K(\Phi, \Phi^\dagger), \tag{2}$$

and as the $d^4\theta$ annihilates purely holomorphic and antiholomorphic functions of chiral superfields, the rigidly supersymmetric theory is automatically invariant under (1).

Supergravity theories are more complicated, and the supergravity action turns out not to be invariant under (1). For example, the superpotential terms in four-dimensional $N = 1$ supergravity are proportional to [21, equ'n (23.3)]

$$\exp(K)\left[g^{ij^*}(D_iW)(D_jW)^* - 3|W|^2\right], \tag{3}$$

[2] It should be noted that supergravity is only meaningful over a large open subset of the full SCFT moduli space, namely the subset 'close' to the weak coupling limit (the large-radius limit, in a Calabi-Yau compactification).

which are clearly not invariant. However, it was noted in [22] that if one combines (1) with an action[3] on the graviton ψ_μ and scalar superpartners χ^i,

$$\psi_\mu \mapsto \exp\left(-\frac{i}{2}\mathrm{Im} f\right)\psi_\mu, \; \chi^i \mapsto \exp\left(+\frac{i}{2}\mathrm{Im} f\right)\chi^i, \tag{4}$$

along with a transformation of the superpotential

$$W \mapsto \exp(-f)W, \tag{5}$$

then under the combined action, the theory is invariant under (1).

Now, let us examine these across elements of an open cover $\{U_\alpha\}$ of $\mathscr{M}$ (formally treating $\mathscr{M}$ as a manifold), following [22]. Let K_α denote the Kähler potential on coordinate patch U_α, and let $f_{\alpha\beta}$ denote the holomorphic function appearing in the transformation (1) across patches U_α, U_β on the intersection $U_\alpha \cap U_\beta$. Then, for example, on the triple intersection $U_\alpha \cap U_\beta \cap U_\gamma$, one finds

$$\begin{aligned} f_{\alpha\beta} + \overline{f}_{\alpha\beta} + f_{\beta\gamma} + \overline{f}_{\beta\gamma} + f_{\gamma\alpha} + \overline{f}_{\gamma\alpha} &= (K_\beta - K_\alpha) + (K_\gamma - K_\beta) + (K_\alpha - K_\gamma), \\ &= 0, \end{aligned} \tag{6}$$

hence

$$f_{\alpha\beta} + f_{\beta\gamma} + f_{\gamma\alpha} = h_{\alpha\beta\gamma}, \tag{7}$$

which are pure imaginary. On quadruple overlaps, it is straightforward to demonstrate that the Cech coboundary $\delta h = 0$.

If there is no gauge symmetry, one can further examine the transformations of the scalar superpartners χ^i, following [22], to argue that on triple overlaps,

$$\exp\left(\frac{i}{2}\mathrm{Im}\, h_{\alpha\beta\gamma}\right) = 1, \tag{8}$$

hence $h_{\alpha\beta\gamma} = 4\pi i n_{\alpha\beta\gamma}$ for integers $n_{\alpha\beta\gamma}$. One then argues that $\exp(-f_{\alpha\beta}/2)$ are transition functions for a line bundle, the "Bagger-Witten line bundle" $\mathscr{L}_{\mathrm{BW}}$, where the $(h_{\alpha\beta\gamma})$ are a Cech representative of c_1 (which is then even).

To be clear, this only defines a line bundle up to a tensor product with a flat line bundle. Consider transition functions, across which the Kähler potential transforms as $K \mapsto K + f + f^*$. Suppose one defines the transition functions to be $\exp(-f/2)$, ala Bagger-witten. Now, K is invariant if one replaces f by $f+g$ for g pure imaginary (hence constant, as follows from holomorphy). After all, $g + g^* = 0$. Hence, we could replace the transition functions $\exp(-f/2)$ by $\exp(-(f+g)/2)$, which in general will define a different line bundle, differing by a flat line bundle. Hence, the supergravity analysis above does not uniquely define a line bundle. However, given a Bagger-Witten line bundle, the analysis above describes how it ties into four-dimensional $N = 1$ supergravity.

---

[3] The action is chiral, and so in the quantum theory, anomalies must be taken into account. We will focus on other aspects here.

From the transformation law (5), we also see that the spacetime superpotential W is a (meromorphic) section of $\mathscr{L}_{\mathrm{BW}}^{\otimes 2}$. Similarly, positivity of the kinetic terms of the scalars and their superpartners, which couple to the pullback of the metric, was used in [22] to argue that $\mathscr{L}_{\mathrm{BW}}^{-1}$ must be ample.

More recently, the Bagger-Witten line bundle has been studied in the context of four-dimensional gauge theories in supergravity. To define the gauge theory, in supergravity, one must specify an action of the gauge group on the scalars and their superpartners – hence, one must specify an action on the moduli space. Now, a group action on a space does not necessarily lift to a line bundle over the same space, and if it does lift, the action will not be unique in general. It was observed in [2, 10] that in the present case, in a gauge theory, if the action of the gauge group on the moduli space does not lift to the Bagger-Witten line bundle, then the action is not (classically) invariant, and hence the gauge theory is well-defined. Furthermore, if a lift does exist, the choice of lift is encoded physically in the Fayet-Iliopoulos parameter, which as a result, must be quantized (as choices of lifts are also quantized).

We can see this more explicitly, following [2, 10]. Under a gauge transformtion, the supergravity scalars ϕ transform as

$$\delta\phi^i = \varepsilon^{(a)} X^{(a)i}, \tag{9}$$

where $\varepsilon^{(a)}$ are gauge transformation parameters and $X^{(a)i}$ the components of a set of holomorphic Killing vectors, the gauge field A^a_μ transforms as

$$\delta A^a_\mu = \partial_\mu \varepsilon^{(a)} + f^{abc}\varepsilon^{(b)} A^c_\mu \tag{10}$$

for f^{abc} the structure constants of the Lie algebra, and the Kähler potential transforms as

$$\delta K = \varepsilon^{(a)} F^{(a)} + \varepsilon^{(a)} \overline{F}^{(a)}, \tag{11}$$

where

$$F^{(a)} = X^{(a)} K + i D^{(a)}. \tag{12}$$

These Kähler potential transformations $F^{(a)}$ also appear in the gauge transformations of the fermions, as for example

$$\delta\chi^i = \varepsilon^{(a)} \left(\frac{\partial X^{(a)i}}{\partial \phi^j} \chi^j + \frac{i}{2} \mathrm{Im}\, F^{(a)} \chi^i \right), \tag{13}$$

$$\delta\psi_\mu = -\frac{i}{2} \varepsilon^{(a)} \mathrm{Im}\, F^{(a)} \psi_\mu, \tag{14}$$

reflecting the fact that if the Kähler potential transforms, then the spinors must also pick up a corresponding phase in order for the theory to remain invariant. This means that the infinitesimal lift of the group action on the moduli space to the Bagger-Witten line bundle is encoded by

$$\frac{i}{2} \varepsilon^{(a)} \mathrm{Im}\, F^{(a)}. \tag{15}$$

Physically, from (12), we see that shifts in the imaginary parts of $F^{(a)}$ are the Fayet-Iliopoulos parameters, so we see that the Fayet-Iliopoulos parameters encode a choice of lift of the group action to the Bagger-Witten line bundle.

Let

$$g = \exp\left(i\varepsilon^{(a)}T^{a}\right) \tag{16}$$

be an element of the Lie group acting on the scalars, where T^{a} is a Lie algebra generator, then from the discussion above, the lift of g to a Bagger-Witten line bundle can be described as

$$\tilde{g} = \exp\left(\frac{i}{2}\varepsilon^{(a)}\mathrm{Im}F^{(a)}\right). \tag{17}$$

We can modify the lift by changing $\tilde{g} \mapsto \tilde{g}\exp(i\theta_g)$ for some collection of phases, subject to the condition that the lift represents the group honestly:

$$\tilde{g}\tilde{h} = \widetilde{gh} \tag{18}$$

for all $g, h \in G$, the gauge group. Such lifts might not always exist – it is not guaranteed that a set $\{\tilde{g}\}$ exist which satisfy (18) even after adding phases. (More formally, if $\tilde{G}$ is the group generated by the $\tilde{g}$, then a priori it is merely an extension of G by $U(1)$, and we need that extension to split in order to be able to satisfy (18).)

Furthermore, if a lift exists satisfying (18), then there are multiple lifts, obtained by deforming by phases $\exp(i\theta_g)$ satisfying

$$\exp(i\theta_g)\exp(i\theta_h) = \exp(i\theta_{gh}), \tag{19}$$

for all $g, h \in G$. A collection of such θ_g defines a group homomorphism $G \to U(1)$, and so we see that if lifts exist, they are classified by $\mathrm{Hom}(G, U(1))$. This means the possible lifts are quantized: for example, if $G = U(1)$, then $\mathrm{Hom}(G, U(1)) = \mathbb{Z}$.

We have already seen that the choice of lift of group action to the Bagger-Witten line bundle is encoded by the Fayet-Iliopoulos parameters. Since those lifts are quantized, we see that the Fayet-Iliopoulos parameters are quantized.

In passing, it was noted in [2] that this story is closely related to the geometric invariant theory (GIT) description of quotients. There, the mathematical analogue of the Fayet-Iliopoulos parameters are also encoded by a choice of group action on a line bundle, which (partially) defines the GIT quotient, in the same way that a choice of point in $\mathfrak{g}^*$ partially determines a symplectic reduction.

Finally, before going on, in cases such as these, when one gauges a group action, the moduli space is best understood as a stack, rather than a space, following e.g. [10, 14]. Revisiting other details of the argument of [22], in [10] it was also pointed out that the transition functions for the bundles in question only close on triple overlaps up to a gauge transformation. The resulting quantities can still be understood as bundles on the corresponding stack, but are not equivalent to ordinary bundles on a space.

3 SCFT description

Bagger-Witten line bundles can also be understood more directly over SCFT moduli spaces, see for example [1, 16]. Consider a family of conformal field theories with some global symmetry G. As one moves across coordinate patches on that family, on overlaps, conformal field theories in one patch are related to those on the overlapping patch by an action of G. This overlap data defines transition functions for a G bundle over the parameter space.

In the case of moduli spaces of two-dimensional $N = 2$ SCFTs, since there is always at least a global $U(1)_R$ symmetry (which exists as part of the $N = 2$ superconformal algebra), the argument above implies that there exists a line bundle over the moduli space.

We can then read off how various CFT operators transform across the moduli space from their $U(1)_R$ charges. Those operators then transform globally as sections of line bundles associated to an underlying principal $U(1)$ bundle by the representations defined by $U(1)_R$ charges.

Two particularly important examples (in the notation of [11]) are as follows:

- The spectral operator $\mathscr{U}_1$. This has the same $U(1)_R$ charge as a holomorphic top-form, and we identify the corresponding line bundle over the moduli space with the Hodge line bundle $\mathscr{L}_{\rm H}$ of holomorphic top-forms.
- The spectral flow operator $\mathscr{U}_{1/2}$. This has half the $U(1)_R$ charge of $\mathscr{U}_1$, and indeed, $(\mathscr{U}_{1/2})^2 = \mathscr{U}_1$. The corresponding line bundle over the moduli space corresponds to a Bagger-Witten line bundle $\mathscr{L}_{\rm BW}$.

The SCFT description illuminates an important relationship. Since $(\mathscr{U}_{1/2})^2 = \mathscr{U}_1$, it is natural to expect that Hodge line bundles and Bagger-Witten line bundles are related by

$$\mathscr{L}_{\rm BW}^{\otimes 2} = \mathscr{L}_{\rm H}. \tag{20}$$

Indeed, this turns out to be the case.

4 Geometric description

Over moduli spaces of Calabi-Yau's, a proposal for a purely geometric description of Bagger-Witten line bundles was presented in [3, 4]. Briefly, it was argued that Bagger-Witten line bundles can be interpreted as bundles of covariantly constant spinors, in the same way that the Hodge line bundle is the bundle of holomorphic top-forms, essentially because of their association with the spectral flow operator $\mathscr{U}_{1/2}$.

On Calabi-Yau threefolds, however, there is a potential issue. A Calabi-Yau threefold has two nowhere-zero covariantly constant spinors. Given the description above, one is led to ask how it can be consistent with the original Bagger-Witten description in four-dimensional $N=1$ supergravity, as only one bundle was discussed there.

To resolve this puzzle, we utilize the fact that the Hodge and Bagger-Witten line bundles over moduli spaces of Calabi-Yau's are believed to be flat. Flatness of the Hodge line bundle over moduli spaces of Calabi-Yau's has been discussed in e.g. [19], [20, theorem 42, corollary 53], and over moduli spaces of two-dimensional SCFTs in[4] [7]. Since Bagger-Witten line bundles are square roots of Hodge line bundles, their flatness follows from that of the Hodge line bundle.

Now, we return to the puzzle of consistency of the four-dimensional $N=1$ supergravity description of Bagger-Witten line bundles with the idea that a Calabi-Yau threefold will host two over its moduli space. As we noted earlier in section 2, the original Bagger-Witten paper [22] only defines the Bagger-Witten line bundle up to tensoring with a flat bundle. Given that the Bagger-Witten line bundle itself is flat, the four-dimensional $N=1$ supergravity description of [22] is completely ambiguous. In particular, in this case one can (and will) have multiple Bagger-Witten line bundles, all of which are consistent with the four-dimensional $N=1$ supergravity description in [22].

5 Example: elliptic curves

Let us begin with elliptic curves. The moduli space of elliptic curves, as frequently advertised, is

$$\mathscr{M}_0 = [\mathfrak{h}/PSL(2,\mathbb{Z})], \tag{21}$$

where $PSL(2,\mathbb{Z}) = SL(2,\mathbb{Z})/\mathbb{Z}_2$, and $\mathfrak{h}$ denotes the upper half plane. This space is the space of τ parameters, which transform as

$$\tau \mapsto \tau' = \frac{a\tau+b}{c\tau+d}, \tag{22}$$

for

$$\begin{bmatrix} a & b \\ c & d \end{bmatrix} \in SL(2,\mathbb{Z}). \tag{23}$$

The $\mathbb{Z}_2$ center of $SL(2,\mathbb{Z})$, namely diagonal matrices $\mathrm{diag}(\pm 1, \pm 1)$, act trivially, and so they are modded out in the description above.

Now, the Hodge line bundle is not defined over $\mathscr{M}_0$. The reason for this is as follows. If we let z denote a local coordinate on T^2, then under $SL(2,\mathbb{Z})$, at the same time that τ transforms [9, section 2.3]

$$z \mapsto \frac{z}{c\tau+d}. \tag{24}$$

[4] In fact, the reference [7] claimed triviality, but as discussed in [8], their arguments only really imply flatness, and indeed, [8] gives examples which are flat but nontrivial. See also [5] for comments on higher-dimensional cases.

As consistency checks, let us consider a pair of special values of z, namely 1 and τ. In principle, they should map to themselves. It is straightforward to compute that under the transformation above,

$$1 \mapsto \frac{1}{c\tau+d} = 1 - c\left(\frac{a\tau+b}{c\tau+d}\right) + (a-1) \sim 1, \tag{25}$$

$$\tau \mapsto \frac{\tau}{c\tau+d} = \frac{a\tau+b}{c\tau+d} + (d-1)\left(\frac{a\tau+b}{c\tau+d}\right) - b \sim \frac{a\tau+b}{c\tau+d}, \tag{26}$$

where $\sim$ denotes the equivalence $z \sim z + m\tau' + n$ for $m,n \in \mathbb{Z}$.

Returning to the actions (22), (24), we see that although τ is invariant under the action of the center $\pm I$, z is not, and in particular, under $\pm I$, $dz \mapsto \pm dz$.

Thus, we see that holomorphic top-forms on T^2 are not invariant under the action of the center of $SL(2,\mathbb{Z})$.

As a result, the Hodge line bundle is defined over a different space, namely

$$\mathscr{M}_1 = [\mathfrak{h}/SL(2,\mathbb{Z})]. \tag{27}$$

The brackets [] indicate that one takes the quotient as a stack, and stacks distinguish quotients by trivial group actions, so even though the difference is merely a trivially-acting group, we see $\mathscr{M}_1 \neq \mathscr{M}_0$.

(As an aside, in physics also, gauging a trivially-acting group results in a different theory than not gauging at all. This was discussed at length in two-dimensional orbifolds and gauge theories in [12, 13, 14], and formed a key component of making sense of string propagation on stacks. In more modern language, a gauge theory in which a subgroup of the gauge group acts trivially has a one-form symmetry, not possessed by the theory in which that same subgroup is not gauged.)

Mathematically, $\mathscr{M}_1$ is a $\mathbb{Z}_2$ gerbe over $\mathscr{M}_0$. The Hodge line bundle of $\mathscr{M}_1$ is the generator of $\mathrm{Pic}(\mathscr{M}_1) = \mathbb{Z}_{12}$.

To define either Bagger-Witten bundle, we must work harder. From the theory of orbifolds, in a supersymmetric sigma model, given a non-R symmetry $\mathbb{Z}_2$ that maps $z \mapsto -z$, its superpartner $\psi \mapsto -\psi$, implying that the Ramond sector vacua map as

$$|\pm\rangle \mapsto \exp(\pm i\pi/2)\,|\pm\rangle = \pm i\,|\pm\rangle. \tag{28}$$

Because of the factors of i, more precisely because $i^2 = -1$ and not $+1$, we see that the center of $\mathbb{Z}_2$ cannot be represented on the Ramond vacua. More generally, under an $SL(2,\mathbb{Z}_2)$ transformation, applying standard rules for orbifolds,

$$|\pm\rangle \mapsto \pm\frac{|\pm\rangle}{\sqrt{c\tau+d}}, \tag{29}$$

the same transformation as $\sqrt{dz}$.

To construct a Bagger-Witten line bundle, we work over the moduli space [8]

$$\mathscr{M}_2 = [\mathfrak{h}/Mp(2,\mathbb{Z})], \tag{30}$$

where $Mp(2,\mathbb{Z})$ is the metaplectic group, the unique nontrivial $\mathbb{Z}_2$ central extension of $SL(2,\mathbb{Z})$:

$$1 \longrightarrow \mathbb{Z}_2 \longrightarrow Mp(2,\mathbb{Z}) \longrightarrow SL(2,\mathbb{Z}) \longrightarrow 1, \tag{31}$$

whose elements can be thought of as pairs

$$\left\{ \begin{bmatrix} a & b \\ c & d \end{bmatrix} \in SL(2,\mathbb{Z}), \pm\sqrt{c\tau+d} \right\}, \tag{32}$$

with product of the form

$$(A, f(-)) \cdot (B, g(-)) = (AB, f(B(-))g(-)), \tag{33}$$

for $A, B \in SL(2,\mathbb{Z})$.

The moduli space $\mathscr{M}_2$ is a $\mathbb{Z}_2$ gerbe over $\mathscr{M}_1$, just as $\mathscr{M}_1$ was itself a $\mathbb{Z}_2$ gerbe over $\mathscr{M}_0$.

The Picard group $\mathrm{Pic}(\mathscr{M}_2) = \mathbb{Z}_{24}$, whose generator g and its inverse g^{-1} can be interpreted as Bagger-Witten line bundles.

This may seem like a rather abstract result, but it has concrete implications. For example, implicitly in the result above is an extension of T-duality for supersymmetric sigma models on T^2, from $SL(2,\mathbb{Z})$ to $Mp(2,\mathbb{Z})$. More generally, in string duality groups, $SL(2,\mathbb{Z})$ is typically extended to $Mp(2,\mathbb{Z})$, after one takes into account sign flips on fermions. This is explored in more detail in [15] (see also [18]).

6 Examples: Calabi-Yau threefolds

Moduli spaces of smooth Calabi-Yau threefolds, constructed from orbifolds of tori, are discussed in [3, 4]. Specifically, these papers studied moduli spaces of toroidal orbifolds by products of $\mathbb{Z}_2$'s, classified in [6], focusing especially on cases in which the orbifolds have $h^{2,1} = 3$.

At some level, the idea of the construction is to utilize results for elliptic curves, in orbifolds of products of the form $E_1 \times E_2 \times E_3$, where each of the E_i is an elliptic curve. For a G orbifold, the moduli spaces are of the form $[\mathfrak{h}^3/H']$, where $H' = H/G$ for H the normalizer of the image of the orbifold group G in the maximal automorphism group of $E_1 \times E_2 \times E_3$ (which includes both $SL(2,\mathbb{Z})^3$ and S_3 from exchanging the three factors).

In most of the cases with $h^{2,1} = 3$, the Picard group contains elements of infinite order; however, in all cases, the Hodge line bundle is of finite order. This is a strong consistency check on the general claim that the Hodge line bundle should always be flat – as the Picard group contains elements of infinite order, not all line bundles are flat, unlike the example of elliptic curves. Also, as a consequence of the finite-order property, a globally-defined Kähler potential exists over the moduli space, which is of the form

$$\log|\eta(\tau_1)\eta(\tau_2)\eta(\tau_3)|^2, \tag{34}$$

where the τ_i are modular parameters associated with the three elliptic curve factors.

Finally, as described in [3], Bagger-Witten bundles exist, and are defined on a $\mathbb{Z}_2$ gerbe over the moduli space which hosts the Hodge line bundle, much as in the case of elliptic curves.

7 Conclusions

We have briefly reviewed progress in understanding Bagger-Witten line bundles over moduli spaces of SCFTs, including recent developments such as applications in supergravities coupled to gauge theories, a proposal for a purely geometric interpretation, and the construction of explicit examples over moduli spaces of Calabi-Yau manifolds.

Acknowledgements We would like to thank J. Distler, R. Donagi, W. Gu, S. Hellerman, and M. Macerato for collaborations on the topics presented here, and also S. Katz, T. Pantev, and J. Tu for useful discussions. E.S. was partially supported by NSF grant PHY-2310588.

References

1. Distler, J.: Notes on N=2 sigma models (1992)
2. Distler, J., Sharpe, E.: Quantization of Fayet-Iliopoulos parameters in supergravity. Phys. Rev. D **83**, 085,010 (2011). DOI 10.1103/PhysRevD.83.085010
3. Donagi, R., Macerato, M., Sharpe, E.: Global aspects of moduli spaces of 2d SCFTs. Commun. Math. Phys. **392**(3), 1063–1098 (2022). DOI 10.1007/s00220-022-04364-3
4. Donagi, R., Macerato, M., Sharpe, E.: On the global moduli of Calabi–Yau threefolds. Asian J. Math. **26**(4), 585–612 (2022). DOI 10.4310/AJM.2022.v26.n4.a4
5. Donagi, R., Morrison, D.R.: Conformal field theories and compact curves in moduli spaces. JHEP **05**, 021 (2018). DOI 10.1007/JHEP05(2018)021
6. Donagi, R., Wendland, K.: On orbifolds and free fermion constructions. J. Geom. Phys. **59**, 942–968 (2009). DOI 10.1016/j.geomphys.2009.04.004
7. Gomis, J., Hsin, P.S., Komargodski, Z., Schwimmer, A., Seiberg, N., Theisen, S.: Anomalies, conformal manifolds, and spheres. JHEP **03**, 022 (2016). DOI 10.1007/JHEP03(2016)022
8. Gu, W., Sharpe, E.: Bagger–Witten line bundles on moduli spaces of elliptic curves. Int. J. Mod. Phys. A **31**(35), 1650,188 (2016). DOI 10.1142/S0217751X16501888
9. Hain, R.: Lectures on moduli spaces of elliptic curves. In: Transformation groups and moduli spaces of curves, *Adv. Lect. Math. (ALM)*, vol. 16, pp. 95–166. Int. Press, Somerville, MA (2011)
10. Hellerman, S., Sharpe, E.: Sums over topological sectors and quantization of Fayet-Iliopoulos parameters. Adv. Theor. Math. Phys. **15**, 1141–1199 (2011). DOI 10.4310/ATMP.2011.v15.n4.a7
11. Lerche, W., Vafa, C., Warner, N.P.: Chiral rings in N=2 superconformal theories. Nucl. Phys. B **324**, 427–474 (1989). DOI 10.1016/0550-3213(89)90474-4
12. Pantev, T., Sharpe, E.: Notes on gauging noneffective group actions (2005)
13. Pantev, T., Sharpe, E.: GLSM's for gerbes (and other toric stacks). Adv. Theor. Math. Phys. **10**(1), 77–121 (2006). DOI 10.4310/ATMP.2006.v10.n1.a4

14. Pantev, T., Sharpe, E.: String compactifications on Calabi-Yau stacks. Nucl. Phys. B **733**, 233–296 (2006). DOI 10.1016/j.nuclphysb.2005.10.035
15. Pantev, T., Sharpe, E.: Duality group actions on fermions. JHEP **11**, 171 (2016). DOI 10.1007/JHEP11(2016)171
16. Periwal, V., Strominger, A.: Kahler geometry of the space of $N = 2$ superconformal field theories. Phys. Lett. B **235**, 261–267 (1990). DOI 10.1016/0370-2693(90)91961-A
17. Silverstein, E., Witten, E.: Criteria for conformal invariance of (0,2) models. Nucl. Phys. B **444**, 161–190 (1995). DOI 10.1016/0550-3213(95)00186-V
18. Tachikawa, Y., Yonekura, K.: Why are fractional charges of orientifolds compatible with Dirac quantization? SciPost Phys. **7**(5), 058 (2019). DOI 10.21468/SciPostPhys.7.5.058
19. Todorov, A.: Ray Singer analytic torsion of Calabi-Yau manifolds I (2000)
20. Todorov, A.: Local and global theory of the moduli of polarized Calabi-Yau manifolds. In: Proceedings of the International Conference on Algebraic Geometry and Singularities (Spanish) (Sevilla, 2001), vol. 19, pp. 687–730 (2003). DOI 10.4171/RMI/365. URL https://doi.org/10.4171/RMI/365
21. Wess, J., Bagger, J.: Supersymmetry and supergravity. Princeton University Press, Princeton, NJ, USA (1992)
22. Witten, E., Bagger, J.: Quantization of Newton's constant in certain supergravity theories. Phys. Lett. B **115**, 202–206 (1982). DOI 10.1016/0370-2693(82)90644-X

G_2-structures with torsion and the deformed Shatashvili–Vafa vertex algebra

Andoni de Arriba de la Hera, Mateo Galdeano and Mario Garcia-Fernandez

Abstract We construct representations of the deformed Shatashvili–Vafa vertex algebra SV_a, with parameter $a \in \mathbb{C}$, as recently proposed in the physics literature by Fiset and Gaberdiel. The geometric input for our construction are integrable G_2-structures with closed torsion, solving the heterotic G_2 system with $\alpha' = 0$ on the group manifolds $S^3 \times T^4$ and $S^3 \times S^3 \times S^1$. From considerations in string theory, one expects the chiral algebra of these backgrounds to include SV_a, and we provide a mathematical realization of this expectation by obtaining embeddings of SV_a in the corresponding superaffine vertex algebra and the chiral de Rham complex. In our examples, the parameter a is proportional to the scalar torsion class of the G_2 structure, $a \sim \tau_0$, as expected from previous work in the semi-classical limit by the second author, jointly with de la Ossa and Marchetto.

1 Introduction

A supersymmetric background of string theory is described by a superconformal field theory on the worldsheet, and the underlying superconformal algebra of chiral symmetries is of special interest. Mathematically, this chiral algebra is identified

Andoni de Arriba de la Hera
Universidad Nacional de Educación a Distancia (UNED), C. de Juan del Rosal, 10, Moncloa-Aravaca, 28040 Madrid, España
e-mail: andoni.dearriba@mat.uned.es

Mateo Galdeano
Fachbereich Mathematik, Universität Hamburg, Bundesstr. 55, 20146, Hamburg, Germany
e-mail: mateo.galdeano@uni-hamburg.de

Mario Garcia-Fernandez
Instituto de Ciencias Matemáticas (CSIC-UAM-UC3M-UCM), Nicolás Cabrera, 13-15, Cantoblanco, 28049 Madrid, España
e-mail: mario.garcia@icmat.es

D. R. Wood et al. (eds.), *2024 MATRIX Annals, Part II*, MATRIX Book Series 8,
https://doi.org/10.1007/978-3-032-16206-9_19

with a supersymmetric (SUSY) vertex algebra, and the correspondence with the background geometry can be made precise by providing an embedding of the vertex algebra in the chiral de Rham complex.

The chiral de Rham complex is a sheaf of vertex algebras on any smooth manifold, introduced by Malikov, Schechtman and Vaintrob [22], which provides a formal quantization of the non-linear sigma model [10]. Vertex algebra embeddings have been previously studied for special holonomy manifolds, see for example [11, 16, 23].

Embeddings of the $\mathcal{N}=2$ vertex algebra for compact non-Kähler complex manifolds have been recently obtained, first in the homogeneous setting [1] and later in more generality from solutions of the Hull-Strominger system [2]. The key assumption in these results is that the underlying manifold admits a solution to the *Killing spinor equations* of heterotic supergravity. In particular, these geometries describe heterotic supersymmetric backgrounds to leading order in the α'-expansion.

Our main motivation in this note is to study the analogous problem for a seven-dimensional Riemannian manifold M. Suppose M admits a solution to the Killing spinor equations with parameter $\lambda \in \mathbb{R}$ and closed NS flux H [25]:

$$\nabla^{+}\eta = 0, \qquad \left(\not{\nabla}^{1/3} - \frac{1}{2}\zeta\right)\cdot\eta = \lambda\,\eta, \qquad dH = 0, \tag{1}$$

for a real spinor η, a three-form $H \in \Omega^3$, and a one-form $\zeta \in \Omega^1$. Here, ∇^+ and $\not{\nabla}^{1/3}$ are the spin connection and Dirac operator of the connections with skew torsion H and $\frac{1}{3}H$, respectively. The equations (1) are equivalent [25] to the *heterotic* G_2 *system* with $\alpha' = 0$ [6, 7]. Consequently, there exists an integrable G_2-structure on M with torsion three-form H (see Proposition 1) satisfying

$$d\varphi = \frac{12}{7}\lambda\,\psi + 3\,\tau_1\wedge\varphi + *\tau_3\,, \qquad d\psi = 4\,\tau_1\wedge\psi\,, \qquad dH = 0\,,$$

where the precise definition of the different torsion classes τ_j can be found in Section 2. Note that, in the present setup, a solution to the first and third equations in (1) determines uniquely a solution of the full system, with the prescription

$$\tau_0 = \frac{12}{7}\lambda, \qquad \tau_1 = \frac{1}{4}\zeta.$$

When τ_1 is exact, the system describes $\mathcal{N}=1$ backgrounds of the form $AdS_3 \times M$ when $\tau_0 \neq 0$, and $\mathbb{R}^{1,2}\times M$ when $\tau_0 = 0$. Remarkably, the eigenvalue λ of the Dirac spinor η is tied to the scalar torsion class $\tau_0 = \frac{12}{7}\lambda$, which is in turn related to the cosmological constant of the spacetime.

In the case of a G_2-holonomy manifold, all torsion classes vanish and the spacetime degenerates to flat Minkowski space. In this situation, the underlying SUSY vertex algebra was described by Shatashvili–Vafa [24], and its embedding in the chiral de Rham complex was found by Rodríguez Díaz [23].

Fiset–Gaberdiel [13] proposed a one-parameter family of algebras, that we call *deformed Shatashvili–Vafa* and denote by SV_a, which depend on a parameter $a \in \mathbb{C}$ and are suitable for describing AdS_3 backgrounds. These algebras belong to the two-parameter family of algebras $\mathscr{SW}(\frac{3}{2}, \frac{3}{2}, 2)$ first constructed by Blumenhagen [4] (see Remark 3 for a more precise statement). By studying the semiclassical limit—where the superconformal field theory induces a Poisson Vertex algebra—it was shown in [8] that the parameter of SV_a is tied to the scalar torsion class, $a = i\frac{7}{6}\tau_0 \ell_s$, where ℓ_s is the string length.

Gathering all the evidence above, we propose the following conjecture:

Conjecture 1. Let M be a seven-dimensional Riemannian manifold admitting a solution to (1). Then, its chiral de Rham complex admits an embedding of the SUSY vertex algebra SV_a, where the value of a is determined by the eigenvalue λ of the Dirac spinor η.

Similarly as in [2], we also expect an analogous result to hold for the more general situation where a gauge bundle equipped with a G$_2$-instanton solving the heterotic Bianchi identity is incorporated. In this case, one has a Killing spinor on a string algebroid describing a solution of the heterotic G$_2$ system with $\alpha' \neq 0$, and the chiral de Rham complex should be the one associated to this algebroid.

In this note, we provide further evidence for the conjecture by showing that it holds for five different families of G$_2$-structures, defined over two different homogeneous manifolds following [1, 12]. Our main result, including the relationship between a and τ_0, is presented in Theorem 1 below.

It is worth pointing out that the torsion class τ_3 does not play a role in our construction. On the other hand, the torsion class τ_1, which is always closed in the examples we consider, appears in an analogous fashion to that of the Lee form in the $\mathcal{N} = 2$ case studied in [1, 2], see Remark 6. Similarly to the results of [1, 2], we also expect that a dilaton-corrected version of Theorem 1 should hold.

2 Integrable G$_2$-structures and their torsion classes

2.1 Torsion of a G$_2$-structure

We begin with a brief summary of G$_2$-structures and introduce the properties which will be relevant to us. For more detailed accounts on the topic and proofs of the statements presented, we refer the reader to [5, 21].

Definition 1. Let M be a seven-dimensional manifold. A G$_2$*-structure* on M is a nowhere-vanishing three-form φ on M such that for every $p \in M$ there exists a basis $\{v^1, \ldots, v^7\}$ of T_p^*M for which

$$\varphi_0 = v^{123} - v^{145} - v^{167} - v^{246} + v^{257} - v^{347} - v^{356},$$

where we are writing $v^{ij} = v^i \wedge v^j$. We call φ the *associative form.*

A G_2-structure determines an orientation and a metric g on M, given by

$$g(X,Y)\,\mathrm{d}vol = \frac{1}{6}(X \lrcorner \varphi) \wedge (Y \lrcorner \varphi) \wedge \varphi\,,$$

for any $X, Y \in TM$. We can use these to define the *coassociative* form $\psi = *\varphi$.

Differential forms on M decompose into irreducible G_2-representations, inducing a decomposition of the spaces $\Omega^k = \Gamma\left(\Lambda^k(T^*M)\right)$. Two of these subspaces which will shortly play an important role are

$$\Omega^2_{14} = \{\beta \in \Omega^2 \,|\, \beta \wedge \psi = 0\}\,, \qquad \Omega^3_{27} = \{\gamma \in \Omega^3 \,|\, \gamma \wedge \varphi = 0\,,\, \gamma \wedge \psi = 0\}\,.$$

We can uniquely characterize the torsion of the G_2-structure by decomposing the exterior derivatives of φ and ψ into irreducible G_2-representations:

Definition 2. Let M be a seven-dimensional manifold and let φ be a G_2-structure on M. The *torsion classes* of the G_2-structure are the differential forms $\tau_0 \in \Omega^0$, $\tau_1 \in \Omega^1$, $\tau_2 \in \Omega^2_{14}$ and $\tau_3 \in \Omega^3_{27}$ uniquely determined by

$$\mathrm{d}\varphi = \tau_0\,\psi + 3\,\tau_1 \wedge \varphi + *\tau_3\,, \qquad \mathrm{d}\psi = 4\,\tau_1 \wedge \psi + *\tau_2\,.$$

We will say that the G_2-structure is *integrable* if $\tau_2 = 0$.

Proposition 1 ([14]). *Let M be a seven-dimensional manifold and let φ be an integrable* G_2*-structure on M. Then, there exists a unique connection with totally skew-symmetric torsion compatible with the* G_2*-structure. Its torsion three-form H is given by*

$$H = \frac{1}{6}\,\tau_0\,\varphi - \tau_1 \lrcorner \psi - \tau_3\,.$$

2.2 G_2-*structures via products*

A simple way of constructing a G_2-structure is via G-structures in lower-dimensional manifolds. To this end, we introduce:

Definition 3. Let X be a four-dimensional manifold. An SU(2)*-structure* on X is a triple $\{\omega_i\}_{i=1,2,3}$ of nowhere-vanishing two-forms on X satisfying

$$\omega_i \wedge \omega_j = 0\,, \qquad \frac{1}{2}\,\omega_i \wedge \omega_i = \frac{1}{2}\,\omega_j \wedge \omega_j \neq 0\,,$$

for each $i, j \in \{1,2,3\}$, $i \neq j$.

Definition 4. Let N be a six-dimensional manifold. An SU(3)*-structure* on N is a pair of well defined, nowhere-vanishing forms (ω, Ω), where ω is a real two-form and Ω is a complex three-form, satisfying that

$$\omega \wedge \Omega = 0, \qquad \frac{1}{6}\omega \wedge \omega \wedge \omega = \frac{i}{8}\Omega \wedge \overline{\Omega} \neq 0,$$

and that the symmetric 2-tensor $\omega(\cdot, J\cdot)$ is positive definite, where J is the almost complex structure determined by (ω, Ω).

We will denote $\Omega_+ := \mathrm{Re}(\Omega)$ and $\Omega_- := \mathrm{Im}(\Omega)$.

The following lemmas describe how to obtain G$_2$-structures from SU(2)- and SU(3)-structures.

Lemma 1. *Let X be a four-dimensional manifold with* SU(2)*-structure* $\{\omega_i\}_{i=1,2,3}$. *Let Y be an orientable three-dimensional manifold and let η^1, η^2, η^3 be a global frame of T^*Y. Then, the three-form*

$$\varphi = \eta^1 \wedge \eta^2 \wedge \eta^3 + \eta^1 \wedge \omega_1 + \eta^2 \wedge \omega_2 + \eta^3 \wedge \omega_3$$

defines a G$_2$*-structure on the product manifold $Y \times X$.*

Lemma 2. *Let N be a six-dimensional manifold with* SU(3)*-structure* (ω, Ω). *Let η be a nowhere-vanishing one-form on the circle S^1. Then, the three-form*

$$\varphi = \eta \wedge \omega + \Omega^+$$

defines a G$_2$*-structure on the product manifold $N \times S^1$.*

3 Examples of manifolds with G$_2$-structures

We now present the manifolds and G$_2$-structures that constitute the main object of study in this note, mostly following [12].

3.1 $S^3 \times T^4$

We first consider the direct product of a three-dimensional sphere and four circles, $S^3 \times T^4$, which we identify with the Lie group $\mathrm{SU}(2) \times \mathrm{U}(1)^4$. We choose a basis of the Lie algebra $\mathbb{R}^3 \oplus \mathfrak{su}(2) \oplus \mathbb{R}$ such that its dual basis $\{v^1, \ldots, v^7\}$ satisfies the following structure equations

$$\mathrm{d}v^4 = v^{56}, \qquad \mathrm{d}v^5 = -v^{46}, \qquad \mathrm{d}v^6 = v^{45}, \qquad \mathrm{d}v^1 = \mathrm{d}v^2 = \mathrm{d}v^3 = \mathrm{d}v^7 = 0.$$

This manifold can be regarded either as a T^3 bundle over the Hopf surface $S^3 \times S^1$, or as an S^3 bundle over T^4. By defining SU(2) structures on $S^3 \times S^1$ and T^4 and applying Lemma 1, we obtain two inequivalent families of integrable G$_2$-structures over $S^3 \times T^4$ with closed torsion three-form. Their torsion classes follow from a straightforward computation:

Proposition 2 ([12]). *Consider the manifold $S^3 \times T^4$ with a basis of its dual Lie algebra as above. Let $\ell, c_1, c_2, c_3, c_7 > 0$ be constant positive numbers. The forms*

$$\omega_1 = \ell v^{45} + \sqrt{c_7 \ell} v^{67}, \quad \omega_2 = \ell v^{46} - \sqrt{c_7 \ell} v^{57}, \quad \omega_3 = -\sqrt{c_7 \ell} v^{47} - \ell v^{56},$$

define an SU(2)*-structure on the Hopf surface $S^3 \times S^1$. Let*

$$\eta^1 = \sqrt{c_1} v^1, \qquad \eta^2 = \sqrt{c_2} v^2, \qquad \eta^1 = \sqrt{c_3} v^3,$$

be a global frame of T^*T^3, then

$$\varphi^1 = \eta^1 \wedge \eta^2 \wedge \eta^3 + \eta^1 \wedge \omega_1 + \eta^2 \wedge \omega_2 + \eta^3 \wedge \omega_3$$

determines an integrable G_2*-structure on $S^3 \times T^4$ with metric*

$$g = \sum_{i=1}^{3} c_i v^i \otimes v^i + \ell \sum_{j=4}^{6} v^j \otimes v^j + c_7 v^7 \otimes v^7, \tag{2}$$

and torsion classes

$$\tau_0 = 0, \quad \tau_1 = \frac{1}{4}\sqrt{\frac{c_7}{\ell}} v^7, \quad \tau_3 = \frac{1}{4}\left(-\eta^{12} \wedge v^4 + \eta^{13} \wedge v^5 + \eta^{23} \wedge v^6 - 3\ell v^{456}\right).$$

The torsion three-form is closed and equal to

$$H = \ell v^{456}. \tag{3}$$

The following G_2-structure corresponds to one of the examples considered in [13] to motivate the SV_a algebra:

Proposition 3. *Consider the manifold $S^3 \times T^4$ with a basis of its dual Lie algebra as above. Let $\ell, c_1, c_2, c_3, c_7 > 0$ be constant positive numbers. The forms*

$$\tilde\omega_1 = \sqrt{c_7 c_1} v^{71} + \sqrt{c_2 c_3} v^{23}, \qquad \tilde\omega_2 = \sqrt{c_7 c_2} v^{72} - \sqrt{c_1 c_3} v^{13},$$
$$\tilde\omega_3 = -\sqrt{c_7 c_3} v^{73} - \sqrt{c_1 c_2} v^{12},$$

define an SU(2)*-structure on T^4. Let*

$$\tilde\eta^1 = \sqrt{\ell} v^4, \qquad \tilde\eta^2 = \sqrt{\ell} v^5, \qquad \tilde\eta^1 = \sqrt{\ell} v^6,$$

be a global frame of T^*S^3, then

$$\varphi^2 = \tilde\eta^1 \wedge \tilde\eta^2 \wedge \tilde\eta^3 + \tilde\eta^1 \wedge \tilde\omega_1 + \tilde\eta^2 \wedge \tilde\omega_2 + \tilde\eta^3 \wedge \tilde\omega_3$$

determines an integrable G_2*-structure on $S^3 \times T^4$ with metric as in* (2) *and torsion classes*

$$\tau_0 = \frac{6}{7}\frac{1}{\sqrt{\ell}}, \qquad \tau_1 = 0, \qquad \tau_3 = \frac{1}{7\sqrt{\ell}}\varphi^2 - \ell v^{456}.$$

The torsion three-form is closed and given by (3).

Remark 1. Note that both G_2-structures in the previous Propositions have the same metric and torsion three-form. This will have interesting implications for our main results regarding vertex algebras in Section 5.

3.2 $S^3 \times S^3 \times S^1$

We now consider the direct product of two three-dimensional spheres and a circle, $S^3 \times S^3 \times S^1$, which we identify with the Lie group $\mathrm{SU}(2) \times \mathrm{SU}(2) \times \mathrm{U}(1)$. We choose a basis of the Lie algebra $\mathfrak{su}(2) \oplus \mathfrak{su}(2) \oplus \mathbb{R}$ such that its dual basis $\{v^1, \dots, v^7\}$ satisfies the following structure equations

$$\begin{aligned}
&\mathrm{d}v^1 = v^{23}, && \mathrm{d}v^2 = -v^{13}, && \mathrm{d}v^3 = v^{12}, && \\
&\mathrm{d}v^4 = v^{56}, && \mathrm{d}v^5 = -v^{46}, && \mathrm{d}v^6 = v^{45}, && \mathrm{d}v^7 = 0.
\end{aligned}$$

Following [12], we construct three different G_2-structures on $S^3 \times S^3 \times S^1$ by choosing different $\mathrm{SU}(3)$-structures on $S^3 \times S^3$ and applying Lemma 2.

Lemma 3 ([12]). *Consider the manifold $S^3 \times S^3$ with a basis of its dual Lie algebra as above. Let $\ell, s > 0$ be constant positive numbers. The forms*

$$\begin{aligned}
\omega &= \sqrt{s\ell}\, v^{14} + \sqrt{s\ell}\, v^{25} - \sqrt{s\ell}\, v^{36}, \\
\Omega_+ &= \sqrt{s^3}\, v^{123} + \ell\sqrt{s}\, v^{156} - \ell\sqrt{s}\, v^{246} - \ell\sqrt{s}\, v^{345}, \\
\Omega_- &= \sqrt{\ell^3}\, v^{456} + s\sqrt{\ell}\, v^{234} - s\sqrt{\ell}\, v^{135} - s\sqrt{\ell}\, v^{126},
\end{aligned}$$

define an $\mathrm{SU}(3)$-structure on $S^3 \times S^3$ with complex volume form $\Omega = \Omega_+ + i\Omega_-$. Similarly, any complex rotation of the complex volume form $\tilde{\Omega} = e^{i\theta}\Omega$ also defines an $\mathrm{SU}(3)$-structure on $S^3 \times S^3$.

Proposition 4. *Consider the manifold $S^3 \times S^3 \times S^1$ with a basis of the dual Lie algebra as above. Let $\ell, s, c_7 > 0$ be constant positive numbers, let ω, Ω be as in Lemma 3 and let $\eta = \sqrt{c_7}\, v^7$. Applying Lemma 2 results in the following:*

- *From (ω, Ω) we obtain an integrable G_2-structure that we denote by φ^3, with torsion classes*

$$\tau_0 = \frac{6}{7}\frac{1}{\sqrt{s}}, \qquad \tau_1 = \frac{1}{4}\sqrt{\frac{c_7}{\ell}}\, v^7, \qquad \tau_3 = \frac{1}{7\sqrt{s}}\varphi^3 + \frac{1}{4\sqrt{\ell}}\Omega_- - s v^{123} - \ell v^{456}.$$

- *From $(\omega, \Omega^1 := e^{i\theta_1}\Omega)$ with $\tan\theta_1 = \sqrt{\frac{\ell}{s}}$ we obtain an integrable G_2-structure that we denote by φ^4, with torsion classes*

$$\tau_0 = 0, \qquad \tau_1 = \frac{\sqrt{c_7}}{4}\sqrt{\frac{1}{\ell} + \frac{1}{s}}\, v^7, \qquad \tau_3 = \frac{1}{4}\sqrt{\frac{1}{\ell} + \frac{1}{s}}\, \Omega^1_- - s v^{123} - \ell v^{456}.$$

- *From* $(\omega, \Omega^2 := e^{i\theta_2}\Omega)$ *with* $\tan\theta_2 = -\sqrt{\frac{s}{\ell}}$ *we obtain an integrable* G_2*-structure that we denote by* φ^5*, with torsion classes*

$$\tau_0 = \frac{6}{7}\sqrt{\frac{1}{\ell}+\frac{1}{s}}, \qquad \tau_1 = 0, \qquad \tau_3 = \frac{1}{7}\sqrt{\frac{1}{\ell}+\frac{1}{s}}\varphi^5 - sv^{123} - \ell v^{456}.$$

All three G_2*-structures have metric*

$$g = s\sum_{i=1}^{3} v^i \otimes v^i + \ell \sum_{j=4}^{6} v^j \otimes v^j + c_7 v^7 \otimes v^7,$$

and closed torsion three-form equal to

$$H = s v^{123} + \ell v^{456}.$$

4 Superaffine algebras and the deformed Shatashvili–Vafa

4.1 SUSY vertex algebras

In this section we review the basics on SUSY vertex algebras and present those that play a role in this note, including the deformed Shatashvili–Vafa vertex algebra SV_a. For a detailed introduction to vertex algebras, we refer the reader to [20]. We will use extensively the formalism introduced by Heluani–Kac [17], although the $\mathcal{N} = 1$ case we employ here was studied earlier by Barron [3].

A *vertex algebra* is a vector superspace V (that is, a vector space with a $\mathbb{Z}_2$-grading) endowed with a non-vanishing even vector $|0\rangle \in V$ (vacuum), an even endomorphism $T : V \to V$ (infinitesimal translation) and a parity-preserving linear map $Y : V \to \mathrm{End}(V)[[z^{\pm}]]$ (state-field correspondence) mapping each vector $a \in V$ to a *field* $a(z)$, defined as a formal sum

$$a(z) := Y(a,z) = \sum_{n\in\mathbb{Z}} a_{(n)} z^{-n-1},$$

in an even variable z, with *Fourier modes* $a_{(n)} \in \mathrm{End}(V)$. This data must satisfy that $Y(a,z)b$ is a formal Laurent series for all $b \in V$, so that the operator product expansions are finite sums

$$a(z)b(w) \sim \sum_{n\geq 0} \frac{(a_{(n)}b)(w)}{(z-w)^{n+1}},$$

along with the vacuum axioms, translation invariance and locality (see e.g. [20]). We will call $a_{(n)}b$ the *n-product* of a and b.

When the vertex algebra is *conformal*, the T-action is enhanced to an action of the Virasoro algebra for some central charge $c \in \mathbb{C}$. We are interested in the case where this action is further enhanced to an $\mathscr{N} = 1$ superconformal symmetry by including an additional odd linear map $S : V \to V$ such that $S^2 = T$. This motivates the definition of a SUSY vertex algebra [17].

A *SUSY vertex algebra* is a tuple $(V, |0\rangle, S, Y(\cdot,z))$, where $(V, |0\rangle, T = S^2, Y(\cdot,z))$ is a vertex algebra and $S : V \to V$ is an odd linear map which is furthermore a derivation for the n-products.

We can now define the λ*-bracket* (the formal Fourier transform by an even formal parameter λ of the operator product expansion) and the *normally ordered product* as follows

$$[a_\lambda b] = \sum_{n\in\mathbb{Z}} \frac{\lambda^n}{n!} a_{(n)} b, \qquad :ab: = a_{(-1)} b.$$

Just as a Lie algebra freely generates the universal enveloping algebra, a $\mathbb{C}[T]$-module with a λ-bracket satisfying a certain set of axioms—that can be found in [17, Definition 2.6]—and an odd derivation S with respect to the λ-bracket such that $S^2 = T$, freely generate a universal enveloping SUSY vertex algebra. The following example is constructed in this way:

Example 1. Let $(\mathfrak{g}, (\cdot|\cdot))$ be a quadratic Lie algebra and $k \in \mathbb{C}$ a scalar. Let $\Pi\mathfrak{g}$ be the corresponding purely odd vector superspace, identifying elements a of $\mathfrak{g}$ with their corresponding odd copies Πa in $\Pi\mathfrak{g}$. The *universal superaffine vertex algebra* with level k associated to $\mathfrak{g}$ is the SUSY vertex algebra $V^k(\mathfrak{g}_{\text{super}})$ freely generated by the $\mathbb{C}[T]$-module $\mathfrak{g} \oplus \Pi\mathfrak{g}$ (and the scalar k), the odd derivation S defined by

$$Sa = T\,\Pi a, \qquad S\,\Pi a = a,$$

for all $a \in \mathfrak{g}$, and the λ-brackets

$$[a_\lambda b] = [a,b] + \lambda(a|b)k, \qquad [a_\lambda \Pi b] = \Pi[a,b], \qquad [\Pi a_\lambda \Pi b] = (b|a)k,$$

for all $a, b \in \mathfrak{g}$, which satisfy the axioms of [17]. Note that there is an explicit construction of $V^k(\mathfrak{g}_{\text{super}})$ as a quotient of the universal enveloping algebra of a central extension of the loop algebra $\mathfrak{g}[t, t^{-1}]$. In this model, the infinitesimal translation T is induced by $-\partial_t$ and the vacuum vector $|0\rangle$ is induced by the unit in the universal enveloping algebra [9, Section 1.7].

Remark 2. For the quadratic Lie algebras we will consider, the universal superaffine vertex algebra with level $k = 2$ embeds into the chiral de Rham complex of the corresponding Lie group. For more details, see Section 5.1.

4.2 The deformed Shatashvili–Vafa vertex algebra

The *deformed Shatashvili–Vafa algebra* with parameter $a \in \mathbb{C}$, which we will denote by SV_a, is the SUSY vertex algebra freely generated by the $\mathbb{C}[T]$-module

$$\mathbb{C}G \oplus \mathbb{C}L \oplus \mathbb{C}\Phi \oplus \mathbb{C}K \oplus \mathbb{C}X \oplus \mathbb{C}M,$$

the odd derivation S defined by

$$SG = 2L\,, \quad 2SL = TG\,, \quad S\Phi = K\,, \quad SK = T\Phi\,, \quad SX = M\,, \quad SM = TX\,,$$

and the λ-brackets

$$
\begin{aligned}
&[L_\lambda L] = (T+2\lambda)L + \left(\frac{21}{2}+3a^2\right)\frac{\lambda^3}{12}\,, \qquad [L_\lambda G] = \left(T+\frac{3}{2}\lambda\right)G\,,\\
&[G_\lambda G] = 2L + \left(\frac{21}{2}+3a^2\right)\frac{\lambda^2}{3}\,, \qquad [L_\lambda \Phi] = \left(T+\frac{3}{2}\lambda\right)\Phi\,,\\
&[L_\lambda X] = -\frac{7}{24}\lambda^3 + (T+2\lambda)X\,, \qquad [G_\lambda \Phi] = a\frac{\lambda^2}{2} + K\,,\\
&[G_\lambda X] = \left(-\frac{1}{2}G + a\Phi\right)\lambda + M\,, \qquad [\Phi_\lambda \Phi] = -\frac{7}{2}\lambda^2 + 6X\,,\\
&[\Phi_\lambda X] = -\left(\frac{5}{2}T + \frac{15}{2}\lambda\right)\Phi\,, \qquad [X_\lambda X] = \frac{35}{24}\lambda^3 - (5T+10\lambda)X\,,\\
&[L_\lambda K] = a\frac{\lambda^3}{4} + (T+2\lambda)K\,, \qquad [L_\lambda M] = \left(-\frac{1}{2}G + a\Phi\right)\frac{\lambda^2}{2} + \left(T+\frac{5}{2}\lambda\right)M\,,\\
&[G_\lambda K] = (T+3\lambda)\Phi\,, \qquad [G_\lambda M] = -\frac{7}{12}\lambda^3 + (L-aK)\lambda + (T+4\lambda)X\,,\\
&[\Phi_\lambda K] = a(3T+6\lambda)\Phi - \left(\frac{3}{2}T + 3\lambda\right)G - 3M\,,\\
&[\Phi_\lambda M] = -a\frac{5}{4}\lambda^3 + a(3T+6\lambda)X + \left(-\frac{1}{2}T + \frac{9}{2}\lambda\right)K + 3\,{:}\Phi G{:}\,,\\
&[X_\lambda K] = -a\frac{5}{4}\lambda^3 + a(3T+6\lambda)X - 3K\lambda - 3\,{:}\Phi G{:}\,,\\
&[X_\lambda M] = +a\frac{4}{7}\,{:}X\Phi{:} + a\left(-\frac{27}{14}T^2 - 3T\lambda - 3\lambda^2\right)\Phi\\
&\qquad + \left(\frac{1}{4}T^2 - \frac{9}{4}T\lambda - \frac{9}{4}\lambda^2\right)G + \left(\frac{1}{2}T - 5\lambda\right) + 4\,{:}XG{:}\,,\\
&[K_\lambda K] = -\frac{7}{2}\lambda^3 + a(3T+6\lambda)K + (3T+6\lambda)X - (3T+6\lambda)L\,,\\
&[K_\lambda M] = a(3T+6\lambda)M - \left(\frac{11}{2}T\lambda + \frac{15}{2}\lambda^2\right)\Phi + 3\,{:}GK{:} - 6\,{:}L\Phi{:}\,,\\
&[M_\lambda M] = -\frac{35}{24}\lambda^4 - a(3T\lambda+3\lambda^2)K + a(\frac{11}{5}T^2 + 10T\lambda + 10\lambda^2)X\\
&\qquad - \left(\frac{5}{2}T^2 - \frac{9}{2}T\lambda - \frac{9}{2}\lambda^2\right) - 6\,{:}GM{:} + 12\,{:}LX{:} + \frac{2}{5}\,{:}XX{:} - {:}KK{:}\,,
\end{aligned}
$$

which satisfy all the required axioms [17] provided that the following condition is imposed:

$$0 = a\left(\frac{9}{7}T^2\Phi - \frac{8}{7}:\Phi X:\right) + 4:GX: - 2:\Phi K: - 4TM - T^2G. \tag{4}$$

In particular, the vectors G and L in SV_a define a Neveu-Schwarz vertex algebra with central charge $c = \frac{21}{2} + 3a^2$.

Remark 3. In the case $a = 0$, the relation (4) was more conceptually understood in [18], where the Shatashvili–Vafa vertex algebra SV_0 was obtained as a quotient of a specific universal superaffine vertex algebra (cf. Example 1). In this setup, one first undertakes quantum Hamiltonian reduction to obtain a two-parameter W-algebra $\mathcal{SW}(\frac{3}{2}, \frac{3}{2}, 2)$, and then quotient by an ideal generated by (4), for $a = 0$ and a suitable choice of parameters, to obtain SV_0. It was then observed in [13] that a similar procedure applies to SV_a, with the more general relation (4).

Remark 4. Note that the parameter a corresponds to $i\sqrt{\frac{2}{\mathrm{k}}}$ in [13], where k is related to the radius of the AdS_3 spacetime via

$$\mathrm{k} = R^2_{\mathrm{AdS}_3}/\alpha'.$$

In the limit $a \to 0$ we formally recover a Minkowski spacetime and the algebra reduces to that of Shatashvili–Vafa [24].

5 Embeddings

5.1 Chiral de Rham complex

We start recalling the coordinate independent description of the chiral de Rham complex using Courant algebroids [15, 19], which we shall apply later to the situation of a Lie group. We refer the reader to [1] and references therein for details.

Let E be an exact Courant algebroid over a smooth manifold M, with exterior differential d, given by a short exact sequence of smooth bundles of the form

$$0 \longrightarrow T^*M \xrightarrow{\pi^*} E \xrightarrow{\pi} TM \longrightarrow 0. \tag{5}$$

We will denote the indefinite pairing and the Dorfman bracket on sections of E by $\langle , \rangle$ and $[,]$, respectively. Recall that a choice of isotropic splitting $s\colon T \to E$ of (5) induces a vector bundle isomorphism $E \cong T \oplus T^*$, such that

$$\langle X+\xi, X+\xi\rangle = \xi(X), \qquad [X+\xi, Y+\eta] = [X,Y] + L_X\eta - i_Y d\xi + i_Y i_X H,$$

where $H \in \Omega^3(M)$ is a closed three-form determined by s.

Let ΠE be the corresponding purely odd super vector bundle. We denote by $\underline{\mathbb{C}}$ the sheaf of locally constant functions on M. The next result provides a coordinate-free description of the chiral de Rham complex on the smooth manifold M.

Proposition 5 ([15, 19]). *There exists a unique sheaf of SUSY vertex algebras $\Omega^{\mathrm{ch}}_M(E)$ over M endowed with embeddings of sheaves of $\underline{\mathbb{C}}$-modules*

$$\iota\colon C^\infty(M) \hookrightarrow \Omega^{\mathrm{ch}}_M(E), \qquad j\colon E \oplus \Pi E \hookrightarrow \Omega^{\mathrm{ch}}_M(E),$$

satisfying the following properties:

1. *ι is an isomorphism of unital commutative algebras onto its image*

$$\iota(fg) = :\iota(f)\iota(g):,$$

2. *ι and j are compatible with the $C^\infty(M)$-module structure of $E \oplus \Pi E$ and the odd endomorphism S of $\Omega^{\mathrm{ch}}_M(E)$*

$$[\iota(f)_\lambda \iota(g)] = 0, \quad j(fA) = :\iota(f)j(A):, \quad S\iota(f) = j(\Pi df),$$

3. *ι and j are compatible with the Dorfman bracket and pairing*

$$[j(A)_\lambda j(B)] = j([A,B]) + 2\lambda\iota(\langle A,B\rangle), \quad [j(A)_\lambda j(\Pi B)] = j(\Pi[A,B]),$$
$$[j(\Pi A)_\lambda j(\Pi B)] = 2\iota(\langle A,B\rangle),$$

4. *ι and j are compatible with the action of $\Gamma(E \oplus \Pi E)$ on $C^\infty(M)$*

$$[j(A)_\lambda \iota(f)] = \iota(\pi(A)(f)), \quad [j(\Pi A)_\lambda \iota(f)] = 0,$$

5. *$\Omega^{\mathrm{ch}}_M(E)$ is universal with these properties,*

for all $f,g \in C^\infty(M)$, $A,B \in \Gamma(E)$.

To apply this to our results in the next section, assume now that our manifold M is a compact Lie group K, and let E be a left-equivariant exact Courant algebroid over K. By [1, Proposition 4.5], we can associate to E a quadratic Lie algebra given by the invariant sections of E

$$\mathfrak{g} = \Gamma(E)^K,$$

endowed with the induced bracket and pairing. Applying the universal construction in Proposition 5, we obtain an embedding of the superaffine vertex algebra $V^2(\mathfrak{g}_{\mathrm{super}})$ of level $k = 2$ (see [1, Proposition 4.17]).

Proposition 6 ([1]). *Let K be a compact Lie group and E a left-equivariant exact Courant algebroid over K. Then there is an embedding*

$$V^2(\mathfrak{g}_{super}) \hookrightarrow \Gamma(K, \Omega^{\mathrm{ch}}_K(E))$$

of the superaffine vertex algebra $V^2(\mathfrak{g}_{super})$ of level $k = 2$ on the space of global sections $\Gamma(K, \Omega^{\mathrm{ch}}_K(E))$ of $\Omega^{\mathrm{ch}}_K(E)$.

5.2 Embeddings of SV_a

For the remainder of the section, we will denote by M either the manifold $S^3 \times T^4$ or $S^3 \times S^3 \times S^1$ and by $\mathfrak{k}$ the corresponding Lie algebra. Let $\{v_i\}$ be a basis of $\mathfrak{k}$ with dual basis $\{v^i\}$. We will denote by φ any of the G_2-structures introduced in Section 3, that is: φ^i with $i = 1,2$ for $S^3 \times T^4$ and $i = 3,4,5$ for $S^3 \times S^3 \times S^1$. By direct application of [1, Proposition 4.5] we obtain the following.

Proposition 7. *Let M, φ and $\mathfrak{k}$ be as above, and let H be the torsion three-form of φ introduced in Proposition 1. The vector space $\mathfrak{g} = \mathfrak{k} \oplus \mathfrak{k}^*$ endowed with the bracket*

$$[v+\alpha, w+\beta] = [v,w] - \beta([v,]) + \alpha([w,]) + i_w i_v H\,,$$

and pairing

$$(v+\alpha \,|\, v+\alpha) = v(\alpha)\,,$$

for any $v,w \in \mathfrak{k}$, $\alpha,\beta \in \mathfrak{k}^*$, has the structure of a quadratic Lie algebra.

For each choice of M and φ we now present explicit embeddings of the SV_a algebra in the corresponding superaffine vertex algebra $V^k(\mathfrak{g}_{\mathrm{super}})$, for arbitrary level $0 \neq k \in \mathbb{C}$. When $k = 2$, applying Proposition 6, we obtain induced embeddings in the global sections of the chiral de Rham complex

$$\mathrm{SV}_a \hookrightarrow V^2(\mathfrak{g}_{\mathrm{super}}) \hookrightarrow \Gamma(M, \Omega_M^{\mathrm{ch}}(E))\,.$$

To the best of our knowledge, these constitute the first explicit embeddings of the SV_0 algebra when the underlying G_2-structure has torsion, and the first ever explicit embeddings of SV_a for $a \neq 0$.

Following [23], we define the following odd vectors of $V^k(\mathfrak{g}_{\mathrm{super}})$

$$e^i = \frac{1}{\sqrt{2}} \Pi \left(g^{ij} v_j + v^i\right)\,.$$

Theorem 1. *Let M, φ be as above, let $\mathfrak{g}$ be as in Proposition 7 and let $0 \neq k \in \mathbb{C}$. Then, the following sections of $V^k(\mathfrak{g}_{super})$*

$$\Phi = \frac{1}{3k}\sqrt{\frac{2}{k}}\, \varphi_{ijk} :e^i :e^j e^k::\,, \qquad K = S\Phi\,,$$

generate an embedding of the SV_a algebra in $V^k(\mathfrak{g}_{super})$ for the choice of parameter

$$a = -\frac{1}{\sqrt{k}}\frac{7}{6}\tau_0\,.$$

The remaining fields are easily recovered via

$$X = \frac{1}{6}\Phi_{(0)}\Phi\,, \qquad M = SX\,, \qquad G = -\frac{1}{3}\Phi_{(1)}K - 2\frac{1}{\sqrt{k}}\frac{7}{6}\tau_0\,\Phi\,, \qquad L = \frac{1}{2}SG\,.$$

Proof. The proof is a straightforward but long verification that the sections defined above satisfy the λ-brackets introduced in Section 4.2. The condition (4) is directly satisfied for this embedding. The computation is more easily performed with the assistance of computer software: we have used the *OPEdefs Mathematica* package by Thielemans [26].

Remark 5. Our results shall be compared with the embedding of SV_0 in the chiral de Rham complex of a G_2-holonomy manifold found by Rodríguez Díaz [23], which uses a local expression for the generator Φ in normal coordinates with an explicit dependence on the Christoffel symbols of the Levi-Civita connection. The simpler expression for Φ in our main result is due to the fact that the connection with skew torsion introduced in Proposition 1 has vanishing Christoffel symbols in the given parallel frames.

Remark 6. The embedding constructed in Theorem 1 is equivalently determined by the sections Φ and G, where the latter can be defined independently of K as

$$G = G_0 + \frac{4\sqrt{2}}{k} T(\tau_{1\,i} e^i),$$

where

$$G_0 = \frac{1}{k}\left(2g_{ij} :(Se^i)e^j: + \frac{1}{3k} 4 g_{ij} g_{mn} :e^i :e^m \left[e^n, e^j\right]::\right).$$

Note that the expression for G_0 only depends on the metric g and the closed torsion three-form H, while G receives a *dilaton correction* from the Lee form τ_1, as found in [1, 2]. By [25, Theorem 4.9], this fulfills our general expectation that a generalized Ricci flat metric with closed divergence should induce an embedding of the $\mathscr{N} = 1$ vertex algebra in the global sections of the chiral de Rham complex. Similarly as above, the generator of supersymmetry G should receive a dilaton correction proportional to T acting on the divergence. Note also that G_0 is formally similar to the $\mathscr{N} = 1$ generator in the Kac–Todorov construction [17], although our formula differs from the standard one, since we are not using a pair of dual basis.

Acknowledgements We would like to thank Andrew Linshaw for help with the program *OPEdefs*, and Jethro van Ekeren for helpful discussions. The first author is grateful to the Department of Mathematics of the University of Denver for the hospitality. The second and third authors would like to thank the mathematical research institute MATRIX in Australia for hospitality and support during the program *The Geometry of Moduli Spaces in String Theory*, where some of the ideas of this note were born. We would specially like to thank the organizers of the program for providing a stimulating environment for discussions.

References

1. Álvarez-Cónsul, L., De Arriba de La Hera, A., Garcia-Fernandez, M.: (0,2) Mirror Symmetry on Homogeneous Hopf Surfaces. Int. Math. Res. Not. **2024**(2) (2024) 1211–1298. DOI 10.1093/imrn/rnad016

2. Álvarez-Cónsul, L., De Arriba de La Hera, A., Garcia-Fernandez, M.: Vertex algebras from the Hull-Strominger system (2023). DOI 10.48550/arXiv.2305.06836
3. Barron, K.D.: N=1 Neveu-Schwarz vertex operator superalgebras over Grassmann algebras and with odd formal variables (1999). DOI 10.48550/arXiv.math/9910007
4. Blumenhagen, R.: Covariant construction of N=1 super W-algebras. Nucl. Phys. **B381** (1992) 641–669. DOI 10.1016/0550-3213(92)90494-V
5. Bryant, R.L.: Some remarks on G_2-structures. In: Proceedings of Gökova Geometry-Topology Conference, pp. 75–109 (2005)
6. Clarke, A., Garcia-Fernandez, M., Tipler, C.: Moduli of G_2 structures and the Strominger system in dimension 7 (2016). DOI 10.48550/arXiv.1607.01219
7. de la Ossa, X., Larfors, M., Svanes, E.E.: The Infinitesimal Moduli Space of Heterotic G_2 Systems. Commun. Math. Phys. **360**(2) (2018) 727–775. DOI 10.1007/s00220-017-3013-8
8. de la Ossa, X., Galdeano, M., Marchetto, E.: $\mathscr{SW}$-algebras and strings with torsion (2024). DOI 10.48550/arXiv.2412.13904
9. De Sole, A., Kac, V.G.: Finite vs affine W algebras. Jpn. J. Math. **1** (2006) 137–261. DOI 10.1088/1751-8121/aa8539.
10. Ekstrand, J., Heluani, R., Kallen, J., Zabzine, M.: Non-linear sigma models via the chiral de Rham complex. Adv. Theor. Math. Phys. **13**(4) (2009) 1221–1254. DOI 10.4310/ATMP.2009.v13.n4.a6
11. Ekstrand, J., Heluani, R., Kallen, J., Zabzine, M.: Chiral de Rham complex on Riemannian manifolds and special holonomy. Commun. Math. Phys. **318** (2013) 575–613. DOI 10.1007/s00220-013-1659-4
12. Fino, A., Martín-Merchán, L., Raffero, A.: The twisted G_2 equation for strong G_2-structures with torsion. Q. J. Pure Appl. Math. **20**(6) (2024) 2711–2767. DOI 10.4310/PAMQ.241112235423
13. Fiset, M.A., Gaberdiel, M.R.: Deformed Shatashvili-Vafa algebra for superstrings on $AdS_3 \times \mathscr{M}_7$. JHEP **05** (2021) 156. DOI 10.1007/JHEP05(2021)156
14. Friedrich, T., Ivanov, S.: Parallel spinors and connections with skew symmetric torsion in string theory. Asian J. Math. **6** (2002) 303–336. DOI 10.4310/AJM.2002.v6.n2.a5
15. Heluani, R.: Supersymmetry of the chiral de Rham complex II: Commuting Sectors. Int. Math. Res. Not **2009**(6) (2009) 953–987. DOI 10.1093/imrn/rnn145.
16. Heluani, R.: Recent advances and open questions on the susy structure of the chiral de Rham complex. J. Phys. A **50**(42), 2017 (2017). DOI 10.1088/1751-8121/aa8539.
17. Heluani, R., Kac, V.G.: Supersymmetric vertex algebras. Commun. Math. Phys. **271** (2007) 103–178. DOI 10.1007/s00220-006-0173-3.
18. Heluani, R., Rodríguez Díaz, L.O.: The Shatashvili-Vafa G_2 superconformal algebra as a Quantum Hamiltonian Reduction of $D(2,1;\alpha)$. Bull. Braz. Math. Soc. (N.S.) **46** (3) (2015) 331–351. DOI 10.1007/s00574-015-0094-x.
19. Heluani, R., Zabzine, M.: Generalized Calabi-Yau manifolds and the chiral de Rham complex. Adv. Math. **223** (2010) 1815–1844. DOI 10.1016/j.aim.2009.10.007
20. Kac, V.G.: Vertex algebras for beginners. University lecture series. American Mathematical Society (1998)
21. Karigiannis, S.: Introduction to G_2 Geometry, pp. 3–50. Springer US, New York, NY (2020). DOI 10.1007/978-1-0716-0577-6_1
22. Malikov, F., Schechtman, V., Vaintrob, A.: Chiral de Rham complex. Commun. Math. Phys. **204** (1999) 439–473. DOI 10.1007/s002200050653
23. Rodríguez Díaz, L.O.: G_2 Holonomy Manifolds are Superconformal. Commun. Math. Phys. **364**(2) (2018) 385–440. DOI 10.1007/s00220-018-3264-z
24. Shatashvili, S.L., Vafa, C.: Superstrings and manifold of exceptional holonomy. Selecta Math. **1** (1995) 347–381. DOI 10.1007/BF01671569
25. Silva, A.A.d., Garcia-Fernandez, M., Lotay, J.D., Earp, H.N.S.: Coupled G_2-instantons (2024). DOI 10.48550/arXiv.2404.12937
26. Thielemans, K.: An Algorithmic approach to operator product expansions, W algebras and W strings. Ph.D. thesis, Leuven U. (1994)

Chapter 5

MATRIX Program: Tensor Categories, Quantum Symmetries, and Mathematical Physics

Low-dimensional indecomposable representations of the braid group B_3

Eric C. Rowell and Yuze Ruan

Abstract In this note we give a complete classification of all indecomposable yet reducible representations of B_3 for dimensions 2 and 3 over an algebraically closed field K with characteristic 0, up to equivalence. We illustrate their utility with an example.

1 Introduction

The problem of classifying representations of the braid groups B_n is unsurprisingly difficult, as is typically the case for infinite non-abelian discrete groups. On the other hand, even for small n and in low dimensions classifications can be invaluable in applications to braided tensor categories [7], especially in questions relevant to quantum computation [4, 8]. Indeed, the classification of irreducible B_3 representations of dimension ≤ 5 by Tuba and Wenzl [9] gave purchase on questions in *loc. cit.*. In particular their classification shows that such irreducible representations are determined, up to finite ambiguity, by the eigenvalues of the image of standard generators of B_3, and conversely shows that irreducible representations always exist for a given list of eigenvalues, provided certain polynomials are non-vanishing.

It is natural to consider indecomposable representations, which play a similar role in non-semisimple settings, such as in the case of the small quantum groups $u_q\mathfrak{sl}_n$. Non-semisimple modular categories have been garnering significant interest recently, for example in [5]. See also [1] for classification questions and [2] for applications to quantum information.

Eric C. Rowell
Texas A&M University, College Station, TX 77843, USA
e-mail: rowell@tamu.edu

Yuze Ruan
Tsinghua University, Beijing, 100084, China
e-mail: ryz22@mails.tsinghua.edu.cn

D. R. Wood et al. (eds.), *2024 MATRIX Annals, Part II*, MATRIX Book Series 8,
https://doi.org/10.1007/978-3-032-16206-9_20

In this paper we give a complete classification of all indecomposable but not irreducible representations (*strictly indecomposable*) of B_3 for dimension up to 3 over an algebraic closed field K of characteristic 0. We find that there are many more forms a strictly indecomposable representation of B_3 may take, which is more complicated than the result of simple representations given by Wenzl and Tuba. More precisely, let σ_1, σ_2 be the generators of B_3, and A is the linear endomorphism by which σ_1 acts on some representation V. If V is simple with dimension less than 3, it is uniquely determined up to equivalence by the eigenvalues of A, whereas for strictly indecomposable representations there may be further parameters and choices. In particular, in some cases there are relations among the eigenvalues of A, and there may be inequivalent indecomposable representations with the same A. Our approach in this paper is computational and elementary: we first reduce A to the Jordan normal form, then determine all possibilities for the common invariant subspace, then compute the solutions, finally we verify all the solutions are indeed inequivalent and indecomposable. In the last section of this paper, we will also discuss an application and outline ways to extend our results. The appendix contains details of the computer calculations.

Acknowledgements The computations for this research were largely carried out while Y.R. was an undergraduate in Mathematics at Texas A&M University, and the hospitality of that institution is gratefully acknowledged. The research of E.C.R. is partially supported by US NSF grant DMS-2205962. Y.R. is supported by Beijing Municipal Science & Technology Commission Z221100002722017.

2 Preliminaries

Definition 1. The Artin braid group B_n is the group generated by $\sigma_1, \sigma_2, \ldots, \sigma_{n-1}$ satisfying relations $\sigma_i\sigma_j = \sigma_j\sigma_i$ if $|i-j| \geq 2$, and

$$\sigma_i\sigma_{i+1}\sigma_i = \sigma_{i+1}\sigma_i\sigma_{i+1}, \; for \; 1 \leq i \leq n-2.$$

The group B_3 which is generated by two elements σ_1, σ_2 satisfying the relation $\sigma_1\sigma_2\sigma_1 = \sigma_2\sigma_1\sigma_2$.

We refer to [3] for more details on the structures of the braid group.

Definition 2. (Direct sums and indecomposable representations) Let G be a group. If (V, φ) and (W, ρ) are both G-representations then:

(a) $(V \oplus W, (\phi, \rho))$ is a G-representation.
(b) if $(V, \varphi) = (W_1 \oplus W_2, (\rho_1, \rho_2))$, and $W_1 \neq V, W_1 \neq 0$, then (V, φ) is **decomposable**. Otherwise, it is said to be **indecomposable**.
(c) (V, φ) is **irreducible** if the only $\varphi(G)$ invariant subspaces of V are 0 and V.
(d) We call a representation **strictly indecomposable** if it is indecomposable but not irreducible.

In what follows, for a B_3 represenation (ρ, V) we set $\rho(\sigma_1) = A$ and $\rho(\sigma_2) = B$. Since B_3 is generated by σ_1, σ_2, we have

Lemma 1. *The representation (V, ρ) is reducible if and only if there exists a proper non-zero subspace $W \subset V$ which is an invariant subspace of both A and B, and (V, ρ) is indecomposable iff there are no proper non-zero invariant subspaces W_i such that $V = W_1 \oplus W_2$.*

The following is synthesized from [9], focusing on the 2 and 3 dimensional cases.

Theorem 1 ([9]). *Any $d = 2$ or $d = 3$ dimensional irreducible representation ρ of B_3 over K is equivalent to one of the form:*

$$A = \begin{pmatrix} \lambda_1 & \lambda_1 \\ 0 & \lambda_2 \end{pmatrix}, \quad B = \begin{pmatrix} \lambda_2 & 0 \\ -\lambda_2 & \lambda_1 \end{pmatrix} \quad \text{for } d = 2,$$

$$A = \begin{pmatrix} \lambda_1 & \lambda_1\lambda_3\lambda_2^{-1} + \lambda_2 & \lambda_2 \\ 0 & \lambda_2 & \lambda_2 \\ 0 & 0 & \lambda_3 \end{pmatrix}, \quad B = \begin{pmatrix} \lambda_3 & 0 & 0 \\ -\lambda_2 & \lambda_2 & 0 \\ \lambda_2 & -\lambda_1\lambda_3\lambda_2^{-1} - \lambda_2 & \lambda_1 \end{pmatrix} \quad \text{for } d = 3.$$

where $\lambda_i \in K^\times$. Moreover, these representations are irreducible if and only if the λ_i do not satisfy $\lambda_1^2 - \lambda_1\lambda_2 + \lambda_2^2 = 0$ for $d = 2$ or $(\lambda_1^2 + \lambda_2\lambda_3)(\lambda_2^2 + \lambda_1\lambda_3)(\lambda_3^2 + \lambda_1\lambda_2) = 0$ for $d = 3$.

It will be useful to assume that A is in Jordan form, so that the standard basis forms a basis of generalized eigenvectors. Since any A-invariant subspace is spanned by generalized eigenvectors, we have:

Lemma 2 ([9]). *If $\{e_i\}_{1 \le i \le n}$ is a basis of generalized eigenvectors of A then $\{b_i\}_{1 \le i \le n}$ is a basis of generalized eigenvectors of B where $b_i = ABAe_i$.*

Proof. This follows from the fact that B is obtained from conjugating A by ABA.

3 Strictly Indecomposable Representations

From now on, we assume (V, ρ) is a strictly indecomposable n-dimensional B_3 representation over an algebraically closed field K of characteristic 0 and W is an invariant space for A and B of minimal dimension.

Lemma 3. *Suppose $\{e_k\}_{1 \le k \le n}$ is a basis of generalized eigenvectors of A. If $W = span\{e_i\}_{i \in I}$ for some $I \subset \{1, \ldots, n\}$ is an invariant subspace for A and B, then we also have $W = span\{b_i := ABAe_i\}_{i \in I}$.*

Proof. Since W is also invariant under ABA, and ABA is invertible, we have $ABA(W) = W$. Thus by Lemma 2, we see $b_i \in W$ for all $i \in I$ and they form a basis of W.

Theorem 2. *If* $\dim(V) = 2$ *then any strictly indecomposable representation is equivalent to one of the following:*

$$(1)\ A = \begin{pmatrix} \lambda_1 & 0 \\ 0 & \lambda_2 \end{pmatrix} \quad B = \begin{pmatrix} \lambda_1 & 1 \\ 0 & \lambda_2 \end{pmatrix}, \quad \lambda_2 = \lambda_1 e^{\pm \pi i/3}$$
$$(2)\ A = B = \begin{pmatrix} \lambda & 1 \\ 0 & \lambda \end{pmatrix}.$$

Remark 1. Note that by Theorem 1 there are no irreducible 2-dimensional representations with eigenvalues as in (1). While the eigenvalues of (2) can appear in an irreducible representation, this representation is clearly reducible.

Proof. In this case the minimal invariant subspace W has $\dim(W) = 1$, so that A and B both act on W by the same scalar λ. We may assume that A is in Jordan form which gives us two cases:

Case 1 : If $A = \begin{pmatrix} \lambda_1 & 0 \\ 0 & \lambda_2 \end{pmatrix}$ then the standard basis vectors e_1, e_2 are eigenvectors, so any 1-dimensional A, B invariant subspace is spanned by either e_1 or e_2. Thus, by Lemma 3, $W = span\{e_i\} = span\{b_i\}$, for either $i = 1$ or $i = 2$. Without loss of generality we may assume $W = span\{e_1, b_1\}$. Thus, since $b_1 = ABAe_1$ we have that $b_1 = \lambda_1^3 e_1$. Set $b_2 = \gamma_1 e_1 + \gamma_2 e_2$. By indecomposability, $\gamma_1 \neq 0$, also $\gamma_2 \neq 0$, by Lemma 3. Then by Lemma 2, $ABA = \begin{pmatrix} \lambda_1^3 & \gamma_1 \\ 0 & \gamma_2 \end{pmatrix}$. Let $M = \begin{pmatrix} \gamma_1 & 0 \\ 0 & 1 \end{pmatrix}$. Since A commutes with M, by conjugation with M we get an equivalent representation with $ABA = \begin{pmatrix} \lambda_1^3 & 1 \\ 0 & \gamma_2 \end{pmatrix}$. Then by computation, we get B by solving the equation $ABA = BAB$ with three unknown variables: $\gamma_2, \lambda_1, \lambda_2$. We find $\gamma_2 = \lambda_2^3$ where $\lambda_2^2 - \lambda_1\lambda_2 + \lambda_1^2 = 0$, and conclude $B = \begin{pmatrix} \lambda_1 & \frac{1}{\lambda_1\lambda_2} \\ 0 & \lambda_2 \end{pmatrix}$. Then conjugating by $\begin{pmatrix} \lambda_1\lambda_2 & 0 \\ 0 & 1 \end{pmatrix}$, we obtain the equivalence class represented by (1).

Case 2 : $A = \begin{pmatrix} \lambda & 1 \\ 0 & \lambda \end{pmatrix}$, then by similar argument, we get $ABA = \begin{pmatrix} \lambda^3 & \gamma_1 \\ 0 & \gamma_2 \end{pmatrix}$. In this case we cannot conjugate any variables away, and we solve the braid equation with three unknown variables $\gamma_1, \gamma_2, \lambda$ directly, obtaining the equivalence class represented by (2).

Clearly these classes are distinct since the Jordan normal forms of A are different in these cases. They are also indecomposable, since for (1) the only way to write V as a direct sum of proper invariant subspaces of A is $Ke_1 \oplus Ke_2$, but Ke_2 is not invariant for B. It is also obvious for (2).

Next we aim to classify 3-dimensional strictly indecomposable representations of B_3. The strategy is similar to that of Theorem 2, which is outlined as follows:

Step 1 Reduce A to the Jordan normal form.
Step 2 Find the possible basis for W and write down ABA with some extra unknown variables with certain conditions.

Step 3 Compute B (which is easy since A^{-1} has a very simple form) and by using Maple [6] (see Appendix) we can solve the equation: $ABA = BAB$, which ensures that we have a B_3 representation, although they may fail to be strictly indecomposable.

Step 4 Verify indecomposability and deduce inequivalence: First determine all the forms of matrices that commute with A (when A's characteristic polynomial equals its minimal polynomial, such matrices are polynomials of A) and conjugate B by them. Thus we can simplify these solutions to guarantee distinctness. Finally we can deduce whether the representation is indecomposable or not by checking finite possibilities of direct sum decomposition (see Appendix).

Theorem 3. *when* $n = 3$, $\dim(W) = 1$, *every strictly indecomposable representation has one of the following forms (up to equivalence), where* λ_i *are mutually different and non-zero:*

$$(1)\ A = \begin{pmatrix} \lambda_1 & 0 & 0 \\ 0 & \lambda_2 & 0 \\ 0 & 0 & \lambda_3 \end{pmatrix}$$

$$(1.1)\ B = \begin{pmatrix} \lambda_1 & 1 & 1 \\ 0 & \lambda_2 & 0 \\ 0 & 0 & \lambda_3 \end{pmatrix} \qquad (\lambda_2 = \lambda_1 e^{\pm\pi i/3},\ \lambda_3 = \lambda_1 e^{\mp\pi i/3})$$

$$(1.2)\ B = \begin{pmatrix} \lambda_1 & 0 & 1 \\ 0 & \lambda_2 & 1 \\ 0 & 0 & \lambda_3 \end{pmatrix} \qquad (\lambda_1 = \lambda_3 e^{\pm\pi i/3},\ \lambda_2 = \lambda_3 e^{\mp\pi i/3})$$

$$(1.3)\ B = \begin{pmatrix} \lambda_1 & 1 & 1 \\ 0 & \lambda_2 & 2\lambda_2 \\ 0 & 0 & \lambda_3 \end{pmatrix} \qquad (\lambda_1 = \lambda_2 e^{\pm\pi i/3},\ \lambda_3 = \lambda_2 e^{\mp\pi i/3})$$

$$(1.4)\ B = \begin{pmatrix} \lambda_1 & 1 & \frac{-\lambda_2^2(\lambda_1^2-\lambda_1\lambda_2+\lambda_2^2)}{\lambda_1(\lambda_1^2+\lambda_2^2)} \\ 0 & \frac{-\lambda_1^4}{\lambda_2(\lambda_1^2+\lambda_2^2)} & \frac{\lambda_1^2(\lambda_1^4+\lambda_1^2\lambda_2^2+\lambda_2^4)}{(\lambda_1^2+\lambda_2^2)^2} \\ 0 & 1 & \frac{\lambda_2^3}{\lambda_1^2+\lambda_2^2} \end{pmatrix} \qquad (\lambda_3 = -\frac{\lambda_1^2}{\lambda_2})$$

$$(2)\ A = \begin{pmatrix} \lambda_1 & 0 & 0 \\ 0 & \lambda_1 & 0 \\ 0 & 0 & \lambda_2 \end{pmatrix} \qquad B = \begin{pmatrix} \lambda_1 & 1 & \frac{\lambda_1\lambda_2}{\lambda_1-\lambda_2} \\ 0 & \frac{\lambda_2^2}{\lambda_2-\lambda_1} & 0 \\ 0 & 1 & \frac{\lambda_1^2}{\lambda_1-\lambda_2} \end{pmatrix} \qquad (\lambda_1 = \lambda_2 e^{\pm\pi i/3})$$

$$(3)\ A = \begin{pmatrix} \lambda_1 & 0 & 0 \\ 0 & \lambda_2 & 1 \\ 0 & 0 & \lambda_2 \end{pmatrix}$$

$$(3.1)\ B = \begin{pmatrix} \lambda_1 & 0 & 1 \\ 0 & \lambda_2 & 1 \\ 0 & 0 & \lambda_2 \end{pmatrix} \qquad (\lambda_1 = \lambda_2 e^{\pm\pi i/3})$$

(3.2) $B = \begin{pmatrix} \lambda_1 & 1 & \frac{\lambda_1+\lambda_2}{\lambda_1^2} \\ 0 & \lambda_2 & 0 \\ 0 & \lambda_1^2 & \lambda_2 \end{pmatrix}$ $(\lambda_2 = \lambda_1 e^{\pm\pi i/2})$

(4) $A = \begin{pmatrix} \lambda_1 & 1 & 0 \\ 0 & \lambda_1 & 0 \\ 0 & 0 & \lambda_2 \end{pmatrix}$

(4.1) $B = \begin{pmatrix} \lambda_1 & 1+\frac{\lambda_1\beta}{\lambda_2^2} & 1 \\ 0 & \lambda_1 & 0 \\ 0 & \beta & \lambda_2 \end{pmatrix}$ $(\lambda_1 = \lambda_2 e^{\pm\pi i/3}$, *$\beta$ an arbitrary parameter.)*

(4.2) $B = \begin{pmatrix} \lambda_1 & -2 & 0 \\ 0 & \frac{\lambda_2}{2} & 1 \\ 0 & \frac{3\lambda_2^2}{4} & -\frac{\lambda_2}{2} \end{pmatrix}$ $(\lambda_1 = -\lambda_2)$

(5) $A = B = \begin{pmatrix} \lambda & 1 & 0 \\ 0 & \lambda & 1 \\ 0 & 0 & \lambda \end{pmatrix}$

Remark 2. Notice that in all of the cases listed in (1) as well as (3.2) and (4.2) no representation with these eigenvalues can be irreducible, as the polynomial $(\lambda_1^2 + \lambda_2\lambda_3)(\lambda_2^2+\lambda_1\lambda_3)(\lambda_3^2+\lambda_1\lambda_2)$ vanishes. Note also that in case (4.1) the equivalence class of representations is not determined up to finite choices by the eigenvalues– there is an additional parameter β.

Proof. The proof splits into cases depending on the Jordan normal forms of A (**Step 1**).

Case 1 : A is a diagonal matrix with three distinct eigenvalues. By Lemma 3, we may assume $W = span\{e_1\} = span\{b_1\}$, then $ABA = \begin{pmatrix} \lambda_1^3 & \beta_1 & \gamma_1 \\ 0 & \beta_2 & \gamma_2 \\ 0 & \beta_3 & \gamma_3 \end{pmatrix}$ with β_1, γ_1 not equal to 0 simultaneously (by indecomposability). Using Maple, we get (1.1)-(1.4).

Case 2 : A is a diagonal matrix with two distinct eigenvalues. We may assume $W_1 = span\{e_1, e_2\}$ is a two-dimensional eigenspace for A with eigenvalue λ_1. Then by Lemma 2, $W_2 = span\{b_1, b_2\}$ is a two-dimensional eigenspace for B with eigenvalue λ_1. Since V is 3-dimensional, we have $W_0 = W_1 \cap W_2 \neq 0$. If $\dim(W_0) = 2$, then $Be_i = \lambda_1 e_i$, which implies $B = \begin{pmatrix} \lambda_1 & 0 & \gamma_1 \\ 0 & \lambda_1 & \gamma_2 \\ 0 & 0 & \lambda_2 \end{pmatrix}$ with $\gamma_1, \gamma_2 \neq 0$. Then conjugating by $M = \begin{pmatrix} \gamma_1 & 0 & 0 \\ 0 & \gamma_2 & 0 \\ 0 & 0 & 1 \end{pmatrix}$ we can assume γ_1, γ_2 both to be 1. Solving the braid equation to get the relation for eigenvalues, we find the equivalence class is represented by $B = \begin{pmatrix} \lambda_1 & 0 & 1 \\ 0 & \lambda_1 & 1 \\ 0 & 0 & \lambda_2 \end{pmatrix}$, However if we let $M_1 = \begin{pmatrix} 1 & 1 & 0 \\ 2 & 1 & 0 \\ 0 & 0 & 1 \end{pmatrix}$, then we have

$M_1^{-1}AM_1 = A,\ M_1^{-1}BM_1 = \begin{pmatrix} \lambda_1 & 0 & 0 \\ 0 & \lambda_2 & 1 \\ 0 & 0 & \lambda_2 \end{pmatrix}$, from which we conclude representation is decomposable.

If $\dim(W_0) = 1$ then there are scalars $t_1, t_2, k_1, k_2 \in K$ such that $t_1e_1 + t_2e_2 = k_1b_1 + k_2b_2$. Without loss of generality, we may assume $t_1 \neq 0$, and change the basis via $(\widetilde{e_1}, \widetilde{e_2}) = (e_1, e_2)\begin{pmatrix} t_1 & 0 \\ t_2 & 1 \end{pmatrix}$. Replacing A and B by their conjugates by $M = \begin{pmatrix} t_1 & 0 & 0 \\ t_2 & 1 & 0 \\ 0 & 0 & 1 \end{pmatrix}$ A is unchanged and $span\{\widetilde{e_1}\}$ is the common invariant space for A, B. Then by Lemma 3, we let $W = span\{\widetilde{e_1}\} = span\{\widetilde{b_1}\}$, we get $ABA = \begin{pmatrix} \lambda_1 & \beta_1 & \gamma_1 \\ 0 & \beta_2 & \gamma_2 \\ 0 & \beta_3 & \gamma_3 \end{pmatrix}$ as usual and this time $\beta_3 \neq 0$ ($\beta_3 = 0$ is the case that $\dim(W_0) = 2$), and we can simplify the form by taking $\beta_3 = 1$, by conjugating by a diagonal matrix, we get (2).

Case 3 : A has one two-dimensional Jordon block, and two distinct eigenvalues, without loss of generality, $\{e_2, e_3\}$ corresponds to eigenvalue λ_2 and e_3 is a generalized eigenvector. Here are two kinds of choices for W, one is $W = span\{e_1\} = span\{b_1\}$, by same argument and computations we get (3.1), (3.2); the other choice is $W = span\{e_2\} = span\{b_2\}$, then we have (4.1), (4.2).

Case 4 : A has one two dimensional Jordon block with only one eigenvalue λ, then without loss of generality, $W_1 = span\{e_1, e_2\}$ is the two dimensional eigenspace for A with respect to λ, $W_2 = span\{b_1, b_2\}$ is the two dimensional eigenspace with respect to λ for B, Let $W_0 = W_1 \cap W_2$, then by similar argument in **Case 2**, when $\dim(W_0) = 2$, we can reduce B to the form $\begin{pmatrix} \lambda & 0 & 1 \\ 0 & \lambda & \gamma \\ 0 & 0 & \lambda \end{pmatrix}$, by easy computation, this case is impossible. When $\dim(W_0) = 1$, we have a similar story with **Case 2**, except the case when $t_1 = 0$; when $t_1 \neq 0$, we follow the trick and reduce it to the case that $W = span\{e_1, b_1\}$, similarly we get a decomposable representation; when $t_1 = 0$, again by Lemma 3, we know $W = span\{e_2, b_2\}$, then we get $ABA = \begin{pmatrix} \alpha_1 & 0 & \gamma_1 \\ \alpha_2 & \lambda^3 & \gamma_2 \\ \alpha_3 & 0 & \gamma_3 \end{pmatrix}$, by solving the equation we get $\gamma_1 = \alpha_2 = \alpha_3 = 0$, then $V = span\{e_1\} \oplus span\{e_2, e_3\}$, A, B are both invariant on each summand, which means it is decomposable as well.

Case 5 : A is a 3-dimensional Jordan block, the only choice for W is $span\{e_1\} = span\{b_1\}$, again by using Maple we get (5).

When A is a scalar matrix, by conjugating $M = ABA$, we get $B = A$, which is obviously decomposable.

Finally, we can easily verify all the solutions we get are distinct and indeed indecomposable but not irreducible (see Appendix), which finishes the proof.

Theorem 4. *when $n = 3$, $\dim(W) = 2$, we have following results: ($\lambda_i \neq 0$ are distinct)*

$$(1)\ A=\begin{pmatrix}\lambda_1 & 0 & 0\\ 0 & \lambda_2 & 0\\ 0 & 0 & \lambda_3\end{pmatrix} \qquad B=\begin{pmatrix}\frac{\lambda_2^3}{\lambda_2^2+\lambda_3^2} & 1 & 1\\ \frac{\lambda_3^2(\lambda_2^4+\lambda_2^2\lambda_3^2+\lambda_3^4)}{(\lambda_2^2+\lambda_3^2)^2} & -\frac{\lambda_3^4}{\lambda_2(\lambda_2^2+\lambda_3^2)} & -\frac{\lambda_3^3(\lambda_2^2+\lambda_2\lambda_3+\lambda_3^2)}{\lambda_2^2(\lambda_2^2+\lambda_3^2)}\\ 0 & 0 & \lambda_3\end{pmatrix}$$

$$(\lambda_1=-\tfrac{\lambda_3^2}{\lambda_2}\ \textit{and}\ \lambda_1^2-\lambda_1\lambda_2+\lambda_2^2\neq 0)$$

$$\textit{Or equivalently}\ A=\begin{pmatrix}\lambda_2 & \lambda_2 & 0\\ 0 & \lambda_1 & 0\\ 0 & 0 & \lambda_3\end{pmatrix} \qquad B=\begin{pmatrix}\lambda_1 & 0 & \frac{\lambda_3^2-(\lambda_1-\lambda_2)\lambda_3+\lambda_2^2}{\lambda_2^2+\lambda_3^2}\\ -\lambda_1 & \lambda_2 & -1\\ 0 & 0 & \lambda_3\end{pmatrix}$$

$$(2)\ A=\begin{pmatrix}\lambda_1 & 1 & 0\\ 0 & \lambda_1 & 0\\ 0 & 0 & \lambda_2\end{pmatrix} \qquad B=\begin{pmatrix}2\lambda_1 & 1 & \frac{2\lambda_1+\lambda_2}{\lambda_2^2}\\ \lambda_2^2 & 0 & 1\\ 0 & 0 & \lambda_2\end{pmatrix} \qquad (\lambda_1=\lambda_2 e^{\pm\pi i/2})$$

$$\textit{Or equivalently}\ A=\begin{pmatrix}\lambda_1 & \lambda_1 & 0\\ 0 & \lambda_1 & 0\\ 0 & 0 & \lambda_2\end{pmatrix} \qquad B=\begin{pmatrix}\lambda_1 & 0 & 1+\frac{\lambda_1}{\lambda_2}\\ -\lambda_1 & \lambda_1 & \frac{\lambda_2}{\lambda_1}\\ 0 & 0 & \lambda_2\end{pmatrix}$$

$$(3)\ A=\begin{pmatrix}\lambda_1 & 0 & 0\\ 0 & \lambda_2 & 1\\ 0 & 0 & \lambda_2\end{pmatrix} \qquad B=\begin{pmatrix}\frac{\lambda_2}{2} & 1 & 0\\ \frac{3\lambda_2^2}{4} & -\frac{\lambda_2}{2} & -2\\ 0 & 0 & \lambda_2\end{pmatrix} \qquad (\lambda_1=-\lambda_2)$$

$$\textit{Or equivalently}\ A=\begin{pmatrix}\lambda_1 & \lambda_1 & 1\\ 0 & \lambda_2 & -2\\ 0 & 0 & \lambda_2\end{pmatrix} \qquad B=\begin{pmatrix}\lambda_2 & 0 & -2\\ -\lambda_2 & \lambda_1 & 4\\ 0 & 0 & \lambda_2\end{pmatrix}$$

Remark 3. Notice that in all cases the polynomial

$$(\lambda_1^2+\lambda_2\lambda_3)(\lambda_2^2+\lambda_1\lambda_3)(\lambda_3^2+\lambda_1\lambda_2)$$

vanishes so that no 3-dimensional B_3 representation with these eigenvalues can be irreducible.

Proof. As usual, we first reduce A to the Jordan normal form. Observe that in each of the cases **Case 2**, **Case 4** and **Case 5** there is a 1-dimensional invariant subspace, contrary to the assumption $\dim(W)=2$. Therefore it suffices to consider the following two cases:

Case 1 : A is a diagonal matrix with three different eigenvalues. In this case the 2-dimensional invariant space for A can only be $span\{e_i,e_j\}$ $i,j\in\{1,2,3\}$. Without loss of generality, we may assume $W=span\{e_1,e_2\}$ is the common invariant subspace of A,B. By Lemma 3, $W=span\{b_1,b_2\}$, then we get $ABA=\begin{pmatrix}\alpha_1 & \beta_1 & \gamma_1\\ \alpha_2 & \beta_2 & \gamma_2\\ 0 & 0 & \gamma_3\end{pmatrix}$, with $\alpha_i,\beta_i\neq 0, i\in\{1,2\}$, otherwise $\dim(W)=1$, and $\gamma_3\neq 0$, by similar computation we get (1).

Case 2 : A has one two-dimensional Jordon block, and two distinct eigenvalues, then the 2-dimensional invariant space for A can only be the generalized eigenspace or the direct sum of two 1-dimensional eigenspace, without loss of generality, we let $span\{e_1,e_2\}$ be the generalized eigenspace, then for one possibility, $W = span\{e_1,e_2\}$, and $W = span\{b_1,b_2\}$ (Lemma 3), therefore we get the same form of ABA as in **Case 1**, except that α_1 or β_1 could be 0 this time, then by computation, we get (2). For the other possibility of W, let $W = span\{e_1,e_3\}$, thus $W = span\{b_1,b_3\}$, by suitably adjusting the basis, we obtain the same form of ABA as in **Case 1**, finally we get (3).

It's easy to verify the solutions are indecomposable and distinct since A satisfies that its characteristic polynomial and minimal polynomial are equal.

4 Applications and Future Work

Analyzing the braid group representations associated with non-semisimple braided categories is an immediate application to our classification. Another key area in which indecomposable braid group representations appear is in the study of Yang-Baxter operators.

Application. Consider the following two Yang-Baxter matrices, which are homogeneous versions of the R-matrices associated with the quantum groups $U_q\mathfrak{sl}_2$ and $U_q\mathfrak{gl}(1|1)$. Here a,b are distinct nonzero complex numbers.

$$R_1 = \begin{pmatrix} a & 0 & 0 & 0 \\ 0 & a+b & -b & 0 \\ 0 & a & 0 & 0 \\ 0 & 0 & 0 & a \end{pmatrix}, \quad R_2 = \begin{pmatrix} a & 0 & 0 & 0 \\ 0 & a+b & -b & 0 \\ 0 & a & 0 & 0 \\ 0 & 0 & 0 & b \end{pmatrix}$$

These yield 8-dimensional representations for B_3 by

$$A_i = \rho_i(\sigma_1) = R_i \otimes \begin{pmatrix} 1 & 0 \\ 0 & 1 \end{pmatrix}, \quad B_i = \rho_i(\sigma_1) = \begin{pmatrix} 1 & 0 \\ 0 & 1 \end{pmatrix} \otimes R_i$$

If $a^2 - ab + b^2 \neq 0$, and also, for $i = 2$,

$$(2a^2 - ab - a + b)\big((6b-1)a^2 + (-8b^2 + b)a + 4b^3\big)(a^2 - ab - b^2) \neq 0$$

these representations are completely reducible, with the following irreducible decompositions:

$$A_1 = a \oplus a \oplus a \oplus a \oplus \begin{pmatrix} a & a \\ 0 & b \end{pmatrix} \oplus \begin{pmatrix} a & a \\ 0 & b \end{pmatrix}, \quad B_1 = a \oplus a \oplus a \oplus a \oplus \begin{pmatrix} b & 0 \\ -b & a \end{pmatrix} \oplus \begin{pmatrix} b & 0 \\ -b & a \end{pmatrix}$$

$$A_2 = a \oplus a \oplus b \oplus b \oplus \begin{pmatrix} a & a \\ 0 & b \end{pmatrix} \oplus \begin{pmatrix} a & a \\ 0 & b \end{pmatrix}, \quad B_1 = a \oplus a \oplus b \oplus b \oplus \begin{pmatrix} b & 0 \\ -b & a \end{pmatrix} \oplus \begin{pmatrix} b & 0 \\ -b & a \end{pmatrix}$$

When a,b satisfy $a^2 - ab + b^2 = 0$ and, for $i = 2$, $(12b^2 - 6b + 1)(b^2 + a - b - 2ab) \neq 0$, we have the following decomposition into indecomposables:

$$A_1 = a \oplus a \oplus \begin{pmatrix} a & 0 & 0 \\ 0 & a & 0 \\ 0 & 0 & b \end{pmatrix} \oplus \begin{pmatrix} a & 0 & 0 \\ 0 & a & 0 \\ 0 & 0 & b \end{pmatrix}, \quad B_1 = a \oplus a \oplus \begin{pmatrix} a & 1 & b-a \\ 0 & a & 0 \\ 0 & 1 & b \end{pmatrix} \oplus \begin{pmatrix} a & 1 & b-a \\ 0 & a & 0 \\ 0 & 1 & b \end{pmatrix}$$

$$A_2 = a \oplus b \oplus \begin{pmatrix} a & 0 & 0 \\ 0 & a & 0 \\ 0 & 0 & b \end{pmatrix} \oplus \begin{pmatrix} b & 0 & 0 \\ 0 & b & 0 \\ 0 & 0 & a \end{pmatrix}, \quad B_2 = a \oplus b \oplus \begin{pmatrix} a & 1 & b-a \\ 0 & a & 0 \\ 0 & 1 & b \end{pmatrix} \oplus \begin{pmatrix} b & 1 & a-b \\ 0 & b & 0 \\ 0 & 1 & a \end{pmatrix}$$

The two distinct 3-dimensional indecomposable representations are both of the form found in Theorem 3 (2).

Extensions It is natural to consider classifying indecomposable but not irreducible representations in higher dimensions. One way is to simply extend our method, which is feasible but much more complicated. The other way is to use the result in paper [9] about irreducible representations we can always reduce A,B to have the following form: $A = \begin{pmatrix} A_1 & \star \\ 0 & A_2 \end{pmatrix}$, $B = \begin{pmatrix} B_1 & \star \\ 0 & B_2 \end{pmatrix}$, where (A_1, B_1) gives an irreducible representation. Although we know their forms from [9], it is inefficient to keep all the unknown variables and directly calculate the equations given by the braid relation. Thus one needs to do some analysis for the forms of A_2 and B_2 to simplify the calculations.

5 Appendix

Here we give more details of the proof of Theorem 3, 4. The key algorithm we use in Maple is to find a Groebner basis. With a fixed Jordan form of A, we start with the polynomial entries of $ABA = BAB$, which are in terms of the eigenvalues of A and the undetermined parameters of ABA. We enforce non-vanishing of the eigenvalues as well the requirement that they be distinct by means of a single polynomial with a new variable k, e.g. $\lambda_1\lambda_2(\lambda_1 - \lambda_2)k - 1$. We use an elimination order to find a set of polynomial consequences that do not use the variable k, and then run the Groebner basis algorithm again with a pure lexicographical order to simplify the equations, which can be analyzed by hand.

To illustrate we will give detailed proof for the case $\dim(W) = 1$ and A is a diagonal matrix with three distinct eigenvalues (Theorem 3, **Case 1**). We begin with the form

$$A = \begin{pmatrix} \lambda_1 & 0 & 0 \\ 0 & \lambda_2 & 0 \\ 0 & 0 & \lambda_3 \end{pmatrix}, \quad ABA = \begin{pmatrix} \lambda_1^3 & \beta_1 & \gamma_1 \\ 0 & \beta_2 & \gamma_2 \\ 0 & \beta_3 & \gamma_3 \end{pmatrix}.$$

Case 1.1 : If $\gamma_2 = \beta_3 = 0$, then by indecomposability and invertibility, all other variables are nonzero, and we may assume $\beta_1 = \gamma_1 = 1$, by conjugating the matrix $\begin{pmatrix} 1 & 0 & 0 \\ 0 & \beta_1 & 0 \\ 0 & 0 & \gamma_1 \end{pmatrix}$. We get a set of polynomials from equation $ABA = BAB$. Using the Groebner basis algorithm in Maple we get the following set of new polynomials that must vanish:

$$\lambda_1 - \lambda_2 - \lambda_3,\ \beta_2 - \gamma_3,\ \lambda_2^2 + \lambda_2\lambda_3 + \lambda_3^2,\ \lambda_2^3 - \gamma_3.$$

Solving them, we obtain:

$B = \begin{pmatrix} \lambda_1 & \frac{1}{\lambda_1\lambda_2} & \frac{1}{\lambda_1\lambda_3} \\ 0 & \lambda_2 & 0 \\ 0 & 0 & \lambda_3 \end{pmatrix}$. Conjugating by $\begin{pmatrix} 1 & 0 & 0 \\ 0 & \frac{1}{\lambda_1\lambda_2} & 0 \\ 0 & 0 & \frac{1}{\lambda_1\lambda_3} \end{pmatrix}$, we have (1.1).

Case 1.2 : If $\gamma_2 \neq 0, \beta_1 = \beta_3 = 0$, again by indecomposability and invertibility all other variables are nonzero. Hence by a similar argument, one can assume $\gamma_1 = \gamma_2 = 1$. Using Maple we get a similar set:

$$\lambda_1 + \lambda_2 - \lambda_3,\ \beta_2 + \gamma_3,\ \lambda_2^2 - \lambda_2\lambda_3 + \lambda_3^2,\ \lambda_2^3 - \gamma_3.$$

Solve them we get (1.2).

Case 1.3 : If $\gamma_2 \neq 0, \beta_1 \neq 0, \beta_3 = 0$, then $\gamma_3 \neq 0$. Using Maple we get the following set:

$$\lambda_1 - \lambda_2 + \lambda_3,\ \beta_2 + \gamma_3,\ \lambda_2^2 - \lambda_2\lambda_3 + \lambda_3^2,\ \beta_1\gamma_2 + 2\gamma_1\gamma_3,\ \lambda_2^3 - \gamma_3.$$

We have $\gamma_1 \neq 0$ from $\beta_1\gamma_2 + 2\gamma_1\gamma_3 = 0$, hence we set $\gamma_1 = \beta_1 = 1$. Using the Groebner command again (or directly plugging in the previous set), we get:

$$\lambda_1 - \lambda_2 + \lambda_3,\ \beta_2 + \gamma_3,\ \lambda_2^2 - \lambda_2\lambda_3 + \lambda_3^2,\ \gamma_2 + 2\gamma_3,\ \lambda_2^3 - \gamma_3.$$

Solving them, we have (1.3).

Case 1.4 : If $\gamma_2, \beta_3 \neq 0$, then we get a longer list of polynomials from the Groebner basis command. The following are five of them:

$$\begin{aligned}&\beta_2 + \gamma_3,\ \gamma_1(\lambda_1^2 + \lambda_2\lambda_3),\ \beta_1(\lambda_1^2 + \lambda_2\lambda_3),\\ &\lambda_1\beta_1\gamma_2 + \lambda_1\gamma_1\gamma_3 + \lambda_2\gamma_1\gamma_3 - \lambda_3\gamma_1\gamma_3,\ -\lambda_1\beta_1\gamma_3 + \lambda_1\beta_3\gamma_1 + \lambda_2\beta_1\gamma_3 - \lambda_3\beta_1\gamma_3.\end{aligned}$$

Together with indecomposability ($(\gamma_1, \beta_1) \neq (0,0)$) implies $\lambda_1^2 + \lambda_2\lambda_3 = 0$ and all the variables are nonzero, moreover $\lambda_1^2 + \lambda_2^2 \neq 0$, since $\lambda_2 \neq \lambda_3$. Now we can set $\beta_1 = \beta_3 = 1$, The Groebner basis command gives a shorter list, it turns out that the following first five relations are enough to get the solution:

$$\beta_2 + \gamma_3,\ \lambda_3\lambda_2 + \lambda_1^2,\ \lambda_1\gamma_1 - \lambda_1\gamma_3 + \lambda_2\gamma_3 - \lambda_3\gamma_3,\ \gamma_1^2 - 2\gamma_1\gamma_3 - \gamma_2,\ \lambda_2\lambda_3\lambda_1 - \gamma_1 + \gamma_3.$$

Solving them we have (1.4).

Since by permuting the basis e_2 and e_3, one can switch the roles played by γ_2 and β_3, **Case 1.2** and **Case 1.3** give all the solutions for the cases when one of γ_2, β_3, is zero or non-zero hence **Case 1.1**-**Case 1.4** give all the possible solutions when A is a diagonal matrix with three distinct eigenvalues.

Moreover, only diagonal matrices commute with A, and conjugating B by a diagonal matrix won't change the positions where the entries are zero or non-zero, therefore the distinctness follows easily. As for indecomposability, since the eigenvalues are all distinct, the only possible decompositions will be $Ke_i \oplus K\{e_j, e_k\}$ for distinct i, j, k and $1 \leq i, j, k \leq 3$, which is impossible for B.

Now for **Case 2** in Theorem 3, we begin with the following matrix

$$A = \begin{pmatrix} \lambda_1 & 0 & 0 \\ 0 & \lambda_1 & 0 \\ 0 & 0 & \lambda_2 \end{pmatrix}, \quad ABA = \begin{pmatrix} \lambda_1^3 & \beta_1 & \gamma_1 \\ 0 & \beta_2 & \gamma_2 \\ 0 & 1 & \gamma_3 \end{pmatrix}.$$

Case 2.1 : $\gamma_2 \neq 0$, using the Groebner basis command with, in addition to $ABA = BAB$, relation $\gamma_2(\lambda_1 - \lambda_2)\lambda_1\lambda_2(\beta_2\gamma_3 - \gamma_2)K - 1$ added, we have the following first several terms:

$$\beta_2 + \gamma_3,\ \lambda_2\beta_1\gamma_3 - \lambda_1\gamma_1,\ \lambda_1^2\lambda_2^2 - \lambda_1\gamma_3 + \lambda_2\gamma_3,\ \lambda_1\lambda_2^2\beta_1 - \beta_1\gamma_3 + \gamma_1.$$

From the third term, we have $\gamma_3 \neq 0$, which implies $\beta_2 \neq 0$ from the first term. Moreover, from the second term $\beta_1 = 0 \Leftrightarrow \gamma_1 \neq 0$, by indecomposability they are both nonzero, hence we can assume $\beta_1 = 1$. Update our set of relations and use the Groebner basis command again, we have a list of relations, the following are the first four terms:

$$\beta_2 + \gamma_3,\ \lambda_1\gamma_1 - \lambda_2\gamma_3,\ \lambda_1\lambda_2^2 + \gamma_1 - \gamma_3,\ \lambda_1\gamma_3^2 + \lambda_2\gamma_1\gamma_3 - \lambda_2\gamma_3^2 + \lambda_2\gamma_2.$$

This gives the solution:

$$B = \begin{pmatrix} \lambda_1 & \frac{1}{\lambda_1^2} & \frac{\lambda_2^2}{\lambda_1 - \lambda_2} \\ 0 & \frac{-\lambda_2^2}{\lambda_1 - \lambda_2} & \frac{\lambda_1^2\lambda_2^2(\lambda_1^2 - \lambda_1\lambda_2 + \lambda_2^2)}{(\lambda_1 - \lambda_2)^2} \\ 0 & \frac{1}{\lambda_1\lambda_2} & \frac{\lambda_1^2}{\lambda_1 - \lambda_2} \end{pmatrix}.$$

Our assumption $\gamma_2 \neq 0$ implies $\lambda_1^2 - \lambda_1\lambda_2 + \lambda_2^2 \neq 0$. Conjugating B by

$$\begin{pmatrix} 1 & 1 & 0 \\ 0 & \frac{\lambda_1^2(\lambda_1^2 - \lambda_1\lambda_2 + \lambda_2^2)}{\lambda_1 - \lambda_2} & 0 \\ 0 & 0 & 1 \end{pmatrix}$$

(which commutes with A), we see the solution is decomposable. (Equivalently, it decomposes as $Ke_1 \oplus K\{e_2 + \frac{\lambda_1^2(\lambda_1^2 - \lambda_1\lambda_2 + \lambda_2^2)}{\lambda_1 - \lambda_2} e_3, e_3\}$)

Case 2.2 : $\gamma_2 = 0$, similarly, we have several terms (part of the list):

$$\beta_2 + \gamma_3,\ \lambda_1^2 - \lambda_1\lambda_2 + \lambda_2^2,\ \lambda_2^3 - \gamma_3,\ \lambda_2\beta_1\gamma_3 - \lambda_1\gamma_1.$$

This implies $\gamma_1, \gamma_3, \beta_1, \beta_2$ are all nonzero, again we assume $\beta_1 = 1$. Then solve a short list of equations, we have (2), its indecomposability will be discussed later.

There are two types of W in **Case 3** of Theorem 3, First one gives the following representation:

$$A = \begin{pmatrix} \lambda_1 & 0 & 0 \\ 0 & \lambda_2 & 1 \\ 0 & 0 & \lambda_2 \end{pmatrix}, \quad ABA = \begin{pmatrix} \lambda_1^3 & \beta_1 & \gamma_1 \\ 0 & \beta_2 & \gamma_2 \\ 0 & \beta_3 & \gamma_3 \end{pmatrix}.$$

Case 3.1 : $\beta_3 = 0$, using the Groebner basis command, we directly get $b_1 = 0$ and $\lambda_1^2 - \lambda_1\lambda_2 + \lambda_2^2 = 0$, then (3.1) is obtained by simple calculations.

Case 3.2 : $\beta_3 \neq 0$, using the Groebner basis command, the following is a subset of the terms we obtain:

$$\gamma_1(\lambda_1^2 + \lambda_2^2), \ \beta_1(\lambda_1^2 + \lambda_2^2), \ -\lambda_1\beta_1\gamma_3 + \lambda_1\beta_3\gamma_1 - \beta_1\beta_3.$$

The second and third terms together with the indecomposability imply $\lambda_1^2 + \lambda_2^2 = 0$. The last term together with the indecomposability implies $\beta_1 \neq 0$, hence we assume $\beta_1 = 1$. Now we simply the set of relations and we obtain the following subset,

$$\beta_2 + \gamma_3, \ \lambda_1^2 + \lambda_2^2, \ \ \lambda_1\lambda_2^2 - \gamma_1\beta_3 + \gamma_3, \ \lambda_2^2\beta_3 - \beta_3\gamma_2 - \gamma_3^2,$$
$$\lambda_1\gamma_1\gamma_3 + \lambda_1\gamma_2 - \beta_3\gamma_1, \ \gamma_1\beta_3\lambda_1 - \gamma_3\lambda_1 - \beta_3.$$

This is enough to get the solution:

$$B = \begin{pmatrix} \lambda_1 & \frac{1}{\lambda_1\lambda_2} & \frac{1}{\lambda_1^3} - \frac{1}{\lambda_2^3} - \frac{\gamma_3}{\lambda_1^5\lambda_2} \\ 0 & \lambda_2 - \frac{\gamma_3}{\lambda_2^2} & \frac{\gamma_3}{\lambda_1^4\lambda_2^2} \\ 0 & \lambda_1^2 & \lambda_2 + \frac{\gamma_3}{\lambda_2^2} \end{pmatrix}.$$

Conjugating by

$$\begin{pmatrix} \frac{1}{\lambda_1\lambda_2} & 0 & 0 \\ 0 & 1 & -\frac{\gamma_3}{\lambda_1^2\lambda_2^2} \\ 0 & 0 & 1 \end{pmatrix}$$

(which commutes with A), we have (3.2).

Case 3.3 : The second type of W gives the following representation:

$$A = \begin{pmatrix} \lambda_1 & 1 & 0 \\ 0 & \lambda_1 & 0 \\ 0 & 0 & \lambda_2 \end{pmatrix}, \quad ABA = \begin{pmatrix} \lambda_1^3 & \beta_1 & \gamma_1 \\ 0 & \beta_2 & \gamma_2 \\ 0 & \beta_3 & \gamma_3 \end{pmatrix}.$$

We first assume $\gamma_2 \neq 0$, and up to equivalence, we may assume $\gamma_2 = 1$. Applying the Groebner basis command, we obtain the following short list of relations,

$$\gamma_1 + \gamma_2, \ \beta_2 + \gamma_3, \ 3\gamma_3^2 - \beta_3, \ 3\gamma_1\gamma_3 + \beta_1, \ \beta_1\gamma_3 + \beta_3\gamma_1, \ \lambda_2^3 + 2\gamma_3,$$

which gives the solution

$$B = \begin{pmatrix} -\lambda_2 & \frac{3\gamma_1\lambda_2-1}{2} & \frac{-\gamma_1\lambda_2-1}{\lambda_2^3} \\ 0 & \frac{\lambda_2}{2} & -\frac{1}{\lambda_2^2} \\ 0 & -\frac{3\lambda_2^4}{4} & -\frac{\lambda_2}{2} \end{pmatrix}.$$

Conjugating by

$$\begin{pmatrix} -\frac{1}{\lambda_2^2} & -\frac{\gamma_1\lambda_2+1}{\lambda_2^3} & 0 \\ 0 & -\frac{1}{\lambda_2^2} & 0 \\ 0 & 0 & 1, \end{pmatrix}$$

we obtain (4.1).

Case 3.4 : $\gamma_2 = 0$, this implies $\gamma_1 \neq 0$, otherwise Ke_3 will be a 1-dimensional common invariant subspace, which is the W of the first type in **Case 3** (see **Case 3.1**, **Case 3.2**). Therefore, we assume $\gamma_1 = 1$. Applying the Groebner basis command, we have the following subset:

$$\beta_2 + \gamma_3,\ \lambda_2^2 - \lambda_1\lambda_2 + \lambda_1^2,\ \lambda_2^3 - \gamma_3,\ \lambda_1\beta_1\gamma_3 - \lambda_1\beta_3 + \lambda_2\beta_3 + 3\gamma_3^2.$$

Solving these equations gives the solution: $B = \begin{pmatrix} \lambda_1 & -2+\frac{\beta_1}{\lambda_1^2} & \frac{1}{\lambda_1\lambda_2} \\ 0 & \lambda_1 & 0 \\ 0 & 3\lambda_1^3 - \beta_1\lambda_1 & -\frac{\lambda_1^3}{\lambda_2^2} \end{pmatrix}$. By simply changing the basis $\tilde{e}_3 := \lambda_1\lambda_2 e_3$ and setting $\beta := \frac{3\lambda_1^2-\beta_1}{\lambda_2}$, we obtain (4.2).

Other cases in Theorem 3, 4 are less complicated and follow similar procedures, so we will now turn to proving our solutions are distinct and indecomposable.

Distinctness: First note that the representations in Theorem 3 and Theorem 4 are inequivalent, and that different Jordan forms of A give inequivalent representations. Moreover, except (4.1), all other representations depend only on eigenvalues and relations among them. Hence only (3.1) and (4.1) need to be analyzed.

In (3.1), eigenvectors corresponding to two eigenvalues are two 1-dimensional common invariant subspaces, which is not the case in (4.1). Moreover different β give different representations, in fact, all the matrices commuting with A in (4.1) are of the form: $M = \begin{pmatrix} a & b & 0 \\ 0 & a & 0 \\ 0 & 0 & c \end{pmatrix}$. Direct calculation shows the β term is invariant under conjugating B by M.

Indecomposability: Except from (2) in Theorem 3, the invariant subspaces decomposition of A in all other cases have only finitely many choices, since all eigenspaces are one dimensional. It is not hard to see they are indecomposable. For (2), since by construction, Ke_1 is a common invariant subspace and $K\{e_1, e_2\}$ is not, we cannot have any other one-dimensional common invariant subspaces. Therefore if the representation is decomposable, it can only decompose as $Ke_1 \oplus K\{ae_1 + be_2, e_3\}$ for some $b \neq 0$, which is impossible for B.

References

1. Chang, L., Kolt, Q.T., Wang, Z., Zhang, Q.: Modular data of non-semisimple modular categories (2024). URL `https://arxiv.org/abs/2404.09314`
2. Iulianelli, F., Kim, S., Sussan, J., Lauda, A.D.: Universal quantum computation using ising anyons from a non-semisimple topological quantum field theory (2025). URL `https://arxiv.org/abs/2410.14860`
3. Kassel, C., Turaev, V.: Braid groups, Graduate Texts in Mathematics, vol. 247. Springer, New York (2008). DOI 10.1007/978-0-387-68548-9. With the graphical assistance of Olivier Dodane
4. Larsen, M.J., Rowell, E.C., Wang, Z.: The N-eigenvalue problem and two applications. Int. Math. Res. Not. (64), 3987–4018 (2005). DOI 10.1155/IMRN.2005.3987
5. Laugwitz, R., Walton, C.: Constructing non-semisimple modular categories with relative monoidal centers. Int. Math. Res. Not. IMRN (20), 15,826–15,868 (2022). DOI 10.1093/imrn/rnab097
6. Maplesoft, a division of Waterloo Maple Inc..: Maple. URL `https://www.maplesoft.com`
7. Rowell, E.C.: Braid representations from quantum groups of exceptional Lie type. Rev. Un. Mat. Argentina **51**(1), 165–175 (2010)
8. Rowell, E.C., Wang, Z.: Mathematics of topological quantum computing. Bull. Amer. Math. Soc. (N.S.) **55**(2), 183–238 (2018). DOI 10.1090/bull/1605
9. Tuba, I., Wenzl, H.: Representations of the braid group B_3 and of $\mathrm{SL}(2,\mathbf{Z})$. Pacific J. Math. **197**(2), 491–510 (2001). DOI 10.2140/pjm.2001.197.491

Part II

Other Contributed Articles

Chapter 6

MATRIX Program: The Geometry of Moduli Spaces in String Theory

Some identities for periods of Hulek-Verrill threefolds

Xenia de la Ossa, Mohamed Elmi

Abstract We study the Hulek-Verril families of Calabi-Yau threefolds[4, 5]. They are birationally equivalent to fibred products of elliptic surfaces so we expect to be able to compute periods on these threefolds by integrating products of elliptic periods over a contour on a $\mathbb{P}^1$. We numerically verify this in several examples. The Hulek-Verril threefolds are interesting because some of them are attractor varieties of rank two. These are modular threefolds where the Hodge structure splits over the rational numbers in a specific way. This note is part of an ongoing effort to better understand attractor varieties of rank two.

1 Hulek-Verril family of 1-folds $\mathscr{E}_{1,1,\frac{1}{\varphi}}$

The Hulek-Verril elliptic surface $\mathscr{E}_{a,b,c}$ is the hypersurface in $\mathbb{P}^2 \times \mathbb{P}^1$, defined by

$$P(X,\lambda) = (X^1 + X^2 + X^3)(aX^2X^3 + bX^1X^3 + cX^1X^2)\lambda^0 - \lambda^1 X^1X^2X^3 = 0 \quad (1)$$

where $[X^1, X^2, X^3] \in \mathbb{P}^2$, $[\lambda^0, \lambda^1] = \mathbb{P}^1$ and $(a,b,c) \in \mathbb{C}^3$ are parameters.[4]

We are interested in the elliptic surface with parameters $\mathscr{E}_{1,1,\frac{1}{\varphi}}$. We choose the affine coordinate $\lambda = \frac{\lambda^0}{\lambda^1}$ on $\mathbb{P}^1$ and work on the coordinate patch $X^3 = 1$ on $\mathbb{P}^2$. With these choices, the holomorphic 1-form on the elliptic fibre at λ is given by

Xenia de la Ossa
Mathematical Institute, Oxford University Oxford, UK
e-mail: delaossa@maths.ox.ac.uk

Mohamed Elmi
Yau Mathematical Sciences Center, Tsinghua University, Beijing 100084, China
e-mail: elmi@mail.tsinghua.edu.cn

D. R. Wood et al. (eds.), *2024 MATRIX Annals, Part II*, MATRIX Book Series 8,
https://doi.org/10.1007/978-3-032-16206-9_21

$$\frac{1}{2\pi i}\oint_C \frac{dX^1 \wedge dX^2}{P(X,\lambda)} \tag{2}$$

where C is a "small contour" that encircles the $P(X,\lambda) = 0$ locus.

The holomorphic period at $\lambda = 0$ is given by

$$\frac{1}{2\pi i}\oint_{|X^1|=\varepsilon}\oint_{|X^2|=\delta} \frac{dX^1 \wedge dX^2}{P(X,\lambda)} = (2\pi i) f_0(\lambda,\varphi) \tag{3}$$

for sufficiently small ε and δ. The function $f_0(\lambda,\varphi)$ given by

$$f_0(\lambda,\varphi) = \sum_{n=0}^{\infty}\left(\sum_{k=0}^{n}\binom{n}{k}^2\binom{2k}{k}\varphi^{-n+k}\right)\lambda^n \tag{4}$$

and the coefficient of λ^n is given by the constant term of

$$\left[(X^1+X^2+1)\left(\frac{1}{X^1}+\frac{1}{X^2}+\frac{1}{\varphi}\right)\right]^n. \tag{5}$$

Both periods of the elliptic curve at λ are annihilated by the Picard-Fuchs differential operator

$$\mathscr{L} = R_2(\lambda,\varphi)(\theta+1)^2 + R_1(\lambda,\varphi)(\theta+1) + R_0(\lambda,\varphi) \tag{6}$$

where $\theta = \lambda\frac{d}{d\lambda}$ and

$$\begin{aligned}
R_2(\lambda,\varphi) &= \left(3+\lambda\left(\frac{1}{\varphi}-4\right)\right)\left(1-\frac{\lambda}{\varphi}\right)\left(1-2\lambda\left(\frac{1}{\varphi}+4\right)+\lambda^2\left(\frac{1}{\varphi}-4\right)^2\right)\\
R_1(\lambda,\varphi) &= -6+6\lambda\left(\frac{1}{\varphi}+6\right)+\\
&\quad 2\lambda^2\left(\frac{1}{\varphi}-4\right)\left(\frac{3}{\varphi}+8\right)-2\lambda^3\left(\frac{1}{\varphi}-4\right)\left(8-\frac{6}{\varphi}+\frac{3}{\varphi^2}\right)\\
R_0(\lambda,\varphi) &= 3-\lambda\left(\frac{1}{\varphi}+14\right)-\\
&\quad \lambda^2\left(\frac{1}{\varphi}+2\right)\left(\frac{3}{\varphi}-10\right)+\lambda^3\left(\frac{1}{\varphi}-4\right)\left(4-\frac{2}{\varphi}+\frac{1}{\varphi^2}\right)
\end{aligned} \tag{7}$$

This differential operator $\mathscr{L}$ is implicitly given by the recursion relations of Verril in [5] and it has the Riemann symbol

$$\mathscr{P}\left\{\begin{array}{ccccc} \frac{3\varphi}{4\varphi-1} & 0 & \varphi & \frac{\varphi}{(1\pm2\sqrt{\varphi})^2} & \infty \\ \hline 0 & 0 & 0 & 0 & 1 \\ 2 & 0 & 0 & 0 & 1 \end{array} ;\lambda\right\}. \tag{8}$$

For generic values of φ, the elliptic periods $\omega(s,\varphi)$ are annhilated by the Picard-Fuchs operator (6). However, at $\varphi = 1$, some of the singularities of (6) coalesce and $\omega(\lambda,1)$ is annihilated by a Picard-Fuchs operator with Riemann symbol

$$\mathscr{P}\left\{\begin{array}{cccc} 0 & \frac{1}{9} & 1 & \infty \\ \hline 0 & 0 & 0 & 1 \\ 0 & 0 & 0 & 1 \end{array} ;\lambda\right\}. \tag{9}$$

This will be important in the following section. We will will also need a basis of periods with integral symplectic monodromy. After some experimentation, we make the following conjecture.

Conjecture: The basis of periods

$$\omega(\lambda,\varphi) = (2\pi i)\begin{pmatrix} 1 & 0 \\ -\frac{\log(\varphi)}{2\pi i} & \frac{3}{2\pi i} \end{pmatrix}\begin{pmatrix} f_0(\lambda,\varphi) \\ f_0(\lambda,\varphi)\log(\lambda) + f_1(\lambda,\varphi) \end{pmatrix} \tag{10}$$

has monodromy in $SL_2(\mathbb{Z})$ for all $\varphi \in \mathbb{C}^*$, where $f_0(\lambda,\varphi)$ is given in (4) and $f_1(\lambda,\varphi)$ is computed from the Picard-Fuchs equation (6) and is normalised so that $f_1(0,\varphi) = 0$. [1]

2 Hulek-Verril family of threefolds as a fibred product of elliptic surfaces

AESZ 34 describes the variation of the Hodge structure of (a quotient of) the one-parameter fibred products of elliptic surfaces

$$\begin{aligned} P_1(X,\lambda) &= (X^1+X^2+X^3)(X^2X^3+X^1X^3+X^1X^2)\lambda^0 - \lambda^1X^1X^2X^3 = 0 \\ P_2(Y,\lambda) &= (Y^1+Y^2+Y^3)\left(Y^2Y^3+Y^1Y^3+\frac{1}{\varphi}Y^1Y^2\right)\lambda^0 - \lambda^1Y^1Y^2Y^3 = 0 \end{aligned} \tag{11}$$

where $\varphi \in \mathbb{C}$ is a parameter. These equations describe the fibred product of Hulek-Verril elliptic surfaces $\mathscr{E}_{1,1,1} \times_{\mathbb{P}^1} \mathscr{E}_{1,1,\frac{1}{\varphi}}$. Equivalently, these equations describe a one-parameter family of complete intersections with configuration matrix

[1] We are grateful to Joseph McGovern for helpful discussions related to this choice of basis.

$$\begin{matrix} \mathbb{P}^2 \\ \mathbb{P}^2 \\ \mathbb{P}^1 \end{matrix} \begin{bmatrix} 3 & 0 \\ 0 & 3 \\ 1 & 1 \end{bmatrix} . \tag{12}$$

On the $X^3 = Y^3 = \lambda^1 = 1$ coordinate patch, the holomorphic 3-form takes the form

$$\Omega = \frac{1}{(2\pi i)^2} \int_C \frac{dX_1 \wedge dX_2 \wedge dY_1 \wedge dY_2 \wedge d\lambda_0}{P_1(X,\lambda) P_2(Y,\lambda)} , \tag{13}$$

where C is a "small contour" that encircles the $P_1(X,\lambda) = P_2(Y,\lambda) = 0$ locus.

Let $\lambda = \frac{\lambda^0}{\lambda^1}$ and $G_i \in H_1(E_1,\mathbb{Z}) \otimes H_1(E_2,\mathbb{Z})$ where E_1 and E_2 are the elliptic fibres defined by $P_1(X,\lambda) = 0$ and $P_2(Y,\lambda) = 0$ respectively. For suitably chosen G_i and contours ℓ_i (closed or open with ends on the singularities) in the complex λ plane, we expect an identify

$$\Gamma^T \Sigma_4 \Pi(\varphi) = \sum_{i=1}^{N} G_i^T \Sigma_{2,2} \int_{\ell_i} \omega(\lambda,1) \otimes \omega(\lambda,\varphi)\, d\lambda , \tag{14}$$

for some $\Gamma \in \mathbb{Q}^4$ and Π that is, up a gauge transformation, a solution of AESZ 34 with integral symplectic monodromy [1]. We have identified G with a vector of its periods in $\mathbb{Z}^4$ and defined

$$\Sigma_n = \begin{pmatrix} 0 & I_{\frac{n}{2}} \\ -I_{\frac{n}{2}} & 0 \end{pmatrix} \tag{15}$$

and

$$\Sigma_{m,n} = \Sigma_m \otimes \Sigma_n . \tag{16}$$

An integral symplectic basis of solutions of AESZ34 is given by

$$\Pi(\varphi) = (2\pi i)^3 \begin{pmatrix} \frac{\zeta(3)}{(2\pi i)^3}\chi & \frac{c_2 \cdot H}{24} & 0 & \frac{H^3}{3!} \\ \frac{c_2 \cdot H}{24} & \frac{\sigma}{2} & -\frac{H^3}{2!} & 0 \\ 1 & 0 & 0 & 0 \\ 0 & 1 & 0 & 0 \end{pmatrix} \begin{pmatrix} 1 & 0 & 0 & 0 \\ 0 & (2\pi i)^{-1} & 0 & 0 \\ 0 & 0 & (2\pi i)^{-2} & 0 \\ 0 & 0 & 0 & (2\pi i)^{-3} \end{pmatrix} \varpi(\varphi) , \tag{17}$$

where $(H^3, c_2 \cdot H, \chi) = (12, 12, -8)$, $\sigma = H^3 \bmod 2 = 0$ and $\varpi = (\varpi_0, \varpi_1, \varpi_2, \varpi_3)$. The components of ϖ are

$$\varpi_j(\varphi) = \varphi \sum_{k}^{3} \binom{j}{k} F_k(\varphi) \log^{j-k}(\varphi) , \tag{18}$$

where $F_j(\varphi)$ are power series that are computed from the Picard-Fuchs equation. They are normalised so that $F_j(0) = \delta_{j,0}$.

We should stress that, when we work with the choice of integral symplectic basis and topological data in (17), we are implicitly working on a free quotient of (11).

As a result of this quotient, we will find that $\Gamma \in \mathbb{Q}^4$ instead of $\mathbb{Z}$ in identities of the form (14).

2.1 Normalisation of Π?

Note that the periods in (18) have been normalised in such a way that

$$\varpi_0(\varphi) = \varphi + \mathscr{O}(\varphi^2). \tag{19}$$

In other words, the Picard-Fuchs equation has indices $(1,1,1,1)$ at the large complex structure point $\varphi = 0$. This normalisation has been included with the benefit of hindsight.

Recall that the holomorphic 3-form at a given value of φ is only defined up to multiplication by a non-zero constant. We are usually free to multiply the periods Π by any non-vanishing holomorphic function of φ without changing anything of consequence. However, a normalisation is implicitly determined by Equation (14), which we can find by analytically computing a single identity of the form (14) as a function of φ. Alternatively, we may fix a number of values of φ, numerically compute a number of identities of the form (14) and then guess what the normalisation of Π must be. This is the quick and dirty approach that has been taken here.

3 Identities

We identify solutions of AESZ34 with identities of the form (14). We compute both $\Pi(\varphi)$ and the right hand side of (14) and then identify the appropriate $\Gamma \in \mathbb{Q}^4$ by using the LLL-algorithm to look for a relation between the numerical values on the left and right hand sides of the identity. All identities in this section are found numerically at the fixed value of $\varphi = \frac{1}{64}$. However, we expect that they continue to hold as φ is varied.

3.1 Monodromy of elliptic periods

We expect identities of the form (14) when we can argue that the 3-chain on the right-hand side is really a 3-cycle. In order to argue that this is the case, we need the monodromy matrices of $\omega(\lambda,1) \otimes \omega(\lambda,\varphi)$ around the various singularities in the complex λ plane.

We use (10) to fix a basis of elliptic periods at $\lambda = 0$ and $\varphi = \frac{1}{64}$. We then analytically continue to the various singularities along the upper half λ-plane and compute anti-clockwise monodromy matrices around the various singularities. Around a singularity λ_*, the elliptic periods transform as

$$\omega(\lambda,1)\otimes\omega(\lambda,\varphi)\to\mu_{\lambda_*}\;\omega(\lambda,1)\otimes\omega(\lambda,\varphi)\,. \tag{20}$$

We find that

$$\mu_0=\begin{pmatrix}1&0\\3&1\end{pmatrix}\otimes\begin{pmatrix}1&0\\3&1\end{pmatrix} \tag{21}$$

$$\mu_{\frac{\varphi}{(1+2\sqrt{\varphi})^2}}=I_2\otimes\begin{pmatrix}1&-2\\0&1\end{pmatrix} \tag{22}$$

$$\mu_\varphi=I_2\otimes\begin{pmatrix}5&-4\\4&-3\end{pmatrix} \tag{23}$$

$$\mu_{\frac{\varphi}{(1-2\sqrt{\varphi})^2}}=I_2\otimes\begin{pmatrix}5&-2\\8&-3\end{pmatrix} \tag{24}$$

$$\mu_{\frac{1}{9}}=\begin{pmatrix}1&-2\\0&1\end{pmatrix}\otimes I_2 \tag{25}$$

$$\mu_1=\begin{pmatrix}7&-6\\6&-5\end{pmatrix}\otimes I_2\,. \tag{26}$$

3.2 Vanishing cycle at $\varphi=\frac{1}{9}$

We expect that the (non-trivial) invariant 1-cycle of a 2×2 monodromy matrix is the vanishing cycle at the associated singularity. As with Hadamard products, we can use this to identify 3-cycles in the fibred product of elliptic surfaces.[3] For example

$$\begin{pmatrix}1&-2\\0&1\end{pmatrix}\begin{pmatrix}1\\0\end{pmatrix}=\begin{pmatrix}1\\0\end{pmatrix}, \tag{27}$$

which tells us that $(1,0)^T$ is a vanishing cycle of the first elliptic curve at $\lambda=\frac{1}{9}$. Similarly, the second elliptic curve has vanishing cycle $(1,1)^T$ at $\lambda=\varphi$. This leads us to the identity

$$\frac{1}{2}(1,0,-5,1)\Sigma_4\Pi(\varphi)=(1,0)\otimes(1,1)\Sigma_{2,2}\int_\varphi^{\frac{1}{9}}\omega(\lambda,1)\otimes\omega(\lambda,\varphi)d\lambda\,. \tag{28}$$

As $\varphi\to\frac{1}{9}$, the integration contour in the above identity vanishes, so we expect that $(1,0,-5,1)^T$ is a vanishing cycle at $\varphi=\frac{1}{9}$. This is indeed the case. However, we must be careful. The precise vanishing cycle we compute depends on the branch that we approach the singularity on. Different choices of branches simply multiply the vanishing cycle by a monodromy matrix. In this case, the vanishing cycle computed in [2] is $(1,0,5,1)^T$, which is simply

$$\begin{pmatrix}1\\0\\5\\1\end{pmatrix} = M_{\frac{1}{25}} \begin{pmatrix}1\\0\\-5\\1\end{pmatrix}, \tag{29}$$

where $M_{\frac{1}{25}}$ is the monodromy matrix around $\varphi = \frac{1}{25}$ that was also computed in [2].

3.3 Vanishing cycle at $\varphi = \frac{1}{25}$

Similarly, we find that

$$\frac{1}{2}(0,0,-5,0)\Sigma_4 \Pi(\varphi) = (1,0)\otimes(1,2)\Sigma_{2,2}\int_{\frac{\varphi}{(1-2\sqrt{\varphi})^2}}^{\frac{1}{9}} \omega(\lambda,1)\otimes\omega(\lambda,\varphi)d\lambda. \tag{30}$$

Again, the contour of integration vanishes as $\varphi \to \frac{1}{25}$, so we would expect that $(0,0,-5,0)^T$ is a vanishing cycle at $\frac{1}{25}$. This is indeed the case. It is just a multiple of the vanishing cycle computed in [2].

3.4 Holomorphic period at $\varphi = 0$ and T^3

As was the case with Hadamard products, we can identify non-trivial 3-cycles by taking the union of some monodromy invariant 2-cycle in the fibre over a contour on the base.[3] In this way, we can identify a 3-torus (T^3) that can be integrated over to obtain the holomorphic period at $\varphi = 0$. More precisely, we find that

$$(-1,0,0,0)\Sigma_4\Pi(\varphi) = (0,1)\otimes(0,1)\Sigma_{2,2}\oint_{\ell_0\ell_{\frac{\varphi}{(1+2\sqrt{\varphi})^2}}\ell_\varphi\ell_{\frac{\varphi}{(1-2\sqrt{\varphi})^2}}} \omega(\lambda,1)\otimes\omega(\lambda,\varphi)d\lambda\,, \tag{31}$$

where the integration contour

$$\ell_0\ell_{\frac{\varphi}{(1+2\sqrt{\varphi})^2}}\ell_\varphi\ell_{\frac{\varphi}{(1-2\sqrt{\varphi})^2}}$$

is a closed contour on the λ plane that first encircles $\lambda = 0$ in an anti-clockwise fashion, then encircles $\lambda = \frac{\varphi}{(1+2\sqrt{\varphi})^2}$ in an anti-clockwise fashion, then encircles... . We see from the above monodromy matrices that $(0,1)\otimes(0,1)$ is invariant under monodromy around this contour. We have therefore identified a T^3 we can integrate over and obtain the holomorphic period at the point of maximal unipotent monodromy

$\varphi = 0$. The results of [3] suggest that this family of T^3s defines an SYZ-like fibration by constant phase contours on the λ-plane.

References

1. Almkvist, G., van Enckevort, C., van Straten, D., Zudilin, W.: Tables of Calabi–Yau equations. arXiv Mathematics e-prints math/0507430 (2005). DOI 10.48550/arXiv.math/0507430
2. Candelas, P., de la Ossa, X., Elmi, M., Van Straten, D.: A One Parameter Family of Calabi-Yau Manifolds with Attractor Points of Rank Two. JHEP **10**, 202 (2020). DOI 10.1007/JHEP10(2020)202
3. Elmi, M.: Hadamard products and BPS networks. JHEP **07**, 076 (2024). DOI 10.1007/JHEP07(2024)076
4. Hulek, K., Verrill, H.: On modularity of rigid and nonrigid Calabi-Yau varieties associated to the root lattice A_4. arXiv Mathematics e-prints math/0304169 (2003). DOI 10.48550/arXiv.math/0304169
5. Verrill, H.A.: Sums of squares of binomial coefficients, with applications to Picard-Fuchs equations. arXiv Mathematics e-prints math/0407327 (2004). DOI 10.48550/arXiv.math/0407327

Mirror symmetry from families of Calabi-Yau manifolds

Shinobu Hosono and Hiromichi Takagi

Abstract We report interesting examples of Calabi-Yau threefolds where birational geometry and geometry of Fourier-Mukai partners of a Calabi-Yau manifold arise naturally from studying its mirror family of Calabi-Yau manifolds. We define mirror symmetry in terms of families of Calabi-Yau manifolds in general. We will find for our examples that all expected geometry of Calabi-Yau manifolds arises from special boundary points, called LCSL points, in the parameter space of mirror Calabi-Yau manifolds.

1 Introduction

Mirror symmetry of Calabi-Yau manifolds has been a central topic in theoretical physics and mathematics since its discovery in the beginning of the 1990s. In particular, its surprising application to enumerative geometry started by Candelas et al [7] draw significant attention from both theoretical physicists and mathematicians. Intensive study over the last three decades has led two main mathematical formulations of mirror symmetry. One is based on the geometry of Calabi-Yau manifolds, called geometric mirror symmetry due to Strominger-Yau-Zaslow [27] and also Gross-Siebert [12, 13], and the other focuses on equivalences of two different categories; the derived category of coherent sheaves and the derived Fukaya category in symplectic geometry for a Calabi-Yau manifold. The latter approach is called homological mirror symmetry [24] and describes mirror symmetry as a derived equivalence interchanging the two different categories defined for Calabi-

Shinobu Hosono
Department of Mathematics, Gakushuin University, Mejiro, Toshima-ku, Tokyo 171-8588, Japan
e-mail: hosono@math.gakushuin.ac.jp

Hiromichi Takagi
Department of Mathematics, Gakushuin University, Mejiro, Toshima-ku, Tokyo 171-8588, Japan
e-mail: hiromici@math.gakushuin.ac.jp

D. R. Wood et al. (eds.), *2024 MATRIX Annals, Part II*, MATRIX Book Series 8,
https://doi.org/10.1007/978-3-032-16206-9_22

Yau manifolds X and its mirror $\check{X}$. While both approaches to mirror symmetry are extensively developed, to fully understand the symmetry, exploring it in many examples of Calabi-Yau manifolds is particularly important since it is still very hard to describe $\check{X}$ explicitly from X by these two approaches.

In this article, based on a series of the present authors' works [18, 19, 20, 21, 22], we will report on mirror symmetry of two interesting classes of Calabi-Yau manifolds; one is Calabi-Yau manifolds related to determinantal quintics in four dimensional projective space, and the other is Calabi-Yau manifolds which are fibered by abelian surfaces. For these Calabi-Yau manifolds, by studying mirror families, it has been observed that mirror symmetry nicely incorporates the problems of birational geometry and also Fourier-Mukai partners of Calabi-Yau manifolds.

2 Families of Calabi-Yau manifolds and mirror symmetry

2.1 Calabi-Yau manifolds

Complex manifolds which admit Ricci-flat Kähler metric are called Calabi-Yau manifolds in general. Abelian varieties or complex tori are simple examples of such Calabi-Yau manifolds. Here in this note, however, we adopt the following definition of Calabi-Yau manifolds in a narrow sense, where abelian varieties and complex tori are excluded.

Definition 1. We call a non-singular projective variety X of dimension d Calabi-Yau manifold if it satisfies $c_1(X) = 0$ and $H^i(X, \mathcal{O}_X) = 0\,(1 \leq i \leq d-1)$.

Given a Calabi-Yau manifold X, we can describe the deformation of its complex structures by the Kodaira-Spencer theory. In particular, due to the theorem by Bogomolov-Tian-Todorov, all infinitesimal deformations are unobstructed and given by the cohomology $H^1(X, TX)$. We denote by $\mathscr{M}$ the deformation space, and write by X_p the Calabi-Yau manifold represented by $p \in \mathscr{M}$. According to the deformation theory, we have a linear map from the tangent space

$$KS_p : T_p\mathscr{M} \to H^1(X_p, TX),$$

which is called Kodaira-Spencer map.

Definition 2. Let X be a Calabi-Yau manifold and $\mathfrak{X}, \mathscr{M}$ be complex manifolds. If there exists a surjective morphism $\pi : \mathfrak{X} \to \mathscr{M}$ satisfying the following conditions: (0) $\pi^{-1}(p_0) \simeq X$, (1) differential π_* is submersive, (2) the fiber $\pi^{-1}(p)$ is compact for all $p \in \mathscr{M}$, (3) the Kodaira-Spencer maps are isomorphic for all $p \in \mathscr{M}$, then we call $\mathfrak{X}$ a deformation family of X.

It is often required that the morphism $\pi : \mathfrak{X} \to \mathscr{M}$ is proper, however, we do not require this property following [23, Thm. 2.8].

2.2 A-structure and B-structure

For a given Calabi-Yau manifold, we introduce two different but similar structures. In what follows, we assume Calabi-Yau manifolds are of dimension three.

2.2.1 A-structure

Since a Calabi-Yau manifold X is projective by definition, we can take a positive line bundle L with its first Chern class $c_1(L) \in H^{1,1}(X,\mathbb{Z})$ being primitive. We take a Kähler form κ satisfying $[\kappa] = [c_1(L)]$. The Kähler form κ is a $(1,1)$ form, and determines the nilpotent linear map $L_\kappa : H^{p,q}(X) \to H^{p+1,q+1}(M)$ in the Hard Lefschetz theorem. For Calabi-Yau manifolds, since it acts trivially on $H^3(X) = \oplus_{p+q=3} H^{p,q}(X)$, we can restrict it on $H^{even}(X) = \oplus_{p+q:\text{even}} H^{p,q}(X)$ and obtain

$$L_\kappa : H^{even}(X,\mathbb{Q}) \to H^{even}(X,\mathbb{Q}), \tag{1}$$

where we use the fact that κ is a real $(1,1)$ form.

For coherent sheaves $\mathscr{E},\mathscr{F}$ on X, we denote the Riemann-Roch (RR) pairing by

$$\chi(\mathscr{E},\mathscr{F}) = \sum_{i=0}^{3} (-1)^i \dim Ext^i(\mathscr{E},\mathscr{F}).$$

This RR pairing is a pairing on the Grothendieck group $K(X)$ of coherent sheaves on X. The numerical K group $K_{num}(X) := K(X)/\equiv$ is defined by the quotient by the radical of RR pairing. $K_{num}(X)$ defines a free abelian group. For Calabi-Yau manifolds, due to the Serre duality, RR pairing introduces a skew symmetric form on $K_{num}(X)$, which we denote by $(K_{num}(X),\chi(-,-))$. Moreover, by Lefshetz (1,1) theorem and Hard-Lefschetz theorem, we can see that the Chern character homomorphism $ch : K_{num}(X) \to H^{even}(X,\mathbb{Q})$ gives rise to an isomorphism $ch(K_{num}(X)) \otimes \mathbb{Q} \simeq H^{even}(X,\mathbb{Q})$ for Calabi-Yau threefolds. Namely, by the homomorphism $ch : K_{num}(X) \to H^{even}(X,\mathbb{Q})$, we have a natural integral structure with a skew symmetric form on $H^{even}(X,\mathbb{Q})$ coming from the structure $(K_{num}(X),\chi(-,-))$.

Definition 3. We define A-structure of a Calabi-Yau manifold X by the nilpotent linear map $L_\kappa : H^{even}(X,\mathbb{Q}) \to H^{even}(X,\mathbb{Q})$ together with the integral structure on $H^{even}(X,\mathbb{Q})$ coming from $(K_{num}(X),\chi(-,-))$.

2.2.2 B-structure

For a Calabi-Yau manifold X, we can define a similar structure to the A-structure defined above, but we need to require some additional assumptions on X. The first assumption is that X has a deformation family $\pi : \mathfrak{X} \to \mathscr{M}$ over a parameter space which is quasi-projective. We assume that there is a compactification $\overline{\mathscr{M}}$ of $\mathscr{M}$ such that the fiber $X_p = \pi^{-1}(p)$ degenerates to a (singular) Calabi-Yau variety $X_{\bar{p}}$

when we take a limit $p \to \bar{p} \in \overline{\mathscr{M}} \setminus \mathscr{M}$. The compactification $\overline{\mathscr{M}}$ is assumed to be a (singular) projective variety whose singular loci is contained in $\overline{\mathscr{M}} \setminus \mathscr{M}$. Also, we assume $\overline{\mathscr{M}} \setminus \mathscr{M}$ is given by a (non-normal crossiong) divisor D in $\overline{\mathscr{M}}$. Let $\bar{o}$ be a singular point on the divisor D and take the following steps:

1. Let $U_{\bar{o}}$ to be an affine neighborhood of $\bar{o}$. By successive blowing-ups starting at $\bar{o}$, we assume the divisor $U_{\bar{o}} \cap D$ is transformed to

$$\hat{U}_{\bar{o}} \setminus U_{\bar{o}}^{sm} = \cup_i D_i,$$

in terms of normal crossing boundary (exceptional) divisors, where $U_{\bar{o}}^{sm}$ represents $U_{\bar{o}} \setminus D$. We focus on a boundary point given by $o = D_{i_1} \cap \cdots \cap D_{i_r} (r = \dim \mathscr{M})$.

2. Recall that for a family of Calabi-Yau manifolds $\pi : \mathfrak{X} \to \mathscr{M}$, we naturally have a locally constant sheaf $R^3\pi_*\mathbb{C}_{\mathfrak{X}}$ over $\mathscr{M}$. This sheaf corresponds to a holomorphic vector bundle whose fibers are $H^3(X_m, \mathbb{C})$ $(m \in \mathscr{M})$ with the holomorphic flat connection (Gauss-Manin connection). Each fiber has (topological) integral structure $H^3(X_m, \mathbb{Z})$ and also a skew symmetric (symplectic) form defined by $(\alpha, \beta) := \int_{X_m} \alpha \cup \beta$. The Gauss-Manin connection is naturally compatible with this symplectic structure.

3. Take a boundary point $o \in \hat{U}_{\bar{o}}$ as described above, and write it $o = D_1 \cap \cdots \cap D_r (r = \dim \mathscr{M})$ by reordering the indices of the (exceptional) divisors. Choose a base point $b_o \in \mathscr{M}$ and fix an integral symplectic basis of $H^3(X_{b_o}, \mathbb{Z})$. With respect to this basis, we represent the monodromy of the Gauss-Manin connection around the boundary divisor D_i by $T_{D_i} : H^3(X_{b_o}, \mathbb{Z}) \to H^3(X_{b_o}, \mathbb{Z})$. We assume that the monodromy matrices T_{D_i} for $i = 1, \cdots, r$ are unipotent, i.e., satisfy

$$(T_{D_i} - \mathrm{id})^{m_i} = O \text{ for some } m_i.$$

Note that this is a condition for the boundary point $o \in \hat{U}_{\bar{o}}$. For each T_{D_i}, we define

$$\begin{aligned} \log T_{D_i} &= \log(\mathrm{id} + (T_{D_i} - \mathrm{id})) \\ &= (T_{D_i} - \mathrm{id}) - \frac{1}{2}(T_{D_i} - \mathrm{id})^2 + \cdots + \frac{(-1)^{m_i-1}}{m_i - 1}(T_{D_i} - \mathrm{id})^{m_i-1} \end{aligned}$$

and denote this by $N_i = \log T_{D_i}$. By definition, N_i determines a nilpotent endomorphism in $\mathrm{End}(H^3(X_{b_o}, \mathbb{Q}))$. Moreover, since the boundary is a normal crossing divisor, we have $N_iN_j - N_jN_i = O$. Now we define $N_\lambda := \sum_i \lambda_i N_i$ in terms of real parameters $\lambda_i > 0$, which gives a nilpotent endomorphism acting on $H^3(X_{b_o}, \mathbb{R})$ due to the commuting relation. The following result is due to Cattani-Kaplan [8, Thm.2]:

Theorem 1. *For general $\lambda_i > 0 (\lambda_i \in \mathbb{Q})$, the nilpotent matrix $N_\lambda := \sum_i \lambda_i N_i$ defines a monodromy weight filtration of the same form on $H^3(X_{b_o}, \mathbb{Q})$.*

As a general result of monodromy theorem, the nilpotent matrix N_λ satisfies $N_\lambda^k = O$ for $k > \dim X (= 3)$. When we have $N_\lambda^3 \neq O$, $N_\lambda^4 = O$, we call the boundary point $o \in \hat{U}_{\bar{o}}$ maximally degenerated. If a boundary point $o \in \hat{M}_{\bar{o}}$ is maximally degenerated and moreover the monodromy weight filtration by N_λ has the form

$$0 \subset W_0 = W_1 \subset W_2 = W_3 \subset W_4 = W_5 \subset W_6 = H^3(X_{b_o}, \mathbb{Q}),$$

then we call o a LCSL (Large Complex Structure Limit) point.

Definition 4. If a Calabi-Yau manifold X has a deformation family $\pi : \mathfrak{X} \to \mathscr{M}$ which gives rise to a LCSL boundary point $o \in \hat{U}_{\bar{o}}$, then we define B-structure of X from o by the nilpotent action $N_\lambda : H^3(X_{b_o}, \mathbb{Q}) \to H^3(X_{b_o}, \mathbb{Q})$ together with the integral structure $(H^3(X_{b_o}, \mathbb{Z}), (\ ,\ ))$.

Remark 1. It is not clear whether every Calabi-Yau manifold admits a family $\mathfrak{X} \to \mathscr{M}$ where we can extract a B-structure defined above. However, for Calabi-Yau hypersurfaces or complete intersections in toric varieties, we can construct explicitly their families where we verify the B-structures [15, 16].

Remark 2. B-structure is not necessarily unique for a Calabi-Yau manifold X. That is, we often observe that many B-structures of X arise from different LCSL points of a family $\mathfrak{X} \to \mathscr{M}$. Such cases are of main interest in this article.

2.2.3 Mirror symmetry

In contrast to A-structures, describing B-structures of Calabi-Yau manifolds are difficult as well as showing their existence. Mirror symmetry of Calabi-Yau manifolds is a conjecture that says for a given Calabi-Yau manifold X, there exists another Calabi-Yau manifold $\check{X}$ such that the A-structure of X is interchanged by the B-structure of $\check{X}$.

To describe the symmetry in more precise, let $\mathscr{K}_X$ be the Kähler cone of X. The Kähler cone is an open convex cone in $H^2(X, \mathbb{R})$ and coincides with the ample cone when $\dim X = 3$ and satisfies $\mathscr{K}_X \cap H^2(X, \mathbb{Z}) \neq \phi$. The shape of the ample cone is complicated in general and is related to the classification problem of algebraic varieties. Take integral elements $\kappa_1, \cdots, \kappa_r \in \overline{\mathscr{K}}_X \cap H^2(X, \mathbb{Z})$ so that these generate the group $H^2(X, \mathbb{Z})$ and define a cone $\sigma_A := \mathbb{R}_{\geq 0}\kappa_1 + \cdots + \mathbb{R}_{\geq 0}\kappa_r$. Cones of this property may be constructed by decomposing the cone $\overline{\mathscr{K}}_X$ into (possibly infinitely many) simplicial cones. We determine a cone σ_A and introduce a Kähler class κ by a general linear combination

$$\kappa = \alpha_1 \kappa_1 + \cdots + \alpha_r \kappa_r$$

with $\alpha_i > 0$. The nilpotent operator L_κ defines an A-structure of X with parameters α_i, which we denote by

$$\left(L_{\kappa(\alpha)}, (K_{num}(X), \chi(-,-)), \Sigma_A\right), \tag{2}$$

where $\Sigma_A := \{\alpha_1 L_{\alpha_1} + \cdots + \alpha_r L_{\alpha_r} \mid \alpha_i > 0\}$ is a cone in $\mathrm{End}(H^{even}(X, \mathbb{R}))$ and corresponds to the cone $\sigma_A \in H^2(X, \mathbb{Z})$.

The nilpotnet matrix $N_\lambda = \sum_i \lambda_i N_i$ to describe B-structure may be regarded as an element in a cone $\Sigma_o := \mathbb{R}_{\geq 0} N_1 + \cdots + \mathbb{R}_{\geq 0} N_m$ $(m = \dim H^1(X, TX))$ for a LCSL boundary point o. This cone Σ_o is called monodromy nilpotent cone in Hodge theory. We denote the B-structure coming from a LCSL boundary point o by

$$\big(N_\lambda, (H^3(X_{b_o}, \mathbb{Z}), (\ ,\ )), \Sigma_o\big). \tag{3}$$

Here $X_{b_o} = \pi^{-1}(b_o)$ represents a smooth fiber over a base point $b_o \in \mathscr{M}$, but this will often be simplified by X.

Definition 5. A-structure of a Calabi-Yau manifold X and B-structure of another Calabi-Yau manifold Y are said isomorphic if the following two properties holds:

(1) There exists an isomorphism

$$\begin{array}{cccc} \varphi : & H^{even}(X,\mathbb{R}) & \xrightarrow{\sim} & H^3(Y,\mathbb{R}) \\ & \cup & & \cup \\ & (K_{num}(X), \chi(-,-)) & & (H^3(Y,\mathbb{Z}), (\ ,\ )), \end{array}$$

which preserves the integral and skew symmetric forms.

(2) When we set $\alpha_i = \lambda_i$, the nilpotent matrices N_λ and $L_{\kappa(\alpha)}$ are related by the isomorphism in (1) as
$$N_\lambda = \varphi \circ L_{\kappa(\alpha)} \circ \varphi^{-1}.$$

Now we are ready to define mirror symmetry (of Calabi-Yau threefolds):

Definition 6. Two Calabi-Yau manifolds X and $\check{X}$ are said mirror to each other if they have A- and B- structures which are exchanged in the following sense:

$$\text{(i) } \varphi : \big(L_{\kappa(\alpha)}, (K_{num}(X), \chi(-,-)), \Sigma_A\big) \xrightarrow{\sim} \big(N_\lambda, (H^3(\check{X},\mathbb{Z}), (\ ,\ )), \Sigma_o\big),$$
$$\text{(ii) } \check{\varphi} : \big(L_{\kappa(\check{\alpha})}, (K_{num}(\check{X}), \chi(-,-)), \Sigma_{\check{A}}\big) \xrightarrow{\sim} \big(N_{\check{\lambda}}, (H^3(X,\mathbb{Z}), (\ ,\ )), \Sigma_{\check{o}}\big).$$

Remark 1. From the isomorphism between A-, B- structures, it holds

$$h^{1,1}(X) = h^{2,1}(\check{X}),\ h^{1,1}(\check{X}) = h^{2,1}(X) \tag{4}$$

when X and $\check{X}$ are mirror symmetric.

Remark 2. When X and $\check{X}$ are mirror symmetric, we have families $\mathfrak{X} \to \mathscr{M}_X$ and $\check{\mathfrak{X}} \to \mathscr{M}_{\check{X}}$ describing their B-structures. We call, for example, the family $\check{\mathfrak{X}} \to \mathscr{M}_{\check{X}}$ a mirror family of X.

Remark 3. A quintic hypersurface $X = (5)$ in $\mathbb{P}^4$ is an example of Calabi-Yau threefold. In this case, we have $h^{1,1}(X) = 1, h^{2,1}(X) = 101$ for the Hodge numbers of X. A Calabi-Yau manifold $\check{X}$ having Hodge numbers $h^{1,1}(\check{X}) = 101, h^{2,1}(\check{X}) = 1$ has been constructed in the pioneering work by Candelas et. al. in 1991 [7], and surprising implications of mirror symmetry have been found there. In this case, while the isomorphism (i) has been shown, verifying the isomorphism (ii) seems difficult since the dimension of the deformation space is large, i.e., $h^{2,1}(X) = 101$. Beyond quintic hypersurfaces in $\mathbb{P}^4$, we can construct pairs $(X, \check{X})$ of Calabi-Yau hypersurfaces or complete intersections satisfying the relation (4) in toric varieties [2, 3]. If the dimensions of the deformation spaces are small enohgh, we can verify either the isomorphism (i) or (ii) [15, 16]. However verifying both relations (i) and (ii) is difficult in general, since the relation (4) implies that the smaller dimensions of the deformation space for X, the larger dimensions for $\check{X}$ when the Euler numbers $e(X) = -e(\check{X})$ have large absolute values. It is believed that both isomorphisms (i) and (ii) hold for the pairs of Calabi-Yau manifolds given in [2, 3].

3 Mirror symmetry from families of Calabi-Yau manifolds

Mirror symmetry is a conjecture which asserts that for a Calabi-Yau manifold X there exists another Calabi-Yau manifold $\check{X}$ which is mirror symmetric in the sense of Definition 6. If $\check{X}$ exists, by definition, it comes with its deformation family $\check{\mathfrak{X}} \to \mathscr{M}_{\check{X}}$. The symmetry of $\check{X}$ to X is described by the B-structure from a LCSL boundary point of $\overline{\mathscr{M}}_{\check{X}}$. By studying families over $\overline{\mathscr{M}}_{\check{X}}$ globally for many examples of X and $\check{X}$, we often observe multiple LCSL boundary points in $\overline{\mathscr{M}}_{\check{X}}$ and recognize the following correspondence:

$$\begin{aligned} &\text{LCSL boundary points in } \overline{\mathscr{M}}_{\check{X}} \\ &\quad \leftrightarrow \quad \text{birational geometry and Fourier-Mukai partners of } X. \end{aligned}$$

We present two interesting examples where we see this correspondence.

3.1 Birational geometry related to determinantal quintics in $\mathbb{P}^4$

Consider the complete intersection of five $(1,1)$ divisors in $\mathbb{P}^4 \times \mathbb{P}^4$,

$$X = \cap_{i=1}^{5}(1,1) \subset \mathbb{P}^4 \times \mathbb{P}^4.$$

X is a smooth Calabi-Yau manifold if we take the five $(1,1)$ divisors general. We can determine its Hodge numbers as $h^{1,1}(X) = 2, h^{2,1}(X) = 52$ [19]. After some works, it turns out that we have a mirror Calabi-Yau manifold $\check{X}$ from a special family X_{sp} of X with parameters a, b;

$$X_{sp} := \{z_i w_i + a z_i w_{i+1} + b z_{i+1} w_i = 0 \ (i = 1, .., 5)\} \subset \mathbb{P}^4_z \times \mathbb{P}^4_w.$$

As described in the following proposition, X_{sp} is singular for general parameters a, b, but there exists a crepant ($c_1 = 0$) resolution of the singularity which gives rise to a mirror Calabi-Yau manifold [19, Thm.5.11,Thm.5.17].

Proposition 1. *When $a, b \in \mathbb{C}$ are general, the following properties hold:*

(1) *X_{sp} is singular along 20 $\mathbb{P}^1$ with the A_1-singularities.*

(2) *There exists a crepant resolution $\check{X} \to X_{sp}$ which gives a Calabi-Yau manifolds with the Hodge numbers $h^{1,1}(\check{X}) = 52$, $h^{2,1}(\check{X}) = 2$.*

Taking the symmetry of the defining equations of X_{sp} into account, we see that the special family is parametrized by a^5 and b^5. Moreover, these affine parameters are projectivised naturally by $[a^5, b^5, 1] \in \mathbb{P}^2$. The family parametrized by $[a^5, b^5, 1]$ has been studied in detail in [21, Prop.3.11].

Proposition 2. *We have a family of Calabi-Yau manifolds $\check{\mathfrak{X}} \to \mathcal{M}_{\check{X}} \subset \mathbb{P}^2$ with the fibers being crepant resolutions $\check{X} \to X_{sp}$. The parameter space $\mathcal{M}_{\check{X}}$ satisfies $\overline{\mathcal{M}}_{\check{X}} = \mathbb{P}^2$, and the fibers over $\mathcal{M}_{\check{X}}$ extend to over $\overline{\mathcal{M}}_{\check{X}} \setminus \mathcal{M}_{\check{X}} = D_1 \cup D_2 \cup D_3 \cup Dis_0$ with some degeneration. Here D_1, D_2, D_3 are the toric divisors in $\mathbb{P}^2$ and Dis_0 represents the discriminant of X_{sp} which has degree 5.*

Since we have the relations $h^{1,1}(X) = h^{2,1}(\check{X}) = 2, h^{2,1}(X) = h^{1,1}(\check{X}) = 52$ for the Hodge numbers, we may expect that the family $\check{\mathfrak{X}} \to \mathcal{M}_{\check{X}}$ is a mirror family of X. In fact, it turns out that not only X but also its birational models arise naturally from the family [21, Prop.3.12].

Proposition 3. *The boundary points $o_1 = D_1 \cap D_2, o_2 = D_2 \cap D_3, o_3 = D_3 \cap D_1$ are all LCSLs, and the B-structures from each point are isomorphic to the A-structures of $X =: X_1$ and its birational models X_2, X_3.*

The birational models X_2, X_3 in the above proposition are described in the following diagram which we can complete starting from $X = X_1$:

$$
\begin{array}{ccccccccccccc}
X_1 & \overset{\rho_{12}}{\dashleftarrow\dashrightarrow} & & & X_2 & \overset{\rho_{23}}{\dashleftarrow\dashrightarrow} & & & X_3 & \overset{\rho_{31}}{\dashleftarrow\dashrightarrow} & & & X_1 \\
& \pi_{21}\searrow & & \swarrow\pi_{22} & & \pi_{32}\searrow & & \swarrow\pi_{33} & & \pi_{13}\searrow & & \swarrow\pi_{11} & \\
& & Z_2 & & & & Z_3 & & & & Z_1 & &
\end{array}
\qquad (5)
$$

Here, Z_2 represents a determinantal quintic hypersurface $(5) \subset \mathbb{P}^4_w$ obtained from X_1 by projecting to the second factor $\pi_{21} : \mathbb{P}^4_z \times \mathbb{P}^4_w \to \mathbb{P}^4_w$. Similarly, we obtain a determinantal quintic hypersurface $Z_1 \subset \mathbb{P}^4_z$ by projecting to the first factor. Both Z_2 and Z_1 are singular quintic hypersurface with 50 ordinary double points. The birational models X_2 and X_3 are small resolutions of Z_2 and Z_1, respectively, see [19] for details. The A-structures of these birational models X_1, X_2, X_3 can be glued together in $H^2(X, \mathbb{R})$ and determines the so-called movable cone. Corresponding to the gluing A-structures, it was found in [21] that the B-structures from the LCSL boundary points o_1, o_2, o_3 can also be glued together by the global analysis of the local solutions over the parameter space $\mathcal{M}_{\check{X}}$. In [21], this gluing B-structures is called gluing of monodromy nilpotent cones contrasting the resulting large cone to the movable cone.

Remark 1. The claim in Proposition 3 is based on the isomorphism of the A-structure of X with the B-structure of $\check{X}$. Verifying the both properties (i) and (ii) in Definition 4 is difficult in this case since the large dimension of the deformation space,

$\dim \mathcal{M}_X = 52$. It should be noted that the Calabi-Yau manifolds in the next section provide examples where we can verify both (i) and (ii).

Remark 2. We came to the above Clabi-Yau manifold $X = \cap_{i=1}^{5}(1,1)$ when studying the three dimensional Reye congruence in [18]. Historically Reye congruence stands for an Enriques surface [9], which we can describe as a $\mathbb{Z}_2$ quotient of complete intersection of four symmetric $(1,1)$ divisors in $\mathbb{P}^3 \times \mathbb{P}^3$. Three dimensional Reye congruence is its generalization [20, 19], which we can describe by a $\mathbb{Z}_2$ quotient of $X_s = \cap_{i=1}^{5}(1,1)_s$ with five symmetric divisors $(1,1)_s$. For symmetric divisors, the determinantal quintic Z_3 in the diagram (5) describes the so-called symmetroid which is singular along a curve. It was found in [18, 20] that the double cover of Z_3 branched along the singular locus defines a Calabi-Yau manifod which is a Fourier-Mukai partner of the three dimensional Reye congruence $X_s/\mathbb{Z}_2$.

3.2 Calabi-Yau manifolds fibered by abelian surfaces

Calabi-Yau manifolds related to determinantal quintics show us interesting implications of mirror symmetry to the birational geometry of Calabi-Yau manifolds. Recall that homological mirror symmetry describes the symmetry in terms of the derived categories of coherent sheaves. Therefore if two Calabi-Yau manifolds are derived equivalent, then these two should have the same mirror Calabi-Yau manifold. As for the derived equivalence, the following result is due to Bridgeland [6]:

Theorem 2. *In dimension three, birational Calabi-Yau manifolds are derived equivalent.*

Combined with homological mirror symmetry, it is natural we have encountered the birational models of X from the study of mirror family $\check{\mathfrak{X}} \to \mathcal{M}_{\check{X}}$. At the same time, Calabi-Yau manifolds which are not birational to a Calabi-Yau manifold but derived equivalent to it come into our interest. Such Calabi-Yau manifolds are called Fourier-Mukai partners of a Calabi-Yau manifold, and the so-called Grassmannian and Phaffian duality provides an interesting example of Fourier-Muaki partners [25, 5, 4, 14]. (For K3 surfaces, we have a general result for the numbers of Fourier-Mukai partners, see [17] and reference therein.) The double cover of Z_3 in the preceding subsection (Remark 2), provides another interesting example of Fourier-Mukai partner [18, 20].

Our second example of this article turns out to be more interesting because we observe there that both birational models and Fourier-Mukai partners arise from its mirror family. To describe it briefly, let us consider an abelian surface A with its polarization $\mathcal{L}$ of $(1,d)$ type. Such an abelian surface A can be embedded into $\mathbb{P}^{d-1}$ by the linear system $|\mathcal{L}|$. Gross and Popescu have studied the ideal $\mathcal{I}(A)$ of the image of this embedding [10]. In particular, when the polarization is of type $(1,8)$, it was found that the ideal contains four quartics of the form;

$$f_1 = \frac{w_0}{2}(x_0^2 + x_4^2) + w_1(x_1x_7 + x_3x_5) + w_2x_2x_6, \; f_{i+1} = \sigma^i f_1 \;\; (i = 1,2,3),$$

where $\sigma : x_i \mapsto x_{i+1}, \tau : x_i \mapsto \xi^{-i}x_i\,(\xi^8 = 1)$ represent the actions of the Heisenberg group $\mathscr{H}_8 = \langle \sigma, \tau \rangle$ on the homogeneous coordinates x_i of $\mathbb{P}^7$. The parameters $w_i = w_i(A)$ are determined by A and represent a point $[w_0, w_1, w_2] \in \mathbb{P}^2$. Thus the correspondence A $\mapsto w(A)$ defines a (rational) map from an open set of the moduli space $\mathscr{A}^{(1,8)}$ of $(1,8)$-polarized abelian surfaces to $\mathbb{P}^2$. The following theorem is due to Gross and Popescu [10]:

Theorem 3. *$\mathscr{A}^{(1,8)}$is birational to a conic bundle over $\mathbb{P}^2$.*

While the ideal $\mathscr{I}(A)$ is related to the moduli space $\mathscr{A}^{(1,8)}$ as above, it defines a complete intersection of four quadrics in $\mathbb{P}^7$ which is a singular Calabi-Yau threefold. We denote this Calabi-Yau variety by V_w. The following properties of V_w has been studied in [10].

Proposition 4. *$V_w = V(f_1, \cdots, f_4) \subset \mathbb{P}^4$ has the following properties:*

(1) *V_w is a pencil of $(1,8)$-polarized abelian surfaces which has 64 base points.*
(2) *Only at the 64 base points, V_w is singular with ordinary double points (ODPs). These singularities are resolved by a blow-up along an abelian surface A, giving a small resolution $X' \to V_w$.*
(3) *By flopping exceptional curves of the small resolution, $X' \to V_w \leftarrow X$, we have a Calabi-Yau threefold with fibers $(1,8)$-polarized abelian surfaces over $\mathbb{P}^1$.*
(4) *Both Calabi-Yau manifolds X and X′ admit free actions of the Heisenberg group $\mathscr{H}_8$ (which actually acts as $\mathbb{Z}_8 \times \mathbb{Z}_8$ on X and X′).*

We can determine the Hodge numbers of X as follows [10]: First we obtain $h^{2,1}(X') = h^{2,1}(X) = 2$ by studying the deformation spaces. Since the singular fibers of X turns out to be translation scrolls of elliptic curves, we see that the topological Euler number of X is zero. This entails that $h^{1,1}(X) = h^{1,1}(X') = 2$.

Using free actions of the Heisenberg group $\mathscr{H}_8$ (which acts on X as $\mathbb{Z}_8 \times \mathbb{Z}_8$), we have quotient Calabi-Yau manifolds

$$\check{X} := X/\mathbb{Z}_8, \; Y := X/\mathbb{Z}_8 \times \mathbb{Z}_8$$

(and also $\check{X}' := X'/\mathbb{Z}_8, Y' := X'/\mathbb{Z}_8 \times \mathbb{Z}_8$). All these are Calabi-Yau manifolds with $h^{1,1} = h^{2,1} = 2$. The following interesting properties are shown in [26, 1].

Proposition 5. *The fiber-wise dual of the abelian-fibered Calabi-Yau manifold $X \to \mathbb{P}^1$ is isomorphic to $Y = X/\mathbb{Z}_8 \times \mathbb{Z}_8$, and X and Y are derived equivalent.*

These properties naturally motivate us studying mirror symmetry of X. Regarding this, Gross and Pavanelli [11] made a conjecture:

Conjecture 1. Mirror of X is $\check{X} = X/\mathbb{Z}_8$, and mirror of $\check{X}$ is $Y = X/\mathbb{Z}_8 \times \mathbb{Z}_8$.

This conjecture came from their study on the Brauer groups and fundamental groups of these Calabi-Yau manifolds. Recently, by constructing families of $\check{X}$ and Y and studying LCSL boundary points carefully, we verified the conjecture affirmatively [22] as follows:

Proposition 6. *If we take a subgroup $\mathbb{Z}_8 \simeq \langle \tau \rangle \subset \mathscr{H}_8$ to define $\check{X} = X/\mathbb{Z}_8$, then there exists a family $\check{\mathfrak{X}} \to \mathscr{M}_{\check{X}}$ which describes a mirror family of X, i.e., the B-structure coming from a LCSL boundary point $o \in \overline{\mathscr{M}}_{\check{X}}$ is isomorphic to the A-structure of X. Also, there is another LCSL boundary point $\tilde{o}$ in $\overline{\mathscr{M}}_{\check{X}}$, the B-structure from which is isomorphic to the A-structure of the Fourier-Mukai partner Y of X.*

The parameter space of the family $\check{\mathfrak{X}} \to \mathscr{M}_{\check{X}}$ is given by a toric variety. Studying the boundary points in $\overline{\mathscr{M}}_{\check{X}} \setminus \mathscr{M}_{\check{X}}$ in more detail, we actually find more LCSL boundary points other than $o, \tilde{o}$ above, and find that the B-structures from them are isomorphic to the birational models X' and Y' [22]. Namely, we observe that the birational models as well as Fourier-Mukai partners of X emerge from the LCSL boundary points in the parameter space of the family $\check{\mathfrak{X}} \to \mathscr{M}_{\check{X}}$. We refer to the recent work [22] for more details.

Acknowledgements One of the author (S.H.) would like to thank the organizers of the MATRIX research program "The Geometry of Moduli Spaces in String Theory" for inviting him to the interesting and fruitful program. He also would like to thank the staff members of MATRIX at the Creswick campus of the University of Melbourne where he enjoyed the wonderful environment in Australia. This work is supported in part by Grants-in-Aid for Scientific Research C 20K03593 and C 24K06743.

References

1. A. Bak, *The spectral construction for a (1,8)-polarized family of Abelian varieties*, arXiv:0903.5488 [math.AG].
2. V. Batyrev, *Dual polyhedra and mirror symmetry for Calabi-Yau hypersurfaces in toric varieties*, Journal of Algebraic Geometry 3 (1994), 493–535.
3. V. Batyrev and L. Borisov, *On Calabi-Yau complete intersections in toric varieties*, Higher-dimensional complex varieties (Trento, 1994), 39–65, de Gruyter, Berlin, 1996.
4. L. Borisov and A. Caldararu, *The Pfaffian-Grassmannian derived equivalence*, J. Algebraic Geom. 18 (2009), no. 2, 201–222.
5. V. Batyrev, I. Ciocan-Fontanine, B. Kim, and D. van Straten, *Conifold transitions and mirror symmetry for Calabi-Yau complete intersections in Grassmannians*, Nuclear Phys. B 514 (1998), 640–666.
6. T. Bridgeland, *Flops and derived categories*, Invent. Math. 147 (2002), 613–632.
7. P. Candelas, X.C. de la Ossa, P.S. Green and L.Parkes, *A pair of Calabi-Yau manifolds as an exactly soluble superconformal theory*, Nucl.Phys. B356(1991), 21-74.
8. E.H. Cattani, *Mixed Hodge structures, compactifications and monodromy weight filtration*, in Topics in Transcendental Algebraic Geometry, ed. P. Griffiths, Princeton Univ. Press (1984).
9. F. Cossec, *Reye congruence*, Transactions of the A.M.S. 280 (1983), 737–751.
10. M. Gross and S. Popescu, *Calabi-Yau threefolds and moduli of abelian surfaces. I*, Compositio Math. 127 (2001), no. 2, 169–228.
11. M. Gross and S. Pavanelli, *A Calabi-Yau threefold with Brauer group* $(\mathbb{Z}/8\mathbb{Z})^2$, Proc. Amer. Math. Soc. 136 (2008), no. 1, 1–9.
12. M. Gross and B. Siebert, *Mirror symmetry via logarithmic degeneration data. I.* J. Differential Geom. 72 (2006), no. 2, 169–338.
13. M. Gross and B. Siebert, *Mirror symmetry via logarithmic degeneration data, II.* J. Algebraic Geom. 19 (2010), no. 4, 679–780.

14. K. Hori and D. Tong, *Aspects of non-abelian gauge dynamics in two-dimensional N=(2,2) theories*, J. High Energy Phys. 2007, no. 5, 079, 41 pp.
15. S. Hosono, A. Klemm, S. Theisen and S.-T. Yau, *Mirror symmetry, mirror map and applications to Calabi-Yau hypersurfaces*, Comm. Math. Phys. 167 (1995), no. 2, 301–350.
16. S. Hosono, A. Klemm, S. Theisen and S.-T. Yau, *Mirror symmetry, mirror map and applications to complete intersection Calabi-Yau spaces*, Nucl. Phys. B433(1995)501–554.
17. S. Hosono, B.H. Lian, K. Oguiso and S.-T. Yau, *Fourier-Mukai number of a K3 surface*, in "Algebraic structures and moduli spaces", CRM Proc. Lecture Notes, 38 (2004), 177 –192.
18. S. Hosono and H. Takagi, *Mirror symmetry and projective geometry of Reye congruences I*, J. Algebraic Geom. 23 (2014), no. 2, 279–312.
19. S. Hosono and H. Takagi, *Determinantal quintics and mirror symmetry of Reye congruences*, Commun. Math. Phys. 329, (2014), 1171–1218.
20. S. Hosono and H. Takagi, *Double quintic symmetroids, Reye congruences, and their derived equivalence*, J. Differential Geom. 104 (2016), no. 3, 443–497.
21. S. Hosono and H. Takagi, *Movable vs monodromy nilpotent cones of Calabi-Yau manifolds*, SIGMA Symmetry Integrability Geom. Methods Appl. 14 (2018), Paper No. 039, 37 pp.
22. S. Hosono and H. Takagi, Mirror symmetry of Calabi-Yau manifolds fibered by (1,8)-polarized abelian surfaces, Commun. Number Theory Phys. 16 (2022), no. 2, 215–298.
23. K. Kodaira, *Complex Manifolds and Deformation of Complex Structures*, Classics in Mathematics, Springer 2005.
24. M. Kontsevich, *Homological algebra of mirror symmetry*, Proceedings of the International Congress of Mathematicians (Zürich, 1994) Birkhäuser (1995) pp. 120 –139.
25. E.A. Rødland, *The Pfaffian Calabi-Yau, its Mirror and their link to the Grassmannian* $G(2,7)$, Compositio Math. 122 (2000), no. 2, 135–149.
26. C. Schnell, *The fundamental group is not a derived invariant*, in "Derived categories in algebraic geometry", 279–285, EMS Ser. Congr. Rep., Eur. Math. Soc., Zürich, 2012.
27. A. Strominger, S.-T. Yau and E. Zaslow, *Mirror symmetry is T-Duality*, Nucl. Phys. B479 (1996) 243-259.

A single point as a Calabi-Yau zerofold

Johanna Knapp and Joseph McGovern

Abstract We give a brief account of a non-abelian GLSM that describes a Calabi-Yau zerofold, in this case a single point, in the "large volume" phase. The other phase is non-regular, i.e. there is no clear separation between the gauge and the matter sectors.

1 Introduction

The construction of the quintic threefold $\mathbb{P}^4[5]$ straightforwardly extends to complex dimension $n-1$ by way of $\mathbb{P}^n[n+1]$. This leads to the obvious realisation that $\mathbb{P}^1[2]$, i.e. a generic quadric in $\mathbb{P}^1$, is a Calabi-Yau zerofold. In other words, two points are Calabi-Yau. This point has been made in [7, §7], where arithmetic methods were applied to CY varieties of various dimensions including 0. While it may seem somewhat pointless to study zero-dimensional Calabi-Yaus, examples like this serve as toy models or motivation for certain constructions. See for instance [19, §2.2] for a GLSM [38] account on $\mathbb{P}^1[2]$ as a preparatory example for the construction of a non-abelian duality in GLSMs.

The point of this short note is two-fold. First, we want to point out the amusing fact that also a single point can be understood as a Calabi-Yau zerofold, in the sense that the axial R-symmetry of the associated GLSM is non-anomalous. We do so by giving a construction in terms of a one-parameter non-abelian GLSM. The construction is the zero-dimensional analogue of a non-abelian GLSM [19] whose two geometric phases are a pair of non-birational Calabi-Yaus first constructed by

Johanna Knapp
School of Mathematics and Statistics, University of Melbourne
e-mail: johanna.knapp@unimelb.edu.au

Joseph McGovern
School of Mathematics and Statistics, University of Melbourne
e-mail: mcgovernjv@gmail.com

D. R. Wood et al. (eds.), *2024 MATRIX Annals, Part II*, MATRIX Book Series 8,
https://doi.org/10.1007/978-3-032-16206-9_23

Hosono and Takagi [25, 24, 27]. The model we present here is already implicit in [19] and other references below.

The second motivation is to produce a simple toy example for non-abelian GLSMs that have been termed "irregular" or "non-regular". First observed in [21] for non-abelian GLSMs associated to Grassmannians, these are GLSMs with phases where there is no scale separation between the gauge sector and the matter sector. A consequence is that field-theoretic methods to describe such phases, such as a Born-Oppenheimer approximation, do not apply and it becomes questionable whether it is possible at all to describe the low-energy behaviour of the GLSM in the phase in terms of a "nice" theory with an action such as a $(2,2)$ non-linear sigma model or a Landau-Ginzburg theory. A similar statement also holds for "bad" or "pseudo-hybrid" models [1] which can also appear in abelian GLSMs, where a direct analysis of the low-energy physics is not straightforward. Since such phases of GLSMs are ubiquitous and may even be generic phases of GLSMs, they should be understood better.

Recent accounts of non-regular GLSMs can be found in [16, 32]. These models are fairly complicated, displaying many interrelated new phenomena, so it may be instructive to first study a simpler example. We hope that the model presented in this note can serve as a toy model for a non-regular GLSM. The non-regular model of [32], studied by the authors, was related to a noncommutative resolution via topological string theory techniques developed in [36, 31]. This demonstrates that such models can contain interesting mathematical structure. We present the non-regular model in this article as a simplest possible entry point in this program.

In the next section we introduce the model following [19]. We show that the "large volume phase" is a single point. We then show that the GLSM is non-regular by demonstrating that there is a Coulomb branch at infinite negative FI-parameter that coexists with a Higgs branch given by a determinantal quadric that cuts out two points in $\mathbb{P}^1$. We go on to consider the hemisphere and sphere partition functions, and study the divergent expressions that this produces in the strongly coupled phase. Moreover, we construct the mirror variety, which is a different description of a point, and compare with GLSM results, in particular the discriminant. In the final section we make suggestions for further research on this example.

2 A non-abelian GLSM for a point

We consider a non-abelian Calabi-Yau GLSM, with gauge group $G = (U(1) \times O(2))/\{\pm 1, \pm \mathbf{1}\} \cong (U(1) \times U(1)) \rtimes \mathbb{Z}_2$ and the following chiral matter content[1]:

ϕ	p^1	p^2	x_1	x_2	y_1	y_2	FI
$U(1)_1$	-1	-1	1	1	0	0	ζ
$U(1)_2$	-1	-1	0	0	1	1	ζ

(1)

[1] By a common abuse of notation we use the same notation for superfields and their scalar components. The $\mathbb{Z}_2$ factor in the semidirect product exchanges the two $U(1)$ factors.

The superpotential is

$$W = \sum_{i,j,k=1}^{2} S_k^{ij} p^k x_i y_j \equiv \sum_{i,j=1}^{2} S^{ij}(p) x_i y_j. \tag{2}$$

The $\mathbb{Z}_2$ exchanges x_i and y_i, and acts trivially on the p-fields. It also swaps the two vector multiplets. Therefore we can only have a single FI parameter ζ. Due to the $\mathbb{Z}_2$-symmetry, the 2×2 matrix $S^{ij}(p)$ is symmetric.

The classical vacua are determined by the solutions of the D-term and F-term equations. The D-terms are

$$-\sum_{i=1}^{2} |p^i|^2 + \sum_{i=1}^{2} |x_i|^2 = \zeta, \qquad -\sum_{i=1}^{2} |p^i|^2 + \sum_{i=1}^{2} |y_i|^2 = \zeta. \tag{3}$$

From this, one concludes that the following deleted sets have to be excluded in the respective phases:

$$F_{\zeta>0} = \{x_1 = x_2 = 0\} \cup \{y_1 = y_2 = 0\}, \qquad F_{\zeta<0} = \{p^1 = p^2 = 0\}. \tag{4}$$

The F-terms are given by $dW = 0$.

2.1 Phases

The $\zeta > 0$ phase is straightforward because the gauge symmetry is broken completely and the vacuum manifold is the Calabi-Yau

$$X = \{(x_i, y_i) \in (\mathbb{P}^1 \times \mathbb{P}^1)/\mathbb{Z}_2 | S_k^{ij} x_i y_j = 0\}, \qquad k = 1,2. \tag{5}$$

This is indeed a single point: A complete intersection of two bilinears in $\mathbb{P}^1 \times \mathbb{P}^1$ would give two points,

$$\begin{aligned} \left(\frac{x_2}{x_1}, \frac{y_2}{y_1}\right) &= \left(\frac{\alpha \pm \sqrt{\nabla}}{2\beta}, \frac{\alpha \mp \sqrt{\nabla}}{2\beta}\right) \\ \alpha = S_1^{11} S_2^{22} - S_2^{11} S_1^{22}, &\qquad \beta = S_2^{12} S_1^{22} - S_1^{12} S_2^{22}, \\ \nabla = \left(S_2^{11} S_1^{22} - S_1^{11} S_2^{22}\right)^2 &+ 4\left(S_2^{11} S_1^{12} - S_1^{11} S_2^{12}\right)\left(S_1^{12} S_2^{22} - S_2^{12} S_1^{22}\right). \end{aligned} \tag{6}$$

However, these two points get identified by the $\mathbb{Z}_2$-action $(x_1, x_2) \leftrightarrow (y_1, y_2)$ that exchanges the two $\mathbb{P}^1$s, which amounts to flipping the sign in the $\pm$ above. This leaves us with just one point.

The $\zeta < 0$ phase turns out to be more complicated due to strong coupling effects and the non-regularity of the model. The vacuum is given by $x_i = y_i = 0$. It has an unbroken continuous symmetry $O(2) = U(1) \rtimes \mathbb{Z}_2$. The p-fields take VEVs in a

$\mathbb{P}^1$. Turning on classical fluctuations of the x- and y-fields generates a potential of the form $W = S^{ij}(\langle p \rangle)x_i y_j$ and the symmetric 2×2-matrix $S^{ij}(p)$ becomes a mass matrix. The x- and y-fields are massive except at the locus where the rank of $S(p)$ drops, i.e. when the determinant vanishes:

$$\begin{aligned}\left(S_1^{11}S_1^{22}-(S_1^{12})^2\right)(p^1)^2+\left(S_2^{11}S_1^{22}+S_1^{11}S_2^{22}-2S_1^{12}S_2^{12}\right)p^1p^2 \\ +\left(S_2^{11}S_2^{22}-(S_2^{12})^2\right)(p^2)^2=0.\end{aligned} \tag{7}$$

Since the p-fields take values in $\mathbb{P}^1$, there can be a rank-1-locus, but there cannot be a rank-0-locus because for generic coefficients the only solution for $S^{ij}(p)=0$ is $p^1=p^2=0$ which is in the deleted set for this phase. So, in analogy to the Hosono-Takagi GLSM [25, 19], one finds a symmetric determinantal variety

$$Y=\{(p^1,p^2)\in\mathbb{P}^1|\mathrm{rk}S(p)=1\}. \tag{8}$$

The solution of the determinantal quadric gives two points rather than one:

$$\begin{aligned}\frac{p_2}{p_1}&=\frac{S_2^{11}S_1^{22}+S_1^{11}S_2^{22}-2S_1^{12}S_2^{12}\pm\sqrt{\Upsilon}}{2(S_2^{12})^2-2S_2^{22}S_2^{11}},\\ \Upsilon&=\left(S_2^{11}S_1^{22}+S_1^{11}S_2^{22}-2S_1^{12}S_2^{12}\right)^2-4\left((S_1^{12})^2-S_1^{11}S_1^{22}\right)\left((S_2^{12})^2-S_2^{11}S_2^{22}\right).\end{aligned} \tag{9}$$

If the $\zeta<0$ phase were just a non-linear sigma model with target Y, we could find a mismatch of Witten indices between the two phases, giving the first piece of evidence that our discussion is incomplete. In our argument we have implicitly made a Born-Oppenheimer approximation, assuming there is an energy scale where the gauge degrees of freedom decouple and the low-energy behaviour is determined by the matter sector. However, this is not true throughout the moduli space, the culprit being a Coulomb branch from the strongly coupled sector located at $\zeta\to-\infty$.

2.2 Coulomb branches

The Coulomb branch analysis for this model carries over from [21] which gives a much more general and comprehensive account and is, to our knowledge, the first reference to point out the existence of non-regular GLSMs. We start with discussing the Coulomb branch of the GLSM itself. The Coulomb branch loci are given in terms of the critical set of the effective superpotential

$$\begin{aligned}\mathcal{W}_{eff}=-t(\sigma_1+\sigma_2)-2(-\sigma_1-\sigma_2)(\log(-\sigma_1-\sigma_2)-1)\\ -2\sigma_1(\log(\sigma_1)-1)-2\sigma_2(\log(\sigma_2)-1),\end{aligned} \tag{10}$$

where $\sigma_{1,2}$ are the vector multiplets. The critical locus is at

$$e^{-t} = \frac{\sigma_1^2}{(\sigma_1+\sigma_2)^2} = \frac{\sigma_2^2}{(\sigma_1+\sigma_2)^2}. \tag{11}$$

From this it follows that $z^2 = \frac{\sigma_2^2}{\sigma_1^2} = 1$. This gives a Coulomb branch at $\varphi \equiv e^{-t} = \frac{1}{4}$ from the locus $\sigma_1 = \sigma_2$. Since this locus is fixed by the $\mathbb{Z}_2$ Weyl group of the non-abelian gauge group, there are actually two Coulomb branches at this point [19]. We expect that this is related to the fact that the quotient action defining the mirror M in §2.5 does not act freely for $\varphi = \frac{1}{4}$. The equations (11) also have a solution for $\sigma_1 + \sigma_2 = 0$, indicating another Coulomb branch at $\zeta \to -\infty$. However, since the p-fields are massless at this locus, it is not allowed to integrate them out, making (10) invalid.

The locus $\sigma_1 + \sigma_2 = 0$, however, makes a reappearance in the strongly coupled phase and eventually does lead to a Coulomb branch at $\zeta \to -\infty$. The condition parametrises the unbroken gauge symmetry in the strongly coupled phase. As was pointed out in [19, §4.2], Hosono-Takagi-type GLSMs with N fundamentals[2] and gauge group $O(k)$ are non-regular when $N-k$ is even, meaning that there is a Coulomb branch for any negative value of the FI parameter. This implies that the gauge degrees of freedom do not decouple. The condition for non-regularity is satisfied in our case, which is $N = k = 2$. It has been further argued in [21, §4.4] in an analysis of models with unbroken $SU(k)$ that this Coulomb branch is lifted almost everywhere by dynamically creating masses for the σ-fields. The discussion carries over almost word by word to our case. We choose a parametrisation

$$\sigma_1 = \sigma - \sigma_0, \qquad \sigma_2 = -\sigma, \tag{12}$$

where σ_0 is associated to the $U(1)$ that is Higgsed in the strongly coupled phase. Inserting this into (10) and integrating out σ_0 leads to

$$\sigma_0 = \frac{1}{1+e^{-\frac{t}{2}}}\sigma =: f(e^{-t})\sigma \tag{13}$$

with

$$t + 2\log(\sigma(1-f)) - 2\log(f\sigma) = 0. \tag{14}$$

Taking this into account and eliminating σ_0 from $\mathcal{W}_{eff}$ gives

$$\mathcal{W}_{eff}(\sigma) = 2\sigma\left(\log\frac{-1}{(1-f)}\right). \tag{15}$$

This has a critical locus at $f = 2$ which is the same as $e^{-t} = \frac{1}{4}$ which is the Coulomb branch locus we have already found. This means there is no Coulomb branch in the $\zeta \ll 0$ phase, except at $\zeta \to -\infty$. This poses a conundrum. We have one point in one phase, and two points plus a Coulomb branch in the other, with an unbroken nonabelian continuous gauge symmetry. It is not obvious how these two phases can

[2] In our model, the fundamentals consist of linear combinations of x- and y-fields.

be equivalent in a way phases of other GLSMs are. We do not know how to resolve this issue but make some comments in §3.

2.3 Structure sheaf and hemisphere partition function

Coming back to the $\zeta \gg 0$ phase, we can make a few further statements about our Calabi-Yau zerofold by computing the hemisphere partition function for the structure sheaf. The hemisphere partition function [37, 18, 20] for the model is[3]

$$Z_{D^2} = \frac{C}{(2\pi)^2} \int d^2\sigma \Gamma(1 - \mathrm{i}\sigma_2 - \mathrm{i}\sigma_2)^2 \Gamma(\mathrm{i}\sigma_1)^2 \Gamma(\mathrm{i}\sigma_2)^2 e^{\mathrm{i}t(\sigma_1+\sigma_2)} f_{\mathscr{B}}(\sigma), \tag{16}$$

where the brane factor $f_{\mathscr{B}} = \mathrm{Tr}_M e^{\mathrm{i}\pi r_*} e^{2\pi\rho(\sigma)}$ is the character of the representations r_* and ρ of the vector R- and gauge symmetries on the Chan-Paton space M of the GLSM brane $\mathscr{B}$ [17, 20]. To evaluate the partition function in the $\zeta \gg 0$ phase we close the integration contour such that the poles of $\Gamma(\sigma_{1,2})$ are enclosed, i.e. we take $\mathrm{i}\sigma_i = -k_i + \varepsilon_i$ for $k_i \in \mathbb{Z}_{\geq 0}$. This guarantees a convergent result in the phase. The integral can be rewritten as

$$Z_{D^2}^{\zeta \gg 0} = \frac{C}{(2\pi \mathrm{i})^2} \sum_{k_1,k_2 \geq 0} \oint d^2\varepsilon \frac{\pi^4}{\sin^2 \pi\varepsilon_1 \sin^2 \pi\varepsilon_2} \cdot \frac{\Gamma(1 + k_1 + k_2 - \varepsilon_1 - \varepsilon_2)^2}{\Gamma(1 + k_1 - \varepsilon_1)^2 \Gamma(1 + k_2 - \varepsilon_2)^2} e^{-t(k_1+k_2-\varepsilon_1-\varepsilon_2)} f_{\mathscr{B}}. \tag{17}$$

The brane that describes the structure sheaf is given by the following matrix factorisation of the superpotential:

$$Q = \sum_{k=1}^{2} p^k \eta_k + S_k^{ij} x_i y_j \overline{\eta}_k, \tag{18}$$

where $\eta_k, \overline{\eta}_k$ satisfy a Clifford algebra. We can associate a GLSM B-brane $\mathscr{B}$ to this which has the brane factor

$$f_{\mathscr{B}} = (1 - e^{2\pi(\sigma_1+\sigma_2)})^2. \tag{19}$$

Inserting this into (17), we find

$$Z_{D^2}^{\zeta \gg 0}(\mathscr{B}) = 2C\varpi_0, \tag{20}$$

where[4]

---

[3] We have fixed the R-charge ambiguity to match with the $\zeta \gg 0$ phase.

[4] Recall that $\varphi = e^{-t}$.

$$\begin{aligned}\varpi_0 =& \sum_{k_1,k_2=0}^{\infty} \frac{\Gamma(1+k_1+k_2)^2}{\Gamma(1+k_1)^2\Gamma(1+k_2)^2}\varphi^{k_1+k_2} \\ =& 1+2\varphi+6\varphi^2+20\varphi^3+70\varphi^4+252\varphi^5+O\left(\varphi^6\right) \\ =& \frac{1}{\sqrt{1-4\varphi}}.\end{aligned} \tag{21}$$

This is annihilated by the Picard-Fuchs operator

$$\mathscr{L} = (1-4\varphi)\theta - 2\varphi = \theta - 2\varphi(2\theta+1), \qquad \theta = \varphi\frac{\mathrm{d}}{\mathrm{d}\varphi}. \tag{22}$$

The Riemann symbol for $\mathscr{L}$ is

$$\left\{\begin{matrix} 0 & \frac{1}{4} & \infty \\ 0 & -\frac{1}{2} & \frac{1}{2} \end{matrix}\right\}. \tag{23}$$

The period $\varpi_0(\varphi)$ is the same as for the mirror of $\mathbb{P}^1[2]$ (see for instance [7, §7]).

We can also obtain ϖ_0 by evaluating the hemisphere partition function for the associated $U(1)$ GLSM, whose $\zeta \gg 0$ phase is $\mathbb{P}^1[2]$, which is easily adapted from the quintic case discussed in [17, 20]. This returns $\varpi_0 = \sum_{m\geq 0}\frac{(2m)!}{m!^2}\varphi^m = \frac{1}{\sqrt{1-4\varphi}}$. Distinguishing two points from one point[5] hence boils down to the choice of the normalisation constant C. We propose that the correct choice for this model is $C=\frac{1}{2}$, where we interpret the factor 2 in the denominator as the rank of the Weyl group, consistent with [20]. In contrast, for the $\mathbb{P}^1[2]$-model we should choose the normalisation in such a way that the central charge of the structure sheaf is $2\varpi_0$.

Finally, we can consider the $\zeta \ll 0$ phase where we find further issues related to the non-regularity of the model. A prescription for evaluating the sphere partition function in the $U(2)$ GLSM describing the Rødland model [35, 21] has been given in [30]. We apply this to our model and take the poles

$$1-\mathrm{i}\sigma_1-\mathrm{i}\sigma_2 = -n_p, \qquad \mathrm{i}\sigma_1 = -n_1, \qquad n_1,n_p\in\mathbb{Z}_{\geq 0}. \tag{24}$$

Then the integral can be written as

$$\begin{aligned}Z_{D^2}^{\zeta\ll 0} = & -\frac{C}{(2\pi\mathrm{i})^2}\sum_{n_1,n_p=0}^{\infty}\oint d\varepsilon_1 d\varepsilon_p \frac{\Gamma(1+n_1+n_p-\varepsilon_1-\varepsilon_p)^2}{\Gamma(1+n_1-\varepsilon_1)^2\Gamma(1+n_p-\varepsilon_p)^2} \\ & \cdot\frac{\pi^4}{\sin^2\pi(-n_1+\varepsilon_1)\sin^2\pi(-n_p-\varepsilon_p)}e^{t(1+n_p-\varepsilon_p)} \\ & \cdot f_{\mathscr{B}}(\mathrm{i}(n_1-\varepsilon_1),-i(1+n_1+n_p-\varepsilon_1-\varepsilon_p)).\end{aligned} \tag{25}$$

[5] Note that there is also a $U(1)\times\mathbb{Z}_2$ model to describe a point, with $\zeta \gg 0$ phase $\mathbb{P}^1[2]/\mathbb{Z}_2$, but that is not the object of discussion here.

The sum in n_1 does not converge for any brane factor[6]. In contrast to the odd-dimensional case, the sin-term gives a $(-1)^{2n_1} = 1$ rather than alternating signs $(-1)^{n_1}$. Hence, the convergence behaviour is worse compared to the threefold and elliptic curve case, but also occurs in the two-dimensional case which is also non-regular. We should not interpret this as the result being divergent. Rather, it tells us that the integration contour we have chosen is not valid for non-regular models. It would be interesting to show if a contour exists that gives a convergent result.

2.4 Sphere partition function

The GLSM can also be placed on a sphere. As shown in [30], the partition function of this theory, which is not topological, can be used to obtain period functions and genus-0 invariants of Calabi-Yau threefolds. Studying this partition function for our model will allow us to further probe the non-regular $\zeta \ll 0$ phase, and gives a complementary view on divergences that arise.

After assigning $U(1)_R$ charges $2q$ to the x_i, y_i fields and $2-4q$ to the p_i fields, we can compute the sphere partition function for our model following [30]. With the GLSM data for our model understood, this takes the form[7]

$$Z_{S^2} = \frac{1}{2}\sum_{m_i\in\mathbb{Z}} \mathrm{e}^{-\mathrm{i}\theta(m_1+m_2)} \int_{-\infty}^{\infty}\int_{-\infty}^{\infty} \frac{\mathrm{d}\sigma_1 \mathrm{d}\sigma_2}{(2\pi)^2} \mathrm{e}^{-2\mathrm{i}\zeta(\sigma_1+\sigma_2)} Z_{\text{matter}}(\sigma_i, m_i) Z_{\text{gauge}}(\sigma_i, m_i), \tag{26}$$

where $Z_{\text{matter}} = \frac{\Gamma(q-\mathrm{i}\sigma_1-\frac{m_1}{2})^2}{\Gamma(1-q+\mathrm{i}\sigma_1-\frac{m_1}{2})^2}\frac{\Gamma(q-\mathrm{i}\sigma_2-\frac{m_2}{2})^2}{\Gamma(1-q+\mathrm{i}\sigma_2-\frac{m_2}{2})^2}$ and $Z_{\text{gauge}} = \frac{\Gamma(1-2q+\mathrm{i}\sigma_1+\mathrm{i}\sigma_2+\frac{m_1+m_2}{2})^2}{\Gamma(2q-\mathrm{i}\sigma_1-\mathrm{i}\sigma_2+\frac{m_1+m_2}{2})^2}$ are prescribed by the 2d $\mathcal{N}=(2,2)$ supersymmetric localisation results of [3, 11]. The prefactor of $\frac{1}{2}$ is taken because our Weyl group is $\mathbb{Z}_2$.

To evaluate Z_{S^2} in the $\zeta \gg 0$ phase we close both σ_i contours in their lower half planes. The integral is performed by residues, with contributing poles at $\sigma_i = -\mathrm{i}q + \mathrm{i}(\frac{m_i}{2} - k_i)$, with $k_i \in \mathbb{Z}_{\geq 0}$. Following [30], we compute

$$\begin{aligned} Z_{S^2}^{\zeta\gg0} &= \frac{(\varphi\overline{\varphi})^{2q}}{2}\oint \frac{\mathrm{d}\varepsilon_1\mathrm{d}\varepsilon_2}{(2\pi\mathrm{i})^2}(\varphi\overline{\varphi})^{-\varepsilon_1-\varepsilon_2}\left(\frac{\pi\sin(\pi(\varepsilon_1+\varepsilon_2))}{\sin(\pi\varepsilon_1)\sin(\pi\varepsilon_2)}\right)^2 \\ &\qquad \left|\sum_{n_i\geq0}\frac{\Gamma(1+n_1+n_2-\varepsilon_1-\varepsilon_2)^2}{\Gamma(1+n_1-\varepsilon_1)^2\Gamma(1+n_2-\varepsilon_2)^2}\varphi^{n_1+n_2}\right|^2 \\ &= (\varphi\overline{\varphi})^{2q}\varpi_0(\varphi)\overline{\varpi_0(\varphi)}, \end{aligned} \tag{27}$$

where, as in [30], the complex conjugation does not act of the variables ε_i, and the contours in ε_i-space are small circles around 0. We reiterate that $\varphi = \mathrm{e}^{-t} = \mathrm{e}^{-\zeta+\mathrm{i}\theta}$.

[6] The dependence on n_1, n_p comes from $e^{2\pi\rho(\sigma)}$ which produces terms of the form $e^{2\pi(a\sigma_1+b\sigma_2)}$ for $a,b\in\mathbb{Z}$. Inserting (24), the n_1,n_p-dependence will always drop out.

[7] Note $\zeta_{\text{here}} = \frac{1}{2\pi}r_{\text{there}}$.

In the phase $\zeta \ll 0$, we close both σ_i contours in the upper half plane, with poles at $\sigma_1+\sigma_2=-2\mathrm{i}q+\mathrm{i}(\frac{m_1+m_2}{2})+k_p$, $\sigma_1=-\mathrm{i}q+\mathrm{i}(1+\frac{m_1}{2}+k_1+k_p)$. We obtain

$$\begin{aligned} Z_{S^2}^{\zeta\gg 0} &= \frac{(\varphi\overline{\varphi})^{2q-1}}{2}\oint \frac{\mathrm{d}\varepsilon_1\mathrm{d}\varepsilon_p}{(2\pi\mathrm{i})^2}(\varphi\overline{\varphi})^{-\varepsilon_p}\left(\frac{\pi\sin(\pi(\varepsilon_1+\varepsilon_p))}{\sin(\pi\varepsilon_1)\sin(\pi\varepsilon_p)}\right)^2 \\ &\qquad\qquad \left|\sum_{n_i\geq 0}\frac{\Gamma(1+n_1+n_p+\varepsilon_1+\varepsilon_p)^2}{\Gamma(1+n_1+\varepsilon_1)^2\Gamma(1+n_p+\varepsilon_p)^2}\varphi^{-n_p}\right|^2 \\ &= (\varphi\overline{\varphi})^{2q-1}\left|\sum_{n_i\geq 0}\frac{\Gamma(1+n_1+n_p)^2}{\Gamma(1+n_1)^2\Gamma(1+n_p)^2}\varphi^{-n_p}\right|^2. \end{aligned} \tag{28}$$

We have taken residues to arrive at the final expression above, however, the series is divergent. Unlike in the $\zeta \gg 0$ phase and consistent with the hemisphere partition function, the sum over n_1 at each fixed n_p is divergent for all φ.

This is reminiscent of the situation concerning the Pfaffian phase of the Rødland model as studied by [30]. There, a convergence factor $\mathrm{e}^{-\delta n_1}$ was incorporated in the sum, so ensuring convergence of the sum for positive δ, and taking the limit $\delta \to 0^+$. However, for this to work one relies on the presence of a $(-1)^{n_1}$ in their summand. There is no such $(-1)^{n_1}$ in our summand. As already observed for the hemisphere, more severe divergences seem to be a feature of non-regular models.

Let us consider more closely the problem of regulating the sum

$$S(\varphi)=\sum_{n_1,n_p\geq 0}\frac{\Gamma(1+n_1+n_p)^2}{\Gamma(1+n_1)^2\Gamma(1+n_p)^2}\varphi^{-n_p}. \tag{29}$$

If we rotate the σ_1-contour counter-clockwise by a small angle δ, so include a convergence factor $\mathrm{e}^{\mathrm{i}\delta\sigma_1}$ in (26), then we instead obtain

$$\begin{aligned} S_\delta(\varphi) &= \sum_{n_1,n_p\geq 0}\frac{\Gamma(1+n_1+n_p)^2}{\Gamma(1+n_1)^2\Gamma(1+n_p)^2}\mathrm{e}^{-\delta n_1}\varphi^{-n_p} \\ &= \sum_{n_p\geq 0} {}_2F_1(1+n_p,1+n_p;1;\mathrm{e}^{-\delta})\varphi^{-n_p}. \end{aligned} \tag{30}$$

The hypergeometric functions account for the divergences we described. Typically one can analytically continue a series ${}_{p+1}F_p$ to argument -1, which is why the models with such $(-1)^{n_1}$ factors in their summands are not problematic. See for instance the expression in [25, eq. 2-5.4], which could also be obtained by a GLSM computation like the ones we are describing. But an argument of 1 in the hypergeometric function is more problematic: there can be a pole. Two useful identities are ${}_2F_1(a,b;c;z)=(1-z)^{c-a-b}{}_2F_1(c-a,c-b;c;z)$ for $|\mathrm{Arg}(1-z)|<\pi$, and ${}_2F_1(a,b;c;1)=\frac{\Gamma(c)\Gamma(c-a-b)}{\Gamma(c-a)\Gamma(c-b)}$ for $\mathrm{Re}(c-a-b)>0$ [9, §15]. These imply

$$ {}_2F_1(1+n_p,1+n_p;1;\mathrm{e}^{-\delta}) = \frac{(2n_p)!}{n_p!^2}\frac{1}{\delta^{2n_p+1}}\left(1+O(\delta)\right). \tag{31} $$

Based on this equation, we will attempt a rescaling. First we multiply S_δ by δ, and then we make a rescaling of our algebraic coordinate, $\varphi = \delta^{-2}\widetilde{\varphi}$. Only then do we take $\delta \to 0^+$, finding

$$ \begin{aligned} \delta S_\delta(\delta^{-2}\widetilde{\varphi})\Big|_{\delta\to 0^+} &= \sum_{n_p\geq 0}\left[\delta^{1+2n_p}\,{}_2F_1(1+n_p,1+n_p;1;\mathrm{e}^{-\delta})\right]_{\delta\to 0^+}\widetilde{\varphi}^{-n_p} \\ &= \sum_{n_p\geq 0}\frac{(2n_p)!}{n_p!^2}\widetilde{\varphi}^{-n_p} = \frac{1}{\sqrt{1-4\widetilde{\varphi}^{-1}}} = \varpi_0(\widetilde{\varphi}^{-1}), \end{aligned} \tag{32} $$

where we employ (31). The appearance of the period function from the $\zeta \gg 0$ phase is reassuring, although we do not have a full interpretation of this manipulation. It would be interesting to better understand the scaling $\varphi = \delta^{-2}\widetilde{\varphi}$, which as a modification of the FI-parameter ζ takes the form

$$ \zeta = \widetilde{\zeta} + 2\log(\delta). \tag{33} $$

Perhaps a better way to obtain the period in the $\zeta \ll 0$ phase is to forego the sphere partition function, and instead obtain an expansion about $\phi = -\frac{1}{16\varphi^2} = 0$ by using the closed form $\varpi_0(\varphi) = \frac{1}{\sqrt{1-4\varphi}}$. This gives

$$ \begin{aligned} \varpi_0(\phi) &= \frac{2\phi}{\sqrt{1+4\phi^2}} = 2\phi\sum_{k=0}^{\infty}\frac{(2k)!}{k!^2}(-1)^k\phi^{2k} \\ &= 2\phi(1-2\phi^2+6\phi^4-20\phi^6+70\phi^8-252\phi^{10}+924\phi^{12}+\ldots). \end{aligned} \tag{34} $$

The number $-1/16$ in our change of variables is chosen such that we get a series with integral coefficients. Our exponent -2 in $\phi = -2^{-4}\varphi^{-2}$ is chosen because the Riemann symbol of $\mathscr{L}$ has an entry $1/2$ at the singularity ∞, see the parallel discussion in [31] concerning their change of variable. In the ϕ-plane, there are two singularities at $\phi = \pm\frac{i}{2}$ equidistant from the singularity $\phi = 0$ about which we are expanding. This seems to be a common property of noncommutative resolution phases of Calabi-Yau moduli spaces, but further study is needed to relate such structures to our model.

2.5 Mirror Symmetry

Here we describe the mirror construction using toric methods. The previous Coulomb branch computation in the GLSM provides the discriminant of the mirror variety. Although the mirror variety is a single point that does not obviously degenerate in any way, we are able to relate the discriminant to the failure of a quotient action to be fixed-point free.

We have described the $\zeta > 0$ geometry X as a $\mathbb{Z}_2$ quotient of the variety

$$\widetilde{X} \;\cong\; \begin{matrix}\mathbb{P}^1\\ \mathbb{P}^1\end{matrix}\begin{bmatrix}1 & 1\\ 1 & 1\end{bmatrix}, \tag{35}$$

where we use the CICY notation [6, 14]. We can use the procedure described in [23, §3] to obtain the mirror variety $\widetilde{M}$ of $\widetilde{X}$ as the locus

$$\widetilde{M}: \qquad U+V=1, \qquad \frac{\varphi_1}{U}+\frac{\varphi_2}{V}=1 \tag{36}$$

in the torus $(\mathbb{C}^*)^2$ with coordinates U,V. This is the zero-dimensional analogue of the threefold families studied by Hulek and Verrill [29]. This mirror variety $\widetilde{M}$ is also two points, given by

$$(U,V)=\left(\frac{1+\varphi_1-\varphi_2\pm\sqrt{\Delta}}{2},\frac{1-\varphi_1+\varphi_2\mp\sqrt{\Delta}}{2}\right). \tag{37}$$

The discriminant is

$$\Delta(\varphi_1,\varphi_2) \;=\; 1-2\varphi_1-2\varphi_2+\varphi_1^2-2\varphi_1\varphi_2+\varphi_2^2 \;=\; \prod_{\varepsilon_i=\pm1}(1+\varepsilon_1\sqrt{\varphi_1}+\varepsilon_2\sqrt{\varphi_2}), \tag{38}$$

which vanishes precisely when $\widetilde{M}$ degenerates into a single point. The fundamental period of $\widetilde{M}$ is

$$\varpi_0(\varphi_1,\varphi_2)=\sum_{m_i\geq 0}\left(\frac{(m_1+m_2)!}{m_1!\;m_2!}\right)^2\varphi_1^{m_1}\varphi_2^{m_2}. \tag{39}$$

The infinite series is provided by the methods of [23]. One can recognise the following two closed forms, using an Appell series F_4 and integrals of products of Bessel functions[8] (for $\mathrm{Re}[\sqrt{\varphi_1}]+\sqrt{\varphi_2}]<1$, so that the integral converges).

$$\varpi_0(\varphi_1,\varphi_2)=F_4(1,1;1,1;\varphi_1,\varphi_2)=\int_0^\infty u\,K_0(u)I_0(u\sqrt{\varphi_1})I_0(u\sqrt{\varphi_2})\,\mathrm{d}u. \tag{40}$$

The mirror M of X is then obtained as the $\mathbb{Z}_2$ quotient of $\widetilde{M}$ by the action $U\leftrightarrow V$, after setting $\varphi_1=\varphi_2=\varphi$. This action is fixed-point free provided $\Delta(\varphi,\varphi)=1-4\varphi$ does not vanish. The fundamental period of M is then

$$\varpi_0(\varphi)=\varpi_0(\varphi,\varphi)=\frac{1}{\sqrt{1-4\varphi}}, \tag{41}$$

[8] Also known as a Bessel moment.

in agreement with the GLSM result (21). Note that the pole of ϖ_0 at $\varphi = 1/4$, where $\Delta = 0$, was explained in the GLSM analysis of §2.2 as the Coulomb branch locus. For threefolds, such Coulomb branch loci are mirrors to conifold points (or their quotient generalisation, hyperconifold points [8]), where three-cycles on the mirror manifold collapse to nodal singularities. In superstring compactifications, branes wrapping these cycles give rise to massless hypermultiplets. The relation between massless hypermultiplets in string compactifications and Coulomb branches in the GLSM was identified in [19, §6.1] . For our present example, the Coulomb branch singularity has the mirror interpretation as the locus in φ-space where the quotient group is not freely acting. From (37), one sees that for $\varphi_1 = \varphi_2 = \varphi = 1/4$ the cover $\widetilde{M}$ is not two points, but instead is a single point $(U,V) = (\frac{1}{2}, \frac{1}{2})$ that is mapped to itself by the $\mathbb{Z}_2$ quotient action $U \leftrightarrow V$. In the GLSM analysis, the Coulomb branch in σ_i-space for $\mathrm{e}^{-t} = \frac{1}{4}$ is similarly fixed by the $\mathbb{Z}_2$ Weyl group action.

3 Where to go from here?

The model we have discussed here, and non-regular GLSMs in general, leave many open questions and possibilities for further research.

- **Witten index.** One expects that phases of GLSMs have matching Witten indices. At $\zeta \gg 0$, the vacuum of the GLSM is a single point, while at $\zeta \ll 0$ we find two points from the Higgs branch and a further point from the Coulomb branch. For non-Calabi-Yau GLSMs, where there can be additional massive Coulomb vacua in addition to Higgs vacua, the Witten index coming from the Higgs branch has to be corrected by adding the number of Coulomb vacua. If we did this for our model we would find a mismatch of 1 vs. 3. We could achieve a matching if the extra Coulomb branch contributes with a negative sign, but we have no explanation why this should be the case. Computing the GLSM elliptic genus [4, 5] or finding a way to compute the Witten index directly in the $\zeta \ll 0$ phase could help clarify this issue.
- **D-brane transport and categorical equivalences.** The B-brane categories of phases of GLSMs are expected to be equivalent. This is understood for abelian GLSMs [17] and for some regular non-abelian GLSMs, including those for the Rødland and Hosono-Takagi models [10, 13, 12]. It would be interesting to generalise these results to non-regular GLSMs. A recent result [16] makes a proposal for a prescription for D-brane transport in a non-regular model which provides valuable guidance.
  Statements on categorical equivalences on our class of models, including non-regular ones, can be found in the mathematics literature [28, 26, 34]. For non-regular models, the statements of categorical equivalences include adding further exceptional collections, similar to what happens in the non-Calabi-Yau case where the exceptional collections can be related to the massive Coulomb vacua. However, in the case of non-regular models like ours, the mathematics result imply that the category associated to the $\zeta \gg 0$ phase would have to be modified

and not the category associated to the $\zeta \ll 0$ phase, where our Coulomb branch appears. While we do not know how to understand this in the context of GLSMs, this may also give hints on what the accurate way to compute Witten indices for non-regular models is. Similar statements have also been made in the mathematics literature for non-regular models associated to Grassmannians [33].

- **Non-abelian duality** A duality for non-abelian GLSMs has been found in [19]. While non-regular models have been excluded from the discussion, our model is covered by the construction of the dual theories. In particular, the statement is that the non-abelian dual of a Hosono-Takagi-type GLSM with gauge group[9] $O(k)_+$ with N fundamentals is $SO(N-k+1)$. For our case of $N=k=2$, the dual non-abelian factor would be trivial and we would expect an abelian dual. It would be interesting to analyse this further and to find out if the duality requires modifications for non-regular GLSMs. There are also other interesting abelian models describing a single point which do not necessarily come from a non-abelian duality. See below for some examples.
- **Non-abelian mirrors** Mirrors for non-abelian GLSMs have been proposed in [22, 15]. It would be interesting to see if these constructions also work for non-regular GLSMs.
- **Divergences** We have not provided a physical interpretation of (33). In this work we have relied on analytic properties of ${}_2F_1$ that were not available to us in [32]. It remains to better understand the divergences of the series provided by GLSM techniques in non-regular models.
- **Other realisations of two points** We have already mentioned the $U(1)$ model describing $\mathbb{P}^1[2]$, studied in [19]. This has three chiral superfields (p, x_0, x_1) with $U(1)$ charges $(-2,1,1)$. At negative FI-parameter the $U(1)$ model flows to a Landau-Ginzburg orbifold.
  Another way to realise two points is through a $U(2)$ GLSM with 4 chiral superfields x_i that transform in the fundamental representation and four chiral superfields p_i transforming in the $\det^{-1}$ representation. Phase analysis at $\zeta > 0$ reveals an intersection of four hyperplanes in $\mathrm{Gr}(2,4)$. This model is non-regular. The fundamental period of the intersection $\mathrm{Gr}(2,4)[1,1,1,1]$ can be obtained by using the "trick with the factorials" from [2], recovering $\varpi_0(\varphi) = \frac{1}{\sqrt{1-\varphi}}$.
- **Other realisations of one point** Motivated by the application in [21] of a simpler dual abelian description to study an non-regular (threefold) model, we describe some other ways that a GLSM can realise a single point, which may help study our main model.
  The $U(1)$ model describing $\mathbb{P}^1[2]$ can be supplemented by an additional gauging, for suitable superpotential. We can replace the gauge group by $U(1) \times \mathbb{Z}_2$, where the $\mathbb{Z}_2$ acts trivially on the single p-field and exchanges $x_1 \leftrightarrow x_2$, whereupon the $\zeta \gg 0$ phase becomes a single point. We expect in this $U(1) \times \mathbb{Z}_2$ model to find a further Landau-Ginzburg orbifold in the $\zeta < 0$ phase.
  Furthermore, we have discussed in the main body of our article the $(U(1) \times U(1)) \rtimes \mathbb{Z}_2$ model with $U(1) \times U(1)$ charge matrix given in (1). One could

[9] We refer to the original paper for the difference between $O(k)_+$ and $O(k)_-$.

consider the same charge matrix, with two unequal FI parameters ζ_1, ζ_2 for an abelian model with gauge group $U(1)^2 \times \mathbb{Z}_2$. Here the $\mathbb{Z}_2$ action has generator $p_1 \leftrightarrow p_2$, $x_1 \leftrightarrow x_2$, $y_1 \leftrightarrow y_2$. For suitable superpotential the vacuum manifold at $\zeta_1 > 0, \zeta_2 > 0$ is a freely acting quotient of two points, again giving a point.

Acknowledgements We thank the participants of the MATRIX program "The geometry of moduli spaces in string theory" for creating a wonderful atmosphere that sparked many interesting discussions, including one on Calabi-Yau zerofolds. We thank Kentaro Hori, Shinobu Hosono, and Hiromichi Takagi for patiently explaining their work to us. We also would like to congratulate MATRIX Institute on the occasion of its tenth anniversary and wish many happy returns. Finally, thanks to the editors of the proceedings volume, Jock McOrist and Sébastien Picard, for accepting our contribution at the very last minute. JK is supported by the Australian Research Council Future Fellowship FT210100514. JM is supported by a University of Melbourne Establishment Grant.

References

1. Aspinwall, P.S., Plesser, M.R.: Decompactifications and Massless D-Branes in Hybrid Models. JHEP **07**, 078 (2010). DOI 10.1007/JHEP07(2010)078
2. Batyrev, V.V., Ciocan-Fontanine, I., Kim, B., van Straten, D.: Conifold transitions and mirror symmetry for Calabi-Yau complete intersections in Grassmannians. Nucl. Phys. B **514**, 640–666 (1998). DOI 10.1016/S0550-3213(98)00020-0
3. Benini, F., Cremonesi, S.: Partition Functions of $\mathcal{N} = (2,2)$ Gauge Theories on S^2 and Vortices. Commun. Math. Phys. **334**(3), 1483–1527 (2015). DOI 10.1007/s00220-014-2112-z
4. Benini, F., Eager, R., Hori, K., Tachikawa, Y.: Elliptic genera of two-dimensional N=2 gauge theories with rank-one gauge groups. Lett. Math. Phys. **104**, 465–493 (2014). DOI 10.1007/s11005-013-0673-y
5. Benini, F., Eager, R., Hori, K., Tachikawa, Y.: Elliptic Genera of 2d $\mathcal{N} = 2$ Gauge Theories. Commun. Math. Phys. **333**(3), 1241–1286 (2015). DOI 10.1007/s00220-014-2210-y
6. Candelas, P., Dale, A.M., Lutken, C.A., Schimmrigk, R.: Complete Intersection Calabi-Yau Manifolds. Nucl. Phys. B **298**, 493 (1988). DOI 10.1016/0550-3213(88)90352-5
7. Candelas, P., de la Ossa, X., Rodriguez-Villegas, F.: Calabi-Yau manifolds over finite fields. 1. (2000). DOI https://doi.org/10.48550/arXiv.hep-th/0012233
8. Davies, R.: Classification and Properties of Hyperconifold Singularities and Transitions (2013). DOI https://doi.org/10.48550/arXiv.1309.6778
9. *NIST Digital Library of Mathematical Functions*. `https://dlmf.nist.gov/`, Release 1.2.4 of 2025-03-15. F. W. J. Olver, A. B. Olde Daalhuis, D. W. Lozier, B. I. Schneider, R. F. Boisvert, C. W. Clark, B. R. Miller, B. V. Saunders, H. S. Cohl, and M. A. McClain, eds.
10. Donovan, W., Segal, E.: Window shifts, flop equivalences and Grassmannian twists. Compos. Math. **150**(6), 942–978 (2014). DOI 10.1112/S0010437X13007641
11. Doroud, N., Gomis, J., Le Floch, B., Lee, S.: Exact Results in D=2 Supersymmetric Gauge Theories. JHEP **05**, 093 (2013). DOI 10.1007/JHEP05(2013)093
12. Eager, R., Hori, K., Knapp, J., Romo, M.: in progress
13. Eager, R., Hori, K., Knapp, J., Romo, M.: Beijing lectures on the grade restriction rule. Chinese Ann. Math. Ser. B **38**(4), 901–912 (2017). DOI 10.1007/s11401-017-1103-8
14. Green, P., Hubsch, T.: Calabi-yau Manifolds as Complete Intersections in Products of Complex Projective Spaces. Commun. Math. Phys. **109**, 99 (1987). DOI 10.1007/BF01205673
15. Gu, W., Sharpe, E.: A proposal for nonabelian mirrors (2018). DOI https://doi.org/10.48550/arXiv.1806.04678
16. Guo, J., Romo, M., Smith, L.: B-brane transport in nonabelian GLSMs for $K_{Gr(2,N)}$ (2025)

17. Herbst, M., Hori, K., Page, D.: Phases Of N=2 Theories In 1+1 Dimensions With Boundary (2008)
18. Honda, D., Okuda, T.: Exact results for boundaries and domain walls in 2d supersymmetric theories. JHEP **09**, 140 (2015). DOI 10.1007/JHEP09(2015)140
19. Hori, K.: Duality In Two-Dimensional (2,2) Supersymmetric Non-Abelian Gauge Theories. JHEP **10**, 121 (2013). DOI 10.1007/JHEP10(2013)121
20. Hori, K., Romo, M.: Exact Results In Two-Dimensional (2,2) Supersymmetric Gauge Theories With Boundary (2013). DOI https://doi.org/10.48550/arXiv.1308.2438
21. Hori, K., Tong, D.: Aspects of Non-Abelian Gauge Dynamics in Two-Dimensional N=(2,2) Theories. JHEP **05**, 079 (2007). DOI 10.1088/1126-6708/2007/05/079
22. Hori, K., Vafa, C.: Mirror symmetry (2000). DOI https://doi.org/10.48550/arXiv.hep-th/0002222
23. Hosono, S., Klemm, A., Theisen, S., Yau, S.T.: Mirror symmetry, mirror map and applications to complete intersection Calabi-Yau spaces. Nucl. Phys. B **433**, 501–554 (1995). DOI 10.1016/0550-3213(94)00440-P
24. Hosono, S., Takagi, H.: Determinantal Quintics and Mirror Symmetry of Reye Congruences. Commun. Math. Phys. **329**, 1171–1218 (2014). DOI 10.1007/s00220-014-1971-7
25. Hosono, S., Takagi, H.: Mirror symmetry and projective geometry of Reye congruences I. J. Alg. Geom. **23**(2), 279–312 (2014). DOI 10.1090/S1056-3911-2013-00618-9
26. Hosono, S., Takagi, H.: Geometry of symmetric determinantal loci (2015). DOI https://doi.org/10.48550/arXiv.1508.01995
27. Hosono, S., Takagi, H.: Double quintic symmetroids, Reye congruences, and their derived equivalence. Journal of Differential Geometry **104**(3), 443 – 497 (2016). DOI 10.4310/jdg/1478138549
28. Hosono, S., Takagi, H.: Derived categories of Artin-Mumford double solids. Kyoto J. Math. **60**(1), 107–177 (2020). DOI 10.1215/21562261-2019-0036
29. Hulek, K., Verrill, H.: On modularity of rigid and nonrigid Calabi-Yau varieties associated to the root lattice A4. Nagoya Mathematical Journal **179**, 103–146 (2005)
30. Jockers, H., Kumar, V., Lapan, J.M., Morrison, D.R., Romo, M.: Two-Sphere Partition Functions and Gromov-Witten Invariants. Commun. Math. Phys. **325**, 1139–1170 (2014). DOI 10.1007/s00220-013-1874-z
31. Katz, S., Klemm, A., Schimannek, T., Sharpe, E.: Topological Strings on Non-commutative Resolutions. Commun. Math. Phys. **405**(3), 62 (2024). DOI 10.1007/s00220-023-04896-2
32. Knapp, J., McGovern, J.: Noncommutative resolutions and CICY quotients from a non-abelian GLSM (2025)
33. Kuznetsov, A.: Homological projective duality for Grassmannians of lines (2006). DOI https://doi.org/10.48550/arXiv.math/0610957
34. Rennemo, J.V.: The homological projective dual of $\mathrm{sym}^2\mathbb{P}(v)$. Compositio Mathematica **156**(3), 476 (2020)
35. Rødland, E.A.: The Pfaffian Calabi–Yau, its mirror, and their link to the Grassmannian G (2, 7). Compositio Mathematica **122**(2), 135–149 (2000)
36. Schimannek, T.: Modular curves, the Tate-Shafarevich group and Gopakumar-Vafa invariants with discrete charges. JHEP **02**, 007 (2022). DOI 10.1007/JHEP02(2022)007
37. Sugishita, S., Terashima, S.: Exact Results in Supersymmetric Field Theories on Manifolds with Boundaries. JHEP **11**, 021 (2013). DOI 10.1007/JHEP11(2013)021
38. Witten, E.: Phases of N=2 theories in two-dimensions. Nucl. Phys. B **403**, 159–222 (1993). DOI 10.1016/0550-3213(93)90033-L

Classically modular points in moduli spaces of Calabi-Yau threefolds

Mohamed Elmi, Joseph McGovern, and Emanuel Scheidegger

Abstract We discuss some aspects of the arithmetic modularity of Calabi-Yau threefolds which relate to constructions in supergravity, or superstring theory.

1 Introduction

A Calabi-Yau threefold X with complex structure modulus $\mathbf{z}$ can be viewed as the fibre $X_{\mathbf{z}}$ over the moduli space of complex structures on X. This moduli space can have dimension $h^{2,1}(X)$ greater than one, so that $\mathbf{z}$ is a vector of moduli space coordinates, but we might also consider the case of a one-parameter family X_z.

There can be interesting points within the moduli space of complex structures, where the fibre takes on certain properties or solves a certain problem.

One such set of points $\mathbf{z}_*$ are of interest in number theory. If X is a variety over a number field then, after reducing the equations that define X modulo p, the set of points on X in each finite field $\mathbb{F}_{p^r}$ can be counted. For each fixed p, the numbers $N_{p^r}(X_{\mathbf{z}})$ of these solutions can be collected in the exponential generating function

$$\zeta_p(X_{\mathbf{z}};T) = \exp\left(\sum_{r=1}^{\infty} \frac{1}{r} N_{p^r}(X_{\mathbf{z}}) T^r\right). \tag{1}$$

Mohamed Elmi
Yau Mathematical Sciences Center, Tsinghua University
e-mail: elmi@tsinghua.edu.cn

Joseph McGovern
School of Mathematics and Statistics, University of Melbourne
e-mail: mcgovernjv@gmail.com

Emanuel Scheidegger
Beijing International Center for Mathematical Research, Peking University
e-mail: esche@bicmr.pku.edu.cn

D. R. Wood et al. (eds.), *2024 MATRIX Annals, Part II*, MATRIX Book Series 8,
https://doi.org/10.1007/978-3-032-16206-9_24

This is the local zeta function, which remarkably is a rational function of T. This rationality property is one of the Weil conjectures [41], proved in the works [18, 27, 14],

$$\zeta_p(X_{\mathbf{z}};T) = \frac{R_p(\mathbf{z};T)}{(1-T)(1-pT)^{h^{1,1}}(1-p^2T)^{h^{1,1}}(1-p^3T)}. \tag{2}$$

Here we have specialised the more general formula to the case of a Calabi-Yau threefold with Hodge numbers $h^{1,1}, h^{2,1}$. The numerator $R_p(\mathbf{z};T)$ is a degree $2h^{2,1}+2$ polynomial in T, whose coefficients contain all dependence of ζ_p on the modulus $\mathbf{z}$. This procedure of point-counting only makes sense for varieties defined over number fields, which means in suitable coordinates that the $\mathbf{z}$ are assumed to take values in a number field. The points $\mathbf{z}_*$ that we wish to draw attention to are those such that for every prime p, $R_p(\mathbf{z}_*;T)$ factors over the number field $\mathbf{Q}(\mathbf{z}_*)$ with a quadratic factor. While for any $\mathbf{z}$ there may be finitely many p such that this factorisation occurs, we are specifically interested in cases where the factorisation occurs for all primes minus those in a certain "bad" finite set. This is said to be "persistent factorisation" of the zeta function.

The varieties $X_{\mathbf{z}}$, for $\mathbf{z}$ in a number field, are expected to encode Siegel modular forms in their zeta function numerators $R_p(\mathbf{z};T)$. This is an area of active research, see for instance [23], which moves in step with advances on Siegel modular forms. The special cases where the zeta function persistently factorises are of interest because the associated modular objects are much simpler, being classical modular forms. The persistently factorising $R_p(\mathbf{z};T)$ may contain factors

$$1-\alpha_p pT+p^3T^2, \qquad \text{or} \qquad 1-\beta_p T+p^3T^2. \tag{3}$$

These are explained as factors in the determinant of the Frobenius action on the middle cohomology of $X_{\mathbf{z}}$. They correspond to reducible blocks contained respectively in $H^{2,1}\oplus H^{1,2}$ or $H^{3,0}\oplus H^{0,3}$, where we use the Hodge filtration.

Through the Euler product formula, these factors are combined into a motivic L-function. Conjecturally this motivic L-function equals an automorphic L-function, which is the Mellin transform of a modular form. The α_p associated to a $H^{2,1}\oplus H^{1,2}$ block are the coefficients of q^p in the q-expansion of a weight-two newform. The β_p associated to a $H^{3,0}\oplus H^{0,3}$ block are the coefficients of q^p in the expansion of a weight-four newform. To summarise, there can be interesting $\mathbf{z}_*$ for which $X_{\mathbf{z}_*}$ are related to classical modular forms via point-counting over finite fields.

The connection to physics comes from the fact that certain problems in supergravity require that the complex structure parameter of a Calabi-Yau threefold (equivalently, the complexified Kähler parameter of a mirror Calabi-Yau threefold) is at a special point in moduli space where the Hodge structure degenerates in some way. We explain these supergravity problems below, and discuss their relation to concepts from arithmetic geometry.

1.1 The attractor mechanism

In a IIB supergravity compactification on X, the $\mathbf{z}$ take on a physical role as the $h^{2,1}$ vector multiplet scalars of the resulting matter-coupled 4d $\mathcal{N}=2$ supergravity.

Static, supersymmetric black hole solutions of the 4d theory have an attractor mechanism [21, 22]. Vector multiplet scalars take some boundary values at spatial infinity, and their dependence on the radial coordinate is then governed by the equations of motion. These take the form of a damped oscillator, and solutions have attractor behaviour. The $\mathbf{z}$ approach asymptotic values at the event horizon which extremise the absolute value of the black hole's central charge. If this central charge does not vanish, the limiting value can be given [20, 39] as the solution of

$$Q = \mathrm{Im}[C\Pi(\mathbf{z})], \tag{4}$$

where Q is the 4d black hole's electromagnetic charge vector and Π is the integral symplectic period vector of $X_{\mathbf{z}}$. The $h^{2,1}$ complex $\mathbf{z}$ and overall complex scale C furnish $2h^{2,1}+2$ real unknowns, which matches the number of real equations in (4). The functions $\mathbf{z}(Q)$ obtained by inverting (4) were analysed in [8].

A given charge vector Q can be interpreted as the periods of an integral 3-cohomology class on a Calabi-Yau threefold and the equation (4) has a solution at a regular point in moduli space precisely when the cohomology class corresponding to Q has vanishing $(2,1)$ and $(1,2)$ parts in its Hodge decomposition.

It may (very rarely) happen that there are two linearly independent charge vectors Q_1 and Q_2 that lead to the same attractor point. Such points are referred to as attractor points of rank two. In this case, there is a rank two sublattice $\Lambda \subset H^3(X,\mathbb{Z})$ with Hodge numbers $(3,0)+(0,3)$. Its symplectically dual lattice then has Hodge number $(2,1)+(1,2)$. When the underlying Calabi-Yau variety at an attractor point of rank two is defined over $\mathbb{Q}$, it has been proven that the Calabi-Yau variety is modular i.e. we expect persistent factorisations of the numerator of the zeta-function and each quadratic factor is determined by the Fourier coefficients of a modular form [24]. This leads to an effective strategy for finding attractor points of rank two [9].

One consequence of modularity is that, at an attractor point of rank two, one may (conjecturally) find exact expressions for the Calabi-Yau periods. These give physically meaningful quantities, such as central charges of D-branes. However, the ultimate role of modularity in Calabi-Yau compactifications is far from understood.

1.2 Supersymmetric flux vacua

We draw on [13, 16], and references therein. The tree-level spacetime superpotential of a 4d $\mathcal{N}=1$ supergravity obtained by reducing IIB supergravity on $X_{\mathbf{z}}$ reads

$$W(\tau,\mathbf{z}) = \int_X (F-\tau H)\wedge\Omega(\mathbf{z}) = (\mathrm{F}-\tau\mathrm{H})\,\Sigma\,\Pi(\mathbf{z}), \tag{5}$$

where τ is the axiodilaton and F, H are the fluxes background values given to the 3-form fields. Let the associated cohomology classes have components F, H in a symplectic basis of $H^3(X,\mathbb{Z})$.

From string theory, there is a requirement for objects of negative tension. These can be provided by O-planes, which necessitate orientifolding the theory. The superpotential remains (5), but only depends on moduli fields that survive the orientifolding. This issue is not important for our present modularity discussions. We will proceed with (5) and discuss implications for the solutions on the threefold, and not on its $\mathbb{Z}_2$ quotient obtained by the orientifolding.

The superpotential defines a potential function that couples all the scalar fields,

$$V(\tau, z_i, t^\alpha) = \mathrm{e}^K \left(G^{AB}(D_A W)(\overline{D_B W}) - 3|W|^2\right), \tag{6}$$

with K the Kähler potential and G_{AB} the metric on the total moduli space. $D_A = \partial_A + (\partial_A K)$ is the Kähler covariant derivative with respect to the field $A \in \{\tau, z_i, t^\alpha\}$. W has no dependence on the Kähler moduli t^α, i.e. $\partial_{t^\alpha} W = 0$, and therefore $D_{t^\alpha} W = (\partial_{t^\alpha} K) W$ vanishes when $W = 0$. The potential V must vanish in a vacuum.

If the vacuum is to preserve supersymmetry, then the superpotential W itself must also vanish. These two conditions $V = W = 0$ can be brought to the form

$$\mathrm{F}\Sigma\,\Pi(\mathbf{z}) = \mathrm{H}\Sigma\,\Pi(\mathbf{z}) = 0, \quad (\mathrm{F} - \tau\mathrm{H})\,\Sigma\,\partial_{z_i}\Pi(\mathbf{z}_i). \tag{7}$$

There is moreover a tadpole cancellation condition, requiring that

$$\mathrm{F}\Sigma\mathrm{H} \neq 0. \tag{8}$$

In particular this implies that the two flux vectors are not a multiple of one another.

Since F and H are integral cohomology classes, the above conditions imply that they span a sublattice of $H^3(X,\mathbb{Z})$ with Hodge numbers $(2,1)+(1,2)$. Just as was the case with attractor points of rank two, this piece of cohomology can be detected as a quadratic factor in the numerator of infinitely many local zeta functions. For Calabi-Yau threefolds with $h^{2,1} = 1$ a rank two attractor is equivalent to a flux vacuum with vanishing superpotential. For $h^{2,1} > 1$, the two are no longer equivalent.

A relation between supersymmetric flux vacua and modularity was conjectured in [34]. [33] makes a striking observation matching the F-theory and modular curves.

These SFV loci are also of use in constructing gauged supergravities, [30]. Here we solely discuss the physical incarnations of modular Calabi-Yau threefolds, for fourfolds see [31, 36].

2 Known 1-parameter examples of split Hodge structures

A seemingly sporadic set has been found by using the methods of [7] to compute the zeta function for a large number of primes and checking for persistent factorisation. We collect as many of these as we are aware of into Table 1. There is not currently a satisfactory explanation as to why any of these varieties are classically modular. We list the AESZ number [1], the special point z_*, the LMFDB labels [40] for the modular forms, and the field K over which the Hodge structure splits[1].

There are many examples of special points with a split Hodge structure that arise from Hadamard products with a symmetry that permutes the singularities at $z = 0$ and $z = \infty$. The special points are typically at apparent singularities, i.e. at points where all the periods and their derivatives are regular, but the Yukawa coupling vanishes. We list examples that are Hadamard squares in Table 1 and a single example is considered in §4, where the method is explained.

Most examples in Table 1 are attractor points of rank two, where the Hodge structure of the middle cohomology of the underlying CY threefold splits as

$$(1,1,1,1) \to (1,0,0,1) + (0,1,1,0) \tag{9}$$

over a real number field. We slightly abuse terminology here when we refer to splittings over non-trivial real number fields (e.g. $\mathbb{Q}(\sqrt{2})$) as rank two attractors. Taken seriously, this would give black holes/D3 branes with charges valued in a number field. The physical meaning here is unclear, but we include them for completeness.

We also point out that a "persistent factorisation" of the numerator of the zeta-function will still take place when the Hodge structure splits over a non-trivial number field for infinitely many primes. For example, in the case of $\mathbb{Q}(\sqrt{2})$, we expect a factorisation of the numerator of the zeta function when $x^2 - 2 = 0$ has a solution in the finite field $\mathbb{F}_p$ and the underlying variety is still smooth when reduced mod p.

There are also points where the Hodge structure splits over an imaginary number field. These cannot be rank two attractor points. Instead, the Hodge structure splits as

$$(1,1,1,1) \to (1,0,1,0) + (0,1,0,1) \tag{10}$$

at these points. Such points support flux vacua with non-vanishing superpotential.

3 Periods and quasiperiods

In [4] the periods of hypergeometric one–parameter families of Calabi–Yau threefolds at an attractor point were expressed in terms of the periods and quasiperiods of the associated modular forms. In this section, we apply this to the two families governed by the Picard–Fuchs operators AESZ17 and AESZ22, discussed in detail in [12]. As discussed in Section 2 at an attractor point the Hodge structure splits as

$$H^3(X_{z_*}, \mathbb{Q}) = \Lambda \otimes \mathbb{Q} \oplus \Lambda^\perp \otimes \mathbb{Q} \tag{11}$$

where $\Lambda \otimes \mathbb{Q}$ is a Hodge structure of type $(1,0,0,1)$ which corresponds to the Hodge structure of a rigid Calabi–Yau threefold, and $\Lambda^\perp \otimes \mathbb{Q}$ is a Hodge structure of type

[1] This is not the same as as the field of definition of z_*. For example, AESZ 34 has rank two attractor points defined over $\mathbb{Q}(\sqrt{17})$. Nevertheless the Hodge structure splits over $\mathbb{Q}$ at these points.

Table 1 Known examples of points z_* in one-parameter moduli spaces where a splitting occurs. We give the AESZ number [1] of the Picard-Fuchs operator, the field K over which the zeta numerator factors, LMFDB labels [40] for the weight-four and weight-two newforms f and g respectively in cases where $K=\mathbb{Q}$, and references that further discuss/discover these examples. $A(\cong B)$ indicates that the operators A and B are equivalent by a change of coordinates.

AESZ#	z_*	K	f	g	References
3	-2^{-8}	$\mathbb{Q}(i)$			
3	$(17\pm 12\sqrt{2})2^{-8}$	$\mathbb{Q}(\sqrt{2})$			
4	$-2^{-3}3^{-6}$	$\mathbb{Q}$	**54.4.a.c**	**54.2.a.a**	[4, 3]
11	$-2^{-4}3^{-3}$	$\mathbb{Q}$	**180.4.a.e**	**36.2.a.a**	[4, 3]
17 ($\cong$ 290)	-1	$\mathbb{Q}$	**14.4.a.b**	**14.2.a.a**	[38, 12]
22 ($\cong$ 118)	-1	$\mathbb{Q}$	**33.4.a.b**	**11.2.a.a**	[38, 12]
34	-7^{-1}	$\mathbb{Q}$	**14.4.a.a**	**14.2.a.a**	[9]
34	$33\pm 8\sqrt{17}$	$\mathbb{Q}$	**34.4.b.a**	**34.2.b.a**	[9]
36	-2^{-6}	$\mathbb{Q}$	**96.4.a.e**	**32.2.a.a**	[3]
49	$-2^{-4}5^{-1}$	$\mathbb{Q}$	**400.4.a.o**	**400.2.a.c**	[3]
55	2^{-4}	$\mathbb{Q}$	**60.4.a.a**	**20.2.a.a**	[3]
84	-2^{-4}	$\mathbb{Q}$	**20.4.a.a**	**20.2.a.a**	[3]
84	$2^{-4}3^{-1}$	$\mathbb{Q}$	**12.4.a.a**	**36.2.a.a**	[3]
100	-2^{-3}	$\mathbb{Q}$	**14.4.a.b**	**14.2.a.a**	[3]
101	1	$\mathbb{Q}$	**22.4.a.a**	**11.2.a.a**	[3]
103	-3^{-2}	$\mathbb{Q}$	**180.4.a.d**	**90.2.a.c**	[3]
107	-2^{-5}	$\mathbb{Q}$	**48.4.a.c**	**48.2.a.a**	[3]
111	$-2^{-8}3^{-1}$	$\mathbb{Q}$	**144.4.a.b**	**144.2.a.b**	[3]
115	-2^{-8}	$\mathbb{Q}$	**32.4.a.c**	**32.2.a.a**	[3]
144	$-2^{-3}3^{-2}$	$\mathbb{Q}$	**306.4.a.c**	**306.2.a.c**	[3]
145	-3^{-6}	$\mathbb{Q}$	**108.4.a.d**	**54.2.a.b**	[3]
154	$-2^{-4}3^{-3}$	$\mathbb{Q}$	**216.4.a.d**	**216.2.a.d**	[3]
155	-2^{-12}	$\mathbb{Q}$	**128.4.a.a**	**128.2.a.b**	[3]
165	-3^{-3}	$\mathbb{Q}$	**54.4.a.d**	**27.2.a.a**	[3]
166	$-2^{-8}3^{-6}$	$\mathbb{Q}$	**864.4.a.b**	**864.2.a.g**	[3]
204	-2^{-10}	$\mathbb{Q}(i)$			
204	-2^{-12}	$\mathbb{Q}(\sqrt{5})$			
204	-2^{-20}	$\mathbb{Q}(\sqrt{41})$			
238	-2^{-4}	$\mathbb{Q}$	**88.4.a.a**	**88.2.a.a**	[3]
277	-2^{-14}	$\mathbb{Q}$	**240.4.a.i**	**80.2.a.b**	[3]
2.32	$2^{-4}3^{-3}$	$\mathbb{Q}$	**324.4.a.a**	**324.2.a.d**	[3]

$(0,1,1,0)$ which (after tensoring by $\mathbb{Q}(1)$) corresponds to the Hodge structure of an elliptic curve. Conjecturally, it is expected that the periods of the modular Calabi-Yau manifold are related to the critical L–values of a modular form whose weight equals the weight of the Hodge structure plus 1. Hence, the standard conjectures imply that there are integers N_1, N_2 and newforms $f\in S_4(\Gamma_0(N_1)), g\in S_2(\Gamma_0(N_2))$ such that there is a factorization $L(X,s)=L(f,s)L(g,s)$. In other words, there is a basis for $H^3(X_{z_*},\mathbb{Q})$ and $H_3(X_{z_*},Q)$ such that period matrix takes the form

$$\mathrm{P} = \begin{pmatrix} \omega_f^+ & \omega_f^- & 0 & 0 \\ \eta_F^+ & \eta_F^- & 0 & 0 \\ 0 & 0 & \tilde{\omega}_g^+ & \tilde{\omega}_g^- \\ 0 & 0 & \tilde{\eta}_G^+ & \tilde{\eta}_G^- \end{pmatrix} \tag{12}$$

where $\omega_f^{\pm}$, $\eta_F^{\pm}$, $\tilde{\omega}_g^{\pm}$, $\tilde{\eta}_G^{\pm}$ are the periods of f and g and the quasiperiods of their meromorphic partners F and G, respectively. Some of the periods are related to critical values of the L–functions of f and g, respectively. For definitions and explanations of these terms see [4].

At the attractor z_* the CY period matrix $\mathrm{T}_{z_*} = (\Pi(z_*), \Pi'(z_*), \Pi''(z_*), \Pi'''(z_*))^T$ can be expressed in terms of the matrix P through $\mathrm{T}_{z_*}^T = \mathrm{APB}$, where A and B are matrices of rational numbers.

3.1 AESZ 17

Here $N_1 = N_2 = 14$ and we have the following modular forms:

$$\begin{aligned} f &= q + 2q^2 - 2q^3 + 4q^4 - 12q^5 - 4q^6 + 7q^7 + 8q^8 - 23q^9 + O(q^{10}) && \in S_4(\Gamma_0(14)), \\ F &= q^{-1} + 38q^2 + 199q^3 + 908q^4 + 2529q^5 + 7796q^6 + O(q^7) && \in S_4^!(\Gamma_0(14)), \\ g &= q - q^2 - 2q^3 + q^4 + 2q^6 + q^7 - q^8 + q^9 + O(q^{10}) && \in S_2(\Gamma_0(14)), \\ G &= q^{-1} + 47q^2 + 199q^3 + 329q^4 + 1023q^5 + 1874q^6 + O(q^7) && \in S_2^!(\Gamma_0(14)), \end{aligned} \tag{13}$$

where $f = f_{14.4.a.b}$, $g = f_{14.2.a.a}$ as in Table 1. The meromorphic modular forms F and G can succinctly be described as

$$\begin{aligned} F &= \frac{1}{h}\Big(\Big(\tfrac{1}{3} - \tfrac{37\sqrt{1969}}{11814}\Big) f_{14.8.a.c} + \Big(\tfrac{1}{3} + \tfrac{37\sqrt{1969}}{11814}\Big)\sigma(f_{14.8.a.c}) \\ &\quad + \tfrac{1}{3} f_{7.8.a.a} - \tfrac{16}{3} f_{7.8.a.a}|_8 V_2\Big) \\ G &= \frac{1}{h}\Big(\Big(-\tfrac{2}{43} - \tfrac{754\sqrt{57}}{2451}\Big) f_{7.6.a.b}|_6 V_2 + \Big(\tfrac{754\sqrt{57}}{2451} - \tfrac{2}{43}\Big)\sigma(f_{7.6.a.b})|_6 V_2 \\ &\quad + \Big(-\tfrac{377\sqrt{57}}{9804} - \tfrac{1}{172}\Big) f_{7.6.a.b} + \Big(-\tfrac{1}{172} + \tfrac{377\sqrt{57}}{9804}\Big)\sigma(f_{7.6.a.b}) + \tfrac{87}{86} f_{14.6.a.a}\Big) \\ h &= \eta(\tau)^2\eta(2\tau)^2\eta(7\tau)^2\eta(14\tau)^2 \end{aligned} \tag{14}$$

where $V_d = \left(\begin{smallmatrix} d & 0 \\ 0 & 1 \end{smallmatrix}\right)$ and σ denotes Galois conjugation. The indices give the label of the cusp form in the LMFDB. Their periods and quasiperiods are

$$
\begin{aligned}
\omega_f^+ &= 3.4463066043046096205\ldots, & \omega_f^- &= 1.0198537741597169074\ldots i,\\
\eta_F^+ &= -12.5614347465463248098\ldots, & \eta_F^- &= -0.7794803062263570521 7\ldots i,\\
\omega_g^+ &= -0.990670978033441617\ldots, & \omega_g^- &= -1.325491239682486714\ldots i,\\
\eta_G^+ &= 15.6872201080337674596 7\ldots, & \eta_G^- &= 8.3043739003453747552 79\ldots i,
\end{aligned} \tag{15}
$$

and $\tilde{\omega}_g^\pm = 2\pi i \tilde{\omega}_g^\pm$, $\tilde{\eta}_G^\pm = 2\pi i \eta_G^\pm$. The periods $\omega_f^\pm$, $\omega_g^\pm$ are related to special values of the L–functions of f and g and their twists by

$$
\begin{aligned}
\omega_f^+ &= -\frac{3}{28}(2\pi i)^2 L(f,1), & \omega_f^- &= -\frac{1}{7}(2\pi i) L(f,2),\\
\omega_g^+ &= -3L(g,1), & \omega_g^- &= -\frac{1}{2}G(\chi)L(\tilde{g},1),
\end{aligned} \tag{16}
$$

where $\chi = \chi_7(6,\cdot)$ with Gauss sum $G(\chi) = \sqrt{-7}$ and $\tilde{g} = g \otimes \chi \in S_2(\Gamma_0(98))$.

3.2 AESZ 22

Here $N_1 = 33$, $N_2 = 11$ and we have the following modular forms:

$$
\begin{aligned}
f &= q - q^2 - 3q^3 - 7q^4 - 4q^5 + 3q^6 - 26q^7 + 15q^8 + 9q^9 + O(q^{10}) & &\in S_4(\Gamma_0(33)),\\
F &= q^{-1} - 4q^2 - 15q^3 - 44q^4 - 126q^5 - 228q^6 - 486q^7 - O(q^8) & &\in S_4^!(\Gamma_0(33)),\\
g &= q - 2q^2 - q^3 + 2q^4 + q^5 + 2q^6 - 2q^7 - 2q^9 + O(q^{10}) & &\in S_2(\Gamma_0(11)),\\
G &= q^{-1} + 116q^2 + 361q^3 + 962q^4 + 2676q^5 + 6172q^6 + O(q^7) & &\in S_2^!(\Gamma_0(11)),
\end{aligned} \tag{17}
$$

with periods and quasiperiods

$$
\begin{aligned}
\omega_f^+ &= 34.669451430295846076\ldots, & \omega_f^- &= 2.3806231256299333180\ldots i,\\
\eta_F^+ &= -15.646511925944874202 3\ldots, & \eta_F^- &= 2.5029717810597002614\ldots i,\\
\omega_g^+ &= -0.6346046521397767108 4\ldots, & \omega_g^- &= 1.4588166169384952293\ldots i,\\
\eta_G^+ &= 11.733403441655492016\ldots, & \eta_G^- &= 17.071571363716644932\ldots i,
\end{aligned} \tag{18}
$$

and $\tilde{\omega}_g^\pm = 2\pi i \tilde{\omega}_g^\pm$, $\tilde{\eta}_G^\pm = 2\pi i \eta_G^\pm$. In this example the expressions for F and G in terms of cusp forms with LMFDB label is too lengthy and is available from the authors on request. The periods $\omega_f^\pm$, $\omega_g^\pm$ are related to special values of the L–functions of f and g. In this case $L(f,2) = 0$ hence we have to replace f with an appropriate twist. Let $\chi = \left(\frac{-3}{\bullet}\right)$ with Gauss sum $G(\chi) = \sqrt{-3}$ and consider

$$
\tilde{f} = f \otimes \chi \in S_4(\Gamma_0(99)), \qquad \tilde{g} = g \otimes \chi \in S_2(\Gamma_0(99)). \tag{19}
$$

Then we have the relations

$$\omega_f^+ = \frac{5}{6}(2\pi i)^2 L(f,1), \quad \omega_f^- = -\frac{1}{132}G(\chi)(2\pi i)^2 L(\tilde{f},1),$$
$$\omega_g^+ = -\frac{5}{2}L(g,1), \qquad \omega_g^- = \frac{1}{2}G(\chi)L(\tilde{g},1). \tag{20}$$

4 Hadamard Products and Symmetries

As explained in previous sections, there are few systematic methods for finding special members of families of Calabi-Yau threefolds where the Hodge structure splits like a rank two attractor (equivalently, in the one-parameter case, like a supersymmetric flux vacuum). At the time of writing, most known examples are Hadamard products of two second order Picard-Fuchs equations, where each second order operator admits a symmetry that permutes its singularities. The Hadamard product will typically inherit this symmetry, and the resulting smooth fixed points will be apparent singularities that are attractor points of rank two. For a geometric explanation and an explicit example, the reader may consult [19].

Let $\mathscr{L}_1$ and $\mathscr{L}_2$ be two second order Picard-Fuchs equations. We say that an operator $\mathscr{L}_1 * \mathscr{L}_2$ is a Hadamard product of $\mathscr{L}_1$ and $\mathscr{L}_2$ if

$$\mathscr{L}_i\left(\sum_{n=0}^{\infty} a_n^{(i)} u^n\right) = 0, \qquad (\mathscr{L}_1 * \mathscr{L}_2)\left(\sum_{n=0}^{\infty} a_n^{(1)} a_n^{(2)} z^n\right) = 0. \tag{21}$$

Note that $\mathscr{L}_1 * \mathscr{L}_2$ is not uniquely defined in this way. There is an entire ideal of differential operators with this property. However, in practice, there is an operator of minimal degree that we refer to as the Hadamard product.

Many second order operators can be found in [2], which can be used to construct attractor points of rank two. In their choice of labeling, operators $\mathscr{L}_a$ to $\mathscr{L}_j$ all admit a symmetry that permutes its singularities. We may take Hadamard products of these operators and produce many examples of fourth order operators with symmetries and smooth fixed points.

We illustrate this approach with the example of AESZ 175 and leave a more thorough search for future work. The operator AESZ 175 has the Riemann symbol

$$\mathscr{P}\left\{\begin{array}{ccccccc} 0 & \frac{1}{81} & \frac{1}{72} & \frac{1}{9} & \frac{1}{8} & \pm\frac{1}{8\sqrt{2}} & \infty \\ \hline 0 & 0 & 0 & 0 & 0 & 0 & 1 \\ 0 & 1 & 1 & 1 & 1 & 1 & 1 \\ 0 & 1 & 1 & 1 & 1 & 3 & 1 \\ 0 & 2 & 2 & 2 & 2 & 4 & 1 \end{array}\right\}. \tag{22}$$

It is the Hadamard product of two operators $\mathscr{L}_c$ and $\mathscr{L}_g$. These have symbols

$$\mathscr{L}_c : \mathscr{P}\left\{\begin{array}{cccc} 0 & \frac{1}{9} & 1 & \infty \\ \hline 0 & 0 & 0 & 1 \\ 0 & 0 & 0 & 1 \end{array}\right\} \quad \text{and} \quad \mathscr{L}_g : \mathscr{P}\left\{\begin{array}{cccc} 0 & \frac{1}{9} & \frac{1}{8} & \infty \\ \hline 0 & 0 & 0 & 1 \\ 0 & 0 & 0 & 1 \end{array}\right\}. \tag{23}$$

AESZ 175 is is invariant under the involution $z \to \frac{1}{648z}$, and we see that both fixed points $\pm\frac{1}{18\sqrt{2}}$ are apparent singularities. We compute the periods at this point and numerically confirm a splitting of Hodge structure of the form (9) over the number field $\mathbb{Q}(\sqrt{2})$. We may fix a basis of periods with rational monodromy by assuming the standard Gamma class formula (see, for example, the appendix of [4]) and search for a choice of the triple intersection number H^3, second Chern-class $c_2 \cdot H$ and Euler characteristic of a smooth mirror manifold that leads to monodromy in $Sp(4,\mathbb{Z})$. Unfortunately, a numerical search suggests that there is no choice of topological data that leads to monodromy in $Sp(4,\mathbb{Z})$ when used with the the standard Gamma class formula. The results below are quoted for the the topological data $(H^3, c_2 \cdot H, \chi) = (72, 72, 216)$. This leads to monodromy matrices in $Sp(4,\mathbb{Q})$, which is good enough for testing splitting of Hodge structure.

We find that

$$\begin{aligned} \Pi\left(\frac{1}{18\sqrt{2}}\right) &= \alpha_1\Gamma_1 + \alpha_2\Gamma_2, & \Gamma_1 &= \left(-\frac{3}{2}, 9+3\sqrt{2}, 1, 0\right), \\ D_z\Pi\left(\frac{1}{18\sqrt{2}}\right) &= \beta_1\overline{\Gamma}_1 + \beta_2\overline{\Gamma}_2, & \Gamma_2 &= \left(9-3\sqrt{2}, -18, 0, 1\right), \end{aligned} \tag{24}$$

and $\overline{\Gamma}_i$ is the Galois conjugate of Γ_i in $\mathbb{Q}\left(\sqrt{2}\right)$. We compute the coefficients to be

$$\begin{aligned} \alpha_1 &= 224.9708237697899355\ldots - i215.5343428532531063\ldots, \\ \alpha_2 &= 165.5115435704756999\ldots + i44.6386239834950043\ldots, \\ \beta_1 &= 3036.246857171535074\ldots + i812.733738853537295\ldots, \\ \beta_2 &= 802.4731812316557819\ldots + i981.0564074692015020\ldots\,. \end{aligned} \tag{25}$$

Note that we normalise the periods so that the holomorphic solution around $z=0$ has leading term equal to $(2\pi i)$. We find similar results at $z = -\frac{1}{18\sqrt{2}}$,

$$\begin{aligned} \Pi\left(-\frac{1}{18\sqrt{2}}\right) &= \alpha_1\Gamma_1 + \alpha_2\Gamma_2, & \Gamma_1 &= \left(-9, 18+6\sqrt{2}, 1, 0\right), \\ D_z\Pi\left(-\frac{1}{18\sqrt{2}}\right) &= \beta_1\overline{\Gamma}_1 + \beta_2\overline{\Gamma}_2, & \Gamma_2 &= \left(18-6\sqrt{2}, -36, 0, 1\right). \end{aligned} \tag{26}$$

As before, $\overline{\Gamma}_i$ is the Galois conjugate of Γ_i in $\mathbb{Q}\left(\sqrt{2}\right)$ and

$$\begin{aligned} \alpha_1 &= -i165.0515735127136317\ldots, \quad \beta_1 = i972.9410328711081378\ldots, \\ \alpha_2 &= 107.07556468687835349\ldots - i82.52578675635681589\ldots, \\ \beta_2 &= -316.5064782118421090\ldots + i486.4705164355540689\ldots. \end{aligned} \tag{27}$$

In each case we find a two dimensional subspace of $H^3\left(X, \mathbb{Q}\left(\sqrt{2}\right)\right)$ with Hodge numbers $(3,0)+(0,3)$, with Galois conjugate subspaces of type $(2,1)+(1,2)$.

5 Supersymmetric flux vacua from $\mathbb{Z}_2$ permutation symmetry

We discuss a solution of the SFV equations (7) from [11], distinct from the method of [15] used in the works [34, 33] that put forward the flux modularity conjecture: the fibres $X_{\mathbf{z}_*}$ over solutions $\mathbf{z}_*$ of the SFV equations are weight-two modular.

The integral symplectic period vector $\Pi(\mathbf{z})$ is a vector of $2h^{2,1}+2$ functions that solve the Picard-Fuchs system of $X_{\mathbf{z}}$. It is known [28, 29] that in an expansion about the large complex structure point $\mathbf{z}=0$ there are $h^{2,1}$ single-log periods Π^i and $h^{2,1}$ double log periods Π_i. Now assume that there exists a pair of moduli z_I, z_J such that

$$\Pi_I(\mathbf{z}) = \Pi_J(\mathbf{z}') \quad \text{and} \quad \Pi^I(\mathbf{z}) = \Pi^J(\mathbf{z}'), \tag{28}$$

where the argument $\mathbf{z}'$ is obtained from $\mathbf{z}$ by permuting z_I and z_J. If a manifold $X_{\mathbf{z}}$ can be found, and coordinates such that these conditions on Π are met, then we can obtain a solution of (7) by choosing flux vectors such that[2]

$$F\Sigma\Pi = \Pi^I - \Pi^J, \qquad H\Sigma\Pi = \Pi_I - \Pi_J. \tag{29}$$

With this choice made, all bar two of the equations (7) are automatically satisfied by restricting $\mathbf{z}$ to the codimension-1 locus $z_I = z_J$. The only condition that is not automatically met is $(F-\tau H)\Sigma\partial_{z_I}\Pi = 0$, which is equivalent to $(F-\tau H)\Sigma\partial_{z_J}\Pi = 0$. This last equation can then be satisfied by constraining the axiodilaton via

$$\tau(\hat{\mathbf{z}}) = -\left.\frac{F\Sigma\partial_{z_I}\Pi(\mathbf{z})}{H\Sigma\partial_{z_I}\Pi(\mathbf{z})}\right|_{z_I=z_J}, \qquad \hat{\mathbf{z}} = \mathbf{z}|_{z_I=z_J}. \tag{30}$$

A large set of examples are informed by the expressions for period functions of mirrors of complete intersection Calabi-Yau manifolds given in [28, 29]. This gives

$$\varpi_0 = \sum_{n_i\geq 0} c(n_1,\dots,n_h) z_1^{n_1}\dots z_h^{n_h}, \qquad h = h^{2,1}(X), \tag{31}$$

we refer to [29, eq.(3.8)] for $c(\mathbf{n})$ in terms of the matrix that defines the mirror Y of X as a complete intersection in a product of projective spaces. We restrict to cases where the number of projective space factors in the ambient space equals $h^{1,1}(Y) = h^{2,1}(X) = h$. All remaining Frobenius periods can be found from

$$\varpi_\varepsilon = \sum_{n_i\geq 0} \frac{c(n_1+\varepsilon_1,\dots,n_h+\varepsilon_h)}{c(\varepsilon_1\dots\varepsilon_h)} z_1^{n_1+\varepsilon_1}\dots z_h^{n_h+\varepsilon_h}, \tag{32}$$

with Gamma functions replacing factorials, and then forming the combinations

[2] See [11, §3] for specific components of F and H, as well as other details of the solution that we omit presently. As discussed in [11], there is the possibility to rescale the flux vectors by integers, which also provides a solution on the same locus in $\mathbf{z}$-space but scales the axiodilaton by a rational number. This is not reflected in the zeta function, but this makes sense because the point counts of isogeneous elliptic curves (related by scaling the modulus τ) are equal.

$$\varpi_{1,i} = \partial_{\varepsilon_i} \varpi_\varepsilon \Big|_{\varepsilon=0}, \quad \varpi_{2,i} = \frac{1}{2} \sum_{j,k} \kappa_{ijk} \partial_{\varepsilon_j} \partial_{\varepsilon_k} \varpi_\varepsilon \Big|_{\varepsilon=0}, \quad \varpi_3 = \frac{1}{6} \sum_{i,j,k} \kappa_{ijk} \partial_{\varepsilon_i} \partial_{\varepsilon_j} \partial_{\varepsilon_k} \varpi_\varepsilon \Big|_{\varepsilon=0}. \tag{33}$$

Here we introduce the triple intersection numbers $\kappa_{ijk} = \int_Y e_i e_j e_k$ of the mirror Y, where the e_i generate $H^2(Y, \mathbb{Z})_{\text{Free}}$. In expansions about the point $\mathbf{z} = 0$ one has

$$\begin{aligned}
\varpi_0 &= 1 + O(z), \qquad\qquad \varpi_i^{(1)} = \varpi_0 \log(z_i) + f_i^{(1)}, \\
\varpi_i^{(2)} &= \frac{\varpi_0}{2} \kappa_{ijk} \log(z_j) \log(z_k) + \kappa_{ijk} f_j^{(1)} \log(z_k) + f_i^{(2)}, \\
\varpi_3 &= \frac{\varpi_0}{6} \kappa_{ijk} \log(z_i) \log(z_j) \log(z_k) + \frac{\kappa_{ijk}}{2} f_i^{(1)} \log(z_j) \log(z_k) + f_k^{(2)} \log(z_k) + f^{(3)},
\end{aligned} \tag{34}$$

where each function f is a power series in $\mathbf{z}$ with constant term 0.

If the configuration matrix is symmetric under exchanging the Ith and J^{th} rows, then (34) will have a $z_I \leftrightarrow z_J$ symmetry. Also, the topological data of the intersection [29, eq.(2.5)] will respect this symmetry. So, the integral symplectic period vector Π will satisfy (28) and an SFV solution is found by setting $z_I = z_J$.

The axiodilaton is constant throughout spacetime, but is a function of the $\mathbf{z}$ on the locus $z_I = z_J$. Using the expressions (34), we can write the axiodilaton from (30) as

$$\begin{aligned}
\tau(\hat{\mathbf{z}}) &= -\frac{1}{2\pi \mathrm{i}} \frac{\partial_{z_I}(\varpi_{2,I} - \varpi_{2,J})}{\partial_{z_I}(\varpi_{1,I} - \varpi_{1,J})}\Bigg|_{z_I = z_J} + \frac{\kappa_{IJJ} - \kappa_{III}}{2} \\
&= -\frac{1}{2\pi \mathrm{i}} (\kappa_{IIk} - \kappa_{IJk}) \log(z_k)\Bigg|_{z_I = z_J} \\
&\qquad - \frac{1}{2\pi \mathrm{i}} \frac{(\kappa_{IIk} - \kappa_{IJk}) f_k^{(1)} + z_I \partial_{z_I}\left(f_I^{(2)} - f_J^{(2)}\right)}{\varpi_0 + z_I \partial_I \left(f_I^{(1)} - f_J^{(1)}\right)}\Bigg|_{z_I = z_J} + \frac{\kappa_{IJJ} - \kappa_{III}}{2}.
\end{aligned} \tag{35}$$

The rightmost term originates in the change of basis from Frobenius ϖ to integral symplectic Π. Note that shifting τ by an integer is a symmetry of the theory, and of the function $j(\tau)$. It was recognised in [33] that this axiodilaton, which defines the F-theory fibre, is also the complex structure modulus of the elliptic curve whose point-counts are reflected in the zeta numerator of the modular threefold $X_{\mathbf{z}_*}$.

Along with explicit checks of the zeta function as computed by the methods of [7, 37, 10], $j(\tau(\mathbf{z}_*))$ was found in a number of examples in [11] to be a rational function of $\mathbf{z}_*$, similarly to the example of [34, 33].

These algebraicity properties of flux vacua were more closely analysed in [26], where families of elliptic curves were identified such that ratios of their period functions gave the axiodilatons of the examples in [11].

We include the two $h^{2,1} = 2$ examples of [11, 26], together with more $\mathbb{Z}_2$ symmetric two-parameter examples that we are aware of, in Table 2. Five of these are mirrors of CICYs [25, 6], and a further two are CICY quotients [5] with $h^{1,1} = 2$, for which the period formulae of [28, 29] can be applied. A further example uses a manifold due to [35], to which we refer for the Picard-Fuchs equations that can be used to obtain the series f in (34) without applying the Frobenius method.

Our last example is mirror to the Gulliksen-Negård threefold, for which no mirror construction is known at present. The PF operators are available from GLSM techniques exploited in [32]. The coordinates therein do not manifestly have a $\mathbb{Z}_2$ symmetry, but the GW invariants found suggest there should be one. By taking their operators [32, eq.(5.21)] and making a coordinate transformation[3] $z_{\text{there}} = z_1^{-2}$, $w_{\text{there}} = -\frac{z_2}{z_1}$, we get a symmetric fundamental period with $\varpi_0^{GN}(z_1,z_2) = \varpi_0^{GN}(z_2,z_1)$. We find an SFV solution, with axiodilaton given in Table 2.

There may be more to understand about this choice of coordinates. We note that $\varpi_0^{GN}(\sqrt{z},\sqrt{z})$ is annihilated by a degree 8 operator whose instanton expansion we do not recognise, while $\varpi_0^{GN}(\mathrm{i}\sqrt{z},-\mathrm{i}\sqrt{z})$ is annihilated by AESZ18.

Table 2 $h^{1,1}=2$ CY threefolds whose mirrors support supersymmetric flux vacua on the locus $z_1=z_2=z$. A number of rational functions $j(\tau(z))$ repeat across examples, sometimes exactly and sometimes only with a coordinate change $z\to z^2$ or $z\to -z^2$.

#	Mirror CY3	κ_{111}	κ_{112}	$j(\tau(z))$
1	$\begin{matrix}\mathbb{P}^2\\ \mathbb{P}^2\end{matrix}\begin{bmatrix}3\\3\end{bmatrix}$	0	3	$-\frac{(1+24z)^3}{z^3(1+27z)}$
2	$\begin{matrix}\mathbb{P}^2\\ \mathbb{P}^2\end{matrix}\begin{bmatrix}3\\3\end{bmatrix}/\mathbb{Z}_3$	0	1	$-\frac{(1-216z)^3}{z(1+27z)^3}$
3	$\begin{matrix}\mathbb{P}^3\\ \mathbb{P}^3\end{matrix}\begin{bmatrix}1&1&2\\1&1&2\end{bmatrix}$	2	6	$\frac{(1+16z+16z^2)^3}{z^4(1+16z)}$
4	$\begin{matrix}\mathbb{P}^3\\ \mathbb{P}^3\end{matrix}\begin{bmatrix}1&2&1\\1&1&2\end{bmatrix}$	2	7	$-\frac{(1+12z+14z^2-12z^3+z^4)^3}{z^5(1+11z-z^2)}$
5	$\begin{matrix}\mathbb{P}^4\\ \mathbb{P}^4\end{matrix}\begin{bmatrix}1&1&1&1&1\\1&1&1&1&1\end{bmatrix}$	5	10	$-\frac{(1+12z+14z^2-12z^3+z^4)^3}{z^5(1+11z-z^2)}$
6	$\begin{matrix}\mathbb{P}^4\\ \mathbb{P}^4\end{matrix}\begin{bmatrix}1&1&1&1&1\\1&1&1&1&1\end{bmatrix}/\mathbb{Z}_5$	1	2	$-\frac{(1-228z+494z^2+228z^3+z^4)^3}{z(1+11z-z^2)^5}$
7	$\begin{matrix}\mathbb{P}^4\\ \mathbb{P}^4\end{matrix}\begin{bmatrix}2&0&1&1&1\\0&2&1&1&1\end{bmatrix}$	4	12	$\frac{(1-16z^2+16z^4)^3}{z^8(1-4z)(1+4z)}$
8	Pfaffian fibration Y_1 of [35, §4.3.2]	0	5	$-\frac{(1+12z+14z^2-12z^3+z^4)^3}{z^5(1+11z-z^2)}$
9	Gulikssen-Negård Threefold, $z_1=z_2=z$	8	24	$\frac{(1-16z^4+16z^8)^3}{z^{16}(1-2z)(1+2z)(1+4z^2)}$

[3] JM thanks Johanna Knapp for discussions on these lines.

We end this section by giving the axiodilaton profiles for two $h^{2,1}=3$ families, whose fully symmetric locus $z_1=z_2=z_3$ has been analysed in [17]. We go to the SFV locus $z_1=z_2=z, z_3=y$.

$$\text{Mirror of} \quad \begin{matrix}\mathbb{P}^2\\ \mathbb{P}^2\\ \mathbb{P}^2\end{matrix}\begin{bmatrix}1&1&1\\1&1&1\\1&1&1\end{bmatrix}^{h^{1,1}=3,\, h^{2,1}=48}, \qquad \kappa_{ijk}=\begin{cases}0, & i=j=k,\\ 6, & i,j,k \text{ distinct},\\ 3, & \text{otherwise},\end{cases}$$

$$j(\tau(z,y)) = -\frac{(1-y+z)^3\left(1-3y+3z+3y^2+18yz+3z^2-y^3+3y^2z-3yz^2+z^3\right)^3}{y^3z^3\left(1-3y+3y^2-y^3+3z+21yz+3y^2z+3z^2-3yz^2+z^3\right)}. \tag{36}$$

$$\text{Mirror of} \quad \begin{matrix}\mathbb{P}^2\\ \mathbb{P}^2\\ \mathbb{P}^2\end{matrix}\begin{bmatrix}1&1&1\\1&1&1\\1&1&1\end{bmatrix}^{h^{1,1}=3,\, h^{2,1}=18}_{/\mathbb{Z}_3}, \qquad \kappa_{ijk}=\begin{cases}0, & i=j=k,\\ 2, & i,j,k \text{ distinct},\\ 1, & \text{otherwise},\end{cases}$$

$$j(\tau(y,z)) = -\frac{(1-y+z)^3\left(1-3y+3z+3y^2-222yz+3z^2-y^3+3y^2z-3yz^2+z^3\right)^3}{yz\left(1-3y+3y^2-y^3+3z+21yz+3y^2z+3z^2-3yz^2+z^3\right)^3}. \tag{37}$$

Acknowledgements We thank the MATRIX Institute for hosting the program "The geometry of moduli spaces in string theory", which was of significant benefit to us and greatly facilitated our projects. We also wish the MATRIX Institute a happy 10th birthday, and hope the best for them with their important future work. We also thank the editors Jock McOrist and Sébastien Picard for their patience with our article. We thank Mainz Institute for Theoretical Physics (MITP) of the Cluster of Excellence PRISMA$^+$ (Project ID 390831469) for hospitality during "The arithmetic of Calabi-Yau manifolds". We thank Philip Candelas, Kilian Bönisch, Xenia de la Ossa, Johanna Knapp, and Pyry Kuusela for discussions. JM is supported by a University of Melbourne establishment grant. ME is supported by the Shuimu Tsinghua Scholar Program.

References

1. Almkvist, G., van Enckevort, C., van Straten, D., Zudilin, W.: Tables of Calabi–Yau equations (2005)
2. Almkvist, G., Zudilin, W.: Differential equations, mirror maps and zeta values (2004)
3. Bönisch, K., Elmi, M., Kashani-Poor, A.K., Klemm, A.: Time reversal and CP invariance in Calabi-Yau compactifications. JHEP **09**, 019 (2022). DOI 10.1007/JHEP09(2022)019
4. Bönisch, K., Klemm, A., Scheidegger, E., Zagier, D.: D-brane Masses at Special Fibres of Hypergeometric Families of Calabi–Yau Threefolds, Modular Forms, and Periods. Commun. Math. Phys. **405**(6), 134 (2024). DOI 10.1007/s00220-024-05006-6
5. Braun, V.: On Free Quotients of Complete Intersection Calabi-Yau Manifolds. JHEP **04**, 005 (2011). DOI 10.1007/JHEP04(2011)005
6. Candelas, P., Dale, A.M., Lutken, C.A., Schimmrigk, R.: Complete Intersection Calabi-Yau Manifolds. Nucl. Phys. B **298**, 493 (1988). DOI 10.1016/0550-3213(88)90352-5

7. Candelas, P., De La Ossa, X., Van Straten, D.: Local Zeta Functions From Calabi-Yau Differential Equations (2021)
8. Candelas, P., Kuusela, P., McGovern, J.: Attractors with large complex structure for one-parameter families of Calabi-Yau manifolds. JHEP **11**, 032 (2021). DOI 10.1007/JHEP11(2021)032
9. Candelas, P., de la Ossa, X., Elmi, M., Van Straten, D.: A One Parameter Family of Calabi-Yau Manifolds with Attractor Points of Rank Two. JHEP **10**, 202 (2020). DOI 10.1007/JHEP10(2020)202
10. Candelas, P., de la Ossa, X., Kuusela, P.: Local Zeta Functions of Multiparameter Calabi-Yau Threefolds from the Picard-Fuchs Equations (2024)
11. Candelas, P., de la Ossa, X., Kuusela, P., McGovern, J.: Flux vacua and modularity for $\mathbb{Z}_2$ symmetric Calabi-Yau manifolds. SciPost Phys. **15**(4), 146 (2023). DOI 10.21468/SciPostPhys.15.4.146
12. Candelas, P., de la Ossa, X., McGovern, J.: Classical Weight-Four L-value Ratios as Sums of Calabi–Yau Invariants (2024)
13. Dasgupta, K., Rajesh, G., Sethi, S.: M theory, orientifolds and G - flux. JHEP **08**, 023 (1999). DOI 10.1088/1126-6708/1999/08/023
14. Deligne, P.: La conjecture de Weil. I. Inst. Hautes Études Sci. Publ. Math. (43), 273–307 (1974)
15. DeWolfe, O.: Enhanced symmetries in multiparameter flux vacua. JHEP **10**, 066 (2005). DOI 10.1088/1126-6708/2005/10/066
16. Douglas, M.R., Kachru, S.: Flux compactification. Rev. Mod. Phys. **79**, 733–796 (2007). DOI 10.1103/RevModPhys.79.733
17. Dücker, J., Klemm, A., Piribauer, J.F.: Calabi-Yau Period Geometry and Restricted Moduli in Type II Compactifications (2025)
18. Dwork, B.: On the Rationality of the Zeta Function of an Algebraic Variety. American Journal of Mathematics **82**(3), 631–648 (1960). URL `http://www.jstor.org/stable/2372974`
19. Elmi, M.: Hadamard products and BPS networks. JHEP **07**, 076 (2024). DOI 10.1007/JHEP07(2024)076
20. Ferrara, S., Kallosh, R.: Supersymmetry and attractors. Phys. Rev. D **54**, 1514–1524 (1996). DOI 10.1103/PhysRevD.54.1514
21. Ferrara, S., Kallosh, R., Strominger, A.: N=2 extremal black holes. Phys. Rev. D **52**, R5412–R5416 (1995). DOI 10.1103/PhysRevD.52.R5412
22. Freedman, D.Z., Van Proeyen, A.: Supergravity. Cambridge Univ. Press, Cambridge, UK (2012). DOI 10.1017/CBO9781139026833
23. Gegelia, N., van Straten, D.: Paramodular forms from Calabi-Yau Operators (2024)
24. Gouvea, F.Q., Yui, N.: Rigid Calabi-Yau Threefolds over Q Are Modular. arXiv e-prints arXiv:0902.1466 (2009). DOI 10.48550/arXiv.0902.1466
25. Green, P., Hubsch, T.: Calabi-yau Manifolds as Complete Intersections in Products of Complex Projective Spaces. Commun. Math. Phys. **109**, 99 (1987). DOI 10.1007/BF01205673
26. Grimm, T.W., van de Heisteeg, D.: Exact flux vacua, symmetries, and the structure of the landscape. JHEP **01**, 005 (2025). DOI 10.1007/JHEP01(2025)005
27. Grothendieck, A.: Formule de Lefschetz et rationalité des fonctions *L*. In: Séminaire Bourbaki, Vol. 9, pp. Exp. No. 279, 41–55. Soc. Math. France, Paris (1995)
28. Hosono, S., Klemm, A., Theisen, S., Yau, S.T.: Mirror symmetry, mirror map and applications to Calabi-Yau hypersurfaces. Commun. Math. Phys. **167**, 301–350 (1995). DOI 10.1007/BF02100589
29. Hosono, S., Klemm, A., Theisen, S., Yau, S.T.: Mirror symmetry, mirror map and applications to complete intersection Calabi-Yau spaces. Nucl. Phys. B **433**, 501–554 (1995). DOI 10.1016/0550-3213(94)00440-P
30. Jockers, H., Kotlewski, S.: Geometry of N=2 Minkowski vacua of gauged N=2 supergravity theories in four dimensions. Phys. Rev. D **110**(4), 046,019 (2024). DOI 10.1103/PhysRevD.110.046019

31. Jockers, H., Kotlewski, S., Kuusela, P.: Modular Calabi-Yau fourfolds and connections to M-theory fluxes. JHEP **12**, 052 (2024). DOI 10.1007/JHEP12(2024)052
32. Jockers, H., Kumar, V., Lapan, J.M., Morrison, D.R., Romo, M.: Two-Sphere Partition Functions and Gromov-Witten Invariants. Commun. Math. Phys. **325**, 1139–1170 (2014). DOI 10.1007/s00220-013-1874-z
33. Kachru, S., Nally, R., Yang, W.: Flux Modularity, F-Theory, and Rational Models (2020)
34. Kachru, S., Nally, R., Yang, W.: Supersymmetric flux compactifications and Calabi-Yau modularity. Commun. Num. Theor. Phys. **18**(3), 653–704 (2024). DOI 10.4310/cntp.240904012658
35. Knapp, J., Scheidegger, E., Schimannek, T.: On genus one fibered Calabi-Yau threefolds with 5-sections (2021)
36. Kotlewski, S.: Flux compactifications and Feynman Integrals - Calabi-Yau geometries in physics. Ph.D. thesis, Mainz U., Mainz U. (2024). DOI 10.25358/openscience-10735
37. Kuusela, P.: Modular Calabi-Yau manifolds, attractor points, and flux vacua. Ph.D. thesis, Oxford University (2022)
38. McGovern, J.: Investigations in Calabi-Yau modularity and mirror symmetry. Ph.D. thesis, Oxford University (2023). DOI 10.5287/ora-w4ykymo0p
39. Strominger, A.: Macroscopic entropy of N=2 extremal black holes. Phys. Lett. B **383**, 39–43 (1996). DOI 10.1016/0370-2693(96)00711-3
40. The LMFDB Collaboration: The L-functions and Modular Forms Database. https://www.lmfdb.org/ (2025). Accessed 22-June-2025
41. Weil, A.: Numbers of solutions of equations in finite fields. Bull. Amer. Math. Soc. **55**, 497–508 (1949). DOI 10.1090/S0002-9904-1949-09219-4.

GPSR Compliance
The European Union's (EU) General Product Safety Regulation (GPSR) is a set of rules that requires consumer products to be safe and our obligations to ensure this.

If you have any concerns about our products, you can contact us on

ProductSafety@springernature.com

In case Publisher is established outside the EU, the EU authorized representative is:

Springer Nature Customer Service Center GmbH
Europaplatz 3
69115 Heidelberg, Germany

www.ingramcontent.com/pod-product-compliance
Ingram Content Group UK Ltd.
Pitfield, Milton Keynes, MK11 3LW, UK
UKHW021614260926
475829UK00010B/16

* 9 7 8 3 0 3 2 1 6 2 0 5 2 *